Techno-Economic Analysis and Optimization for Energy Systems

Techno-Economic Analysis and Optimization for Energy Systems

Editors

Konstantinos Aravossis
Eleni Strantzali

Basel • Beijing • Wuhan • Barcelona • Belgrade • Novi Sad • Cluj • Manchester

Editors
Konstantinos Aravossis
School of Mechanical
Engineering
National Technical University
of Athens
Athens
Greece

Eleni Strantzali
Department of Naval
Architecture
University of West Attica
Athens
Greece

Editorial Office
MDPI
St. Alban-Anlage 66
4052 Basel, Switzerland

This is a reprint of articles from the Special Issue published online in the open access journal *Energies* (ISSN 1996-1073) (available at: www.mdpi.com/journal/energies/special_issues/7J53C91GKC).

For citation purposes, cite each article independently as indicated on the article page online and as indicated below:

Lastname, A.A.; Lastname, B.B. Article Title. *Journal Name* **Year**, *Volume Number*, Page Range.

ISBN 978-3-7258-0836-6 (Hbk)
ISBN 978-3-7258-0835-9 (PDF)
doi.org/10.3390/books978-3-7258-0835-9

Contents

About the Editors

Konstantinos Aravossis

Konstantinos Aravossis is Professor and UNESCO Chairholder of Green Innovation and Circular Economy at the NTUA (National Technical University of Athens), as well as a Professor at the ATHENS MBA. He is a mechanical engineer (Technical University of Aachen, Germany) with an M.Sc. in Management Science from Imperial College, University of London. He has further been awarded a PhD by the National Technical University of Athens, focusing on operations research. He is the Coordinator of the "Environmental Economics and Sustainable Development Research Group" of the Sector of Industrial Management and Operations Research at the Mechanical Engineering School.

Prof. Konstantinos Aravossis is a former Secretary General for Forestry at the Greek Ministry of Environment and Energy. Prior to that, he was also the Secretary General of Natural Environment & Water. He has also been President of the Greek Branch of the Waste to Energy Research and Technology Council (www.wtert.gr), coordinated by Columbia University, and an Associate Researcher at the Earth Engineering Center, Columbia University, as well as a visiting Professor at Imperial College, University of London. Moreover, he has been President of the Board of the Institute of Production and Operations Management / Hellenic Management Association.

His research interests include environmental management, circular economy, environmental economics, sustainability, green innovation, entrepreneurial ventures, techno-economic evaluations of projects and investments, engineering economics, waste management, renewable energy resources and forest management.

Eleni Strantzali

Eleni Strantzali is an Assistant Professor at the Department of Naval Architecture at the University of West Attica, with a specialization in techno-economic analysis and supply chain in maritime transport.

She is a mechanical engineer, with a M.Sc. in MBA Techno-Economic Systems, NTUA (Management of Technological Systems) and has a PhD in the Section of Industrial Management and Operational Research from the School of Mechanical Engineering, NTUA.

Her research interests include techno-economic analysis, maritime transport and environment, maritime transport supply chain, naval engineering, energy planning, design and optimization of circular economy supply chains, decision support systems, operational research and environmental risk management.

Preface

The challenges of climate change and energy transition require fundamental changes in energy systems. Significant progress is necessary for the current energy systems to satisfy the ambitious targets of the Energy Roadmap 2050 of the European Commission, aiming to fully decarbonize the European economy by reducing GHG emissions in developed countries to below 80–95% of the 1990 levels. A circular economy has risen high in the agendas of policymakers as a way of enhancing the decarbonization approach.

This Special Issue presents articles related to techno-economic analysis and optimization approaches, especially for energy systems, and collates the latest research and advancements in their application.

Nineteen papers were submitted to this Special Issue, of which fifteen were published. Although each paper covers a different topic, three categories can be identified based on the following focus areas of the published papers:

- Decarbonization: One of the main current challenges is decarbonization, with energy systems being significant contributors to the transition. The published articles encompass the implementation of decarbonization from the generation of electricity to the transport sector, proposing cost-effective solutions and optimization approaches. The authors discuss all types of renewable energy sources and their different production methodologies, with both on-land and shipping applications.

- Energy Demand: Estimating and managing energy demand is an issue that has occupied researchers throughout the years. It is fully linked to the need to meet demands and the parallel need to reduce greenhouse gas emissions, making it a constant timely challenge.

- Optimization of Energy Systems: In achieving the optimization of energy systems, the optimization of their individual components is a fundamental condition.

Konstantinos Aravossis and Eleni Strantzali
Editors

Article

Assessment of Co-Gasification Methods for Hydrogen Production from Biomass and Plastic Wastes

Jonah M. Williams [1,*] and A. C. (Thanos) Bourtsalas [1,2,*]

1 Department of Earth and Environmental Engineering, Columbia University, New York, NY 10027, USA
2 Earth Engineering Center, Columbia University, New York, NY 10027, USA
* Correspondence: jmw2291@columbia.edu (J.M.W.); ab3129@columbia.edu (A.C.B.)

Abstract: In recent decades, economic development and population growth has been accompanied by the generation of billions of tonnes of solid residues or municipal "wastes", a substantial portion of which is composed of plastics and biomass materials. Combustion-based waste-to-energy is a viable and mature method of extracting calorific value from these end-of-life post-recyclable materials that are otherwise landfilled. However, alternative thermochemical methods, such as gasification, are becoming attractive due to the ability to synthesize chemical precursors for supply chain recirculation. Due to the infancy of gasification technology deployment, especially in the context of anthropogenic CO_2 emission reduction, additional systems engineering studies are necessary. Herein, we conduct an attributional life cycle analysis to elucidate the syngas production and environmental impacts of advanced thermochemical gasification methods for the treatment of biomass and plastic wastes obtained from municipal solid wastes, using a comprehensive thermodynamic process model constructed in AspenTech. Feedstock composition, process parameters, and gasification methods are varied to study the effects on syngas quality, yield, power generation potential, and overall greenhouse gas emissions. Steam-based gasification presents up to 38% reductions in CO_2 emissions when compared to conventional thermochemical methods. Using gasifier-active materials, such as metal hydroxides, can also further reduce CO_2 emissions, and realizes a capture load of 1.75 tonnes of CO_2 per tonne of plastic/stover feedstock. This design alteration has implications for reductions in CAPEX due to the mode of CO_2 capture utilized (e.g., solid sorbent vs. liquid SELEXOL). The use of renewable energy to provide a method to generate steam for this process could make the environmental impact of such MSW gasification processes lower by between 60–75% tonnes of CO_2 per tonne of H_2. Overall, these results can be used to inform the guidance of advanced waste gasification methods as a low-carbon transition towards a circular economy.

Keywords: municipal waste treatment; hydrogen production; gasification; carbon capture utilization and storage; thermochemical conversion; plastic pollution

Citation: Williams, J.M.; Bourtsalas, A.C. Assessment of Co-Gasification Methods for Hydrogen Production from Biomass and Plastic Wastes. *Energies* **2023**, *16*, 7548. https://doi.org/10.3390/en16227548

Academic Editor: Dmitri A. Bulushev

Received: 21 October 2023
Revised: 8 November 2023
Accepted: 10 November 2023
Published: 13 November 2023

1. Introduction

Humanity's dependence on plastic since the oil boom in the first half of the 20th century has gone hand-in-hand with many of the technological advancements society enjoys today [1]. Although methods of recycling exist to re-purpose waste plastics into virgin materials, a majority of used plastics are either disposed of in a managed fashion (e.g., combustion-based waste to energy or landfilling) or improperly disposed of (open dumping or ocean dumping) [2–4]. In the U.S., in 2018 alone, about 35,680K tonnes of plastic were generated and of that amount, 26,970K U.S. tons were landfilled (75.5%) [2]. Globally, the amount landfilled is closer to 350 million tonnes per year, and global plastic production is projected to grow at a rate of about 15 million tonnes per year due to increased industrialization [5,6]. The staggering issue of plastics generation and their longevity has led to the development of recycling programs to offset the usage of virgin polymers. However, not all plastics can be recycled, as their relative complexity, contamination, accessibility, heavy metal content, etc., make it more challenging [5,7,8]. For example, plastic food containers can be challenging to mechanically recycle due to the presence of residual food wastes. This leaves few alternatives for dealing with the issues of growing plastic pollution. Additionally, China recently enacted its "National Sword" policy (2017), decreeing that it will no longer import Western recycling materials [9]. This has severely impacted the global flows of recyclables, and furthers the need for interim technologies to mitigate the sheer volume of plastics that are landfilled. One such method is Waste-to-Energy (WtE) processes, which allow the recovery of calorific value in the form of power and/or fuels from waste streams through combustion, gasification, or pyrolysis [10].

Ciuffi et al. (2020) enumerate many disposal pathways for MSW plastics. The first two, primary and secondary, include mechanical recycling, which is applicable only with pure, point-source separated feedstocks. The tertiary recycling method, WtE, provides a solution for contaminated bulk MSWs that can be continually processed and incinerated [11]. Among the WtE processes listed, gasification shows clear advantages, as a high quality syngas rich in CO/H_2 can be recovered (Table 1) [12]. This syngas can then be utilized in heat-recovery operations or as a precursor to downstream fuels and polymer synthesis processes (e.g., Fischer–Tropsch). Although the technology maturity is currently low, gasification facilities for the treatment of MSWs show future promise, especially in the context of a circularized and constrained carbon economy for the production of chemical precursors for supply chain recirculation. Currently, there are many types of gasification reactors and process schemes that can be used generate synthesis gas. For high-carbon feedstocks such as coal or biomass, there is much flexibility in the type of gasifier that can be used. For MSW streams, which usually have a medium gross calorific value, the best type of gasifier to use is likely a moving-bed type, where the waste is pre-pulverized and fluidized in the gasifier with an oxidant. This allows for lower residence times, increased char and tar cracking, higher temperatures, and overall better conversions and volatilization. Gasifiers typically run at elevated temperatures (>1000 °C) and pressures greater than 50 bar to effectively convert the feedstocks. Although H_2 and CO are the primary syngas components, CO_2 is also produced during gasification along with partially oxidized sulfur species (H_2S and COS), chlorides (HCl, Cl^-), and trace heavy metals. The gasification of MSWs is poised to generate more hazardous metals and species, such as dioxins and chloro-compounds due to the wide variability of the feed based on the addition refuse components found with the plastics in the feed [13–15].

The gasifier can be operated in many different oxidant modes such as air-blown, oxygen-blown, steam, and sorbent-based gasification (Figure 1). Air-blown gasifiers are the most widely used, are generally inexpensive relative to the other methods; they have a simplistic reactor design but produce low-value syngas (low LHV) with larger carbon emissions. Oxygen-blown gasifiers produce high-purity syngas with a high LHV with lower pollutant levels; however, a full-scale Air Separation Unit (ASU) is needed to provide the oxidant charge. Steam gasification uses superheated or supercritical steam as the oxidant, which can produce a high hydrogen content in the syngas, increased char and tar cracking, which leads to higher gas yields; however, greater energy needs are necessary as the overall process is endothermic when steam is used [11,16]. Lastly, sorbent-based gasification utilizes catalytically active gasifier bed materials that aid in both feedstock conversion and CO_2 sequestration in the form of metal (M) carbonates $M(CO_3)_x$ [17]. In situ carbon capture is very attractive as it has implications for the cost reduction of syngas cleanup downstream; however, external heat needs to be supplied to the gasifier due to the slightly endothermic nature of the reactions. Additionally, due to the large volumes and reserves of alkaline waste materials that can act as gasifier-active species (e.g., olivine, serpentine, portlandite, etc.), the motivation to run these gasification systems is increasing. Greater hydrogen yields can theoretically be obtained due to the participation of the OH^- ions in the reaction, as exemplified with the alkaline thermal treatment of portlandite below: [18,19]

$$Ca(OH)_2 + Carbon\ (C) + H_2O \rightarrow CaCO_3 + 2H_2 \tag{1}$$

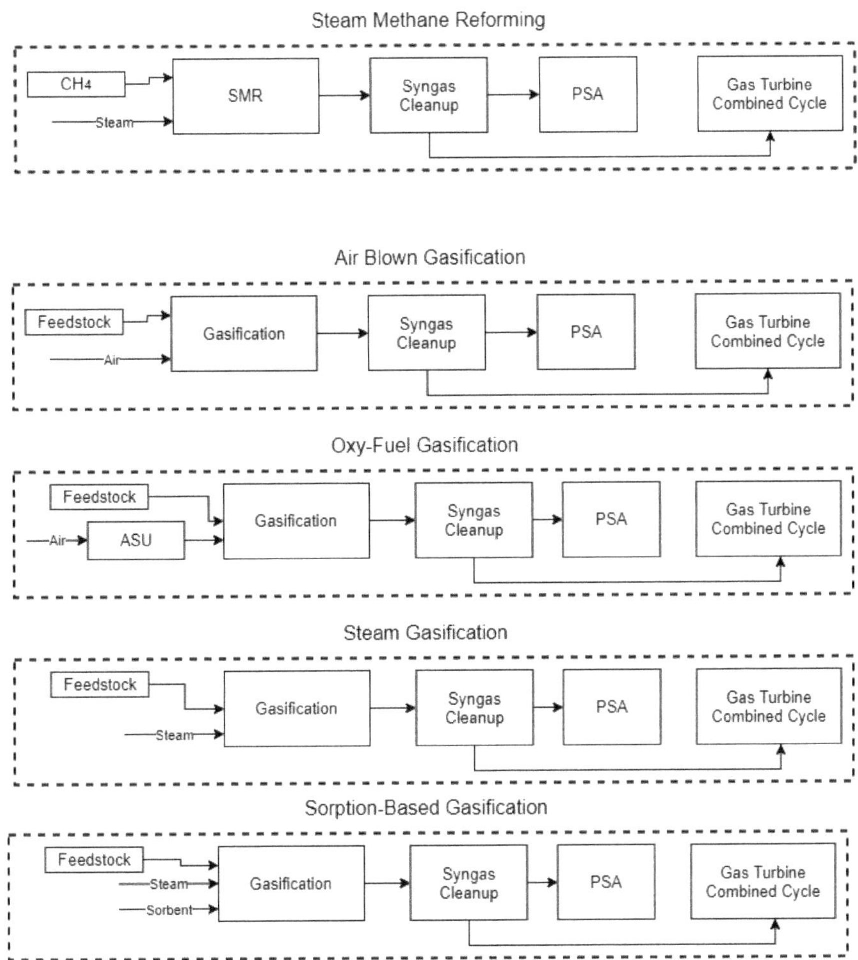

Figure 1. Different gasification cases compared in this study, analyzed in the context of hydrogen and power generation capacities compared to the benchmark of SMR.

Table 1. Comparison of the three major thermochemical treatment methods to recovery calorific value from MSW streams.

	Pyrolysis	Gasification	Combustion
Air provided to the system	No air	Sub stoichiometric air	Excess air
Feedstock	Source separated plastic materials	Source separated high calorific value materials, e.g., plastics, and paper, and biomass	Mixed wastes
Products	Liquid fuels, e.g., oil	Syngas (CO and H_2)	Energy—electricity and/or heat
By Products	High char, unconverted solid will remain Pollutants in reduced form (H_2S, COS)	Char @ low Temp; Vitrified slag @ high Temp Lower fly ash carries over, compared to combustion Pollutants in reduced form (H_2S, COS)	Bottom ash (inert), fly ash (hazardous) Pollutants in oxidized form (SO_x, NO_x, etc.)

Table 1. *Cont.*

	Pyrolysis	Gasification	Combustion
Temperature	<500 °C	700–1200 °C	>1100 °C
Maturity	Not proven—small scale, ~10 tonne per day	Not proven—failures reported, e.g., Tees Valley in the UK and PyroGenesis in Florida, USA	Proven and dominant, ~1000 plants worldwide with capacities from 100 tonnes per day up to 5000 tonnes per day. Flexible and optimized system

Previous studies have tried to compare different gasification models in the context of either biomass, coal, or petroleum coke gasification, but few examine it in the context of MSWs, especially with an underlying assessment of emissions metrics [20–23]. Additionally, most do not consider models that combine upstream gasification with downstream gas scrubbing, sulfur removal, and heat recovery units [24,25]. In order to address this need, this study compares these four viable promising gasification methods in the context of MSW treatment and examines the ramifications of each from a comprehensive thermodynamic system engineering perspective. Hydrogen production and purification is carefully examined with respect to these thermochemical conversion cases as an alternative to the most conventional method of producing H_2, steam methane reforming (SMR), which is CO_2-intense. Due to the growing interest in and the need to reduce anthropogenic CO_2 emissions, this model also considers cases with advanced point-source carbon capture systems using commercialized physisorption thermal-swing processes. Materials balances were produced from the AspenTech gasification combined cycle simulations for six individual cases. All cases considered the same general process configuration, where the main difference was the type of feedstock or method of gasification. Overall, this study represents the development and assessment of a complex thermodynamic MSW gasification model, which considers the production of hydrogen while capturing and producing a pure stream of CO_2 for utilization or storage.

2. Materials and Methods

2.1. Block Flow Diagram and Boundary Conditions

An attributional cradle-to-gate Life Cycle Analysis (LCA) was conducted, examining the conversion of raw materials (biomass and MSW-derived plastics) through an integrated gasification combined cycle (IGCC) facility which produces high-purity hydrogen gas and electricity. The hydrogen will be produced via pressure swing absorption of syngas, and the purge/reject gas will be further oxidized in a gas turbine (GC) equipped with a heat recovery steam generator (HRSG).

Figure 2 shows a block flow diagram of the considered process along with the input and exit boundaries. The boundaries of this project include the feedstocks arriving at the gasifier block, and exclude pretreatment and transportation. The raw material inputs are air, water, feedstock (MSW-plastic or biomass), energy (in the form of either electricity or steam), and SELEXOL charge (fresh ethyl ethers of polyethylene glycol) for CO_2 capture. The materials exiting the plant are energy, in the form of power, stack emissions, from the gas turbine combined cycle (GTCC) and heat recovery steam generator (HRSG), pure hydrogen, char, and ash from the gasifier, and process waste-water. Although not included in Figure 2, pure hydrogen sulfide gases and pure carbon dioxide gases, to feed either a Claus Unit to make elemental sulfur or to be stored in a saline aquifer, respectively, are also exiting the process. For the purpose of this study, the plant efficacy is determined solely based on hydrogen product value and energy produced.

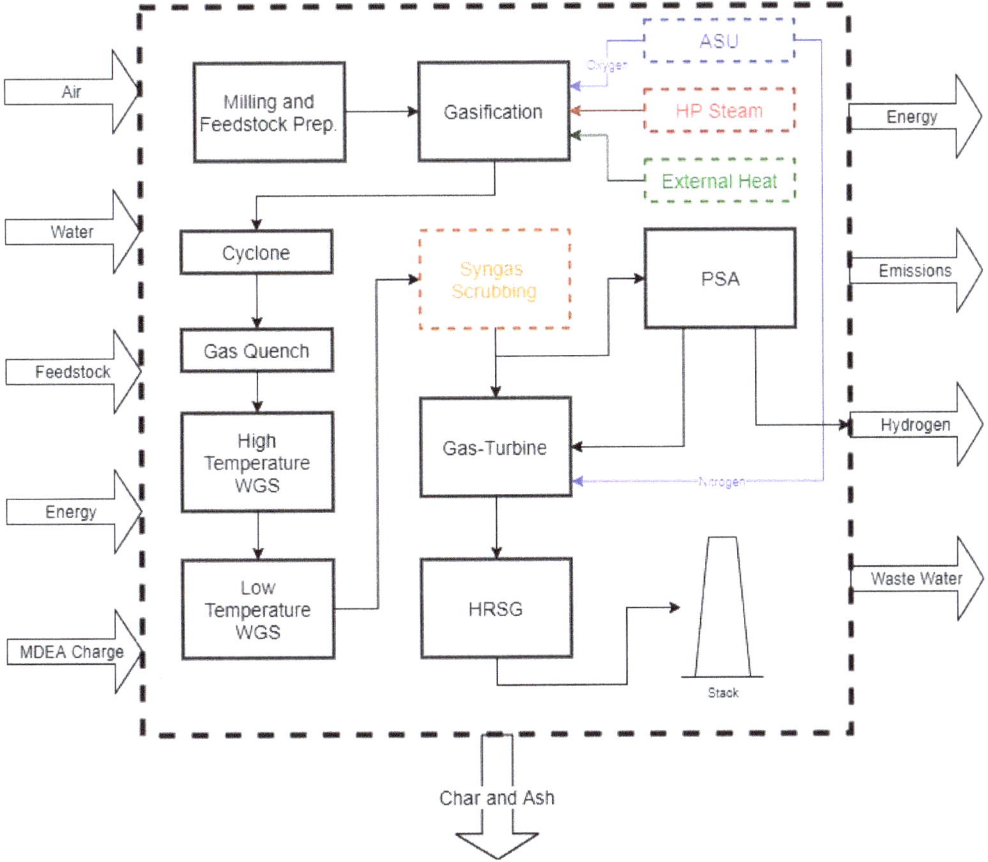

Figure 2. Block flow diagram (BFD) of the proposed LCA showing the boundary conditions and the inputs/outputs from the plant. Near the gasifier, the ASU, HP Steam, and External heat represent different gasification operation modes. The ASU is only used in the case of oxygen-blown gasification, HP steam to the gasifier is for the case of sorbent-based gasification and steam gasification, and external heat is for sorbent-based gasification.

2.2. Functional Unit

Two different metrics were used to assess the flows at the boundary condition, one being hydrogen produced per metric tonne of feedstock basis and MWh produced per metric tonne of feedstock. The first basis will be useful in comparing the amount of energy produced, hydrogen purified, CO_2-captured, etc., per tonne of feedstock when different feedstock slates are compared. Additionally, with the potential to co-gasify plastics and biomass, this metric is important to compare potential synergies between the feedstocks. The second metric will be useful for assessing the functionality of the plant at scale, and will provide plant-wide data relative to power generation, which is an important parameter in the success of WtE facilities.

2.3. Assumptions

There are many assumptions present throughout this comparison and they are used to provide greater support and motivation for the subsequent data. As established by ISO 14040 [26], assumptions allow for less misinterpretation of LCA study results (Table S1).

2.4. Thermodynamic Model of MSW Gasification Combined Cycle

For almost all of the simulation models, the Predictive Soave-Redlich-Kwong (PSRK) property set method from the AspenTech Property Set Database was used. The PSRK method is useful for predicting chemical equilibria for high-pressure gas systems, which is the majority of this process [27,28]. However, to model the liquid–liquid gas equilibria in the SELEXOL process the NRTL-Electrolyte method from the AspenTech Property Set Databased was chosen [29]. These methods were selected to yield the most robust and accurate process model possible. A full process flow diagram (PFD) to showcase the AspenTech model developed is available in Figure S1 and a description to accompany this PFD is present in Supplementary Information Description S3.

2.5. Systems Assessed and Sensitivity Analysis

This study assesses four methods of industrial gasification in the context of both biomass and MSW-derived plastic wastes. The six test cases studied and variations of the thermodynamic model are described below:

Case 1 is a coal-fired oxyfuel plant (benchmark), Case 2 is a methane fired autothermal steam methane reforming plant (benchmark), and Case 3 is an MSW-plastic/corn stover (1:1 ratio by mass) oxyfuel plant. These three cases are almost identical, aside from the different feedstocks utilized. Case 4 is an MSW-plastic/corn stover (1:1 ratio by mass) steam gasification plant, which does not have an ASU but instead uses very high-pressure steam as the oxidant in the gasifier. Case 5 is the same as Case 4, but utilizes a mixture of steam, oxygen, and a sorbent (in this case $Ca(OH)_2$—portlandite) to perform in-situ carbon capture and hydrogen generation. Lastly, Case 6 is an air-blown gasifier which recovered hydrogen from an MSW-plastic/biomass mixture (1:1 ratio by mass), but does not have an ASU nor gas turbine nor gas cleanup process due to the presence of large amounts of nitrogen circulating throughout. The general differences in the gasification islands are highlighted in Figure 1.

2.6. Feedstocks Considered and Heating Values

The feedstock considered for the MSW/Biomass gasification processes was selected from a group of known WtE feedstocks and their associated ultimate analyses (Table S2). A representative mix of corn stover and MSW-derived plastics, in equal mass portions. This was done as a proof-of-concept to simulate ideal conditions for MSW plastics contaminated with biomass wastes (e.g., agricultural) from landfilling. Additionally, since the full ultimate elemental analysis is known for each component, they were ideal candidates for the development of the thermodynamic mode. For the purpose of this analysis, the MSW-derived plastics was considered to be a poly-ethylene derivative. Sub-bituminous coal and a shale-derived natural gas were also considered as reference feedstocks to compare the proposed theoretical process to industrially available gasification processes. Table S3 showcases the different energetic values in Btu/lb for the four feedstocks. Natural gas, due to the abundance of methane, shows the highest heating value while the MSW-derived plastic comes in second due to the assumption that is a poly-ethylene derivative; the energetic value of C-H bonds and relative abundance in the $(CH_2\text{-}CH_2)_n$ backbone is quite high. Corn stover, with a heating value of ~7590 Btu/lb, is much lower than the other feedstocks, however quite in line with other lignocellulosic biofuel materials [30]. The low heating value of biomass feedstocks makes them quite challenging in gasification/power-generation activities; generally, more feedstock is needed to produce the same electricity/chemical yields in reference to coal, for example, due to the high water and oxygen content. This can lead to the presence of tars and waxes which can also foul up the gasifier.

2.7. Indicators Examined

Environmental impact indicators were considered in the assessment of the gasification thermodynamic models developed from AspenTech. The most important metrics considered were carbon dioxide (CO_2), nitrogen oxides (e.g., NO_x), carbon monoxide (CO), reduced sulfur species (H_2S, COS), chloride emissions (Cl^-), ashes from the gasifier block, and predicted process waste-water (from steam). Each metric is reported in the most salient metric possible. For instance, gaseous emissions, such as NO_x, CO, and H_2S/COS, were reported according to U.S. powerplant standards (e.g., lbs contaminant per MMBtu fired in turbines). Other metrics were reported as metric tonnes per hour for simplicity.

2.8. Sensitivity to Examine CO_2 Reduction Using Renewables to Generate Steam

The use of steam gasification and steam-sorbent gasification (Case 4 and Case 5, respectively) for the disposal and recovery of value from MSW plastics and biomass show potential advantages over oxyfuel and air gasification processes. This is due to the use of water as an oxidant, which allowed for enhanced methanation and greater hydrogen yields per feedstock charge. However, high energy penalties are incurred as a function of the external boiler firing duty needed to vaporize BFW to produce VHP steam to supply to the gasifier. This can hamper the overall CO_2 emissions of the process and the energetics, making steam gasification less desirable than oxyfuel gasification. However, if the energy needed to maintain the gasifier temperature of the energy needed to vaporize water into steam for Case 4 and 5 could be sourced by alternative energy systems, the carbon balance of these plants may fall into a more desirable range. A sensitivity analysis was performed to examine the difference in carbon emissions if the energy to charge the gasifier with steam could be sourced from renewables (e.g., solar thermal).

3. Results and Discussion

3.1. Material and Energy Balances

Table 2 shows the overall material balances for the processes examined. Case 2, steam methane reforming (SMR) produces the most hydrogen of all the processes. This is due to the clean nature of the feed which readily reacts with steam to yield H_2 much more efficiently than the solid gasification processes. Case 3 (plastic/biomass oxyfuel) is very similar to Case 1 (coal oxyfuel), however the yield of hydrogen is about half. This suggests that due to the lower combined heating value of the fuel, an MSW-cogasification facility needs to input more feedstock per desired unit hydrogen [30]. The biomass and organic fraction of the feed brings down the intrinsic heating value. Plastics, however, boost the calorific value of the feed due to the abundance of C-C and C-H bonds (e.g., LDPE is about 85% carbon by mass) [31]. The potential for the creation of tars and waxes is very prevalent during the gasification of plastics, and thus the gasifier temperature needs to be kept constantly elevated [11]. Case 4 elucidates that using steam as an oxidant can more readily produce hydrogen in an MSW-cogasification facility and unload power requirements from the ASU; however, the energy requirements for producing the steam charge to the gasifier are large. Case 5 showcases that the use of a gasifier bed material, in this case portlandite, can seriously alleviate both carbon dioxide emission and the penalties associated with the physisorption SELEXOL process [32]. Finally, the air gasification shows little merit (Case 6), with low hydrogen yields and high emissions. This further supports a growing consensus that either pure oxygen or steam must be used as a gasifier oxidant to avoid penalties from circulating large amounts of nitrogen.

Energy balances, in terms of the main users and producers, were tabulated to evaluate the efficacy of the plants from a full-scale power generation standpoint. Overall, similar trends as discussed for the material balances exist for the energy balances as well (Table 3). Case 1 and Case 3 are very similar from an energy balance, as expected, showcasing the predictability of oxyfuel fired gasification processes. The ability to generate roughly the same amount of power, but different hydrogen yields, shows that the differences in the energetic value of the feedstocks are most important in the yield of hydrogen. In Case 4 and Case 5, steam gasification and sorption enhanced steam gasification, significantly boosting the yield of net power, almost 4- and 3.2-fold, respectively. This is impressive, but likely due to the large presence of reformation and methanation reactions that are driven by the steam. These reactions are less likely to occur in the presence of oxygen, as partial oxidation will be dominant. However, it should be noted that preparation of the steam to feed the gasifier requires large amounts of energy, almost 1024 MMKcal/h in Case 4 and 512.1 MMKcal/h in Case 5. This is because in the absence of oxygen, steam gasification is mostly endothermic, so the heat for gasification must come from an ancillary boiler. This boiler will likely be fired with natural gas, and thus will incur additional CO_2 penalties. However, from an absolute yield Case 4 and Case 5 are the most efficient in terms of power generation.

Another important metric in assessing the efficacy of these gasification processes is the feedstock conversion capacities, expressed in both per tonne of hydrogen produced and per MW of power generated. Figure 3 showcases these LCA metrics for all six of the gasification cases studies. As observed, the process utilizing natural gas SMR to generate hydrogen and power is the most efficient from per tonne of methane utilized and per tonne of hydrogen produced. This is attributed to the energetic value of the feedstocks, with methane being the highest. Similarly, coal oxyfuel gasification (Case 2) and the plastic/stover oxyfuel gasification (Case 3) follow the same trend in tonne feedstock/tonne hydrogen, with Case 3 being the most at about 22 tonnes of feedstock per tonne of hydrogen produced. As seen in Case 4 and Case 5, the use of steam and steam plus a gasifier bed sorbent can actually reduce the required tonnage of feedstock per tonne of hydrogen and tonne of feedstock per MW of power. The use of steam as an oxidant increases the amount of methane and hydrogen relative to the other gases, while oxygen in the gasifier increases the relative amounts of CO due to the partial oxidation. Using a gasifier-bed sorbent brings down the tonne of feedstock required per tonne of hydrogen produced by further increasing the amount of hydrogen through in situ carbon capture. By capturing carbon dioxide that is being generated in the gasifier, the sorbent effectively shifts the equilibria of the water gas shift by removing CO_2 in solid form as a carbonate salt, thereby allowing more CO to react with water to yield increased fractions of hydrogen [18]. Thus, the use of bed-active gasifier materials could be an attractive way to further enhance the production of hydrogen from low-calorific-value feedstocks.

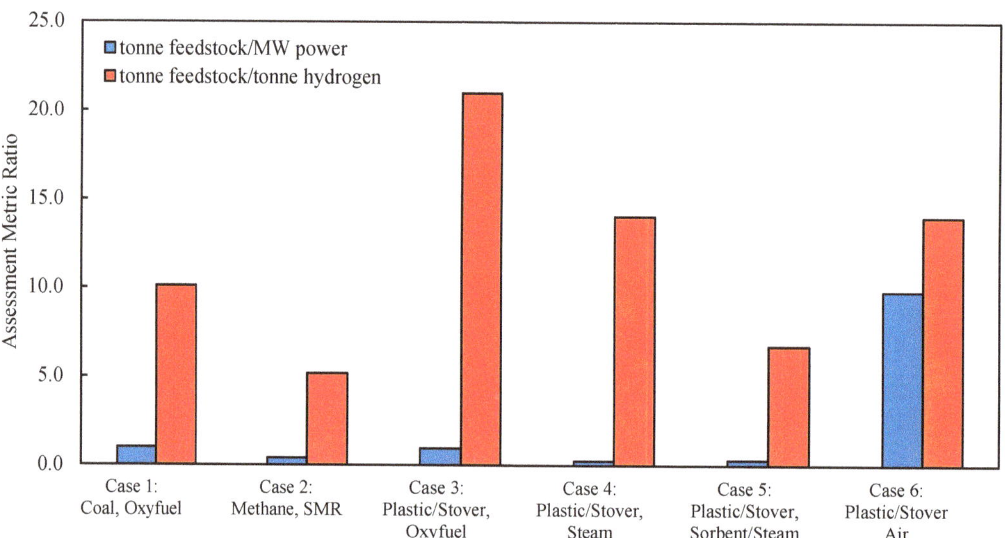

Figure 3. Feedstock production capacity LCA metric for six gasification cases in terms of tonne of feedstock needed for tonne of hydrogen produced (red) and tonne of feedstock needed per MW of power generated (blue).

Table 2. Material balances for the six gasification cases studied as produced by AspenPlus V9.

OVERALL BALANCE	Case 1: Coal, Oxyfuel	Case 2: Methane, SMR	Case 3: Plastic/Stover, Oxyfuel	Case 4: Plastic/Stover, Steam	Case 5: Plastic/Stover, Sorbent/Steam	Case 6: Plastic/StoverAir
FEED						
Feedstock (t/h)	100	100	100	100	100	100
Air to ASU/Gasifier (t/h)	463.4	700	477.9	5.12	202.70	514
Gas Turbine Air (t/h)	297.6	550	350	500	560	0
Cooling Water (t/h)	7115.8	15,105.2	6939.7	6961.9	10,387.8	227
HRSG Water Feed (t/h)	312.7	673	317.9	1247	713.4	328.7
Gasifier Feed Water (for internal steam demand, t/h)	113.4	300	72.06	1157	501.4	208
Diluent Water GTCC	0	0	0	216.1	54	0
Portlandite Feed (t/h)	0	0	0	0	366.7	0
SUM	**8403**	**17,428**	**8258**	**10,187**	**12,886**	**1378**
PRODUCTS						
Hydrogen (t/h)	9.893	19.3	4.765	7.12	14.845	7.14
CO_2 Prod (t/h)	155.4	286.8	155	97.5	0	0
Acid Gas for Claus (t/h)	2.641	5.01	2.44	3.64	1.866	0
Nitrogen Release (t/h)	68.64	159.7	74.1	0	0.85	0
HRSG Stack Gas (t/h)	680.6	1072	705.5	871.9	800.4	599.5
Process Waste-Water (t/h)	57.6	106.5	58	997.6	471.1	215.8
HRSG to Cooling Tower	312.7	673	317.9	1247	713.4	328.7
Cooling Water (t/h)	7115.8	15,105.2	6939.7	6961.9	10,387.8	227
Calcium Carbonate/ Chloride Mass (t/h)	0	0	0	0	495.4	0
SUM	**8403**	**17,428**	**8257**	**10,187**	**12,886**	**1378**

Table 3. Energy balances for the six gasification processes studied as produced by AspenTech simulation.

SUMMARY	Case 1: Coal, Oxyfuel	Case 2: Methane, SMR	Case 3: Plastic/Stover, Oxyfuel	Case 4: Plastic/Stover, Steam	Case 5: Plastic/Stover, Sorbent/Steam	Case 6: Plastic/StoverAir
Gross Total Power Generated (MW)	240.7	479.9	255.3	481.2	451.6	84.2
ASU Air/Oxygen/Nitrogen Compressors (MW)	91.2	136.0	94.3	0.0	49.3	0.0
Gas Turbine Compressor (MW)	39.4	72.8	45.7	66.2	74.1	0.0
Carbon Dioxide Flash Compressor (MW)	8.4	15.9	8.8	5.8	0.2	0.0
SELEXOL Recycle Gas Comp. (MW)	0.3	0.6	0.4	0.3	0.0	0.0
Gasifier Air Compressor (MW)	0.0	0.0	0.0	0.0	0.0	74.0
Net Power (MW)	**101.3**	**254.5**	**106.2**	**408.9**	**327.9**	**10.2**
Refrigeration Load (MMKcal/h)	20.5	37.9	16.5	14.0	0.02	0.0
Steam Firing Duty (MMkcal/h)	-	-	-	1024	512.1	0

3.2. Environmental Impact Assessment

Table 4 shows the environmental impact of the six gasification cycles studied. The major environmental pollutants produced from this site are: carbon dioxide emissions, nitrogen oxide (NOx) emissions from the GTCC, CO emissions from non-combusted material in the GTCC, hydrogen sulfide emissions from the gasification process, chloride emissions from the gasifier, ash emissions (both bottom and fly ash), and process waste-water [33]. Carbon dioxide is produced in both the gasification process and the gas turbine as well. In Cases 4 and 5, where very-high-pressure (VHP) steam is injected into the gasifier, additional CO_2 emissions are incurred from a boiler producing the steam. NOx emissions stem from the gas turbine, where nitrogen can itself become oxidized. CO is emitted from incompletely combusted materials in the gas turbine, and has to be controlled as it is poisonous in large quantities. Diluents can be injected into the gas turbine to mitigate NOx emissions by lowering the combustion temperature; however, lowering the combustion temperature will also increase the amount of CO [34]. Thus, the addition of a diluent must be carefully tuned. Chlorides and ash wastes are significant problems in MSW gasification operations, and need to be carefully controlled to prevent the emission of hazardous pollutants. Chlorides must be removed early on in the process to prevent corrosion downstream via chloride-induced corrosion stress cracking [35]. As seen in Table 4, Case 3 (plastic/biomass, oxyfuel) actually generates the least amount of carbon dioxide and correspondingly low amounts of NOx and CO relative to all other scenarios. Due the presence of the ASU, water was only used to make a feedstock slurry in this case, and thus process waste-water requirements are also reduced. Case 4 yields the highest amount of carbon emissions due to the firing of almost 1124 MMKcal/h of VHP steam as an oxidant in the gasifier, higher CO emissions, and large amounts of waste-water produced. The addition of a sorbent, as shown in Case 5, can assist in reducing CO_2, CO, Cl, Ash, and waste-water emissions relative to Case 4. Thus, the use of a catalytically active gasifier bed material could help at reducing the emissions profile of steam-based gasification processes, which can also produce more hydrogen.

3.3. Emission Potential (CO$_2$-Equivalents)

Overall, it was determined that the air-blown gasification of an MSW plastic/corn stover is the worst from a carbon emission standpoint per tonne of hydrogen produced and MW power (Figure 4). The use of the SELEXOL-based carbon capture process and its advantages can clearly be observed in all other gasification cases relative to Case 6. It should be noted that most of the carbon dioxide emissions for Cases 1–5 used in these LCA metrics come from the HRSG stack, and are generated during the combined cycle. Additional carbon capture systems could be installed for the stack gas; however, for the purpose of this study it was not considered due to the dilute amounts of CO_2 present in the flue gas. Cases 1–5 are quite similar in their carbon emissions relative to power generated, however using steam as an oxidant does incur a greater CO_2 penalty per MW power and an extreme penalty in terms of tonnes of CO_2 per tonne of hydrogen produced. This can be explained by the fact that water is a lesser oxidant in gasification systems. A penalty is incurred in steam gasification because of the large gas firing duties from the auxiliary boilers to supply steam. The chief reactions occurring during steam-based gasification are methanation coupled with water–gas shift, yielding high amounts of methane and hydrogen. The sheer mass of methane allows the significant recovery of energy by firing in the gas turbine, suggesting that from a power perspective, steam gasification could make energetic sense, however from a commodity chemical standpoint, it may not be the best method of recovering value from MSW and biomass. The data suggest that full-scale oxyfuel MSW-plastic/biomass co-gasification facility equipped with a GTCC-HRSG is competitive enough to compete in the range of conventional SMR and coal gasification plants on all emissions LCA metrics as indicated by Figure 4.

Table 4. Environmental impact data for the six gasification cases studied in terms of the major pollutants of this process: carbon dioxide emissions, NOx emissions, CO emissions, H_2S emissions, Chloride emissions, Ash, and Waste-Water. NOx, CO, and H_2S emissions are reported to EPA standards, which is commonly reflected as lbs generated per MMBtu of gasified feedstock.

Environmental Impact Parameters	Case 1: Coal, Oxyfuel	Case 2: Methane, SMR	Case 3: Plastic/Stover, Oxyfuel	Case 4: Plastic/Stover, Steam	Case 5: Plastic/Stover, Sorbent/Steam	Case 6: Plastic/StoverAir
Gross CO_2 emissions (t/h)	111.4	204.9	77.8	286.8	175.3	171.2
NOx emisions (lbs/MMBtu fired)	0.87	1.10	0.47	0.00	0.63	0.00
CO emisions (lbs/MMBtu fired)	0.13	0.34	0.27	4.09	0.35	9.94
H2S emissions (lbs/MMBtu fired)	0.00	0.00	0.00	0.00	0.00	0.01
Chloride Emissions (t/h)	0.379	0.001	0.31	0.31	0.15	0.31
Ash Emissions (Bottom + Fly) (t/h)	15.4	0.0154	8.78	8.78	5.98	8.78
Waste Water Flow (t/h)	57.6	106.5	58	997.6	471.1	215.8

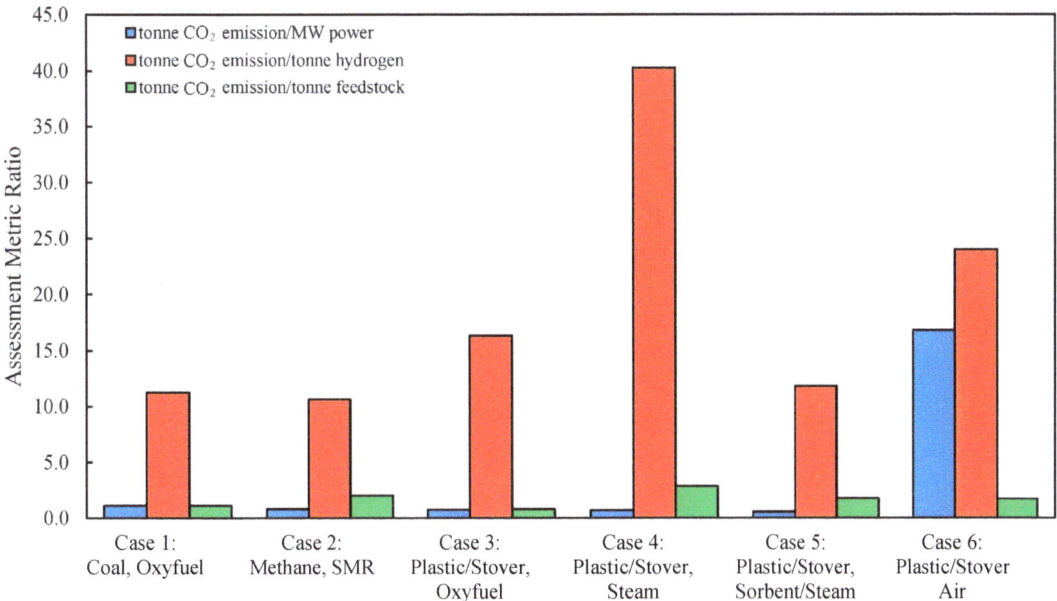

Figure 4. Carbon dioxide LCA emission metric for six gasification cases in terms of tonnes of carbon dioxide emitted per tonne of feedstock (green), tonnes of carbon dioxide emitted per tonne of hydrogen produced (red), and tonnes of carbon dioxide emitted per MW of power generated (blue).

Carbon Capture Potential

As a component of the LCA, a full carbon balance was performed to assess the overall environmental greenhouse gas release potential of the various gasification cycles studies herein. Overall, all gasification cases, with the exception of the air-blown gasifier, were tuned such that almost all of the carbon dioxide produced during gasification was captured using the SELEXOL process. Thus, the intrinsic syngas carbon capture efficiency for all cases (except 6) was approximately 100%. In all cases, as a component of the combined cycle process, a gas turbine was operated to fully combust the PSA offgas. As visualized in Figure 5, the carbon capture efficiency for Case 1 (coal, oxyfuel) and Case 2 (autothermal methane reforming) is quite similar, however the overall carbon emissions for Case 2 are the highest of all the processes. The large amount of CO_2 released in the autothermal reforming process is due to the highly energetic value of the methane feed, producing a PSA off-gas which is quite rich in methane and CO with a great combustible value. The MSW-plastic/biomass oxyfuel, steam, and steam-sorbent gasification series showed relatively similar total CO_2 emissions, suggesting that the PSA off-gas has roughly the same energetic composition to feed the gas turbine. Interestingly, the sorption-enhanced gasification process yielded the highest degree of potential carbon capture due to the presence of the portlandite, which performs in situ capture of CO_2 in the gasifier (producing carbonates), yielding a higher-purity hydrogen and methane stream for downstream use. Portlandite or Ca-bearing phases can be sourced from alkaline industrial wastes, such as steel slag, construction and demolition waste, and mine tailings, which could be used as sorbent materials in this process [36,37]. Regeneration of Ca gasifier-bed active materials could be achieved through conventional calcination and slaking cycles (e.g., calcium looping) which would produce pure gaseous CO_2 for storage, but this could be energy-intense, require additional unit operations, and require piping and access to CO_2 storage wells [38]. The produced calcium carbonates from such a process could be re-used in carbon utilization applications, especially within the context of the built environment. Recent studies have attempted to understand the processes and mechanisms that are the most important in the crystallization of $CaCO_3$ for carbon utilization [39,40] in addition to how these carbonates modify cement hydration [41] and rheology [42] when they are reincorporated as new built environment building blocks. Alternatively, if the carbonate materials are not re-utilized within industrial applications, they can be stored as thermodynamically stable CO_2 sinks underground (e.g., reclaimed mines) for deep and permanent sequestration.

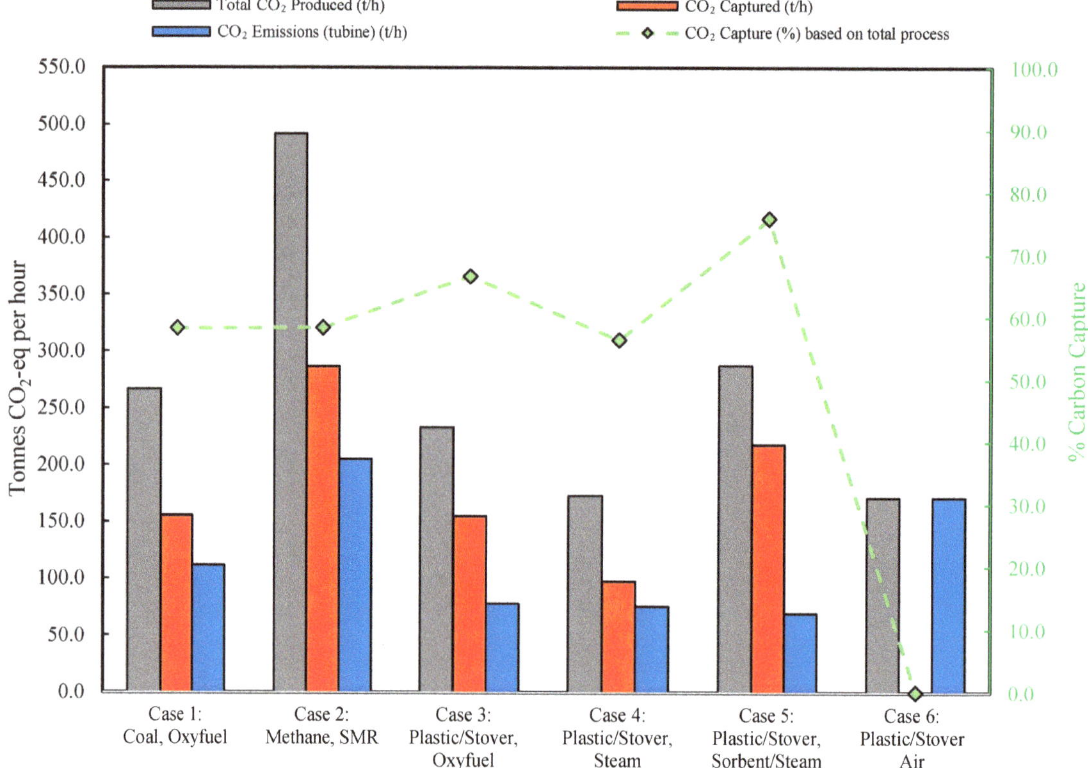

Figure 5. Carbon capture potential for the six gasification cycles considered in this study. The chart shows the total carbon emissions produced (grey), the total carbon captured (red), and the carbon emissions from the turbine (blue). On the right axis (in green), the carbon capture efficiency was tabulated for all the processes.

3.4. Sensitivity Analysis: Use of Renewable Energy Systems

In order to reduce carbon emissions during the generation of steam for the steam gasification of MSWs, a sensitivity test was conducted to investigate offsets using renewable steam generation methods. One such promising technology is the use of concentrating solar power (CSP) plants to supply thermal energy. These facilities utilize thousands of heliostats and power towers to concentrate solar light to a point source boiler at high-pressure and high-temperature conditions (>60 bar, >550 °C). A comparison between the different emission profiles of Case 4 and Case 5 was tabulated on both a per power basis (Figure 6a) and on a per tonne of hydrogen basis (Figure 6b). Overall, significant reductions in CO_2 emissions can be achieved by using CSP to provide heat to the gasifier, especially in Case 4, where no sorbent was utilized in the gasifier itself. The potential hybridization of these steam gasification processes with alternative energy systems may make them more attractive for both near-term and long-term investment prospects.

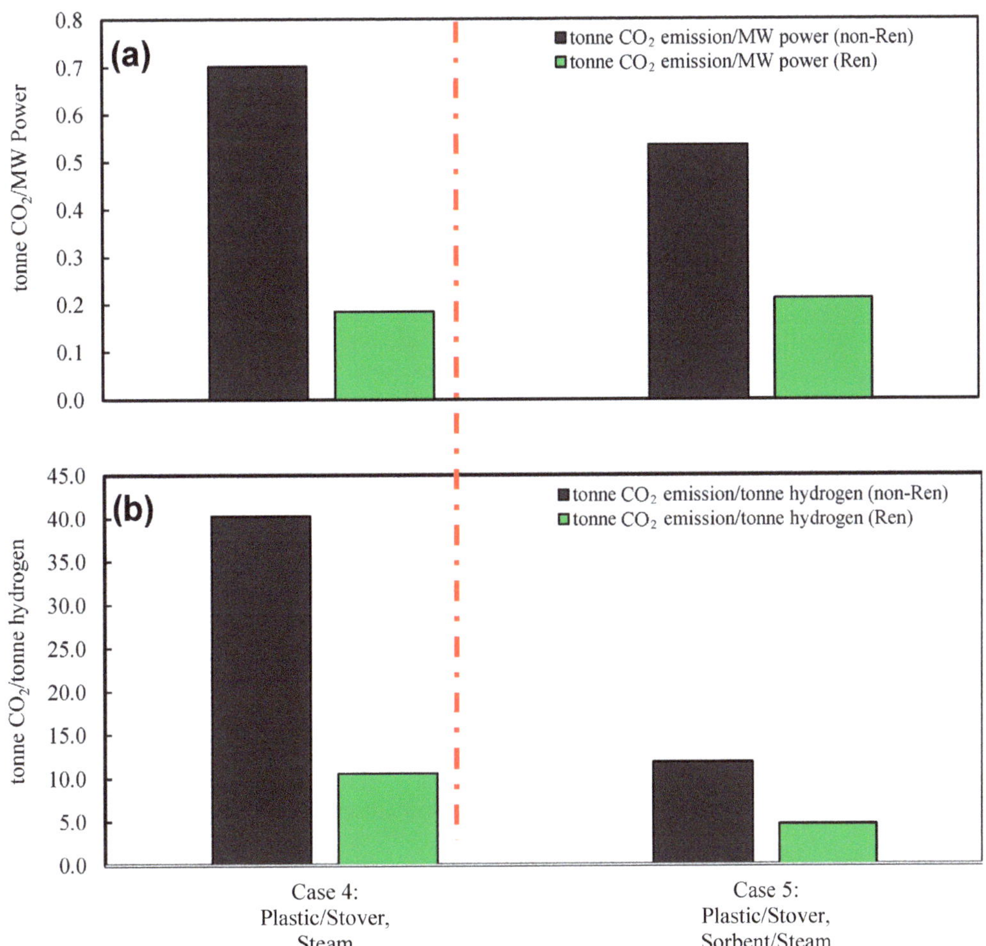

Figure 6. Analysis of using a solar thermal-based power process to produce the steam required for the steam gasification cases, Case 4 and Case 5 in the context of MSW plastic/biomass gasification. Offsetting carbon emissions that are incurred through boiler firing to generate the steam for the gasifier would yield significant benefits from carbon balance standpoint. This figure showcases the reductions (green) versus the base case (black) in tonne of CO_2 per MW power (**a**) and tonne of CO_2 per tonne of hydrogen (**b**) produced.

3.5. Limitations of the Study

While the methodology used in this study provides valuable insights into the conversion of biomass and plastic materials via integrated gasification, several limitations should be noted:

1. **Scope Limitation:** The attributional cradle-to-gate LCA primarily focuses on the conversion process and excludes pretreatment and transportation of feedstocks. This might not capture the complete environmental footprint of the entire lifecycle.

2. **Data Quality and Reliability:** Assumptions made throughout the study, as high-lighted by ISO 14040, can impact the results. While they are meant to provide clarity, they may also introduce biases or inaccuracies.

3. **Thermodynamic Models:** The choice of the Predictive Soave–Redlich–Kwong (PSRK) and NRTL-Electrolyte methods for thermodynamic modeling, while robust, may not account for all possible chemical interactions or unforeseen process deviations.

4. **Feedstock Representation:** The study considers a mix of corn stover and MSW-derived plastics as a representation. The variability in actual feedstock compositions in real-world scenarios might result in different outcomes.

5. **Limitation of Indicators:** While multiple environmental indicators were assessed, other potential environmental impacts might not have been captured in this study.

6. **Scaling Limitations:** The results obtained are based on the described process configurations and may not directly scale or apply to different setups or larger industrial scenarios.

7. **Sensitivity Analysis:** The study assumes that the energy for certain processes could be sourced from renewables. In real-world scenarios, the availability, consistency, and reliability of renewable sources can vary, impacting the outcomes.

8. **External Factors:** External factors like policy changes, technological advancements, or economic factors that might influence the feasibility and efficiency of the described processes in the future were not considered.

Future research should consider addressing these limitations for a more comprehensive understanding of the gasification process and its environmental impacts.

4. Conclusions

Gasification systems applied to mixture of biomass and MSW-derived plastic wastes comprise a technology that has immense potential. As seen by the carbon and energy balances presented herein, steam gasification is a promising method to dispose of these plastics/biomass feedstocks and yields 75% greater fraction of power per tonne of feedstock and 33% greater fraction of hydrogen per tonne of feedstock as compared to air or oxyfuel gasification. Although steam gasification could be a better gasification pathway, extreme heat penalties are incurred via an auxiliary boiler to supply the steam charge. This decreased the total carbon capture ability of the plant by about 10%. However, sourcing alternative energy resources to supply the gasifier with the steam (e.g., CSP), could become practical in a carbon-constrained world, allowing the hybridization of alternative energy, waste disposal, and commodity chemical production. Lastly, using gasifier active bed materials, such as sorbents like portlandite [$Ca(OH)_2$], can dramatically reduce the need for complicated scrubbing systems (e.g., SELEXOL) and the produced calcium carbonate could be safely stored or reused for carbon utilization efforts. Overall, this study provides a thermodynamic metric assessment of emerging gasification technologies to deal with the growing problem of MSWs, especially those rich in biogenic and plastic fractions. As society moves closer towards the development of circularized commodities economies, these advanced gasification facilities, especially using sorbent materials, become quite attractive.

Supplementary Materials: The following supporting information can be downloaded at: https://www.mdpi.com/article/10.3390/en16227548/s1, Table S1: Assumptions for the gasification LCA presented in this study adhering to the ISO 14040. Table S2: Various WtE feedstocks and their associated elemental ultimate analysis (if known) and corresponding Gross Calorific Value (MJ/kg). Table S3: The estimated feedstock energetic values for the gasifier thermodynamic block based on the aforementioned collected feedstock data from Table S2. Figure S1: Overall Process Flow Diagram (PFD) of the modeled IGCC showing all of the major unit operations and the flow of materials throughout. Refs. [43–47] are cited in the Supplementary Materials.

Author Contributions: J.M.W. and A.C.B. conceptualized and planned this study. A.C.B. performed preliminary research and gathered relevant background information. J.M.W. developed the process model and performed the data collection in addition to constructing and writing the main manuscript. All authors have read and agreed to the published version of the manuscript.

Funding: This research received no external funding.

Data Availability Statement: Data are contained within the article and Supplementary Materials.

Acknowledgments: This work was supported by the Earth Engineering Center (Columbia University, NY, 10027).

Conflicts of Interest: The authors declare that they have no known competing financial interest or personal relationship that could have appeared to influence the work reported herein.

References

1. Abdel-Shafy, H.I.; Mansour, M.S.M. Solid Waste Issue: Sources, Composition, Disposal, Recycling, and Valorization. *Egypt. J. Pet.* **2018**, *27*, 1275–1290. [CrossRef]
2. Plastics: Material-Specific Data | Facts and Figures about Materials, Waste and Recycling | US EPA. Available online: https://www.epa.gov/facts-and-figures-about-materials-waste-and-recycling/plastics-material-specific-data (accessed on 13 April 2021).
3. Eriksen, M.; Lebreton, L.C.M.; Carson, H.S.; Thiel, M.; Moore, C.J.; Borerro, J.C.; Galgani, F.; Ryan, P.G.; Reisser, J. Plastic Pollution in the World's Oceans: More than 5 Trillion Plastic Pieces Weighing over 250,000 Tons Afloat at Sea. *PLoS ONE* **2014**, *9*, e111913. [CrossRef] [PubMed]
4. Ritchie, H. Plastic Pollution. Available online: https://ourworldindata.org/plastic-pollution (accessed on 13 April 2021).
5. Geyer, R.; Jambeck, J.R.; Law, K.L. Production, Use, and Fate of All Plastics Ever Made. *Sci. Adv.* **2017**, *3*, 5. [CrossRef] [PubMed]
6. Di, J.; Reck, B.K.; Miatto, A.; Graedel, T.E. United States Plastics: Large Flows, Short Lifetimes, and Negligible Recycling. *Resour. Conserv. Recycl.* **2021**, *167*, 105440. [CrossRef]
7. Hopewell, J.; Dvorak, R.; Kosior, E. Plastics Recycling: Challenges and Opportunities. *Philos. Trans. R. Soc. B Biol. Sci.* **2009**, *364*, 2115–2126. [CrossRef] [PubMed]
8. Nikiema, J.; Asiedu, Z. A Review of the Cost and Effectiveness of Solutions to Address Plastic Pollution. *Environ. Sci. Pollut. Res.* **2022**, *29*, 24547–24573. [CrossRef] [PubMed]
9. Wen, Z.; Xie, Y.; Chen, M.; Dinga, C.D. China's Plastic Import Ban Increases Prospects of Environmental Impact Mitigation of Plastic Waste Trade Flow Worldwide. *Nat. Commun.* **2021**, *12*, 425. [CrossRef]
10. Dong, J.; Tang, Y.; Nzihou, A.; Chi, Y.; Weiss-Hortala, E.; Ni, M.; Zhou, Z. Comparison of Waste-to-Energy Technologies of Gasification and Incineration Using Life Cycle Assessment: Case Studies in Finland, France and China. *J. Clean. Prod.* **2018**, *203*, 287–300. [CrossRef]
11. Ciuffi, B.; Chiaramonti, D.; Rizzo, A.M.; Frediani, M.; Rosi, L. A Critical Review of SCWG in the Context of Available Gasification Technologies for Plastic Waste. *Appl. Sci.* **2020**, *10*, 6307. [CrossRef]
12. Mukherjee, C.; Denney, J.; Mbonimpa, E.G.; Slagley, J.; Bhowmik, R. A Review on Municipal Solid Waste-to-Energy Trends in the USA. *Renew. Sustain. Energy Rev.* **2020**, *119*, 109512. [CrossRef]
13. Wender, I. Reactions of Synthesis Gas. *Fuel Process. Technol.* **1996**, *48*, 189–297. [CrossRef]
14. Maurstad, O. Integrated Gasification Combined Cycle. 2005. Available online: https://sequestration.mit.edu/pdf/LFEE_2005-0 02_WP.pdf (accessed on 20 October 2023).
15. Wall, T.F. Combustion Processes for Carbon Capture. *Proc. Combust. Inst.* **2007**, *31*, 31–47. [CrossRef]
16. Communication, S. Effect of Air Separation Unit Integration on Integrated Gasification Combined Cycle Performance and NO Emission Characteristics. *Korean J. Chem. Eng.* **2007**, *24*, 368–373.
17. Zhang, K.; Ouassil, N.; Campo, C.A.O.; Rim, G.; Kim, W.J.; Park, A.H.A. Kinetic and Mechanistic Investigation of Catalytic Alkaline Thermal Treatment of Xylan Producing High Purity H2 with In-Situ Carbon Capture. *J. Ind. Eng. Chem.* **2020**, *85*, 219–225. [CrossRef]
18. Zhang, K.; Kim, W.J.; Park, A.H.A. Alkaline Thermal Treatment of Seaweed for High-Purity Hydrogen Production with Carbon Capture and Storage Potential. *Nat. Commun.* **2020**, *11*, 3783. [CrossRef] [PubMed]
19. Ferguson, T.E.; Park, Y.; Petit, C.; Park, A.H.A. Novel Approach to Hydrogen Production with Suppressed CO x Generation from a Model Biomass Feedstock. In Proceedings of the Energy and Fuels, Philadelphia, PA, USA, 19–23 August 2012; Volume 26, pp. 4486–4496.
20. Al-Zareer, M.; Dincer, I.; Rosen, M.A. Production of Hydrogen-Rich Syngas from Novel Processes for Gasification of Petroleum Cokes and Coals. *Int. J. Hydrogen Energy* **2020**, *45*, 11577–11592. [CrossRef]
21. Pauls, J.H.; Mahinpey, N.; Mostafavi, E. Simulation of Air-Steam Gasification of Woody Biomass in a Bubbling Fluidized Bed Using Aspen Plus: A Comprehensive Model Including Pyrolysis, Hydrodynamics and Tar Production. *Biomass Bioenergy* **2016**, *95*, 157–166. [CrossRef]
22. Zaman, S.A.; Ghosh, S. A Generic Input–Output Approach in Developing and Optimizing an Aspen plus Steam-Gasification Model for Biomass. *Bioresour. Technol.* **2021**, *337*, 125412. [CrossRef]

23. Tian, W.; Yan, F.; Liang, R. Simulation Analysis of Steam Gasification of Petroleum Coke with CaO. *Pet. Sci. Technol.* **2018**, *36*, 2170–2183. [CrossRef]
24. Indrawan, N.; Mohammad, S.; Kumar, A.; Huhnke, R.L. Modeling Low Temperature Plasma Gasification of Municipal Solid Waste. *Environ. Technol. Innov.* **2019**, *15*, 100412. [CrossRef]
25. Hasanzadeh, R.; Mojaver, P.; Azdast, T.; Chitsaz, A.; Park, C.B. Low-Emission and Energetically Efficient Co-Gasification of Coal by Incorporating Plastic Waste: A Modeling Study. *Chemosphere* **2022**, *299*, 134408. [CrossRef] [PubMed]
26. *ISO 14040*; Environmental Management—Life Cycle Assessment—Principles and Framework. ISO: Geneva, Switzerland, 2006.
27. Gutiérrez Ortiz, F.J.; Ollero, P.; Serrera, A.; Sanz, A. Thermodynamic Study of the Supercritical Water Reforming of Glycerol. *Int. J. Hydrogen Energy* **2011**, *36*, 8994–9013. [CrossRef]
28. Gutiérrez Ortiz, F.J. Biofuel Production from Supercritical Water Gasification of Sustainable Biomass. *Energy Convers. Manag. X* **2022**, *14*, 100164. [CrossRef]
29. Dyment, J.; Watanasiri, S. *Acid Gas Cleaning Using Amine Solvents: Validation with Experimental and Plant Data*; Aspen Technology Inc.: Bedford, MA, USA, 2015; pp. 2–10.
30. Fuels—Higher and Lower Calorific Values. Available online: https://www.engineeringtoolbox.com/fuels-higher-calorific-values-d_169.html (accessed on 31 August 2022).
31. Kartik, S.; Balsora, H.K.; Sharma, M.; Saptoro, A.; Jain, R.K.; Joshi, J.B.; Sharma, A. Valorization of Plastic Wastes for Production of Fuels and Value-Added Chemicals through Pyrolysis—A Review. *Therm. Sci. Eng. Prog.* **2022**, *32*, 101316. [CrossRef]
32. Dai, J.; Whitty, K.J. Chemical Looping Gasification and Sorption Enhanced Gasification of Biomass: A Perspective. *Chem. Eng. Process.—Process Intensif.* **2022**, *174*, 108902. [CrossRef]
33. Glarborg, P. Detailed Kinetic Mechanisms of Pollutant Formation in Combustion Processes. *Comput. Aided Chem. Eng.* **2019**, *45*, 603–645. [CrossRef]
34. Stiehl, B.; Morales, A.; Genova, T.; Otero, M.; Martin, S.; Yoon, C.; Ahmed, K. Controlling Pollutant Emissions in a High-Pressure Combustor with Fuel-Diluent Blending. *Fuel* **2022**, *317*, 123481. [CrossRef]
35. Tsubouchi, N.; Fukuyama, K.; Matsuoka, N.; Mochizuki, Y. Removal of Hydrogen Chloride from Simulated Coal Gasification Fuel Gases Using Honeycomb-Supported Natural Soda Ash. *Fuel* **2022**, *317*, 122231. [CrossRef]
36. Rim, G.; Roy, N.; Zhao, D.; Kawashima, S.; Stallworth, P.E.; Greenbaum, S.G.; Park, A.-H.A. CO_2 Utilization in Built Environment via the PCO2 Swing Carbonation of Alkaline Solid Wastes with Different Mineralogy. *Faraday Discuss.* **2021**, *230*, 187–212. [CrossRef]
37. Zhang, N.; Moment, A. Upcycling Construction and Demolition Waste into Calcium Carbonates: Characterization of Leaching Kinetics and Carbon Mineralization Conditions. *ACS Sustain. Chem. Eng.* **2023**, *11*, 866–879. [CrossRef]
38. Bosoaga, A.; Masek, O.; Oakey, J.E. CO_2 Capture Technologies for Cement Industry. *Energy Procedia* **2009**, *1*, 133–140. [CrossRef]
39. Williams, J.M.; Zhao, D.; Moon, S.; Kawashima, S.; Park, A.-H.A.; Moment, A.J. Stabilization of Pure Vaterite During Carbon Mineralization: Defining Critical Activities, Additive Concentrations, and Gas Flow Conditions for Carbon Utilization. *Cryst. Growth Des.* **2023**, *23*, 8103–8115. [CrossRef]
40. Williams, J.M.; Zhao, D.; Zhang, N.; Chin, A.; Kawashima, S.; Moment, A.J. Directed Synthesis of Aragonite through Semi-Continuous Seeded Crystallization Methods for Carbon Utilization. *CrystEngComm* **2023**, *25*, 6050–6066. [CrossRef]
41. Zhao, D.; Williams, J.M.; Li, Z.; Park, A.-H.A.; Radlińska, A.; Hou, P.; Kawashima, S. Hydration of Cement Pastes with Calcium Carbonate Polymorphs. *Cem. Concr. Res.* **2023**, *173*, 107270. [CrossRef]
42. Zhao, D.; Williams, J.M.; Park, A.H.A.; Kawashima, S. Rheology of Cement Pastes with Calcium Carbonate Polymorphs. *Cem. Concr. Res.* **2023**, *172*, 107214. [CrossRef]
43. Mukherjee, S.; Kumar, P.; Hosseini, A.; Yang, A.; Fennell, P. Comparative Assessment of Gasification Based Coal Power Plants with Various CO_2 Capture Technologies Producing Electricity and Hydrogen. *Energy Fuels* **2014**, *28*, 1028–1040. [CrossRef] [PubMed]
44. Al Lagtah, N.M.A.; Onaizi, S.A.; Albadarin, A.B.; Ghaith, F.A.; Nour, M.I. Techno-Economic Analysis of the Effects of Heat Integration and Different Carbon Capture Technologies on the Performance of Coal-Based IGCC Power Plants. *J. Environ. Chem. Eng.* **2019**, *7*, 103471. [CrossRef]
45. Smith, R.W. Steam Turbine Cycles and Cycle Design Optimization: Combined Cycle Power Plants. In *Advances in Steam Turbines for Modern Power Plants*; Woodhead Publishing: Cambridge, UK, 2017; pp. 57–92. ISBN 9780081003251.
46. Ibrahim, T.K.; Kamil, M.; Awad, O.I.; Rahman, M.M.; Najafi, G.; Basrawi, F.; Abd Alla, A.N.; Mamat, R. The Optimum Performance of the Combined Cycle Power Plant: A Comprehensive Review. *Renew. Sustain. Energy Rev.* **2017**, *79*, 459–474. [CrossRef]
47. Cheng, M.; Verma, P.; Yang, Z.; Axelbaum, R.L. Single-Column Cryogenic Air Separation: Enabling Efficient Oxygen Production with Rapid Startup and Low Capital Costs—Application to Low-Carbon Fossil-Fuel Plants. *Energy Convers. Manag.* **2021**, *248*, 114773. [CrossRef]

Article

Techno-Economic Analysis of Hydrogen–Natural Gas Blended Fuels for 400 MW Combined Cycle Power Plants (CCPPs)

Ju-Yeol Ryu [1,2], Sungho Park [1,*], Changhyeong Lee [1], Seonghyeon Hwang [1] and Jongwoong Lim [1]

[1] Institute for Advanced Engineering, Yongin 17180, Republic of Korea; juyeol.ryu@iae.re.kr (J.-Y.R.); chlee@iae.re.kr (C.L.); shhwang@iae.re.kr (S.H.); limjw97@iae.re.kr (J.L.)
[2] Department of Mechanical Engineering, Sungkyunkwan University, Suwon 16419, Republic of Korea
* Correspondence: sh.park@iae.re.kr

Abstract: Various research and development activities are being conducted to use hydrogen, an environmentally friendly fuel, to achieve carbon neutrality. Using natural gas–hydrogen blends has advantages such as the usage of traditional combined cycle power plant (CCPP) technology and existing natural gas piping infrastructure. Therefore, we conducted CCPP process modeling and economic analysis based on natural gas–hydrogen blends. For process analysis, we developed a process model for a 400 MW natural gas CCPP using ASPEN HYSYS and confirmed an error within the 1% range through operation data validation. For economic analysis, we comparatively reviewed the levelized cost of electricity (LCOE) of CCPPs using hydrogen blended up to 0.5 mole fraction. For LCOE sensitivity analysis, we used fuel cost, capital expenditures, capacity factor, and power generation as variables. LCOE is 109.15 KRW/kWh when the hydrogen fuel price is 2000 KRW/kg and the hydrogen mole fraction is increased to 0.5, a 5% increase from the 103.9 KRW/kWh of CCPPs that use only natural gas. Economic feasibility at the level of 100% natural gas CCPPs is possible by reducing capital expenditures (CAPEX) by at least 20%, but net output should be increased by at least 5% (20.47 MW) when considering only performance improvement.

Keywords: hydrogen–natural gas blends; economic analysis; levelized cost of electricity; total revenue requirement; low-carbon fuels

Citation: Ryu, J.-Y.; Park, S.; Lee, C.; Hwang, S.; Lim, J. Techno-Economic Analysis of Hydrogen–Natural Gas Blended Fuels for 400 MW Combined Cycle Power Plants (CCPPs). *Energies* **2023**, *16*, 6822. https://doi.org/10.3390/en16196822

Academic Editors: Konstantinos Aravossis and Eleni Strantzali

Received: 1 September 2023
Revised: 19 September 2023
Accepted: 25 September 2023
Published: 26 September 2023

1. Introduction

The use of fossil energy in various industries generated 37.1 billion tons of CO_2 emissions worldwide in 2021, which has been causing environmental problems, such as global warming and ocean acidification [1]. Moreover, since CO_2 is a major cause of climate change, in February 2021, 124 countries pledged to make joint efforts to eliminate carbon using carbon reduction technologies to become carbon-neutral by 2050 or 2060 [2]. The plan is to continue to reduce CO_2 emissions through various kinds of research and development activities, but transition to environmentally friendly fuels is crucial at this point to achieve zero emissions. When hydrogen, which is a typical environmentally friendly fuel, is used as a fuel for power generation, only oxygen (O_2) is created as the by-product in the process, and hence it can be the ultimate solution to problems related to energy and the climate crisis. According to market research by the International Energy Agency (IEA), the global demand for hydrogen was 75 million tons in 2019, but it is expected to increase sharply to approximately 1200 million tons by 2070 as its application scope expands to industries, transport, and fuels. Moreover, using hydrogen or hydrogen-based fuels is expected to reduce CO_2 emissions by 8% per year, which is why it is necessary to use hydrogen for sustainable energy industries [3]. To use hydrogen in various industries, it is necessary to establish the entire process of production, storage, and supply. First, hydrogen is classified into three types according to the production method. Gray hydrogen is produced through steam methane reforming (SMR) of fossil fuels (coal, oil, or gas), blue hydrogen is produced by additionally applying carbon capture and storage (CCS) equipment, and green hydrogen produces hydrogen through renewable energy [4]. Of all the hydrogen produced worldwide, 48% is produced using natural gas, 30% using oil, and 18% using coal; only about 4% is produced using water electrolysis [5]. Moreover, less than 1% is produced using renewable energy, which suggests a need to increase production of green hydrogen through continuous R&D and demonstration [6]. Next, the storage and supply method of hydrogen is addressed. To use hydrogen as a fuel, countries with insufficient hydrogen production are considering phase-converting and storing gaseous hydrogen in a liquid state and then supplying it through transport. Liquid hydrogen has an extremely low melting point, 20 K, and it generates boil-off gas (BOG) even with a small heat input from the outside, which limits long-distance transport. Hence, continuous efforts are being made to establish a hydrogen ecosystem by developing technologies, such as slush hydrogen production for zero boil-off application [7,8] or methods to transport hydrogen using catalytic reactions of organic liquids, such as toluene/methylcyclohexane and ammonia (NH3) [9–11].

It Is difficult to ensure economic feasibility with existing technologies, considering the production, storage, and transport process of hydrogen, but it will be possible to produce grey hydrogen for USD 1.0–USD 2.1/kg, blue hydrogen for USD 1.5–USD 2.9/kg, and green hydrogen for USD 3.0–USD 7.5/kg [12]. As 7.5–8 kg of oxygen is generated per kg of hydrogen through electrolysis when a hydrogen electrolyzer is used, a plan has also been suggested to ensure economic feasibility by lowering the cost of produced oxygen to USD 2.98–USD 3.2/kg-H_2 in connection with biomass gas and the process [13]. Moreover, the method of blending natural gas and hydrogen has been receiving attention for using business infrastructure that is already established, and many studies are currently being conducted on this method [14]. Blending hydrogen into a natural gas pipeline network can reduce greenhouse gas emissions more than using just natural gas alone. An experiment proved that blending 20% hydrogen into natural gas for combustion can reduce CO_2 by up to 9.33% per year [15]. Other experiments have also confirmed that blending as much as 20% hydrogen into the engine using natural gas results in lower emissions such as hydrocarbon and carbon monoxide than recommended by European emission standards such as Euro-5 (Euro V) and Euro-6 (Euro VI) [16,17]. Furthermore, it is possible to ensure economic feasibility and increase supply by using the natural gas pipeline networks established in each country, and the demand and supply of hydrogen can be adjusted by gradually increasing the amount of blended hydrogen from 0.1% to 10%, until a large amount of hydrogen production is secured [18,19]. Countries such as the UK, Netherlands, and France have studied ways to blend 2–20% hydrogen into the existing natural gas pipelines and reviewed the applicability by changing the method of combustion control and reinforcing safety equipment [20–22]. However, an experiment regarding the effect of operating pressure on piping when blending natural gas and hydrogen proved that fatigue life rapidly decreased when the amount of hydrogen blended into high-pressure 12 MPa natural gas piping was increased up to 50%, which suggests the need for additional research on materials [23]. A combustor design to prevent flashbacks is important since hydrogen combusts faster than natural gas. Cameretti et al. suggested a method that does not cause flashbacks even when blending more than 10% hydrogen into natural gas using computational fluid dynamics (CFD) [24]. The well-known problem of flashback at higher hydrogen concentrations can be prevented by using water dilution [25]

Recently, gas turbines have been developed, such as the distributed electric and thermal energy generation to avoid any possible waste [26]. Combustor development is one of the key technologies of gas turbines, and the GE DLN-2.6 combustor is capable of 15% hydrogen cofiring, which is limited to 5% in actual operation. There is ongoing research and demonstration to apply high-concentration hydrogen of more than 50% [27]. Siemens is capable of up to 15% hydrogen blending without significantly changing the current natural gas combustor for natural gas–hydrogen cofiring and is currently validating the performance of the gas turbine combustor to apply up to 50% [28]. An examination of fuel characteristics and review of the performance of diaphragm gas meters to accurately measure the flow rate of blended gas revealed that the error is small when 0–15% hydrogen is blended into natural gas [29].

Meanwhile, many studies anticipate several benefits from using natural gas–hydrogen blends, but there are several problems. Italy has a natural gas pipeline network of approximately 300,000 km, so economic benefits are expected from blending hydrogen. However, the lower heating value (LHV) per unit mass of hydrogen is 120.1 MJ/kg, which is higher than that of natural gas (49.3 MJ/kg), but the heating value per unit volume is 10.8 MJ/Nm^3, which is lower than that of natural gas (39.08 MJ/Nm^3). Hence, the volume of hydrogen should be at least 3.6 times that of natural gas to produce the same heating value [30]. Therefore, when using blended fuel, it is important to design the combustor according to the increase in volume. Moreover, various studies have been conducted on the levelized cost of hydrogen (LCOH) in which hydrogen is produced and stored using a hydrogen electrolyzer associated with renewable energy, but many studies are still needed to ensure economic feasibility at the level of USD 37.9–USD 52.9/kg when applying a 200–300 kW hydrogen electrolyzer [31]. Therefore, this study validated a process model for a combined cycle power plant (CCPP) using natural gas–hydrogen blends as fuels and examined the economic benefits of using natural gas–hydrogen blends through economic analysis. First, we validated the analytical model by comparing the simulation results of the existing CCPP process that uses 100% natural gas as fuel with actual operation data. Then, using the validated model, we calculated the change rate in power generation and temperature character depending on the amount of hydrogen blended. Therefore, we verified the fuel costs of adequate hydrogen by comparing the levelized cost of electricity (LCOE) expected from operating a 400 MW CCPP with natural gas–hydrogen blends. In addition, we proposed proper operation conditions to secure competitiveness with natural gas CCPPs by comparing LCOE according to changes in hydrogen fuel cost, capacity factor, and facilities investment cost, namely, capital expenditures (CAPEX).

2. Methodology

2.1. Process Model

2.1.1. Assumption of Combined Cycle Power Plant (CCPP)

The CCPP process generates power using natural gas as fuel, and it is a system that operates at more than 60% efficiency by generating power from a gas turbine while recovering the heat from the high-temperature exhaust gas discharged simultaneously, which is supplied to the steam turbine [32]. CCPPs mainly comprise a compressor, gas turbine, heat recovery steam generator, steam turbine, deaerator, condenser, boiler feedwater pump (BFP), and condensate extraction pump (CEP).

Figure 1 shows the schematic diagram for the performance review of a CCPP, which mostly comprises 1 gas turbine, 1 heat recovery steam generator, 1 steam turbine, and balance of plant (BOP) equipment. The net power output of the process is 393.58 MW, and the net power efficiency at higher heating value (HHV) and lower heating value (LHV) is 53.6% and 58.8%, respectively. Conditions such as ambient relative humidity of 60%, ambient dry bulb temperature of 15°C, and atmospheric pressure of 1.013 bar(a) were considered, and the HHV and LHV of the natural gas supplied were 54,136 kJ/kg and 49,300 kJ/kg, respectively [33]. In addition, the following conditions were set for process analysis.

- The flow is in a steady state.
- Air and combustion products are assumed as ideal gas.
- The gas turbine and steam turbine models are operated at a steady state.
- Heat transfer between the components of the plant and the environment is negligible.

We used ASPEN HYSYS V 12.0 for the CCPP process modeling and applied the Peng–Robinson (PR) equation of state (EOS) for analysis. The values provided by the HYSYS database were used for material properties. The composition of Natural gas is shown in Table 1.

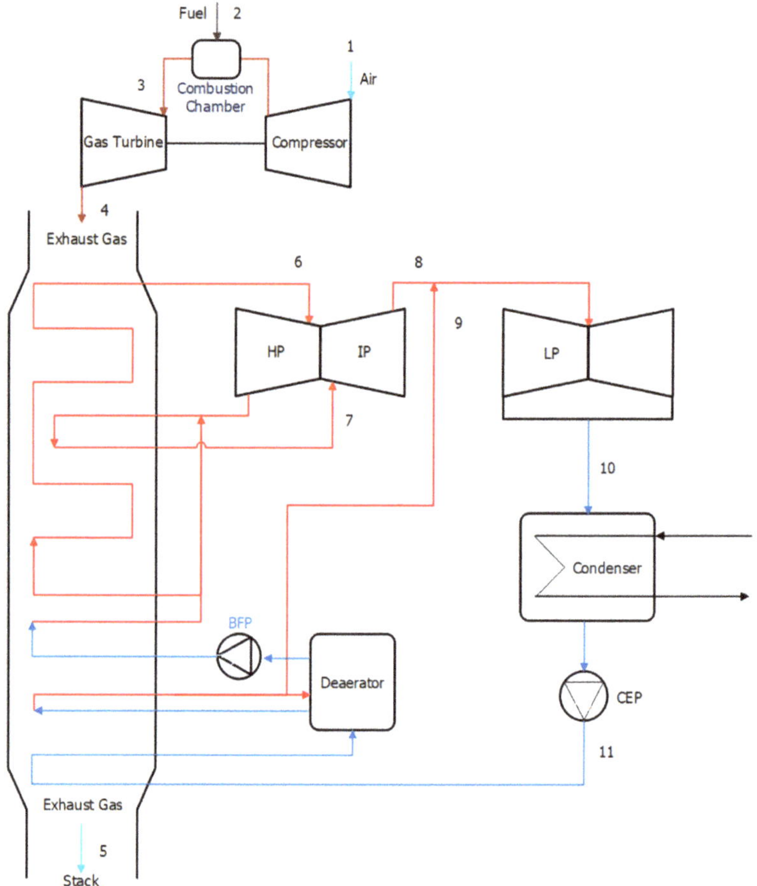

Figure 1. Schematic diagram of a combined cycle power plant.

Table 1. Natural gas fuel composition.

Gas	CH_4	C_2H_6	C_3H_8
Vol (%)	89.5	8.8	1.7

2.1.2. Model simulation

For CCPP process modeling, we used the 400 MW CCPP heat and mass balance diagram operated by Korea South-East Power Co., Ltd. (Jinju, Republic of Korea). To perform block modeling including the gas turbine, the combustion efficiency of the combustor was set at 100% and the heat loss that may occur in the combustion process was set at 3%. The efficiency of the gas turbine and compressor was set at 85% and 89.3%, respectively, and it was modeled so that 11% of the compressed air flow would be used for cooling the gas turbine. Required equations for the calculation of components of compressor and gas turbine are given below [34].

Compressor

$$T_{out} = T_{in}\left(1 + \frac{1}{\eta_{AC}}\left(r_{AC}^{\frac{k-1}{k}} - 1\right)\right)$$

(1)

Gas turbine

$$T_{out} = T_{in}\left(1 - \eta_{GT}\left(1 - \left(\frac{P_{in}}{P_{out}}\right)^{\frac{k-1}{k}}\right)\right) \tag{2}$$

The heat recovery steam generator of the steam turbine block was modeled by arranging 4 economizers, 3 evaporators, and 7 superheaters, and the minimum approach temperature was set at 5 K. We conducted a comparative review on temperature, pressure, and flow rate at the major points, and the differences between the actual heat and mass balance diagram and the simulation model are as shown in Table 2.

Table 2. Thermophysical property comparison of actual and simulation data.

Point	Stream	Temperature (°C)		Pressure (Bar)		Mass Flow Rate (t/h)	
		Actual	Simulation	Actual	Simulation	Actual	Simulation
1	Air	15	15	1.013	1.013	2,122	2132
2	Natural gas	200	200	39	39	48.83	48.83
3	Combustion gas	1500	1,514	39	39	2,170	2181
4	Exhaust gas	611.8	616.0	1.039	1.09	2,170	2181
5	Exhaust gas	83.0	83.6	1.013	1.07	2,170	2181
6	Steam	596.4	596.0	129.7	129.7	257.5	288.8
7	Steam	582.3	582.0	27.2	31.3	283.1	317.4
8	Steam	235.5	238.2	2.0	2.5	289.5	289.6
9	Steam	244.2	245	4.0	4.2	47.8	49.2
10	Steam	29.4	31.2	0.041	0.094	340.1	342
11	Water	29.5	29.5	9.5	9.5	340.8	345

The model analysis results revealed a difference in flow rate at certain points, and there were some errors in the process since the LP sealing steam and the steam fumed intermittently to the condenser. However, we confirmed that the maximum error was around 1% by similarly controlling the rates of fuel consumption and total power produced in the steam turbine and gas turbine blocks. The thermodynamic efficiency of the CCPP was evaluated by net efficiency ($\eta_{net,CCPP}$) based on the power produced, and it is defined as shown in Equation (3).

$$\eta_{net,\,CCPP} = \frac{P_{net,\,GT} + P_{net,\,ST}}{(\dot{m}_{NG}) \times LHV} \times 100 \tag{3}$$

Here, $P_{net,GT}$ is the net power of the gas turbine, excluding the auxiliary power generated in the compressor from the gross power produced in the gas turbine. $P_{net,ST}$ is the net power of the steam turbine, excluding power such as BFP and CEP from the gross power produced in the steam turbine, and $\dot{m}_{NG}$ is the fuel supply based on LHV.

We compared the change in the amount of hydrogen blended with natural gas by increasing the amount from 0 to 0.5 in mole fraction. Equation (4) shows the natural gas-hydrogen blend ratio in mole fraction [35], and the amount of natural gas–hydrogen blends injected is as shown in Table 3.

$$\text{Mole fraction}_{H2} = \frac{\chi_{H2}}{\chi_{H2} + \chi_{NG}} \times 100 \tag{4}$$

2.2. Economic Model

2.2.1. Methodology of Levelized Cost of Electricity (LCOE)

Connecting the processes or converting fuels can improve the efficiency of the CCPP system, but it generally involves a complicated system or reduces economic feasibility. Hence, a newly proposed process or a process altered by fuel conversion requires a comparative review between different power generation systems through economic evaluation.

The LCOE can quantitatively evaluate the economic feasibility of the source of power through the process of converting the costs required for constructing and operating the equipment in the CCPP into the present value and levelizing them. The total revenue requirement (TRR) methodology used by the US Electric Power Research Institute (EPRI) was applied to calculate the LCOE of the 400 MW natural gas–hydrogen CCPP [36].

Table 3. Flow rate of blended fuel based on mole fraction.

Fuel Composition			
H_2 Mole Fraction	H_2 Flow Rate (t/h)	NG Mole Fraction	NG Flow Rate (t/h)
0	0	1.0	48.83
0.1	0.662	0.9	47.38
0.2	1.435	0.8	45.68
0.3	2.352	0.7	43.67
0.4	3.455	0.6	41.24
0.5	4.808	0.5	38.26

The TRR calculates the cost of system construction and other expenditures with the cost that must be recovered annually by selling electric power. Hence, it requires the calculation of TCI (total capital investment), which consists of FCI (fixed capital investment) and OO (other outlay). FCI is divided into DC (direct cost) and IC (indirect cost) and is expressed as shown in Equation (5).

$$TCI = FCI + OO = DC + IC + OO \tag{5}$$

DC includes purchased equipment cost (PEC), piping, land, and service facilities, and IC includes engineering cost, construction cost, and contingency. OO includes startup cost, working capital, and allowance for funds used during construction.

Meanwhile, TRR is calculated as the sum of annual expense and CC (carrying charge) required for facility operation. Expenses comprise electricity cost (or fuel cost, FC) and O&M cost (OMC), and CC includes capital recovery, return on equity, return on debt, income taxes, other taxes, and insurance. Figure 2 shows the diagram for calculating TRR [37].

CR_j (capital recovery) is calculated as the sum of BD_j (book depreciation), $DITX_j$ (differed income taxes), and $RCEAF_j$ (recovery of common-equity AFUDC), as shown in Equation (6).

$$CR_j = BD_j + DITX_j + RCEAF_j \tag{6}$$

$DITX$ is the tax incurred owing to the difference between TXD (tax depreciation) and BD (book depreciation), and it is as shown in Equation (7), considering $f_{MARCS,\,j}$ (rate of depreciation), t (tax rate), and TL (taxation period).

$$\begin{aligned}
TXD &= TDI + f_{MARCS,\,j} & j &= 1, \ldots, \text{TL} + 1 \\
TXD &= 0 & j &= \text{TL} + 2, \ldots, n \\
DITX &= (TXD - BD) \times t & j &= 1, \ldots, \text{TL} + 1 \\
DITX &= -\frac{\sum_{k=1}^{TL+1} DITX_k}{n - (TL-1)} & j &= \text{TL} + 2, \ldots, n
\end{aligned} \tag{7}$$

Meanwhile, CC (carrying charge) is calculated as shown in Equation (8), using variables such as ROI (return on investment), BBY (balance beginning of year), f_x (funding ratio), ADJ (adjustment), and BD (book depreciation).

$$
\begin{aligned}
ROI &= BBY_{j,x} \times i_x & x &= \mathrm{d, ps, ce} \\
BBY &= TCI \times f_x & x &= \mathrm{d, ps, ce} \\
BBY_j &= BBY_{j-1} - (BD_{j-1} + ADJ_{j-1}) & j &= 2, \ldots, \mathrm{n} \\
ADY_{j,d} &= DITX_j \times f_x & j &= 2, \ldots, \mathrm{n},\ x = \mathrm{d, ps} \\
ADY_{j,d} &= DITX_j \times f_{ce} + RCEAF_j & j &= 1, \ldots, \mathrm{n} \\
ITX &= \tfrac{t}{1-t}\left(ROI_{ce} \times ROI_{ps} + RCEAF_j\right) - DITX & & \\
CC &= TCR + ROI_{ce} + ROI_{ps} + ROI_d + ITX + OXTI & & \\
\text{Expense} &= FC + OMC & &
\end{aligned}
\tag{8}
$$

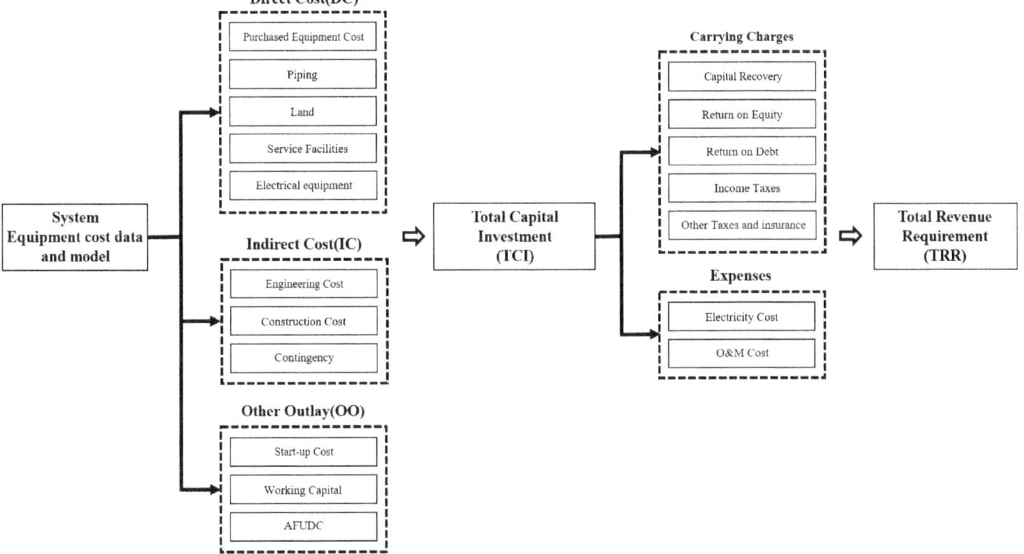

Figure 2. The concept of the TRR method for a CCPP.

Then, to levelize TRR, i.e., the sum of CC and expense, the constant escalation levelization factor (CELF) is applied to the cost incurred for overall system operation, as shown in Equation (9).

$$
\begin{aligned}
CC_L &= CRF \times \sum_{1}^{n} \frac{CC_j}{\left(1+i_{eff}\right)^j} \\
CRF &= \frac{i_{eff}\left(1+i_{eff}\right)^n}{\left(1+i_{eff}\right)^n - 1} \\
FC_L &= FC_O \times CELF_{FC} \\
CELF_{FC} &= \frac{k_{FC}\left(1-k_{FC}^n\right)}{1-k_{FC}} \times CRF, \quad k_{FC} = \frac{1+r_{FC}}{1+i_{eff}} \\
OMC_L &= OMC_O \times CELF_{OMC} \\
CELF_{OMC} &= \frac{k_{OMC}\left(1-k_{OMC}^n\right)}{1-k_{OMC}} \times CRF, \quad k_{OMC} = \frac{1+r_{OMC}}{1+i_{eff}} \\
TRR_L &= CC_L + FC_L + OMC_L
\end{aligned}
\tag{9}
$$

The LCOE is calculated by subtracting BPV (by-product value) from TRR and dividing the result by annual power, as shown in Equation (10) [38].

$$LCOE \; [\$/MWh] = \frac{TRR_L - BPV}{Annual \; Power} \tag{10}$$

2.2.2. Capital Cost Calculation

To calculate the LCOE of a natural gas–hydrogen CCPP, a levelization process is required to prepare cash flows, calculate annual costs to be recovered, and convert them into present values. Hence, a few necessary conditions for economic analysis were assumed, as shown in Table 4. For the annual inflation rate, nominal inflation rate, and exchange rate, 1.5%, 1.5%, and KRW 1100 were applied, respectively, with the consideration of the means from 2012 to 2020 for each [39,40]. The first and second FPI supply refer to the facilities investment cost of each year, assuming that the construction period is two years, and this is randomly assumed to convert the interest incurred during the construction period into allowance for funds used during construction. Total income tax rate was set as 22%, and other tax rate as 2.0%, which is 10% of total income tax rate [41]. Since the lifetime of a turbine, which is a major facility, is generally about 30 years, plant life was set as 30 years and tax years, as 20 years [42]. In addition, capacity factor was set as 28.6% based on the actual utilization rate, and for fuel cost, the actual fuel cost of the CCPP operated by Korea South-East Power Co., Ltd. was applied. Regarding the combustor replacement cost for natural gas–hydrogen cofiring, we used the data provided by a gas turbine company in Korea. The results obtained from the experiment and analysis of CFD can be used to adjust the amount of hydrogen blended or replace the combustor for application to the existing CCPP system [43].

Table 4. Economic assumptions and index input for economic analysis.

		Contents	**Unit**	**Value**
Overall economic index		Annual inflation rate [39]	%	1.5
		Nominal inflation rate [39]	%	1.5
		Fuel escalation	%	1.0
		Levelized interest rate	%	4.7
		First FPI supply	%	40.0
		Second FPI supply	%	60.0
		Won–dollar exchange rate [40]	KRW	1100
System financing		Plant design start year	year	2020
		Plant construction start year	year	2020
		Plant operation start year	year	2022
	Common equity	Financing fraction	%	50.7
		Required annual return	%	7.0
	Preferred stock	Financing fraction	%	0.0
		Required annual return	%	8.0
	Debt	Financing fraction	%	49.3
		Required annual return	%	2.4
		Resulting average cost of money	%	4.7
		Total income tax rate [41]	%	22.0
		Other tax income rate [41]	%	2.0
Plant operation index		Plant life [42]	year	30
		Tax life	year	20
		Capacity factor (or plant operation rate)	%/year	28.6
		Power plant net power	kW	406,211
	Fuel cost	Natural gas unit price	USD/MJ	20,488
		Hydrogen unit price	USD/t	7273
	Combustor	Number of combustors	ea.	14
		Unit cost per combustor	USD/ea.	272,727
		Lifetime of combustor	h	25,000
		Total combustor cost for repair	USD	26,757,818.2
		Total combustor cost for repair per year	USD/year	1,337,891

Meanwhile, for the total facilities investment cost, land cost, and other utility costs of the natural gas CCPP, we used the data provided by Korea South-East Power Co., Ltd., and other main equipment manufacturers to calculate DC, IC, and OO. Table 5 summarizes the total net outlay and total facilities investment cost that is not depreciated from the total investment cost calculated.

Table 5. Capital cost calculation summary.

Contents				Cost (USD)
Fixed capital investment	Direct cost	Onsite costs	Purchased equipment cost	209,090,909
		Offsite costs	Land cost	20,909,901
			Civil, structural and supervision	118,181,818
		Total cost		348,181,818
	Indirect cost	Engineering and supervision		27,854,545
		Construction cost		52,227,273
		Contingency		64,239,545
		Total cost		144,321,364
	Total cost			492,503,182
Other outlay	Startup cost	Fuel and O&M for startup		9,543,459
		Escalated startup cost		288,451
		Total cost		9,831,910
	Working capital	Working capital cost		23,233,479
		Escalated working capital cost		1,061,267
		Total cost		24,294,746
	AFUDC	Allowance for funds used during construction		30,372,455
		Total AFUDC after 2 years		34,883,398
Total capital investment (TCI)	Total net outlay		Land cost	20,909,091
			Plant facilities investment	490,220,655
			Startup cost	9,831,910
			Working capital	24,294,746
			Total net outlay	545,256,402
		Total cost		580,139,800
Total depreciable capital Investment	Total net capital investment	Total capital investment		580,139,800
		Total cost		580,139,800
	Total nondepreciable capital investment	Land cost		20,909,091
		Working capital		24,294,746
		Common equity AFUDC		25,948,579
		Total cost		71,152,416
		Total depreciable capital investment		508,987,384

2.2.3. Model Development

Before calculating the LCOE of the 400 MW natural gas–hydrogen CCPP system, it is necessary to validate the TRR model. Hence, validation was conducted to determine whether the same level of LCOE is calculated by applying the actual facilities investment cost of Bundang CCPP Unit 2, which has been in operation since 1997. Table 6 summarizes the variables applied for validation. The results showed that the LCOE calculated using the suggested TRR method was 96.5 KRW/kWh, indicating a 1.4% difference from that of Bundang CCPP Unit 2 (95.16 KRW/kWh), which confirmed that the proposed model has sufficient reliability.

Table 6. Evaluation of TRR method model.

Contents	Unit	Bundang CCPP-2	TRR Method Simulation
Total capital investment	KRW	162,900,000,000	162,900,000,000
Common equity financing fraction	%	50.73	50.73
Cost of equity capital	%	7.02	7.02
Debt financing fraction	%	49.27	49.27
Cost of debt capital	%	2.36	2.36
Weighted average cost of capital	%	4.7	4.7
Income tax rate	%	22	22
Plant lifetime	Year	30	30
Capacity factor	%	28.6	28.6
Plant net power	MW	368	368
Fuel cost/year, only NG	KRW	80,200,000,000	80,200,000,000
Levelized cost of electricity	KRW/kWh	95.16	96.5

3. Analysis Results

3.1. Process Simulation Results

Based on the model that has been validated using actual CCPP data, we checked for a change in performance according to the natural gas–hydrogen blend ratio. For performance comparison, we applied the same condition by setting the heat energy of the natural gas–hydrogen blend supplied to the gas turbine at 743.3 MW and consistently supplying air at a flow rate of 2132 t/h by replacing only the combustor in the existing gas turbine. Moreover, the amount of air required according to the increase in hydrogen cofiring rate increased gradually when the hydrogen volume was 80% or higher, but the ratio was around 0.5%, proving that there was almost no change in the characteristics of the compressor [33]. Table 7 shows the comparison of the process analysis results and efficiency.

Table 7. Results of the thermodynamic analysis.

Contents		Unit	Actual	Simulation	Error (%)
Gas turbine block	NG flow rate	t/h	48.83	48.83	-
	Air flow rate	t/h	2122	2132	0.46
	GT inlet temperature	°C	1500	1500	-
	GT outlet temperature	°C	611.8	616	0.65
	GT exhaust gas flow rate	t/h	2170	2181	0.46
	Net power	kW	263,180	263,197	0.01
Steam turbine block	BFP flow rate	t/h	340	345	1.47
	HRSG inlet temperature	°C	83	83.6	0.72
	Net power	kW	130,400	130,968	0.43
Total net power generation		kW	393,580	394,165	0.14
$\eta_{net.CCPP}$ (LHV)		%	58.86	58.94	0.13

An increase in the ratio of hydrogen blended into the fuel led to an increase in the output of the gas and steam turbines. At 0.5 mole fraction, the gas turbine block generated 271.17 MW of power, showing that power generation increased by 3.03%, while the steam turbine block generated 135.36 MW, showing that power generation increased by approximately 3.1%. Thus, a total of 406.53 MW was generated. Figure 3 shows the characteristics of the increase in an enthalpy change and output due to the increase in the partial pressure of water as the hydrogen blend ratio increased.

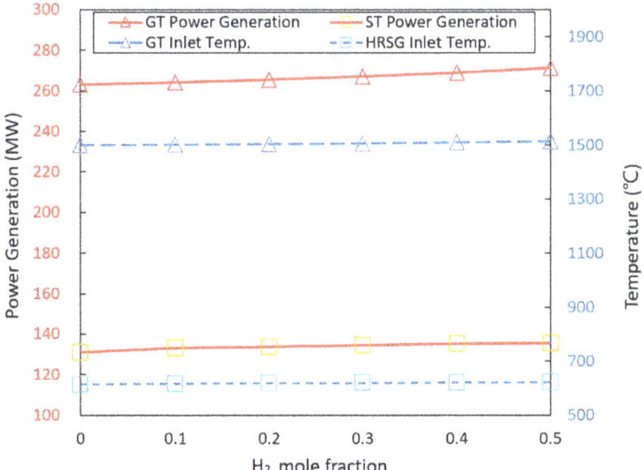

Figure 3. The performance results of the CCPP based on blended fuels.

Figure 4 shows that when the mole fraction of hydrogen in the fuel increases to 0.5, the net efficiency improves by 1.86% from 58.94% to 60.8%, proving that fuel supply in mass decreases by 11.8%. This is because the per unit mass LHV of hydrogen is 2.43 times greater than that of LNG, but the analysis was conducted assuming the heating value of the fuel supplied to the CCPP is the same. Therefore, higher efficiency can be expected by increasing the hydrogen blend ratio in the fuel. The inlet volume flow increased by 63.9% from that when supplying 100% natural gas. This proves that combustor design is important for using hydrogen fuel blends.

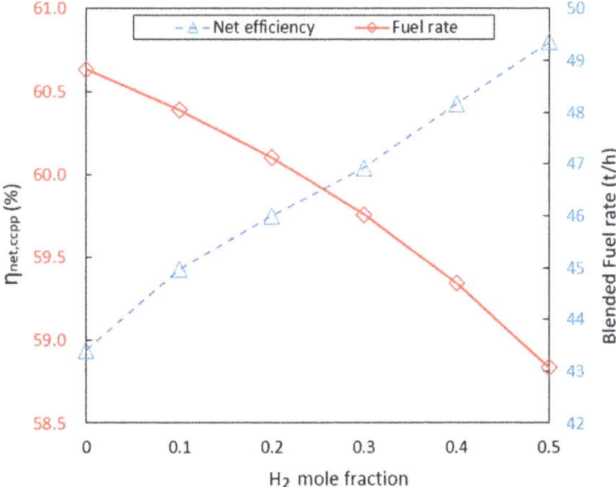

Figure 4. Correlation of net efficiency with fuel consumption based on blended fuels.

3.2. LCOE of Natural Gas–Hydrogen CCPP

We reviewed the expected LCOE in case a natural gas–hydrogen CCPP is operated by changing or replacing the existing combustor in the LCOE model based on the TRR method. The target hydrogen supply price was 6000 KRW/kg in 2022 according to the Korean government's Hydrogen Economy Roadmap, but as the distribution was 7000–8800 KRW/kg in 2022, the hydrogen fuel supply price was set as 8000 KRW/kg [44]. Moreover, we calculated the combustor replacement cost and TCI based on the expected cost of constructing a 400 MW

CCPP provided by a turbine manufacturer and used the values listed in Table 6 for other variables. The LCOE when only natural gas is used versus when natural gas–hydrogen blends are used is as shown in Table 8.

Table 8. Index input for LCOE of a natural gas–hydrogen CCPP.

	Unit	Natural Gas	Natural Gas and Hydrogen
Total capital investment	KRW	360,000,000,000	360,000,000,000
Hydrogen mole fraction	-	0	0.5
Plant lifetime	Year	20	20
Plant operating rate	%	28.6	28.6
Plant net power	MW	394.165	406.53
Combustor repair cost	KRW/year	-	1,470,000,000
Fuel cost/year	KRW	72,000,000,000	152,800,000,000
Levelized cost of electricity	KRW/kWh	103.9	180.67

For the natural gas CCPP, fuel and maintenance costs play a bigger role than capital expenditures (CAPEX) when calculating LCOE. Consequently, it was found that the cost increased up to 180.67 KRW/kWh when operating a natural gas–hydrogen CCPP owing to hydrogen fuel cost.

3.3. Sensitivity Analysis of LCOE

The key variables that affect the LCOE calculation of the CCPP include fuel cost, capacity factor, CAPEX, and power generation. Hence, we reviewed ways to achieve price competitiveness by conducting a sensitivity analysis of the variables that affect LCOE.

The IEA has predicted that the price of hydrogen in China will decrease to USD 2–USD 5/kg by 2030, and the price of hydrogen in the global market is expected to be USD 1.5–USD 2.5/kg [45]. For fuel cost, since LCOE may fluctuate greatly depending on hydrogen supply price, LCOE was reviewed at the price range of 2000–8000 KRW/kg (USD 1.8–USD 7.2/kg). Figure 5 shows the analysis results based on a hydrogen fuel supply of up to 50% in terms of mole fraction. The range of LCOE when using 50% blends is 109.15–180.67 KRW/kWh, and when the supply fuel is converted 100% to hydrogen, the expected LCOE could be 432.08 KRW/kWh (8000 KRW/kg), 280.20 KRW/kWh (5000 KRW/kg), or 128.32 KRW/kWh (2000 KRW/kg).

Capacity factor, which is the utilization rate of the natural gas–hydrogen CCPP, can also be a key variable. Increasing the capacity factor from 28.6% to 35% or more can lower the LCOE to 103.76 KRW/kWh, down to the LCOE level (103.9 KRW/kWh) of a CCPP using only natural gas. Here, hydrogen fuel price must be lowered to 2000 KRW/kg, and the results of the economic analysis on hydrogen supply price and capacity factor are as shown in Figure 6.

The investment cost associated with hydrogen production is expected to be reduced by approximately 30% by 2050 with a learning rate of 17–23% due to technology development and learning effects [46,47]. Even in the case of combustors and related equipment for using hydrogen fuel, it is necessary to review LCOE according to an approximately 30% change in CAPEX, considering a case in which CAPEX decreases owing to technology development or the cost increases owing to increased technical difficulty. Figure 7 shows the change in LCOE according to the increase and decrease in CAPEX. Even when CAPEX decreases by up to 30%, the LCOE changes only by around 5%. When the hydrogen supply price is 8000 KRW/kg, the LCOE is 171 KRW/kWh even when CAPEX is reduced by 30%. However, it is 2000 KRW/kg, the LCOE is 118.17 KRW/kWh even when CAPEX is increased by 30%, which is significantly lower. Therefore, a change in fuel price is much more important than a change in CAPEX.

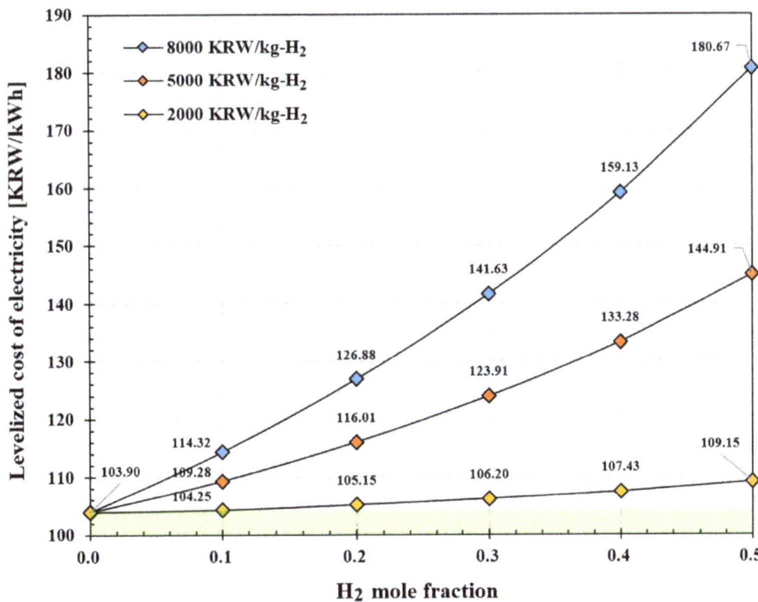

Figure 5. LCOE analysis results considering fuel blend.

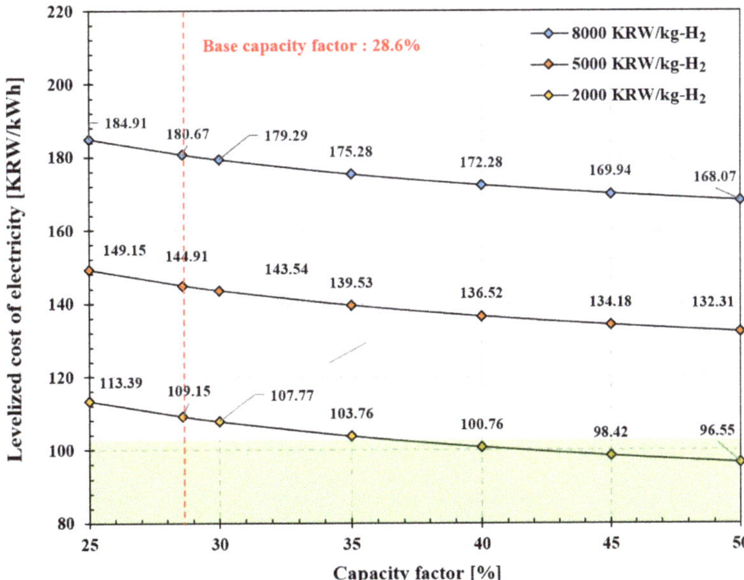

Figure 6. LCOE analysis results considering capacity factor.

Finally, Figure 8 shows the power generation required for a natural gas–hydrogen CCPP to achieve competitiveness with a CCPP that uses only natural gas as fuel. If the output of a CCPP using 2000 KRW/kg hydrogen blended in 0.5 mole fraction is 427 MW or more, it achieves competitiveness with a CCPP that only uses natural gas. The results of the process analysis conducted earlier show that the output (power generation) of a natural gas–hydrogen CCPP is 406.53 MW. If the output is increased by at least 20.47 MW by optimizing the process and improving performance, it will be possible to achieve a similar level of LCOE as a natural gas CCPP.

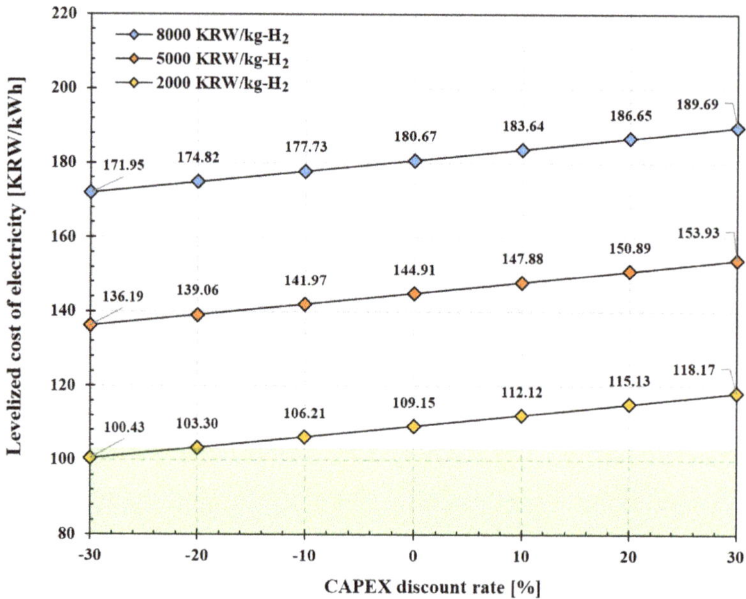

Figure 7. LCOE analysis results considering CAPEX discount rate.

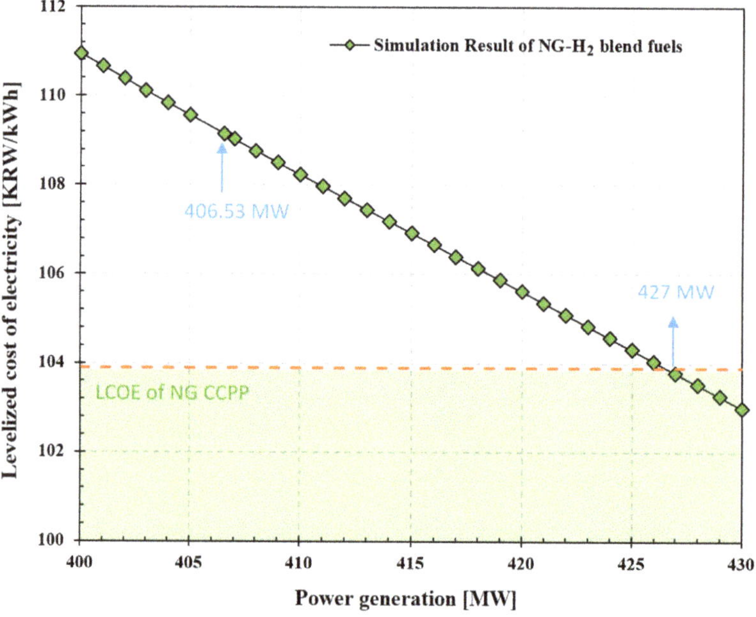

Figure 8. LCOE analysis results considering power generation.

Thus, the LCOE sensitivity analysis showed that, for a natural gas–hydrogen CCPP to secure competitiveness, it is more effective to reduce the hydrogen fuel supply price than

CAPEX and increase utilization rate. Furthermore, performance must be improved by at least 5% (20.47 MW) to secure an LCOE at the level of a natural gas CCPP.

4. Discussion

Various R&D and demonstration projects are underway to build a hydrogen ecosystem within the energy industry, from hydrogen production and storage to its transport and use. This study economically evaluated the gradual increase in hydrogen use as well as the use of natural gas–hydrogen blends that can be linked to natural gas-based CCPPs, which are used as a key power source in various countries. First, we simulated the natural gas CCPP process with 400 MW of output using ASPEN HYSYS to evaluate the benefits of using hydrogen fuel. The simulated model showed an error of around 1% by comparing the material properties of the key points of actual operation data, thereby confirming the excellence of the validation and implementation model.

Based on the validated process model, we reviewed ways to secure the economic feasibility of natural gas–hydrogen CCPPs compared with natural gas CCPPs. We completed the validation of the LCOE calculation model based on the TRR method using the commercialization costs of the operational Bundang CCPP-2. A sensitivity analysis was conducted with fuel cost, capacity factor, CAPEX, and power generation as the variables to evaluate the LCOE of natural gas–hydrogen CCPPs. The results showed that the change in LCOE was most significant according to hydrogen fuel prices, revealing that when hydrogen supply price decreases to 2000 KRW/kg, the LCOE does not change much even if hydrogen is blended into the fuel by up to 0.5 mole fraction. Hence, it will be possible to obtain an LCOE at a similar level as that of natural gas CCPPs by optimizing the process and improving performance while gradually increasing the ratio of hydrogen fuel. The capacity factor is expected to gradually increase more than 28.6% as the ratio of coal-fired power plants decreases and that of natural gas CCPPs increases within the power system in Korea. Hence, using natural gas–hydrogen blends will help to improve economic feasibility. Finally, regarding CAPEX, according to the use of hydrogen fuel, a reduction in cost is expected owing to the expansion of hydrogen-related industries and continuous technology development. Consequently, even if the amount of blended hydrogen is increased by up to 50%, natural gas–hydrogen CCPPs will be able to achieve sufficient competitiveness owing to technology development and green energy policies.

This study has a few limitations. First, the process model was validated using the operation data of natural gas CCPPs, but there is no operation data of the CCPP model using natural gas–hydrogen blends. Therefore, it is necessary to validate the reliability of the model through actual operation and experimental data in the future. Next, LCOE was analyzed by limiting the scope of variables used in economic analysis to certain values, but it is necessary to consider additional variables based on the ones confirmed in this study. Finally, it was assumed that hydrogen is blended into natural gas in certain ratios, but it is necessary to also consider specific hydrogen supply plans for future economic analysis.

Despite several limitations, this study suggested a method to secure economic feasibility of CCPP by using natural gas–hydrogen blended fuels instead of using only natural gas. In further research, we intend to analyze the probabilistic effects using methodologies such as Monte Carlo simulation for extensive economic analysis while connecting variables such as CAPEX, and capacity factor with learning rate. The cumulative probability curve using Monte Carlo will show the optimal LCOE conditions by reflecting price fluctuations in the equipment and electricity costs.

5. Conclusions

This study examined ways to secure the economic feasibility of using hydrogen fuel by simulating the process of a CCPP that uses natural gas–hydrogen blends and calculating LCOE. We increased the ratio of hydrogen in natural gas from 0 to 0.5 mole fraction and analyzed LCOE according to changes in the values of variables, such as fuel cost, capacity factor, CAPEX, and power generation. The results are as follows.

- We developed a process model for natural gasbased CCPPs and compared the material properties of each key point with operation data, which revealed an error range of around 1%, thereby completing the validation of the process model.
- When hydrogen fuel is supplied at 2000–8000 KRW/kg, the LCOE is 103.9–180.67 KRW/kWh. When it is supplied at under 2000 KRW/kg, the LCOE is 109.15 KRW/kWh even if the ratio of hydrogen blending is increased to 50%, showing a 5.0% increase from the LCOE of existing natural gas CCPPs (103.9 KRW/kWh).
- When the capacity factor of the CCPP is increased from 28.6% to at least 35% after blending 50% hydrogen at the price of 2000 KRW/kg with natural gas, the LCOE falls under 103.76 KRW/kWh, thereby ensuring price competitiveness over CCPPs using only natural gas.
- Even when CAPEX is reduced by up to 30%, the LCOE is reduced by only around 5%, not showing much of a reduction effect. However, when it is reduced by 20%, the LCOE is 103.3 KRW/kWh, which is lower than that of a CCPP that uses only natural gas.
- The process analysis showed that blending 50% hydrogen is expected to result in power generation of 406.53 MW and an LCOE of 109.15 KRW/kWh, suggesting that the same LCOE as that of existing natural gas CCPPs can be secured when net power generation is increased by 20.47 MW by optimizing the process and improving efficiency.

Author Contributions: Conceptualization, J.-Y.R. and S.P.; methodology, S.P.; software, J.-Y.R. and J.L.; validation, S.H. and S.P.; investigation, C.L. and S.H.; resources, S.P.; writing—original draft preparation, J.-Y.R.; writing—review and editing, J.-Y.R. and S.P.; visualization, C.L.; supervision, J.-Y.R. and S.P.; project administration, S.P.; funding acquisition, S.P. All authors have read and agreed to the published version of the manuscript.

Funding: This research was funded by Korean Energy Technology Evaluation and Planning, grant number 20213030020280.

Institutional Review Board Statement: Not applicable.

Informed Consent Statement: Not applicable.

Data Availability Statement: The data presented in this study are available upon request from the corresponding author.

Conflicts of Interest: The authors declare no conflict of interest.

Nomenclature

ADJ	Adjustment
AFUDC	Allowance for funds used during construction
BBY	Balance beginning of year
BD	Book depreciation
BFP	Boiler feedwater pump
BPV	Byproduct value
CC	Carrying charge
CCPP	Combined cycle power plant
CEP	Condensate extraction pump
CP	Cumulative probability
CRF	Capital recovery factor
DC	Direct cost
DITX	Differed income taxes
ESS	Energy storage system
FCI	Fixed capital investment
FOM	Fixed operating and maintenance
IC	Indirect cost
LCOE	Levelized cost of electricity

MACRS	Modified accelerated cost recovery system
OO	Other outlay
OTXI	Other taxes and insurance
PEC	Purchased equipment cost
PEI	Plant facilities investment
RCEAF	Recovery of common-equity AFUDC
ROI	Return of investment
SRHF	Standing reserve hourly fee
SRP	Standing reserve payment
SRSC	Standing reserve scheduled capacity
TCI	Total capital investment
TCR	Total capital recovery
TDI	Total depreciable investment
TRR	Total revenue requirement
TRRL	Total revenue requirement levelized
TXD	Tax depreciation
Subscript	
a	Annualized
ce	Common equity
d	Debt
FC	Fuel cost
j	J th year
k	Ratio of specific heats
L	Levelized
η	Net efficiency
n	Operating year
OMC	Operating and maintenance cost
ps	Preferred stock
R	Replacement
r	Pressure ratio
t	Tax rate

References

1. Friedlingstein, P.; O'Sullivan, M.; Jones, M.W.; Andrew, R.M.; Gregor, L.; Hauck, J.; Le Quéré, C.; Luijkx, I.T.; Olsen, A.; Peters, G.P.; et al. Global carbon budget 2022. *Earth Syst. Sci. Data* **2022**, *14*, 4811–4900.
2. Chen, J.M. Carbon neutrality: Toward a sustainable future. *Innovation* **2021**, *2*, 100127. [CrossRef] [PubMed]
3. IEA. Energy Technology Perspectives 2020. Special Report on Carbon Capture, Utilisation and Storage. 2020. Available online: https://webstore.iea.org/download/direct/4191 (accessed on 7 August 2023).
4. Yu, M.; Wang, K.; Vredenburg, H. Insights into low-carbon hydrogen production methods: Green, blue and aqua hydrogen. *Int. J. Hydrogen Energy* **2021**, *46*, 21261–21273.
5. Ryi, S.K.; Han, J.Y.; Kim, C.H.; Lim, H.; Jung, H.Y. Technical Trends of Hydrogen Production. *Clean Technol.* **2017**, *23*, 121–132.
6. International Energy Agency. *The Future of Hydrogen: Seizing Today's Opportunities*; OECD: Paris, Franace, 2019. [CrossRef]
7. Park, S.; Lee, C.; Ryu, J.-Y.; Hwang, S. An Economic Analysis on Slush Hydrogen Containing Liquid and Solid Phase for Long-Term and Large-Scale Storage. *Trans. Korean Hydrog. New Energy Soc.* **2022**, *33*, 247–254. [CrossRef]
8. Lamb, K.E.; Webb, C.J. A quantitative review of slurries for hydrogen storage–Slush hydrogen, and metal and chemical hydrides in carrier liquids. *J. Alloys Compd.* **2022**, *906*, 164235.
9. Hamayun, M.H.; Maafa, I.M.; Hussain, M.; Aslam, R. Simulation Study to Investigate the Effects of Operational Conditions on Methylcyclohexane Dehydrogenation for Hydrogen Production. *Energies* **2020**, *13*, 206. [CrossRef]
10. Andersson, J. Application of liquid hydrogen carriers in hydrogen steelmaking. *Energies* **2021**, *14*, 1392.
11. Meille, V.; Pitault, I. Liquid Organic Hydrogen Carriers or Organic Liquid Hydrides: 40 Years of History. *Reactions* **2021**, *2*, 94–101.
12. Mneimneh, F.; Ghazzawi, H.; Abu Hejjeh, M.; Manganelli, M.; Ramakrishna, S. Roadmap to Achieving Sustainable Development via Green Hydrogen. *Energies* **2023**, *16*, 1368.
13. Park, S.; Ryu, J.Y.; Sohn, G. Techno-Economic Analysis (TEA) on Hybrid Process for Hydrogen Production Combined withBiomass Gasification Using Oxygen Released from the Water Electrolysis Based on Renewable Energy. *J. Korean Inst. Gas* **2020**, *24*, 65–73.
14. Chae, M.J.; Kim, J.H.; Moon, B.; Park, S.; Lee, Y.S. The present condition and outlook for hydrogen-natural gas blending technology. *Korean J. Chem. Eng.* **2022**, *39*, 251–262. [CrossRef]

15. Sun, M.; Huang, X.; Hu, Y.; Lyu, S. Effects on the performance of domestic gas appliances operated on natural gas mixed with hydrogen. *Energy* **2021**, *244*, 122557.
16. Tangoz, S.; Kahraman, N.; Akansu, S.O. The effect of hydrogen on the performance and emissions of an SI engine having a high compression ratio fuelled by compressed natural gas. *Int. J. Hydrogen Energy* **2017**, *42*, 25766–25780. [CrossRef]
17. Makaryan, I.A.; Sedov, I.V.; Salgansky, E.A.; Arutyunov, A.V.; Arutyunov, V.S. A Comprehensive Review on the Prospects of Using Hydrogen–Methane Blends: Challenges and Opportunities. *Energies* **2022**, *15*, 2265. [CrossRef]
18. Erdener, B.C.; Sergi, B.; Guerra, O.J.; Chueca, A.L.; Pambour, K.; Brancucci, C.; Hodge, B.M. A review of technical and regulatory limits for hydrogen blending in natural gas pipelines. *Int. J. Hydrogen Energy* **2023**, *48*, 5595–5617.
19. Melaina, M.W.; Antonia, O.; Penev, M. *Blending Hydrogen into Natural Gas Pipeline Networks: A Review of Key Issues*; Technical Report; National Renewable Energy Laboratory: Golden, CO, USA, 2013.
20. Quarton, C.J.; Samsatli, S. Power-to-gas for injection into the gas grid: What can we learn from real-life projects, economic assessments and systems modelling? *Renew. Sustain. Energy Rev.* **2018**, *98*, 302–316.
21. Zachariah-Wolff, J.L.; Egyedi, T.M.; Hemmes, K. From natural gas to hydrogen via the Wobbe index: The role of standardized gateways in sustainable infrastructure transitions. *Int. J. Hydrogen Energy* **2007**, *32*, 1235–1245. [CrossRef]
22. Janès, A.; Lesage, J.; Weinberger, B.; Carson, D. Experimental determination of minimum ignition current (MIC) ratio of hydrogen/methane (H2NG) blends up to 20 vol.% of hydrogen. *Process Saf. Environ. Prot.* **2017**, *107*, 299–308. [CrossRef]
23. Meng, B.; Gu, C.; Zhang, L.; Zhou, C.; Li, X.; Zhao, Y.; Zheng, J.; Chen, X.; Han, Y. Hydrogen effects on X80 pipeline steel in high-pressure natural gas/hydrogen mixtures. *Int. J. Hydrogen Energy* **2016**, *42*, 7404–7412. [CrossRef]
24. Cameretti, M.C.; De Robbio, R.; Tuccillo, R. CFD Study of a Micro—Combustor Under Variable Operating Conditions. In Proceedings of the ASME Turbo Expo 2017, Charlotte, NC, USA, 26–30 June 2017. Paper No: GT2017-63661.
25. Pappa, A.; Bricteux, L.; Benard, P.; De Paepe, W. Can water dilution avoid flashback on a hydrogen enriched micro gas turbine combustion?—A large eddy simulation study. *J. Eng. Gas Turbines Power* **2021**, *143*, 041008-1. [CrossRef]
26. De Robbio, R. Micro gas turbine role in distributed generation with renewable energy sources. *Energies* **2023**, *16*, 704. [CrossRef]
27. Goldmeer, J. *Power to Gas: Hydrogen for Power Generation*; General Electric Company: Boston, MA, USA, 2019.
28. Larfeldt, J.; Andersson, M.; Larsson, A.; Moell, D. Hydrogen Co-Firing in Siemens Low NOX Industrial Gas Turbines. In Proceedings of the POWER-GEN Europe, Cologne, Germany, 27–29 June 2017. Available online: https://pdfs.semanticscholar.org/37fd/8e07212bf1e60f6db535d6e422b11880b816.pdf (accessed on 7 August 2023).
29. Jaworski, J.; Blacharski, T. Study of the effect of addition of hydrogen to natural gas on diaphragm gas meters. *Energies* **2020**, *13*, 3006. [CrossRef]
30. Saccani, C.; Pellegrini, M.; Guzzini, A. Analysis of the existing barriers for the market development of power to hydrogen (P2H) in Italy. *Energies* **2020**, *13*, 4835. [CrossRef]
31. Sorgulu, F.; Dincer, I. Analysis and Techno-Economic Assessment of Renewable Hydrogen Production and Blending into Natural Gas for Better Sustainability. *Int. J. Hydrogen Energy* **2022**, *47*, 19977–19988. [CrossRef]
32. Kotowicz, J.; Brzęczek, M. Analysis of increasing efficiency of modern combined cycle power plant: A case study. *Energy* **2018**, *153*, 90–99. [CrossRef]
33. Shin, Y.; Cho, E.S. Numerical Study on H$_2$ Enriched NG Lean Premixed Combustion. *J. Korean Soc. Combust.* **2021**, *26*, 51–58. [CrossRef]
34. Kurt, H.; Recebli, Z.; Gedik, E. Performance analysis of open cycle gas turbines. *Int. J. Energy Res.* **2009**, *33*, 285–294. [CrossRef]
35. Khan, A.R.; Ravi, M.R.; Ray, A. Experimental and chemical kinetic studies of the effect of H$_2$ enrichment on the laminar burning velocity and flame stability of various multicomponent natural gas blends. *Int. J. Hydrogen Energy* **2019**, *44*, 1192–1212. [CrossRef]
36. Bejan, A.; Tsatsaronis, G.; Moran, M. *Thermal Design and Optimization*; Wiley: New York, NY, USA, 1996.
37. Sohn, G.; Ryu, J.Y.; Park, H.; Park, S. Techno-economic analysis of oxy-fuel power plant for coal and biomass combined with a power-to-gas plant. *Energy Sustain. Dev.* **2021**, *64*, 47–58. [CrossRef]
38. Park, S.H.; Lee, Y.D.; Ahn, K.Y. Performance analysis of an SOFC/HCCI engine hybrid system: System simulation and thermoeconomic comparison. *Int. J. Hydrogen Energy* **2014**, *39*, 1799–1810. [CrossRef]
39. Statistics Korea. Consumer Price Index. Available online: https://https://kostat.go.kr/cpi (accessed on 7 August 2023).
40. Statistics Korea. Exchange Rate of World Bank. Available online: https://kosis.kr/statHtml/statHtml.do?orgId=101&tblId=DT_2AQ4516&conn_path=I2 (accessed on 7 August 2023).
41. National Tax Service South Korea. Income Tax. Available online: https://www.nts.go.kr/nts/cm/cntnts/cntntsView.do?mi=2372&cntntsId=7746 (accessed on 7 August 2023).
42. Ji, D.-M.; Sun, J.-Q.; Sun, Q.; Guo, H.-C.; Ren, J.-X.; Zhu, Q.-J. Optimization of start-up scheduling and life assessment for a steam turbine. *Energy* **2018**, *160*, 19–32.
43. Nazari, M.A.; Alavi, M.; Salem, M.; El Haj Assad, M. Utilization of hydrogen in gas turbines: A comprehensive review. *Int. J. Low-Carbon Technol.* **2022**, *17*, 513–519. [CrossRef]
44. Shin, J.-E. Hydrogen Technology Development and Policy Status by Value Chain in South Korea. *Energies* **2022**, *15*, 8983. [CrossRef]
45. International Energy Agency (IEA). The Future of Hydrogen 2019. Available online: https://www.iea.org/reports/the-future-of-hydrogen (accessed on 17 August 2023).

46. Ajanovic, A.; Haas, R. Economic prospects and policy framework for hydrogen as fuel in the transport sector. *Energy Pol.* **2018**, *123*, 280–288. [CrossRef]
47. Abdin, Z.; Khalilpour, K.; Catchpole, K. Projecting the levelized cost of large scale hydrogen storage for stationary applications. *Energy Convers. Manag.* **2022**, *270*, 116241. [CrossRef]

Article

Economic Assessment of Polypropylene Waste (PP) Pyrolysis in Circular Economy and Industrial Symbiosis

Anastasia Zabaniotou * and Ioannis Vaskalis

Bioenergy, Circular Economy and Sustainability Group, Department of Chemical Engineering, Engineering School, Aristotle University of Thessaloniki, Un. Box 455, University Campus GR, 54124 Thessaloniki, Greece
* Correspondence: azampani@auth.gr

Abstract: Plastic waste has a high energy content and can be utilized as an energy source. This study aims to assess the economic feasibility of polypropylene plastic waste (PP) pyrolysis. A literature review was carried out to determine the optimal pyrolysis conditions for oil production. The preferred pyrolysis temperature ranges from 450 °C to 550 °C, where the oil yields vary from 82 wt.% to 92.3 wt.%. Two scenarios were studied. In the first scenario, pyrolysis gas is used for the pyrolysis heating needs, whereas in the second scenario, natural gas is used. An overview of the economic performance of a pyrolysis plant with a capacity of 200,000 t/year is presented. Based on the results, the plant is economically viable, as it presents high profits and a short payback time for both scenarios considered. Although the annual revenues are smaller in scenario 1, the significant reduction in operating costs makes this scenario preferable. The annual profits amount to 37.3 M€, while the return on investment is 81% and the payback time is 1.16 years. In scenario 2, although the plant is still feasible and shows high profitability, the annual profits are lower by about 1.5 M€, while the payback time is 1.2 years.

Keywords: pyrolysis; plastic waste; polypropylene; feasibility assessment; circular economy; industrial symbiosis

Citation: Zabaniotou, A.; Vaskalis, I. Economic Assessment of Polypropylene Waste (PP) Pyrolysis in Circular Economy and Industrial Symbiosis. *Energies* **2023**, *16*, 593. https://doi.org/10.3390/en16020593

Academic Editors: Konstantinos Aravossis and Eleni Strantzali

Received: 6 December 2022
Revised: 22 December 2022
Accepted: 28 December 2022
Published: 4 January 2023

1. Introduction

The production of plastics on a global scale is on an upward trajectory, due to their extensive use in agriculture, construction, packaging, the automobile industry, and electrical equipment manufacturing. In 2020, global production reached 367 million t, displaying a 25% increase compared to 2010. It is estimated that by 2050, the production could potentially exceed 1 billion t if the current production and consumption trends persist [1]. The improper disposal of plastic waste leads to soil and groundwater pollution, and thus poses a serious threat to the environment and human health. Presently, approximately only 10% of plastic waste is recycled properly, while the bulk of it is either landfilled, incinerated, or generally left untreated and mismanaged [2].

In Greece, approximately 700 thousand t of plastic waste, or 68 kg per capita, is generated annually. Currently, the majority (i.e., 84%) is landfilled and only 8% is recycled [3]. The improper management of plastic waste is mostly attributed to low collection rates, highly mixed waste streams and limited recycling infrastructure [4]. The accumulation of plastic waste poses an important issue for the country, as more than 40 thousand t of plastic leaks into nature and local ecosystems each year. Additionally, there are negative implications on the national economy, with annual losses amassing to 26 M€, affecting the tourism, shipping, and fishing sectors [3,5].

There is still significant room for improvement in terms of the diversion of plastic waste from landfills and incineration plants and managing them in an efficient and environmentally sound manner [6]. The potential of utilizing plastic waste and feeding it into a forward supply chain, within the model of circular economy, is significant. A circular economy model focuses on waste management and resource recovery, through reuse, recycling, and energy utilization. Additionally, it aids in the development of new industries and jobs, reducing emissions, and promoting the efficient use of resources [7].

There are several pathways for the proper management and utilization of plastic waste within a circular economy concept. Plastic waste can be recycled and converted into other useful products. Mechanical recycling, which is also referred to as secondary recycling, involves a plastic waste recovery process based on mechanical means. Moreover, plastic waste has a high energy content and can, therefore, be utilized as an energy source.

An efficient method of utilizing these materials is through the process of pyrolysis, which involves indirect energy recovery from the feedstock. During pyrolysis, the feedstock is heated in the absence of oxygen, and the molecular chains are deconstructed. There are three main products of this process, which are pyrolysis oil, gas, and char. Pyrolysis oil can be used as a fuel, and it has properties that are similar to those of conventional fuels. The gas can be used to partially cover the energy demands of the process, while the solid product (char) can either be sold or used to produce activated carbon and other useful products. The conversion of plastic waste into valuable products and energy carriers through the pyrolysis process contributes to the reduction in the negative environmental impacts of the waste and to the reduction in fossil fuel use [6–8].

The utilization of plastic waste through pyrolysis can promote the transition to a circular economy that emphasizes industrial symbiosis. Industrial symbiosis is a part of industrial ecology, and it aims to foster cooperation between enterprises through the physical exchange of materials, energy, and/or by-products by using neighboring geographical advantages. Material symbiosis includes the use of by-products or waste generated by upstream production units as raw materials for downstream production. Energy symbiosis promotes the improvement of energy efficiency in industry, through the optimization of energy exchange networks in line with the overall supply–demand relationship. In the concept of energy symbiosis, industrial plants are urged to adopt the model of energy cascading and cogeneration, and thus improve their energy utilization efficiency [8,9].

A graphical conceptual representation of plastic waste treatment via pyrolysis in a circular economy and industrial symbiosis is presented in Figure 1. Several industrial plants that produce plastic waste can offer their waste as feedstock to a pyrolysis plant, ensuring an efficient and environmentally safe utilization route for their waste. The pyrolysis oil can be sold, providing a source of revenue for the pyrolysis plant. As mentioned, it can be used to generate heat and electricity, or it can be upgraded to produce fuels. Apart from the oil, a pyrolysis plant can generate two additional products, which can be used efficiently within the concept of industrial symbiosis. The gaseous product can be utilized to cover the energy demands of the pyrolysis plant. It can potentially also be used by neighboring industrial plants as a source of thermal energy, thus reducing their operating costs and dependence on fossil fuels, such as natural gas. Lastly, the char can be offered to neighboring plants and used as fuel. It can also be used as feedstock in a plant that produces activated carbon.

Countries (governments and companies) must implement circular economy pathways to reduce waste, conserve biodiversity, maintain environmental quality, and achieve economic sustainability. Investment in alternative energy sources including bioenergy, must be prioritized [10].

Different approaches can be used to mitigate and reduce the global environmental impacts of plastic waste. These include taxes on plastic products, especially plastic packaging, incentives to reuse and repair, target values for recycled products, extended producer responsibility, improved waste management infrastructure and schemes, and increased litter collection rates [11]. However, the successful implementation of such policies on a larger scale is still a significant challenge. Furthermore, the lack of plastic waste treatment infrastructure is an important issue, as efficient large-scale plastic waste recycling pathways are still scarce [12].

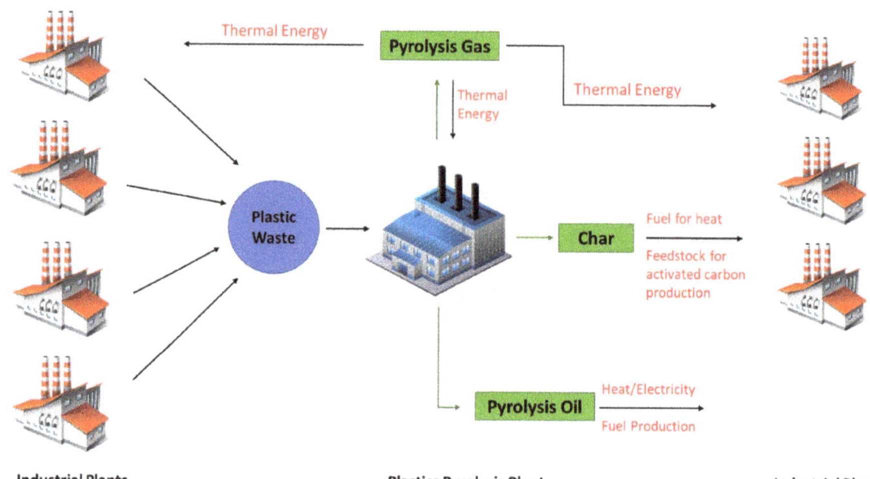

Figure 1. Conceptual representation of plastic waste pyrolysis in a circular economy and industrial symbiosis model.

In the European context, the EU has proposed several policies and actions towards the more sustainable management of plastic waste, where the focus is plastic reuse and recycling. The Action Plan for a Circular Economy, which was adopted in December 2015, identified the management of plastic waste as an area of high priority and focused on combating the potential challenges that arise from plastics throughout their value chain and their entire life cycle. The Plastics Strategy of 2018 outlined the transition to a circular plastics economy, made commitments for action at the EU level and recommended measures to national authorities and industry, to make plastic waste recycling profitable for businesses. The Revised Waste Framework Directive, which was passed in May 2018, updated the rules for waste management in the EU, including the management of plastic waste. The European Green Deal was presented in December 2019 and set out a roadmap for no net emissions of greenhouse gases by 2050. The Plastics Strategy aims at implementing new legislation, as well as specific targets and measures for tackling over-packaging and waste generation. Moreover, it promotes the strengthening of legal requirements to boost the market for secondary raw materials with mandatory recycled content, as well as guarantee that all packaging in the EU market is reusable or recyclable in an economically viable manner by 2030 [13,14].

In Greece, there are efforts underway to minimize the generation of plastic waste and promote its proper management and utilization. The Extended Producer Responsibility (EPR) Law 2939/01 obliges producers to finance the collection and recycling of waste through EPR schemes. The New Recycling Law 4496/2017 introduced a national plan for a four-stream collection system, including paper, glass, metals, and plastics. It also sets new targets, as 74% of the waste produced must be diverted from landfills. Moreover, the Landfill tax 4042/2012 proposed a 35 €/t tax for landfilling untreated waste, aiming to reach up to 60 €/t, but it was never rolled out. The government has also made active efforts to cease operations of illegal dump sites or convert them [3]. However, despite the introduction of such initiatives, there are still several challenges that hinder their proper implementation, such as low capacity and stakeholder pushback [4,5]. Additionally, the country still has limited infrastructure for recycling or utilizing plastic waste to generate useful products.

In order to promote the establishment of pyrolysis and other installations for the treatment of plastic waste, it is vital to determine the optimal conditions for their operation and assess the feasibility of such systems [4].

The aim of this study is to estimate the economic feasibility of polypropylene (PP) plastic waste utilization using pyrolysis for closing loops in energy and materials. PP is a thermoplastic polymer with properties such as fire resistance, simplicity, high heat distortion temperature, and dimensional solidity and accounts for 16% of the worldwide plastics market [1]. Petrochemical companies have generated an increasing demand for PP products and have raised environmental concerns related to PP waste.

This study is part of a Greek project entitled "Utilization of plastic and rubber waste for the production of alternative liquid fuels and adsorbent materials with innovative processes within the framework of the circular economy and industrial symbiosis model - ACTOIL". It is, therefore, focused on the development of a PP waste pyrolysis plant in Greece, within the context of circular economy and industrial symbiosis, by utilizing the industrial sector's plastic waste, such as PP waste from HELLENIC OILS, which is a Greek company responsible for the refining, supply and sales of petroleum and petrochemical products. The company produces considerable PP waste, which could be used as feedstock for a pyrolysis plant in Greece. Through economic assessment, it is possible to evaluate whether such installations can be profitable, to identify potential areas of improvement and to outline the optimal plant capacities. Overall, the aim of this study is to contribute to supporting developers and investors in the establishment of plastic waste pyrolysis plants. The findings of the study stress the potential of the pyrolysis of plastic waste and can, thus, play a pivotal role in the promotion of circular economies of plastic waste in the Greek context and in other countries.

In this paper, comprehensive information on the study's data, theoretical background, and methodology is provided. A literature review is provided in Section 2 to help position the paper. In Section 4, the economic analysis provides detailed insights and useful data for decision-makers and investors and in Section 5, the main findings derived from the economic evaluation are depicted.

2. Literature Review

Through a literature review, this paper identifies the most important parameters that affect the quality and yields of pyrolysis products. It also provides a critical overview on the optimal pyrolysis conditions required to produce oil. Thus, the results of the literature review were used as guidelines for the selection of pyrolysis operating parameters for different types of plastic waste feedstocks. Furthermore, the feasibility of a PP pyrolysis plant is examined through a preliminary techno-economic assessment.

The selection of relevant papers for the literature review was made (Chapter 2). In Chapter 3 of the report, the effect of the different parameters that affect pyrolysis product yields is presented, as well as an overview of the optimal conditions for oil production based on the literature review. Chapter 4 includes a feasibility assessment of a polypropylene pyrolysis plant located in Greece, while Chapter 5 provides conclusions and suggestions.

The feedstocks that were considered were as follows: polypropylene (PP), polystyrene (PS), polyethylene terephthalate (PET), poly-vinyl chloride (PVC), high-density polyethylene (HDPE), low-density polyethylene (LDPE) and mixed plastic waste, but the focus was on PP.

Internet search engines and electronic libraries were used for the review of relevant articles and journals. Scopus, ScienceDirect and Google Scholar were used to research plastic waste pyrolysis, focusing on publications between 2012 and 2022. The following keywords were used: 'plastic waste' AND 'pyrolysis', 'polypropylene' AND 'pyrolysis', 'polystyrene' AND 'pyrolysis', 'polyethylene terephthalate' AND 'pyrolysis', 'poly-vinyl chloride' AND 'pyrolysis', 'high-density polyethylene' AND 'pyrolysis', 'low-density polyethylene' AND 'pyrolysis', 'mixed plastics' AND 'pyrolysis'. The contribution of the published articles for each search term is presented in Figure 2. An increasing trend in the number of articles can be observed, with the highest number of articles of interest published in 2021 and 2022. Mixed plastic waste was the most common type of feedstock encountered in the articles, followed by polypropylene and polystyrene.

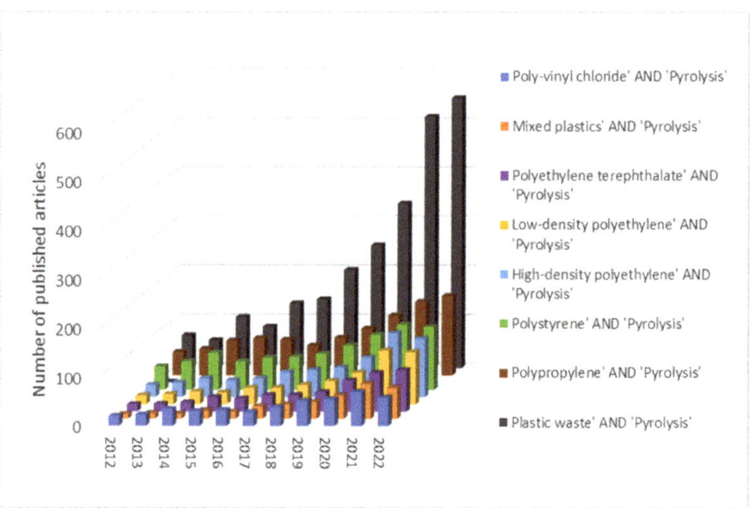

Figure 2. Annual contribution of the published articles of interest in the period 2012–2022.

The number of articles was 7656 and after an initial screening process based on the titles and abstracts, the number decreased to 295. A second screening process was carried out, in order to define the most relevant articles related to the effect of pyrolysis parameters on product yields and properties. Eventually, 44 articles were chosen for the plastic waste pyrolysis literature review. The distribution of the selected articles per year of publication and per type of feedstock is presented in Figure 3 and Table 1, respectively.

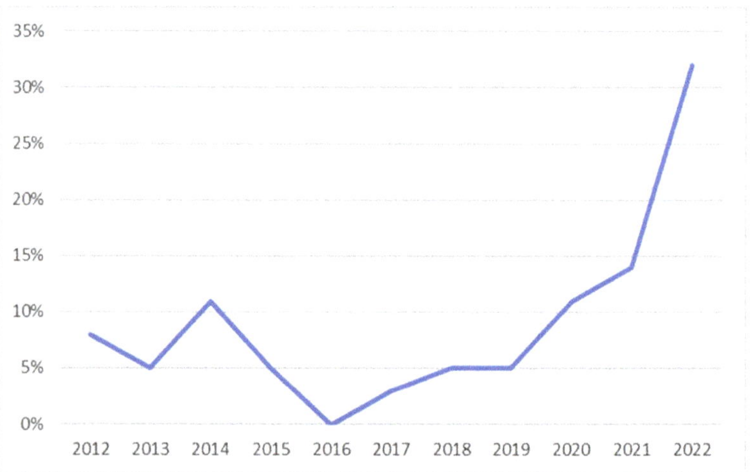

Figure 3. Share of the articles used in the study per year of publication.

Table 1. Share of the articles used in the study per type of feedstock.

Type of Feedstock	Share of Articles Used in the Study (%)
PP	24%
LDPE	17%
HDPE	15%
PS	15%
Mixed plastics	12%
PET	7%
PE	5%
PVC	5%

Most of the articles considered within this study were published in 2022 (32%), followed by 2021 (14%), 2020 and 2014 (11% each). Polypropylene (PP) and low-density polyethylene (LDPE) were the most common feedstocks in the relevant articles.

3. Literature Review Findings on Plastic Pyrolysis Maximizing the Oil's Yield

Pyrolysis is a thermochemical process that involves the deconstruction of molecular chains of materials by heating them in an oxygen-free environment. Typically, the temperature of pyrolysis varies between 300 °C and 800 °C. The products of pyrolysis are a liquid product, called pyrolysis oil, a solid product called char and a gaseous product. Pyrolysis oil can be used for the production of electricity or thermal energy. It can also undergo further processing to produce a fuel with similar properties to conventional fuels, i.e., diesel, gasoline or kerosene [15,16]. The gaseous product, which consists mainly of carbon monoxide and carbon dioxide, hydrogen, and hydrocarbons, has a relatively high calorific value and can, therefore, be exploited to meet part of the energy needs of the process [17]. The solid product of pyrolysis can be used as a fuel or as raw material for the production of activated carbon and other useful products [18]. In the pyrolysis process, an inert gas is always used, which does not participate in the reaction and in most cases, this gas is nitrogen [15].

The type of raw material, the conditions of pyrolysis (i.e., temperature, heating rate, duration and pressure), the type of reactor and the use of a catalyst are parameters that affect the yield and properties of the products, as well as their final composition [19,20].

3.1. Temperature and Heating Rate Effect on Pyrolysis

The temperature of pyrolysis significantly affects the product yields. Temperatures between 300 °C and 600 °C are preferred if the desired product is pyrolysis oil, while temperatures above 600 °C favor the formation of gaseous products. In addition, the heating rate and the duration also affect the pyrolysis process. In general, oil production is favored by intermediate temperatures, short residence times and relatively high heating rates, while the optimal yields of liquid products can be observed at temperatures between 450 °C and 600 °C [21]. On the contrary, gaseous products are favored by very high temperatures and long residence times [22], while low heating rates at low temperature with long residence times lead to char production [18]. Temperature, residence time and heating rate determine the type of pyrolysis (Table 2).

Table 2. Types of pyrolysis [15,20,21].

Type of Pyrolysis	Temperature (°C)	Residence Time	Heating Rate (°C/s)	Feedstock Size (mm)
Slow	300–650	5–60 min	0.1–1	5–50
Fast	450–600	0.5–10 s	10–200	<1
Flash	450–1000	<0.5 s	>1000	<0.2

3.2. Impact of the Type of Plastic Waste on Pyrolysis Process

The type of feedstock, as well as its characteristics and composition, have a significant impact on the pyrolysis process. It affects the yields, as well as the properties of the products. In the context of this study, different types of plastic waste were examined as feedstocks in the pyrolysis process.

A significant advantage of pyrolysis compared to other methods of plastic waste management is that no sorting process is required, and different types of plastics can be used simultaneously as feedstocks of the process [23]. Nevertheless, the type of plastic that is used can affect product yields, as well as their properties and quality. A high volatile content in the raw material favors the formation of pyrolysis oil, while on the contrary, a high content of ash leads to an increased yield of gas and solid products [15]. The list of plastics used in pyrolysis processes, as well as some of their applications, are presented in Table 3. The proximate analysis of the different types of plastic is presented in Table 4.

Table 3. Types of plastic and their applications [19,24,25].

Type of Plastic	Symbol	Uses/Applications
Polyethylene terephthalate	PET	Plastic beverage packaging, electrical insulation, magnetic tapes; printing sheets
Polypropylene	PP	Packaging, stationery, reusable containers, textiles, auto parts; laboratory equipment
High-density polyethylene	HDPE	Bottles for packaging, oil cans, toys; insulating material
Low-density polyethylene	LDPE	Plastic bags, wrapping sheets, insulating material; household goods
Polystyrene	PS	Food packaging, electronics, medical equipment, appliances; toys
Polyvinyl chloride	PVC	Pipes, insulation material, flooring, medical equipment; construction material

Table 4. Proximate analysis of different types of plastic [19,21].

Type of Plastic	Moisture (wt.%)	Fixed Carbon (wt.%)	Volatile Matter (wt.%)	Ash (wt.%)
PET	0.45–0.7	6–14	85–92	0–0.1
HDPE	0–0.3	0.01–0.03	94–99.8	0.2–1.5
LDPE	0–0.3	0	99–99.8	0–0.4
PVC	0.7–0.8	5–7	85–95	0–0.1
PP	0.15–0.4	0.15–1.2	95–99	1–4
PS	0.25–0.3	0.12–0.2	99–99.8	0–0.5

PET and PVC have the lowest volatile content; therefore, it is expected that their pyrolysis will favor the formation of gaseous and solid products. In addition, the pyrolysis of PVC results in the release of harmful products, such as HCl, as well as the formation of chlorobenzene in pyrolysis oil, which is toxic to the environment. The use of PVC pyrolysis oil requires further processing, adding further costs to the process [26]. PP, PS, HDPE and LDPE have a high volatile content, reaching up to 99.8 wt. %, indicating their suitability for the production of pyrolysis oil.

As mentioned, the properties of pyrolysis oil are significantly influenced by the type of plastic used as feedstock, as shown in Table 5. The higher heating value (HHV) of the pyrolysis oil from HDPE, LDPE, PS and PP is usually within the range of 39–43 MJ/kg, which is similar to conventional fuels, such as gasoline and diesel. In contrast, the thermal content of PVC and PET pyrolysis oil is significantly lower (21.8 MJ/kg and 28.2 MJ/kg, respectively).

Table 5. Properties of plastic pyrolysis-derived oil and comparison with conventional fuel-derived oil [15,20,21].

Feedstock	HHV (MJ/kg)	Density, 15 °C (g/cm^3)	Elemental Composition (wt.%)				
			C	H	N	S	Other
PET	28.2	0.087–0.9	62.1	4.21	-	-	33.7
HDPE	43.3	0.8–0.92	85.4	14.2	-	0.28	0.12
LDPE	40.6	0.77–0.8	85.4	14.2	-	0.25	0.15
PVC	21.8	0.84	39.2	4.9	-	0.58	55.3
PP	39.6	0.77–0.86	84.7	14.1	-	0.33	0.87
PS	41.5	0.85–0.86	91.5	7.4	-	0.19	0.91
Gasoline	42.5	0.78					
Diesel	44	0.81	85.6	14.1	0.3	-	-

3.3. Pyrolysis Reactor's Effect on Pyrolysis

The type of reactor used for pyrolysis significantly affects the performance and duration of the process, as well as the properties of the final products. Based on the conditions, the scale of production and the desired end products, the selection of the appropriate type of reactor can be made.

Batch reactors are closed systems where there is no flow of reactants or products during the reaction. Semi-batch reactors allow the addition of reagents and the removal of products during operation, giving them greater flexibility. An important advantage of these types of reactors is the ease of controlling the parameters of pyrolysis, leading to high yields and conversion rates. However, the high operating costs make this type of reactor better for small- and laboratory-scale applications [19].

Fixed-bed reactors are used extensively, especially in the case of catalytic pyrolysis. Despite their simple design, they have some disadvantages since the available surface of the catalyst during the reaction is limited. Fluidized-bed reactors use a fluidizing gas that ensures better mixing of the catalyst with the raw material, and thus a more efficient reaction. This type of reactor ensures good temperature control and is more flexible than batch and semi-batch reactors. In general, fluidized-bed reactors are preferred for large-scale applications, mainly from an economic point of view [17,19].

Rotary kiln reactors present several advantages, such as simple construction, low purchase costs, simple operation, and feedstock flexibility. These reactors are heated using solid heat carriers. In addition, the rotational movement of the furnace improves the mixing between materials and heat dissipation, while at the same time preventing the formation of agglomerates. It should be noted that the filling rate of this type of reactor significantly affects the performance and quality of the finished products. The ideal filling rate for oil production is about 15–20%, ensuring optimal heat transfer [27].

3.4. Catalyst's Effects on Pyrolysis

Pyrolysis is an energy-intensive process and catalysts contribute significantly to its optimization, as they limit the energy requirements. For example, optimal yields in pyrolysis oil can be achieved even at temperatures below 450 °C using the appropriate catalyst [17,28].

Catalysts are used in the pyrolysis of plastic waste to accelerate the rate of reactions. Their presence significantly enhances the efficiency of the process and reduces the required reaction time and degradation temperature of the raw material, while at the same time improving the quality of the fuel.

Several studies have reported that the ratio of raw material to catalyst significantly affects both the yield and the composition of pyrolysis products. The increase in the amount of catalyst is not linearly related to process efficiency. Usually, an increase in efficiency is observed up to a certain amount of catalyst, while further addition does not affect the reaction [20]. Various catalysts, such as natural zeolite (NZ), FCC, $Cu-Al_2O_3$, Fe_2O_3, MCM-41, ZSM5, HZSM-5 and $Al(OH)_3Ca(OH)_2$, have been used in plastic waste pyrolysis processes [15,17,19,28].

3.5. Pyrolysis Oil

The results of the literature review on plastic waste pyrolysis are disaggregated based on the type of plastic used. The parameters studied are the type of reactor, the use of catalyst, and the temperature of the pyrolysis process.

3.5.1. Polypropylene (PP) Pyrolysis Oil Yields

Polypropylene is used extensively as feedstock for pyrolysis, as typically very high yields in pyrolysis oil are achieved. Optimal oil production can usually be achieved at temperatures ranging from 450 °C to 550 °C.

Generally, very low pyrolysis temperatures are not preferable, since in these cases, the formation of gaseous products is favored. Ahmadis et al. [29] performed PP pyrolysis at 300 °C, with a heating rate of 20 °C/min, and the process yield of oil reached 69.9 wt.%. The increase in temperature resulted in the improved production of oil, as at 380 °C, the yield reached 80.1 wt.% [30]. Similarly, Kusenberg et al. [31] reported an oil yield of 87 wt.% at 450 °C, using a CSTR reactor. Pyrolysis at 500 °C [32] and 550 °C [33] resulted in oil yields of 82.1 wt.% and 92 wt.%, respectively. It should be noted that excessively high temperatures can negatively affect the process. As demonstrated by Demirbas [34], the oil yield fell to only 48.8 wt.% at 740 °C, as gas production was favored (49.9 wt.%).

Catalytic pyrolysis of PP has several advantages since optimal yields are achieved at lower temperatures. In a semi-batch reactor, an FCC catalyst was used, achieving the optimal oil yield (85 wt.%) at 400 °C [35]. Abbas-Abadi et al [36], by using the same type of catalyst and reactor, achieved improved oil production (92.3 wt.%), due to the higher operating temperature (450 °C). The use of natural zeolite for catalytic pyrolysis at 430 °C resulted in an oil yield of 72.2 wt.% in a semi-batch reactor [37].

3.5.2. Polyvinyl Chloride (PVC) Pyrolysis Oil Yields

PVC is not generally used as feedstock for pyrolysis, due to its relatively low oil yield and the production of toxic by-products. Thermal degradation of polyvinyl chloride is estimated to take place at temperatures between 220 °C and 520 °C [19]. During PVC pyrolysis at 500 °C and in a fixed-bed reactor, very low oil production was observed, with a yield of 12.3 wt.%, while the main product of the process was gas (87.8 wt.%) [24]. Marino et al. [38] used a fixed-bed reactor with a ZSM-5 catalyst at 450 °C with significantly improved results. The process yield of oil was 60 wt.% and the yield of gas was 35 wt.%, in which a high HCl content was observed.

3.5.3. Polystyrene (PS) Pyrolysis Oil Yields

Unlike PET and PVC, polystyrene pyrolysis displays very high yields of oil. For its thermal breakdown, temperatures between 350 °C and 500 °C are required, while it is estimated that the pyrolysis temperature should not exceed 550–600 °C to achieve maximum oil yields [19]. Generally, PS pyrolysis oil yields vary between 90 wt.% and 99 wt.%. In a batch reactor, an oil yield of 89.5 wt.% was observed at 580 °C [34], while at a lower temperature (450 °C), the yield of liquid products was 56 wt.% [39].

The use of a catalyst during the pyrolysis of PS reduces the required reaction time, but there is a slight improvement in the oil yields of the process. Terapalli et al. [40] used PS as feedstock for pyrolysis at 600 °C in a microwave reactor, using KOH as a catalyst. The oil yield reached 95.2 wt.%, using a heating rate of 31 °C/min and 7.5 gr of KOH for 27.5 gr of PS. In these conditions, the gas and char yields were 3.5 wt.% and 1.3 wt.%, respectively. Adnan, Shah and Jan [41] used a Zn catalyst in a batch reactor at 500 °C. The yields of the oil and gas products were 96.7 wt.% and 3.3 wt.%, respectively. A similarly high oil yield was observed with the use of an MgO catalyst in a fixed-bed reactor. PS pyrolysis took place at 400 °C and 500 °C. The final yield of oil was 93 wt.% in both cases, while gas production increased from 2 wt. % to 5 wt.% [35]. Finally, Miandad et al. [28] studied PS pyrolysis at 450 °C using the following two different catalysts: (i) natural zeolite with heat treatment at 500 °C for 5 h (TA-NZ) and (ii) natural zeolite with treatment with 0.1 m HNO_3 for 48 h (AA-NZ). The highest oil yield observed was 70 wt.% and this was achieved using the TA-NZ catalyst. However, while the yield of the liquid product with the AA-NZ catalyst was lower (60 wt.%), the HHV of the oil produced (42.1 MJ/kg) was higher.

3.5.4. Polyethylene Terephthalate (PET) Pyrolysis Oil Yields

As mentioned, the use of PET as a pyrolysis feedstock is not preferred, mainly due to its low content of volatile components and its generally low oil yield compared to other plastics. However, there have been some studies that have examined the utilization of PET via pyrolysis.

It has been observed that the thermal breakdown of PET takes place in a temperature range between 350 °C and 520 °C [19]. Çepelioğullar and Pütün [17] used a fixed-bed reactor at a temperature of 500 °C with a heating rate of 10 °C/min. The main product was gas, with a yield of 76.9 wt.%, while the yield for oil was found to be 23.1 wt.%. Furthermore, a high content of benzoic acid was observed in the oil produced (49.93 wt.%), giving it a strongly acidic character. Additionally, Shahbaz et al. [42] studied PET pyrolysis at 450 °C, with a heating rate of 10 °C/min, achieving an oil yield of 18 wt.%. Finally, PET pyrolysis, using a fixed-bed reactor at 500 °C and at a heating rate of 6 °C/min, resulted in oil production of 39.89 wt.%, while at the same time, yields for the gas and solid products were 52.13 wt.% and 7.98 wt.% [32], respectively.

3.5.5. Low Density Polyethylene (LDPE) Pyrolysis Oil Yields

The use of low-density polyethylene as feedstock for pyrolysis has been studied extensively. In general, it is estimated that the thermal breakdown of LDPE takes place from 360 °C up to 550 °C [19] and optimal yields of oil are achieved between 500 °C and 550 °C. Based on the study by Bagri and Williams [43], LDPE pyrolysis at 500 °C and in a fixed-bed reactor produces 95 wt.% oil and 5 wt.% gas. In addition, FakhrHoseini and Dastanian [30] observed an oil yield of 80.4 wt.% at 500 °C, while in another study in a fluidized-bed reactor, it was observed that a very high temperature (600 °C) led to a significant decrease in oil yields to 51 wt.% [44]. It should also be noted that the pressure of the process affects the quantity of the products. Odejobi et al. [33], while performing pyrolysis at 450 °C, managed to achieve oil yields of approximately 41 wt.%, with gas production reaching 57 wt.%.

According to Wu et al. [45], the use of HZSM5 as a catalyst helped to increase the process oil yields in a fixed-bed reactor. At a temperature of 550 °C and a heating rate of 20 °C/min, oil production reached 93.42 wt.%. Similarly, in another study, the combination of an HZSM5 catalyst with a relatively high temperature (550 °C) again resulted in a very high oil yield of 93.1 wt.% [46].

3.5.6. High-Density Polyethylene (HDPE) Pyrolysis Oil Yields

Like LDPE, HDPE is used extensively for pyrolysis oil production. In the absence of a catalyst, Dzol et al. [46] achieved an oil yield of 90 wt.% at 500 °C in a fixed-bed reactor. On the other hand, when using HZSM5 or waste chicken eggshells (WCE) as catalysts, optimal results were obtained with WCE, providing an oil yield of 80 wt.% [47]. According to Mastral et al. [48], oil production dropped to 68.5 wt.% when the process was performed at 650 °C and with a duration of 20 min. Using an MIL-53 (Cu)-derived zeolite Y catalyst at 500 °C [49], the oil yield reached 95.3 wt.%, while with the HZSM5 catalyst at 550 °C [46], oil production was 85 wt.%. Abbas-Abadi et al. [41] studied HDPE pyrolysis, where the process took place in a semi-batch reactor at 450 °C and with an FCC catalyst. In this case oil, the gas and solid product yields were 91.2 wt.%, 4.1 wt.% and 4.7 wt.%, respectively.

3.5.7. Mixed Plastic Waste Pyrolysis Oil Yields

In many cases, a mixture of plastics can be used as the feedstock of pyrolysis. For example, Donaj et al. [50] performed pyrolysis of an LDPE/HDPE/PP mixture in a fluidized-bed reactor. Without the use of catalyst, the optimal oil yield, 48.4 wt.%, was reached at 650 °C, while using a Ziegler–Natta catalyst, oil production increased to 89 wt.% at the same temperature. It should also be noted that the same series of experiments were carried out at 730 °C with lower yields.

Moreover, the PE/PP/PS mixture has been extensively studied, showing generally low yields of oil. With the PE-PP/PS mixture (75 wt.%/25 wt.%), Kaminsky, Schlesselmann and Simon [51] achieved a yield of 48.4 wt.% of oil, performing pyrolysis in a 730 °C fluidized-bed reactor. Similar results, with an oil yield of 46.6 wt.%, were reported by Demirbas [32] in a batch reactor. Based on the findings of the literature review, the use of a catalyst does not significantly affect the oil yields of the process in this case. In a batch reactor, at 450–500 °C and using natural zeolite as a catalyst, Nugroho, Pratama and Saptoadi [52] reported that oil production reached 45.1 wt.%, as the gaseous product was favored (50 wt.%). Miandad et al. [28] used the PS/PE/PP mixture (50 wt.%/25 wt.%/25 wt.%) at 450 °C with a natural zeolite catalyst. The yields of oil, gas and solid products were 44 wt.%, 37 wt.% and 19 wt.%, respectively.

3.5.8. Selection of Pyrolysis Operating Conditions for Maximizing Oil Production

Based on the results of the literature review, pyrolysis parameters were selected on the basis of maximizing oil yields. An overview of all the results with the optimal conditions for oil production are depicted in Table 6.

In the study of Miandad et al. [28], where various mixtures of plastics were examined, oil yields of less than 55 wt.% were obtained. Optimal results were reported for the following two cases of pyrolysis at 450 °C: (i) a PS/PE mixture (50 wt.%/50 wt.%) with a catalyst of natural zeolite treated with HNO3, producing an oil yield of 52 wt.%, and (ii) a PS/PP mixture (50 wt.%/50 wt.%) with a natural zeolite catalyst and heat treatment, resulting in an oil yield of 54 wt.%.

Table 6. Literature review results of catalytic and non-catalytic plastics pyrolysis.

Feedstock	Reactor	T (°C)	Catalyst	Yield (wt.%)			Ref.
				Oil	Gas	Char	
PP	-	450	TA-NZ	40	41.1	18.9	[28]
			AA-NZ	54	26.1	19.9	
	-	300	-	69.82	28.84	1.34	[29]
	Batch	380	-	80.1	6.6	13.3	[30]
	CSTR	450	-	87	9	3	[31]
	-	500	-	82.1	17.8	0.1	[32]
	-	550	-	92	8	0	[33]
	Batch	740	-	48.8	49.6	1.6	[34]
	-	400	FCC	85	13	2	[35]
	-	450	FCC	92.3	4.1	3.6	[36]
	Batch	430	NZ	72.17	27.83	0	[37]
PVC	Fixed-bed	500	-	12.3	87.7	0	[24]
	Fixed-bed	450	ZSM-5	60	35	5	[38]

Table 6. *Cont.*

Feedstock	Reactor	T (°C)	Catalyst	Yield (wt.%)			Ref.
				Oil	Gas	Char	
PS	-	450	TA-NZ	70	14.2	15.8	[28]
			AA-NZ	60	24.6	15.4	
	Batch	581	-	89.5	9.9	0.6	[34]
	Fixed-bed	400	MgO	93	2	5	[35]
		500		93	5	2	
	Batch	450	-	56	45	1	[39]
	-	600	KOH	95.2	3.5	1.3	[40]
	Batch	500	Zn	96.73	3.27	0	[41]
PET	-	400–500	-	26-28	-	-	[24]
	Fixed-bed	500	-	39.89	52.13	7.98	[32]
	-	450	-	18	33	49	[42]
LDPE	-	500	-	80.4	19.4	0.2	[30]
	Batch	450	-	41	57	2	[33]
	Fixed-bed	500	-	95	5	0	[43]
	Fluidized-bed	600	-	51	24.2	0	[44]
	Fixed-bed	550	HZSM5	93.4	6.4	0.2	[45]
	Batch	550	HZSM5	93.1	14.6	0	[46]
HDPE	-	550	-	80	20	0	[53]
	-	450	FCC	91.2	4.1	4.7	[41]
	Batch	550	HZSM5	84.7	15.3	0	[46]
	Fixed-bed	500	WCE	90	9	1	[47]
			-	80	13	7	
	Fluidized-bed	650	-	68.5	31.5	0	[48]
	Fixed-bed	500	MIL-53(Cu) Y zeolite	95.3	-	-	[49]
	CSBR	500–900	-	14.7	84.5	0.8	[54]
PE	-	450	TA-NZ	40	47	13	[28]
	-		AA-NZ	42	50.8	7.2	
	CSTR	450	-	85	10	5	[39]
PS/PE (50/50)	-	450	AA-NZ	52	29.2	18.8	
PS/PP (50/50)	-		TA-NZ	54	25.7	20.3	
PP/PE (50/50)	-		TA-NZ	44	44.6	11.4	[28]
PS/PE/PP (50/25/25)	-		TA-NZ	44	37	19	
PS/PP/PE/PET (40/20/20/20)	-		AA-NZ	30	38.4	31.6	
PP/PE/PS	Batch		-	46.6	35	2.2	[32]

Table 6. *Cont.*

Feedstock	Reactor	T (°C)	Catalyst	Yield (wt.%)			Ref.
				Oil	Gas	Char	
LDPE/HDPE/PP	Fluidized-bed	650	Ziegler–Natta	89	6.5	4.5	[50]
			-	48.4	36.9	15.7	
PE-PP/PS (75/25)	Fluidized-bed	730	-	48.4	35	16.6	[51]
PE/PP/PS (50/40/10)	Batch	450–500	NZ	45.1	50	4.9	[52]

Polypropylene (PP) pyrolysis displays very high yields of pyrolysis oil. The properties of the oil are similar to those of conventional fuels. The optimal temperature range for PP pyrolysis is from 450 °C to 550 °C, where oil yields from 82 wt.% to 92.3 wt.% were reported. For temperatures below 450 °C, lower yields were observed (<80 wt.%), while at much higher temperatures (740 °C), the gas production increases significantly, limiting the formation of oil. The use of an FCC catalyst significantly affects the process, achieving an oil production value of 92 wt.% at 450 °C.

The pyrolysis of PVC and PET favors the production of gaseous products, with yields ranging between 55 wt.% and 88 wt.%. Oil production is limited to the range of 12-40 wt.%. The process is carried out at 450–500 °C. During the pyrolysis of PVC, harmful by-products are released, such as HCl in the gas product and chlorobenzene in the oil. Their removal is vital and significantly adds to the overall cost of the process.

The pyrolysis of PS, regardless of the type of reactor and the presence of a catalyst, typically has oil yields ranging from 60 wt.% up to 98.7 wt.% for temperatures of 450–600 °C. Optimal oil yields are reported at temperatures from 550 °C up to 600 °C, while the use of catalysts such as FCC and Zn allows the process to be carried out at lower temperatures with high oil production. In general, polystyrene showed the highest oil yields compared to other plastics.

Both LDPE and HDPE pyrolysis result in high oil yields, typically exceeding 80 wt.% for temperatures of 450 °C to 550 °C. In the case of LDPE, the increase in pressure and the presence of a catalyst has a positive effect on the oil yield; oil production of 93.1 wt.% can be achieved at 550 °C, using an HZSM5 catalyst. The use of an FCC, Si-Al or HZSM5 catalyst during HDPE pyrolysis resulted in oil yields higher than 85 wt.% for temperatures ranging from 450 °C to 550 °C. The yields of oil for LDPE pyrolysis (73.6 wt.%) are higher than PP (73 wt.%) and HDPE (71.5 wt.%).

Compared to the pyrolysis of a single type of plastic, the use of a mixture of plastics shows much lower oil yields, which are usually in the range of 30–50 wt.%. Optimal pyrolysis temperatures are in the range of 450 °C–650 °C. When using a Ziegler–Natta catalyst at 650 °C, the oil yields are 90 wt.%. The properties of the final product show many similarities with those of conventional fuels.

4. Feasibility Study for a PP Pyrolysis Plant in Greece

A preliminary economic assessment was carried out for a PP pyrolysis plant. The assessment was conducted for different plant capacities. The economic performance of the plant was evaluated based on the following economic indicators: the initial investment, the operating costs, annual cash inflows, gross and net profits, return on investment (R.O.I) and pay-out time (P.O.T).

R.O.I is an indicator used to measure the profitability of a particular investment and is used to express the percentage of the initial investment that can be recovered over one year. Generally, a positive R.O.I denotes an investment that is profitable. However, if similar investments occur with a higher R.O.I, these are preferable [55].

P.O.T is an indicator used to measure the period required for the profit or other benefits of an investment to equal the cost of the investment. A smaller P.O.T is a sign of a more attractive investment opportunity [55].

By studying the effect of the plant's capacity on the two indicators, it was possible to determine a range for the preferred capacity of the pyrolysis plant. The present study examines the following two scenarios:

- **Scenario 1:** Pyrolysis gas is used to meet the energy requirements of the process. If there is further availability, then the remaining amount of gas is sold.
- **Scenario 2:** the entire amount of pyrolysis gas is sold.

4.1. Hypotheses of the Study

The pyrolysis plant utilizes PP waste from HELLENIC OILS, which is a Greek company responsible for the refining, supply and sales of petroleum and petrochemical products. The company produces considerable PP waste, which could be used as feedstock for a pyrolysis plant in Greece. In the context of circular economy and industrial symbiosis models, the pyrolysis plant receives the waste from HELLENIC OILS and utilizes it to generate the following three main products: pyrolysis oil, pyrolysis gas and char. Liquid and solid products are sold and are an important source of revenue for the unit. Pyrolysis gas can either be sold or used to meet the energy requirements of the process either partially or completely.

The hypotheses of the study are as follows:

- ❖ The capacity selected for the PP pyrolysis plant is 200,000 t/year; the process followed to make this selection is shown in Chapter 4.3.7.
- ❖ Based on the literature review, the optimal pyrolysis temperature for oil production is 500 °C; the process takes place at atmospheric pressure.
- ❖ According to the literature review findings, the oil yield is to be considered equal to 86 wt.%, the gas yield is equal to 13.9 wt.% and the char yield is equal to 0.1 wt.%.
- ❖ A rotary kiln reactor was chosen, due to its extensive use and advantages, such as simple construction, low purchase costs and simple operation.
- ❖ It was assumed that product yields are not affected by the increase in plant capacity.
- ❖ The unit operates for 330 days and 24 h/day.
- ❖ The facility is in Greece.
- ❖ PP waste transportation costs are negligible, assuming that this cost is integrated into the price of the PP waste.

A general overview of the characteristics of the studied plant is given in Table 7.

Table 7. Characteristics of PP pyrolysis plant under study.

Days of Operation Per Year	330
Operating hours (h/d)	24
Feedstock	Industrial PP
PP input (t/day)	606
Temperature (°C)	500
Pressure (atm)	1
Pyrolysis oil production (t/day)	521
Pyrolysis gas production (t/day)	84
Pyrolysis char production (t/day)	0.6

4.2. Description of the PP Pyrolysis Plant

Figure 4 shows the generalized flowchart for a PP pyrolysis unit. A brief description of the devices used, as well as the process currents, is also presented.

The main types of equipment used in the process are as follows:

- M-101: shredder—used to shred the feedstock and facilitate its energy utilization.
- M-102: feed hopper.

- R-101: rotary kiln reactor where the pyrolysis process takes place.
- C-101: Condenser—the gases produced in the reactor are fed into C-101. The resulting condensate contains a percentage of water and is, therefore, led to a separator to be removed.
- C-102: Condenser—the gaseous product that was not condensed in C-101 is fed into C-102. The condensate does not contain water and is driven into an oil collection container.
- S-101: oil/water separator—used to remove the water from the pyrolysis oil.
- T-101: pyrolysis oil collection container.
- G-101: gas storage container—the gas can either be sold or used to meet the reactor's energy requirements.

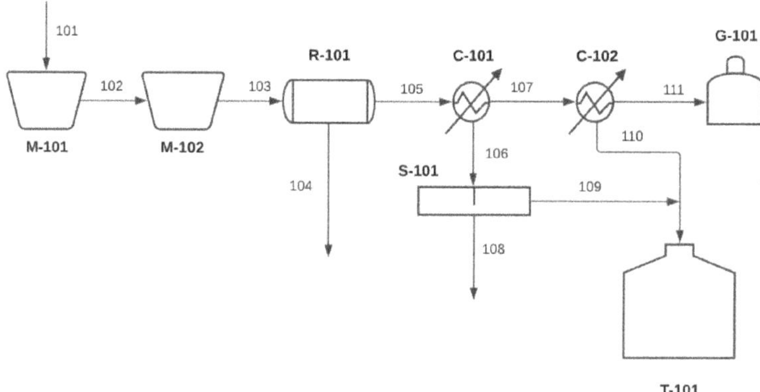

Figure 4. Generalized flowchart of the pyrolysis process [56–58].

The main process streams, as depicted in the flowchart, are as follows:

- **101:** feedstock (industrial PP).
- **102:** shredded feedstock.
- **103:** feed stream of the pyrolysis reactor.
- **104:** solid residue of pyrolysis (char), which can either be stored for further processing or sold directly.
- **105:** main product of pyrolysis, which contains the products of the process in gaseous form and is led to the condensers.
- **106:** contains water and pyrolysis oil and is led to a separation device.
- **107:** contains gases and pyrolysis oil in gaseous form and is led to the second condenser for further separation.
- **108:** contains water that is removed.
- **109:** contains pyrolysis oil for storage.
- **110:** contains pyrolysis oil resulting from the condenser; the product is fed to the storage tank.
- **111:** contains the gaseous product of pyrolysis that is either stored or used to meet the energy requirements of the process.

4.3. Preliminary Economic Assessment

A preliminary economic assessment was carried out, considering the initial investment, the operating costs, annual cash inflows and gross and net profits, as well as two economic indicators, which are R.O.I (return on investment) and P.O.T (pay-out time). Moreover, the effect of plant capacity on the economic performance was also considered.

4.3.1. Initial Investment (CAPEX)

The Chemical Engineering Plant Cost Index (CEPCI) was used to calculate the investment requirements for the PP pyrolysis plant. The initial investment of one unit for 2022 can be calculated from the following equation:

$$I_{F,\,2022} = I_{F,X} \times \frac{CEPCI_{2022}}{CEPCI_X} \tag{1}$$

where $I_{F,2022}$ is the investment for the year 2022 and $I_{F,X}$ the investment for the year X, for which bibliographic data were found. As no values are available for the $CEPCI$ index for 2022, it is possible to make an estimate based on the data presented in Figure 5, showing that $CEPCI_{2022}$ = 651.4.

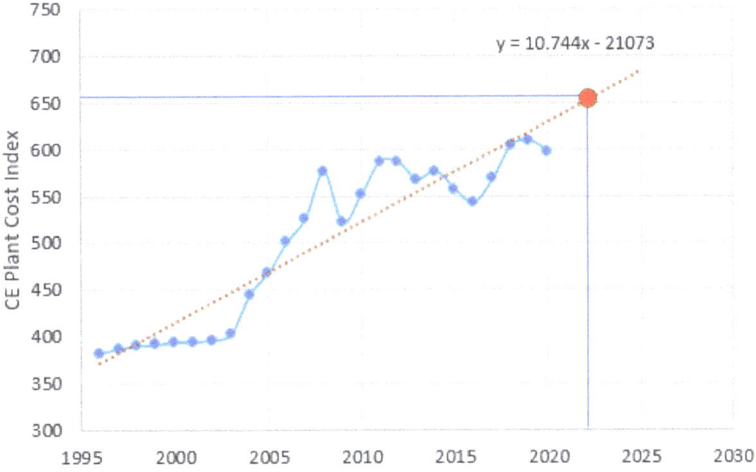

Figure 5. CEPCI indicators for 1996–2020 [59].

The initial investment is directly related to the capacity of the pyrolysis plant. Table 8 shows the investment requirement for units of different capacities, based on the literature review. Using the CEPCI indicators, all values are transposed and refer to 2022.

Table 8. Initial investment for pyrolysis units of different capacities.

$I_{F,X}$ (M EUR)	Capacity (T/Day)	Year	CEPCI	$I_{F,2022}$ (M€)	Ref.
1.17	2.5	2018	603.1	1.35	[60]
3.12	40	2019	608	3.34	[61]
26.1	110	2021	640.6	26.5	[62]
33.1	120	2018	603.1	35.8	[63]
8.07	133	2021	640.6	8.2	[64]
32.4	876	2019	608	34.7	[65]
102.9	1000	2021	640.6	104.6	[66]

Using the data in Table 8, it is possible to make an estimate of the relationship between capacity and the investment requirement for a plastics pyrolysis plant. This can be achieved based on the equation obtained from Figure 6, which is as follows:

$$I_{F,2022} = 667,520 \times Q^{0.6603} \tag{2}$$

where $I_{F,2022}$ is the initial investment for 2022 and Q is the plant capacity (t/day). Consequently, the initial investment for the PP pyrolysis plant with the selected capacity of 200,000 t per year is I_F = 45.9 M€. It should be noted that the investment is the same for both PP pyrolysis scenarios considered.

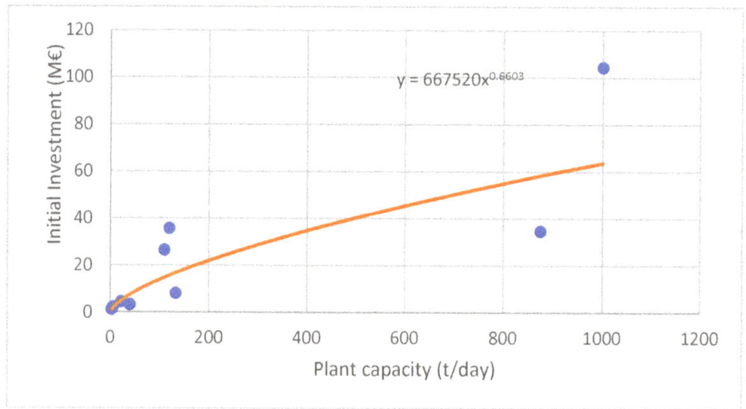

Figure 6. Relation of plant capacity and initial investment.

4.3.2. Operating Costs (OPEX)

The next step in the economic assessment of the pyrolysis plant is to determine the operating costs. This category of expenses includes labor, raw materials, and the cost of utilities, as well as some additional costs.

Operating expenses differ between the two scenarios, due to the difference in costs for utilities. In scenario 1, where pyrolysis gas is used to meet the energy requirements of the process, natural gas is not used, thus reducing costs by approximately 3 M€. The overall operating costs for the two scenarios are presented in Table 9 [67]. The detailed calculation methodology is presented below.

Table 9. Overview of PP pyrolysis operating costs.

Type of Cost	Cost Estimation	EUR/year
I. Production Costs		
A. Direct Costs		
i. Feedstock		35×10^6
ii. Labor		831,140
iii. Supervision	15% A(ii)	124,670
iv. Utilities	Scenario 1	6.28×10^6
	Scenario 2	9.45×10^6
v. Maintenance	5% I_F	2.3×10^6
vi. Materials	0.75% I_F	344,210
vii. Lab costs	10% A(ii)	83,115
B. Fixed Costs		
i. Insurance	1% I_F	458,950
ii. Taxes	1% I_F	458,950
C. Additional Costs	60% (A(ii)+A(iii)+A(v))	1.65×10^6
II. General Costs		
A. Administration costs	5% A (ii)	41,560
B. Distribution costs	5% I_F	2.29×10^6
Total Operating Costs C	Scenario 1	**51.3 M€**
	Scenario 2	**54.8 M€**

Labor Costs

The man-hours are calculated using the Wessels method with the following equation [67]:

$$\frac{manhours}{(days) \times (stages)} = \alpha \times \left(\frac{tonnes}{day}\right)^{0.24} \tag{3}$$

The days of operation were considered to be 330, the stages to be 3, the parameter α equal to 11, and the capacity of the unit to be 606 t/day. Using the man-hour price for Greece [68], the total labor costs were defined, as observed in Table 10.

Table 10. Labor costs calculation.

Days of Operation	330
Operation stages	3
Capacity (t/day)	606
Number of man-hours	50,679
Man-hour cost (€/hour, Greece)	16.4
Total Labor Costs (€/year)	**831,143**

Feedstock Costs

To determine the cost of the raw material, the price of PP scraps was taken as 175 €/t [69]. Based on the capacity of the unit and the feedstock price, it was calculated that the total annual feedstock costs amount to 35 M€.

Utilities Cost

The utilities required for the operation of the pyrolysis plant are electricity, cooling water and natural gas for heating the reactor.

It is estimated that a pyrolysis plant requires 28 kWh of electricity and 13 m^3 of cooling water per t of feedstock [63]. Considering the cost of electricity of 0.13 €/kWh [70] and the cost of water of 0.98 €/m^3 [71], the annual costs were 728,000 €/year and 2.55 M€/year respectively. The electricity requirements (28 kWh/t) do not include the cost of operating the shredder. It is estimated that this expenditure amounts to approximately 15 €/t of raw material [72], adding 3 M€ to the annual costs of utilities.

In addition, for the calculation of the annual cost of natural gas, it is required to determine the energy requirements of the process. For this, the following equation was used:

$$Q_{pyro} = m_{PP,dry} \times C_{p,PP} \times \left(T_{final} - T_{in}\right) + m_{water} \times h_{evap} + m_{water} \times \left(h_{w,final} - h_{w,in}\right) \tag{4}$$

All the required data for this calculation are presented in Table 11.

Table 11. Calculation of PP pyrolysis energy requirements.

Characteristics	Value	Ref.
Feedstock moisture (wt.%)	0.6	-
Feedstock quantity $m_{PP,dry}$ (kg/s)	6.97	-
Specific heat $C_{p,PP}$ (kJ/kgK)	1.92	[73]
T_{in} (°C)	20	-
T_{out} (°C)	500	-
Water quantity m_{water} (kg/s)	0.04	-
h_{evap} (kJ/kg)	2454	[74]
$h_{w,20}$ (kJ/kg)	83.9	[75]
$h_{w,500}$ (kJ/kg)	3488	[75]
Energy Requirements Q		**6.67 MW**

The energy requirements for the pyrolysis reactor (Q = 6.67 MW) can be met either using pyrolysis gas (scenario 1) or by natural gas (scenario 2).

Based on bibliographic data, it is estimated that the calorific value of pyrolysis gas is 26 MJ/kg [76–78]. Given that the amount of pyrolysis gas produced is 84 t/day, the gas can provide about 25 MW, thus fully meeting the energy requirements of the reactor. Consequently, the cost to heat the pyrolysis reactor is zero for scenario 1.

In scenario 2, natural gas is used. The calorific value of natural gas is 31.7 MJ/m^3 [79], while its price is 0.53 EUR/m^3 [80]. The total cost of the PP pyrolysis plant utilities is shown in Table 12. The utilities costs are significantly lower for scenario 1.

Table 12. Utility costs for PP pyrolysis unit for scenarios 1 and 2.

Type of Utility		Quantity	Cost (M€/year)
Electricity		5.6 GWh/year	0.73
Cooling water		2.6 Mm^3/year	2.55
Shredding (electricity)			3
Natural Gas	**Scenario 1**	0	0
	Scenario 2	6 Mm^3/year	3.17
Total Utilities Cost	**Scenario 1**		**6.28**
	Scenario 2		**9.45**

4.3.3. Cash Inflow

The revenue of the PP pyrolysis plant originates from the sale of the three products. Based on the quantities produced and their respective selling prices, it is possible to calculate the annual cash inflow of the installation, as shown in Table 13.

In scenario 1, part of the pyrolysis gas is used to meet the energy requirements of the processes and the remaining amount is sold. In scenario 2, the entire amount of pyrolysis gas is sold, leading to higher cash inflows for the plant.

Table 13. Total cash inflows of the PP pyrolysis unit.

Product		Quantity (t/d)	Selling Price (€/t)	Cash Inflow (M€/year)	Source
Pyrolysis oil		521	652	112	[81,82]
Pyrolysis gas	**Scenario 1**	62	176	3.6	[83]
	Scenario 2	84	176	4.89	[83]
Char		0.6	580	0.12	[84]
Total Cash Inflows S	**Scenario 1**			**115.86 M€**	
	Scenario 2			**117.15 M€**	

4.3.4. Net Profits

Based on the total cash inflows (S) and operating costs (C), it is possible to calculate the annual gross profit of the PP pyrolysis plant for both scenarios, as shown by the equations below:

$$R_{Scenario\ 1} = S - C = 64.5\ M€/y$$

$$R_{Scenario\ 2} = S - C = 62.3\ M€/y$$

The net annual profit of the installation is calculated with the following equation:

$$P_{Scenario\ 1} = R - e \times I_f - \left(R - d \times I_f\right) \times t = 37.3\ M€/y$$

$$P_{Scenario\ 2} = R - e \times I_f - \left(R - d \times I_f\right) \times t = 36\ M€/y$$

The following assumptions were made for this calculation:

- The economic life of the plant is N = 15 years.
- Depreciation is linear.
- The uniform tax rate is t = 0.4.
- The depreciation rate for tax purposes is d = 1/N = 0.06.
- The depreciation rate of the fixed investment is e = d.

4.3.5. Financial Indicators

To assess the viability of the PP pyrolysis plant, two economic indicators were also examined, namely the return based on the initial investment (R.O.I) and payback time (P.O.T). These indicators are calculated as follows:

$$ROI = \frac{P}{Cost\ of\ Investment}$$

$$POT = \frac{I_f}{P + e \times I_f}$$

Initially, the return based on the initial investment (R.O.I) is calculated as follows for the two scenarios:

$$ROI_{Scenario\ 1} = 81\%$$

$$ROI_{Scenario\ 2} = 78\%$$

In addition, the payback time (P.O.T) is as follows:

$$POT_{Scenario\ 1} = 1.16\ years$$

$$POT_{Scenario\ 2} = 1.2\ years$$

4.3.6. Overview of Feasibility

An overview of the economic performance of the PP pyrolysis plant with a capacity of 200,000 t/year is presented in Table 14. Based on the results of the analysis, it appears that the plant is economically viable, as it presents high profits and a short payback time for both scenarios considered.

More specifically, the use of pyrolysis gas to supply heat to the reactor (scenario 1) has a positive effect on the feasibility of the plant. Although the annual revenues are smaller in scenario 1, the significant reduction in the operating costs makes this scenario preferable. The annual profits amount to 37.3 M€, while the return on investment is quite high at 81%. Finally, the payback time of the unit is only 1.16 years. In contrast, in scenario 2, although the plant is still feasible and shows high profitability, the annual profits are lower by about 1.5 M€, while the payback time is a little higher at 1.2 years.

Table 14. Overview of the feasibility of the PP pyrolysis plant.

	Scenario 1	Scenario 2
Capacity (t/year)	200,000	
Investment (M€)	45.9	
Operating costs (M€ /year)	51.3	54.8
Annual cash inflows (M€/year)	115.86	117.15
Net annual profit (M€/year)	37.3	36
R.O.I	81%	78%
P.O.T (years)	1.16	1.2

Pyrolysis oil is the most important product of the process, as more than 95% of the annual revenue of the plant is due to its sale. The solid product (char) is produced in very small quantities; therefore, it does not contribute significantly to the profitability of the plant.

The annual operating costs amount to 51.3 and 54.8 M€, respectively, for the two scenarios, 1 and 2. The cost of purchasing the raw material constitutes about 60% of the total. The cost of utilities is a significant part of the total operating costs, with shredding costs amounting to 3 M€. Some additional operating costs, such as supervision, maintenance, local taxes, administration, and distribution costs, make up about 15% of the unit's total annual operating costs.

4.3.7. Sensitivity Analysis Effect of Plant Capacity on Economic Indicators

Based on the process described in the previous sections, the economic performance of the PP pyrolysis plant can be examined for a range of capacities. The impact of plant capacity on the initial investment is depicted in Figure 7. The investment increases with the plant's capacity, from 4.02 M€ for 5000 t/year to 59.98 M€ for 300,000 t/year.

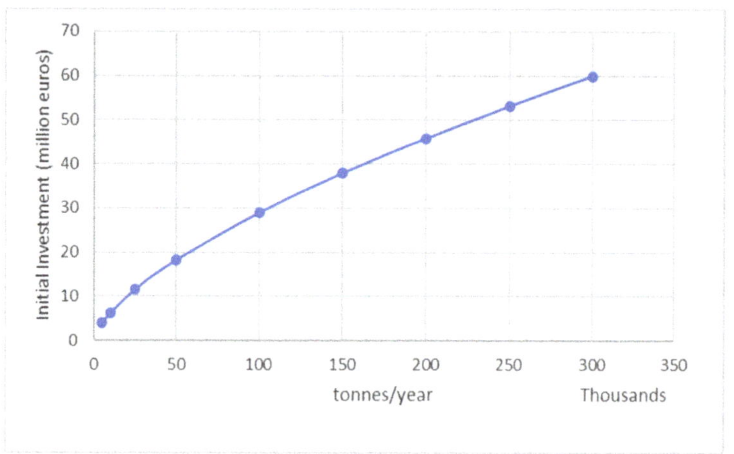

Figure 7. Effect of PP pyrolysis plant capacity on the initial investment (scenario 1).

Nevertheless, this is not a sufficient criterion for comparing the economic viability of facilities of different scales; therefore, the indicators R.O.I and P.O.T are preferred, as shown in Figure 8, to assess optimal capacity.

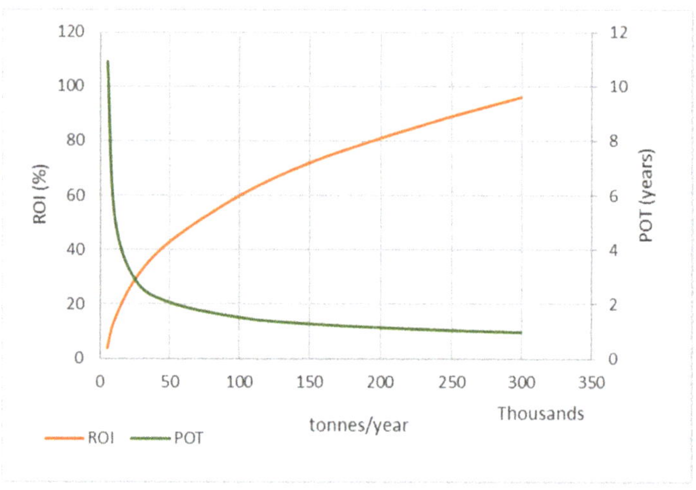

Figure 8. Effect of PP pyrolysis plant capacity on R.O.I and P.O.T (scenario 1).

For small capacities (<50,000 t/year), the PP pyrolysis plant seems profitable, but does not appear to be a very viable or attractive investment. The payback period is between 2 and 10 years, which means that it takes a very long time to recover the initial investment. However, the R.O.I index is less than 45%, which is significantly lower compared to higher capacity installations.

In plants with a capacity of >100,000 t per year, the investment is much more advantageous. It seems that the initial investment can be recovered in less than two years, as the time decreases with increasing capacity. Accordingly, the higher the capacity of the unit, the more efficient the investment.

Nevertheless, the initial investment increases significantly. For example, for a pyrolysis plant with a capacity of 300,000 t/year, the R.O.I index is at its maximum (96%) and the payback time is at its minimum (1 year), which means that it presents the optimal financial performance. However, due to the size of the facility, the initial investment amounts to 60 M€, which can be a deterrent. Accordingly, to determine the optimal capacity, a profitable investment with a reasonably high initial investment must be found.

As shown in Figure 8, the P.O.T index decreases with increasing capacity. When the capacity has significantly increased, there is a very small reduction in payback time. Especially for capacities >200,000 t/year, the payback time is insignificantly reduced, meaning that while the initial investment significantly increases, the economic performance of the plant is only improved slightly. The R.O.I index is constantly increasing, but its increase is less significant for capacities >200,000 t/year.

Based on these observations, it can be assumed that while the unit is viable for all the capacities that were considered, optimal economic efficiency can be achieved for capacities in the range of 150,000–200,000 t per year.

5. Conclusions

Pyrolysis is an effective method of managing plastic waste mainly for oil production, which exhibits similar properties to conventional fuels, such as gasoline or diesel. Many factors affect the yields of the process, as well as the properties and quality of the final products. The temperature, the type of feedstock and the presence of a catalyst are the most important parameters to consider, although the type of reactor, the pressure and the heating rate are also important.

Polypropylene (PP) is very often used as feedstock for pyrolysis, as very high yields of pyrolysis oil are observed. The properties of the oil are similar to those of conventional fuels. The preferred temperature for PP pyrolysis ranges from 450 °C to 550 °C, where oil yields vary from 82 wt.% to 92.3 wt.%. For temperatures below 450 °C, lower oil yields can be observed (<80 wt.%), while at higher temperatures (740 °C), the gas yield increases significantly, limiting oil production.

Based on the results of the feasibility assessment, a PP-based pyrolysis plant can be economically viable, with its optimal capacity being in the range of 150,000 to 200,000 t per year. Pyrolysis oil is the main product of the plant, and it is responsible for generating a significant part of the total revenue, while the gaseous product also contributes to the economic viability. The pyrolysis plant shows significant profits both when the pyrolysis gas is used to cover the energy requirements of the process (scenario 1) and when the gas is sold as a final product (scenario 2). The solid (char), despite its relatively high selling price, constitutes a very low percentage of the annual profits, due to its low yields.

For scenario 1, the pyrolysis plant with a capacity of 200,000 t PP/year is estimated to have annual profits of 37.3 M€, while the payback time of the initial investment is very low, at 1.16 years. Using part of the gas to meet the energy needs of the pyrolysis reactor significantly reduces the operating costs, as the use of natural gas is not required. However, the remaining amount of gas is sold and contributes to the annual revenue. In the case where the entire amount of gas is sold (scenario 2), the profitability of the plant remains, but this option is less preferable, due to the higher operating costs associated with the use of natural gas.

The results of this study show that the PP pyrolysis unit is economically viable for a wide range of capacities. However, this is an approximate analysis based on certain assumptions; thus, a detailed feasibility study is proposed to produce more accurate results.

Author Contributions: Conceptualization, A.Z.; methodology, A.Z. and I.V.; formal analysis, I.V.; investigation, A.Z. and I.V.; resources, I.V. and A.Z.; data curation, I.V.; writing—original draft preparation, I.V.; writing—review and editing, A.Z.; visualization, I.V. and A.Z.; supervision, A.Z.; project administration, A.Z.; funding acquisition. All authors have read and agreed to the published version of the manuscript.

Funding: This research was funded by the action «Investment Plans for Innovation' of the region of Central Macedonia, Greece; grant number KMP6-0283869.

Institutional Review Board Statement: Not applicable.

Informed Consent Statement: Not applicable.

Data Availability Statement: Not applicable.

Acknowledgments: We acknowledge the administrative support provided by CAO HELLAS.

Conflicts of Interest: The authors declare no conflict of interest.

References

1. Xayachak, T.; Haque, N.; Parthasarathy, R.; King, S.; Emami, N.; Lau, D.; Pramanik, B.K. Pyrolysis for Plastic Waste Management: An Engineering Perspective. *J. Environ. Chem. Eng.* **2022**, *10*, 108865. [CrossRef]
2. Zhu, H.; He, D.; Duan, H.; Yin, H.; Chen, Y.; Chao, X.; Zhang, X.; Gong, H. Study on coupled combustion behaviors and kinetics of plastic pyrolysis by-product for oil. *Energy* **2023**, *262A*, 125452. [CrossRef]
3. WWF. *Plastic Pollution in Greece: How to Stop It—A Practical Guide for Policy Makers*; Worldwide Fund for Nature: Vaud, Switzerland, 2019.
4. Elliott, T.; Xirou, H.; Stergiou, V.; Bapasola, A.; Gillie, H. *Policy Measures on Plastics in Greece*; Eunomia Research & Consulting: London, UK, 2020.
5. YPEKA; EOAN; GIZ. Citizens Leaflet—Practical Tips for a More Sustainable and Cleaner Greece. Available online: https://ypen.gov.gr/wp-content/uploads/2021/09/Guidelines_Citizens_FINAL_EN.pdf (accessed on 19 December 2022).
6. Babaremu, K.O.; Okoya, S.A.; Hughes, E.; Tijani, B.; Teidi, D.; Akpan, A.; Igwe, J.; Karera, S.; Oyinlola, M.; Akinlabi, E.T. Sustainable plastic waste management in a circular economy. *Heliyon* **2022**, *8*, e09984. [CrossRef] [PubMed]
7. Hossain, R.; Islam, M.T.; Ghose, A.; Sahajwalla, V. Full circle: Challenges and prospects for plastic waste management in Australia to achieve circular economy. *J. Clean. Prod.* **2022**, *368*, 133127. [CrossRef]
8. Mirkarimi, S.M.R.; Bensaid, S.; Chiaramonti, D. Conversion of mixed waste plastic into fuel for diesel engines through pyrolysis process: A review. *Appl. Energy* **2022**, *327*, 120040. [CrossRef]
9. Chen, X.; Dong, M.; Zhang, L.; Luan, X.; Cui, X.; Cui, Z. Comprehensive evaluation of environmental and economic benefits of industrial symbiosis in industrial parks. *J. Clean. Prod.* **2022**, *354*, 131635. [CrossRef]
10. Hassan, S.T.; Wang, P.; Khan, I.; Zhu, B. The impact of economic complexity, technology advancements, and nuclear energy consumption on the ecological footprint of the USA: Towards circular economy initiatives. *Gondwana Res.* **2023**, *113*, 237–246. [CrossRef]
11. OECD. *Global Plastics Outlook: Policy Scenarios to 2060*; OECD: Paris, France, 2022.
12. Kusenberg, M.; Eschenbacher, A.; Delva, L.; De Meester, S.; Delikonstantis, E.; Stefanidis, G.D.; Ragaert, K.; Van Geem, K.M. Towards high-quality petrochemical feedstocks from mixed plastic packaging waste via advanced recycling: The past, present and future. *Fuel Process. Technol.* **2022**, *238*, 107474. [CrossRef]
13. European Commission. Circular Economy: Commission Takes Action to Reduce waste from Single-Use Plastics. (Online) European Commission. 2022. Available online: https://ec.europa.eu/commission/presscorner/detail/en/ip_22_5731 (accessed on 19 October 2022).
14. CMS. Plastics and Packaging Laws in the European Union. (Online) CMS. 2021. Available online: https://cms.law/en/int/expert-guides/plastics-and-packaging-laws/european-union (accessed on 20 November 2022).
15. Eze, W.; Umunakwe, R.; Obasi, H.; Ugbaja, M.; Uche, C.; Madufor, I. Plastics waste management: A review of pyrolysis technology. *Clean Technol. Recycl.* **2021**, *1*, 50–69. [CrossRef]
16. Rehan, M.; Miandad, R.; Barakat, M.A.; Ismail, I.M.I.; Almeelbi, T.; Gardy, J. Effect of zeolite catalysts on pyrolysis liquid oil. *Int. Biodeterior. Biodegrad* **2017**, *119*, 162–175. [CrossRef]
17. Armenise, S.; SyieLuing, W.; Ramírez-Velásquez, J.; Launay, F.; Wuebben, D.; Ngadi, N.; Rams, J.; Muñoz, M. Plastic waste recycling via pyrolysis: A bibliometric survey and literature review. *J. Anal. Appl. Pyrolysis* **2021**, *158*, 105265. [CrossRef]

18. Jamradloedluk, J.; Lertsatitthanakorn, C. Characterization and Utilization of Char Derived from Fast Pyrolysis of Plastic Wastes. *Procedia Eng.* **2014**, *69*, 1437–1442. [CrossRef]
19. Anuar Sharuddin, S.; Abnisa, F.; Wan Daud, W.; Aroua, M. A review on pyrolysis of plastic wastes. *Energy Convers. Manag.* **2016**, *115*, 308–326. [CrossRef]
20. Maqsood, T.; Dai, J.; Zhang, Y.; Guang, M.; Li, B. Pyrolysis of plastic species: A review of resources and products. *J. Anal. Appl. Pyrolysis* **2021**, *159*, 105295. [CrossRef]
21. Erdogan, S. Recycling of Waste Plastics into Pyrolytic Fuels and Their Use in IC Engines. *Sustain. Mobil.* **2020**. [CrossRef]
22. Basu, P. *Biomass Gasification and Pyrolysis*; Academic Press: Amsterdam, The Netherlands, 2010.
23. Abnisa, F.; Wan Daud, W. A review on co-pyrolysis of biomass: An optional technique to obtain a high-grade pyrolysis oil. *Energy Convers. Manag.* **2014**, *87*, 71–85. [CrossRef]
24. Çepelioğullar, Ö.; Pütün, A. Thermal and kinetic behaviors of biomass and plastic wastes in co-pyrolysis. *Energy Convers. Manag.* **2013**, *75*, 263–270. [CrossRef]
25. Pandey, U.; Stormyr, J.; Hassani, A.; Jaiswal, R.; Haugen, H.; Moldestad, B. Pyrolysis of plastic waste to environmentally friendly products. *Energy Prod. Manag. 21st Century IV* **2020**, *246*, 61–74.
26. López, A.; de Marco, I.; Caballero, B.; Laresgoiti, M.; Adrados, A. Dechlorination of fuels in pyrolysis of PVC containing plastic wastes. *Fuel Process. Technol.* **2011**, *92*, 253–260. [CrossRef]
27. Wang, H.; Hu, H.; Yang, Y.; Liu, H.; Tang, H.; Xu, S.; Li, A.; Yao, H. Effect of high heating rates on products distribution and sulfur transformation during the pyrolysis of waste tires. *Waste Manag.* **2020**, *118*, 9–17. [CrossRef]
28. Miandad, R.; Rehan, M.; Barakat, M.A.; Aburiazaiza, A.S.; Khan, H.; Ismail, I.M.I.; Dhavamani, J.; Gardy, J.; Hassanpour, A.; Nizami, A.S. Catalytic Pyrolysis of Plastic Waste: Moving Toward Pyrolysis Based Biorefineries. *Front. Energy Res* **2019**, *7*, 27. [CrossRef]
29. Ahmad, I.; Khan, M.; Khan, H.; Ishaq, M.; Tariq, R.; Gul, K.; Ahmad, W. Pyrolysis Study of Polypropylene and Polyethylene Into Premium Oil Products. *Int. J. Green Energy* **2014**, *12*, 663–671. [CrossRef]
30. Sakata, Y.; Uddin, M.; Muto, A. Degradation of polyethylene and polypropylene into fuel oil by using solid acid and non-acid catalysts. *J. Anal. Appl. Pyrolysis* **1999**, *51*, 135–155. [CrossRef]
31. Kusenberg, M.; Zayoud, A.; Roosen, M.; Thi, H.; Abbas-Abadi, M.; Eschenbacher, A.; Kresovic, U.; De Meester, S.; Van Geem, K. A comprehensive experimental investigation of plastic waste pyrolysis oil quality and its dependence on the plastic waste composition. *Fuel Process. Technol.* **2021**, *227*, 107090. [CrossRef]
32. FakhrHoseini, S.; Dastanian, M. Predicting Pyrolysis Products of PE, PP, and PET Using NRTL Activity Coefficient Model. *J. Chem.* **2013**, *2013*, 487676. [CrossRef]
33. Paraschiv, M.; Kuncser, R.; Tazerout, M.; Prisecaru, T. New energy value chain through pyrolysis of hospital plastic waste. *Appl. Therm. Eng.* **2015**, *87*, 424–433. [CrossRef]
34. Demirbas, A. Pyrolysis of municipal plastic wastes for recovery of gasoline-range hydrocarbons. *J. Anal. Appl. Pyrolysis* **2014**, *72*, 97–102. [CrossRef]
35. Inayat, A.; Fasolini, A.; Basile, F.; Fridrichova, D.; Lestinsky, P. Chemical recycling of waste polystyrene by thermo-catalytic pyrolysis: A description for different feedstocks, catalysts and operation modes. *Polym. Degrad. Stab.* **2022**, *201*, 109981. [CrossRef]
36. Abbas-Abadi, M.; Haghighi, M.; Yeganeh, H.; McDonald, A. Evaluation of pyrolysis process parameters on polypropylene degradation products. *J. Anal. Appl. Pyrolysis* **2014**, *109*, 272–277. [CrossRef]
37. Erawati, E.; Hamid and Permatasari, R. Pyrolysis of polypropylene waste with natural zeolite as catalyst. Exploring resources, process and design for sustainable urban development. In Proceedings of the 5th International Conference on Engineering, Technology, and Industrial Application (ICETIA), Surakarta, Indonesia, 12–13 December 2018.
38. Marino, A.; Aloise, A.; Hernando, H.; Fermoso, J.; Cozza, D.; Giglio, E.; Migliori, M.; Pizzaro, P.; Giordano, G.; Serrano, D.P. ZSM-5 zeolites performance assessment in catalytic pyrolysis of PVC-containing real WEEE plastic wastes. *Catal. Today* **2022**, *390–391*, 210–220. [CrossRef]
39. Odejobi, O.J.; Oladunni, A.A.; Sonibare, J.A.; Abegunrin, I.O. Oil yield optimization from co-pyrolysis of low-density polyethylene (LDPE), polystyrene (PS) and polyethylene terephthalate (PET) using simplex lattice mixture design. *Fuel Commun.* **2020**, *2–5*, 100006. [CrossRef]
40. Terapalli, A.; Kamireddi, D.; Srivedi, V.; Tukarambai, M.; Suriapparao, D.V.; Rao, C.S.; Gautam, R.; Modi, P.R. Microwave-assisted in-situ catalytic pyrolysis of polystyrene: Analysis of product formation and energy consumption using machine learning approach. *Process Saf. Environ. Prot.* **2022**, *166*, 57–67. [CrossRef]
41. Adnan; Shah, J.; Jan, M. Thermo-catalytic pyrolysis of polystyrene in the presence of zinc bulk catalysts. *J. Taiwan Inst. Chem. Eng.* **2014**, *45*, 2494–2500. [CrossRef]
42. Shahbaz, M.; AlNouss, A.; Mckay, G.; Mackey, H.; Ansari, T.A. Techno-economic and environmental analysis of pyrolysis process simulation for plastic (PET) waste. *Comput. Aided Chem. Eng.* **2022**, *51*, 115–120.
43. Bagri, R.; Williams, P. Catalytic pyrolysis of polyethylene. *J. Anal. Appl. Pyrolysis* **2022**, *63*, 29–41. [CrossRef]
44. Williams, P.T.; Williams, E.A. Fluidized bed pyrolysis of low density polyethylene to produce petrochemical feedstock. *J. Anal. Appl. Pyrolysis* **1999**, *51*, 107–126. [CrossRef]
45. Wu, Y.; Wang, K.; Wei, B.; Yang, H.; Jin, L.; Hu, H. Pyrolysis behavior of low-density polyethylene over HZSM-5 via rapid infrared heating. *Sci. Total Environ.* **2021**, *806*, 151287. [CrossRef]

46. Marcilla, A.; Beltrán, M.; Navarro, R. Thermal and catalytic pyrolysis of polyethylene over HZSM5 and HUSY zeolites in a batch reactor under dynamic conditions. *Appl. Catal. B Environ.* **2009**, *86*, 78–86. [CrossRef]

47. Dzol, M.A.A.M.; Balasundram, V.; Shameli, K.; Ibrahim, N.; Manan, Z.A.; Isha, R. Catalytic pyrolysis of high-density polyethylene over nickel-waste chicken eggshell/HZSM-5. *J. Environ. Manag.* **2022**, *324*, 116392. [CrossRef]

48. Mastral, F.; Esperanza, E.; García, P.; Juste, M. Pyrolysis of high-density polyethylene in a fluidised bed reactor. Influence of the temperature and residence time. *J. Anal. Appl. Pyrolysis* **2012**, *63*, 1–15. [CrossRef]

49. Mousavi, S.A.H.S.; Sadrameli, S.M.; Dehaghani, A.H.S. Energy recovery from high density polyethylene plastic via pyrolysis with upgrading of the product by a novel nano MIL-53 (Cu) derived@Y zeolite catalyst using response surface methodology. *Fuel Process. Technol.* **2022**, *231*, 107257. [CrossRef]

50. Donaj, P.; Kaminsky, W.; Buzeto, F.; Yang, W. Pyrolysis of polyolefins for increasing the yield of monomers' recovery. *Waste Manag.* **2012**, *32*, 840–846. [CrossRef] [PubMed]

51. Kaminsky, W.; Schlesselmann, B.; Simon, C. Thermal degradation of mixed plastic waste to aromatics and gas. *Polym. Degrad. Stab.* **1996**, *53*, 189–197. [CrossRef]

52. Nugroho Pratama, N.; Saptoadi, H. Characteristics of Waste Plastics Pyrolytic Oil and Its Applications as Alternative Fuel on Four Cylinder Diesel Engines. *Int. J. Renew. Energy Dev. (IJRED)* **2014**, *3*, 13–20. [CrossRef]

53. Tian, X.; Zeng, Z.; Liu, Z.; Dai, L.; Xu, J.; Yang, X.; Yue, L.; Liu, Y.; Ruan, R.; Wang, Y. Conversion of low-density polyethylene into monocyclic aromatic hydrocarbons by catalytic pyrolysis: Comparison of HZSM-5, Hβ, HY and MCM-41. *J. Clean. Prod.* **2022**, *358*, 131989. [CrossRef]

54. Artetxe, M.; Lopez, G.; Elordi, G.; Amutio, M.; Bilbao, J.; Olazar, M. Production of Light Olefins from Polyethylene in a Two-Step Process: Pyrolysis in a Conical Spouted Bed and Downstream High-Temperature Thermal Cracking. *Ind. Eng. Chem. Res.* **2012**, *51*, 13915–13923. [CrossRef]

55. Kagan, J.; Drury, A. Payback Period Definition. (Online) Investopedia. Available online: https://www.investopedia.com/terms/p/paybackperiod.asp (accessed on 10 September 2022).

56. Bezergianni, S.; Dimitriadis, A.; Faussone, G.; Karonis, D. Alternative Diesel from Waste Plastics. *Energies* **2017**, *10*, 1750. [CrossRef]

57. Kodera, Y.; Yamamoto, T.; Ishikawa, E. Energy- and economic-balance estimation of pyrolysis plant for fuel-gas production from plastic waste based on bench-scale plant operations. *Fuel Commun* **2021**, *7*, 100016. [CrossRef]

58. Mathieson, J.; Somerville, M.A.; Deev, A.; Jahanshahi, S. *Utilization of Biomass as an Alternative Fuel in Ironmaking*; Woodhead Publishing: Soston, UK, 2015; pp. 581–613.

59. Turton, R.; Shaeiwitz, J.; Bhattacharyya, D.; Whiting, W. *Analysis, Synthesis and Design of Chemical Processes*, 5th ed.; Prentice Hall: Boston, MA, USA, 2018.

60. Fivga, A.; Dimitriou, I. Pyrolysis of plastic waste for production of heavy fuel substitute: A techno-economic assessment. *Energy* **2018**, *149*, 865–874. [CrossRef]

61. Ghodrat, M.; Abascall Alonso, J.; Hagare, D.; Yang, R.; Samali, B. Economic feasibility of energy recovery from waste plastic using pyrolysis technology: An Australian perspective. *Int. J. Environ. Sci. Technol.* **2019**, *16*, 3721–3734. [CrossRef]

62. Riedewald, F.; Patel, Y.; Wilson, E.; Santos, S.; Sousa-Gallagher, M. Economic assessment of a 40,000 t/y mixed plastic waste pyrolysis plant using direct heat treatment with molten metal: A case study of a plant located in Belgium. *Waste Manag.* **2021**, *120*, 698–707. [CrossRef] [PubMed]

63. Yang, Y.; Wang, J.; Chong, K.; Bridgwater, A. A techno-economic analysis of energy recovery from organic fraction of municipal solid waste (MSW) by an integrated intermediate pyrolysis and combined heat and power (CHP) plant. *Energy Convers. Manag.* **2018**, *174*, 406–416. [CrossRef]

64. Rogers, C.; Means, P.; Gonzalez, R.; Sheets, K.; Townsend, H. Economic Feasibility of Mixed Plastic Waste Pyrolysis Using Twin Reactor System in Northwest Arkansas. Bachelor's Thesis, University of Arkansas, Fayetteville, AR, USA, 2021.

65. Petersson, I.; Svensson, A. Regional Plastic Waste Recycling through Pyrolysis—A Technoeconomic Evaluation. Ph.D. Thesis, Chalmers University of Technology, Department of Space, Earth and Environment, Gothenburg, Sweden, 2019.

66. Almohamadi, H.; Alamoudi, M.; Ahmed, U.; Shamsuddin, R.; Smith, K. Producing hydrocarbon fuel from the plastic waste: Techno-economic analysis. *Korean J. Chem. Eng.* **2021**, *38*, 2208–2216. [CrossRef]

67. Peters, S.M.; Timmerhaus, D.C. *Plant Design and Economics for Chemical Engineers*; McGraw Hill International Editions; McGraw Hill: New York, NY, USA, 2003.

68. Eurostat. Hourly Labour Costs—Statistics Explained. (Online) Ec.europa.eu. Available online: https://ec.europa.eu/eurostat/statistics-explained/index.php?title=Hourly_labour_costs#Hourly_labour_costs_ranged_between_.E2.82.AC6.5_and_.E2.82.AC45.8_in_2020 (accessed on 25 September 2022).

69. Bora, R.; Wang, R.; You, F. Waste Polypropylene Plastic Recycling toward Climate Change Mitigation and Circular Economy: Energy, Environmental, and Technoeconomic Perspectives. *ACS Sustain. Chem. Eng.* **2020**, *8*, 16350–16363. [CrossRef]

70. DEI. Electricity Selling Price. (Online) Dei.gr. 2021. Available online: https://www.dei.gr/Documents/xt.tim.1.7.08.pdf (accessed on 14 September 2022).

71. EYDAP. Pricing EYDAP. (Online) Eydap.gr. 2022. Available online: https://www.eydap.gr/userfiles/47614413-661a-4fba-ba7c-a14f00cfa261/Timologio_EYDAP_2.pdf (accessed on 25 September 2022).

72. REC. Shredding. (Online) Recyclingequipment.com. 2020. Available online: https://recyclingequipment.com/shredders/398-tire-shredding (accessed on 15 September 2022).

73. Engineering Toolbox. Specific Heat of Common Substances. (Online). 2003. Available online: https://www.engineeringtoolbox.com/specific-heat-capacity-d_391.html (accessed on 20 August 2022).

74. EngineeringToolbox. Water—Heat of Vaporization. (Online) Engineeringtoolbox.com. 2021. Available online: https://www.engineeringtoolbox.com/water-properties-d_1573.html (accessed on 2 September 2022).

75. NIST. Compressed Water and Superheated Steam. (Online) Nist.gov. Available online: https://www.nist.gov/system/files/documents/srd/NISTIR5078-Tab3.pdf (accessed on 25 August 2022).

76. Abdallah, R.; Juaidi, A.; Assad, M.; Salameh, T.; Manzano-Agugliaro, F. Energy recovery from waste tires using pyrolysis: Palestine as a case study. *Energies* **2020**, *13*, 1817. [CrossRef]

77. Chavando, J.A.M.; de Matos, E.C.J.; Silva, V.B.; Tarelho, L.A.C.; Cardoso, J.S. Pyrolysis characteristics of RDF and HPDE blends with biomass. *Int. J. Hydrog. Energy* **2021**, *47*, 19901–19915. [CrossRef]

78. Jaafar, Y.; Abdelouahed, L.; El Hage, R.; El Samrani, A.; Taouk, B. Pyrolysis of common plastics and their mixtures to produce valuable petroleum-like products. *Polym. Degrad. Stab.* **2022**, *195*, 109770. [CrossRef]

79. EngineeringToolbox. Fuel Gases Heating Values. (Online) Engineeringtoolbox.com. 2021. Available online: https://www.engineeringtoolbox.com/heating-values-fuel-gases-d_823.html (accessed on 12 September 2022).

80. Aerio Attikis. Natural Gas Price and Savings. (Online) Aerioattikis.blob.core.windows.net. 2021. Available online: https://aerioattikis.blob.core.windows.net/wp-uploads/2021/07/%CE%A4%CE%B9%CE%BC%CE%AD%CF%82_%CE%99%CE%B4%CE%B9%CF%8E%CF%84%CE%B5%CF%82_%CE%99%CE%BF%CF%8D%CE%BB%CE%B9%CE%BF%CF%82_2021-60f1866822637.pdf (accessed on 26 August 2022).

81. Desai, S.B.; Galage, C.K. Production and Analysis of pyrolysis oil from waste plastic in Kolhapur city. *Int. J. Eng. Res. Gen. Sci.* **2015**, *3*, 590–595.

82. Hamid, K.; Sabir, R.; Hameed, K.; Waheed, A.; Ansari, M.U. Economic Analysis of Fuel Oil Production from Pyrolysis of Waste Plastic. *Austin Environ Sci.* **2021**, *6*, 1053. [CrossRef]

83. Yahya, S.; Iqbal, T.; Omar, M.; Ahmad, M. Techno-Economic Analysis of Fast Pyrolysis of Date Palm Waste for Adoption in Saudi Arabia. *Energies* **2021**, *14*, 6048. [CrossRef]

84. GVR. Recovered Carbon Black Market Size, Share & Trends Analysis Report by Application (Tires, Rubber, High Performance Coatings, Plastics), by Region (North America, Europe, APAC, CSA, MEA), and Segment Forecasts, 2020—2027. (ebook) Grand View Research. 2020. Available online: https://www.grandviewresearch.com/industry-analysis/recovered-carbon-black-market/toc (accessed on 26 January 2022).

Article

Comparative Techno-Economic Evaluation of a Standalone Solar Power System for Scaled Implementation in Off-Grid Areas

Muhammad Sadiq [1], Phimsupha Kokchang [2,*] and Suthirat Kittipongvises [3]

[1] Environment Development and Sustainability (EDS), Graduate School, Chulalongkorn University, Bangkok 10330, Thailand; engr.sadiq95@gmail.com
[2] Energy Research Institute, Chulalongkorn University, Bangkok 10330, Thailand
[3] Environmental Research Institute, Chulalongkorn University, Bangkok 10330, Thailand; suthirat.k@gmail.com
* Correspondence: phimsupha.k@chula.ac.th

Abstract: The increasing environmental concerns and dependence on fossil fuel-based energy sectors necessitate a shift towards renewable energy. Off-grid communities can particularly benefit from standalone, scaled renewable power plants. This study developed a comprehensive techno-economic framework, analyzed the objective metrics, and assessed the influence of economies of scale in solar PV power plants to electrify off-grid communities, taking Baluchistan, Pakistan, as a pilot case. Simulations and analyses were performed using the System Advisor Model (SAM). The results indicate a noteworthy reduction in the levelized cost of energy (LCOE) with increased power generation capacity. It was observed that utilizing bi-facial modules with single-axis tracking leads to a more cost-effective LCOE compared to the relatively expensive dual-axis trackers. The main cost factors identified in the analysis were capital costs, installed balance of plant (BOP), mechanical, and electrical costs. Notably, the disparity between the highest and lowest LCOE values across the six different power generation pathways amounted to approximately 38.5%. The average LCOE was determined to be 2.14 USD/kWh for fixed-mounted plants, 1.79 USD/kWh for single-axis plants, and 1.74 USD/kWh for dual-axis plants across the examined power generation capacity range. The findings can serve as a valuable benchmark, specifically for regional key stakeholders, in making informed investment decisions, formulating effective policies, and devising appropriate strategies for off-grid electrification and the development of renewable energy value chains.

Keywords: solar PV; techno-economic analysis; economy of scale; off-grid electrification

Citation: Sadiq, M.; Kokchang, P.; Kittipongvises, S. Comparative Techno-Economic Evaluation of a Standalone Solar Power System for Scaled Implementation in Off-Grid Areas. *Energies* **2023**, *16*, 6262. https://doi.org/10.3390/en16176262

Academic Editors: Konstantinos Aravossis and Eleni Strantzali

Received: 29 July 2023
Revised: 23 August 2023
Accepted: 24 August 2023
Published: 28 August 2023

1. Introduction

The renewable energy sector is continuously evolving, with technological advancements improving the efficiency and cost-effectiveness of renewable energy systems (RES) [1]. As technology progresses, off-grid communities can benefit from more affordable and improved renewable energy solutions [2]. Reliance on centralized power grids can be challenging in remote or off-grid areas where infrastructure development is limited. The adoption of RES would enable these communities to achieve self-sufficiency and reduce dependence on external energy sources.

Like many other developing nations, Pakistan faces significant electricity challenges access, affordability, and shortfall. Around 40% of the population lacks access to reliable electricity, while even those with access encounter issues such as electricity shortages and high prices [3]. Moreover, 37.2% of Pakistan's population live in extreme poverty in 2023, earning an average of USD 3.65 per day [4]. This means that they may struggle to afford its usage even if they have access to the national electricity grid. As of August 2022, the electricity deficit in Pakistan has escalated to 7461 MW [5]. Considering an anticipated surge in electricity demand, this deficit is anticipated to escalate significantly, reaching a staggering 40,000 MW by the year 2030 [6]. These statistics highlight the need for Pakistan to address the electricity shortfall issues while considering the accessibility and affordability challenges faced by a significant portion of the population.

Baluchistan is the largest province in terms of area in Pakistan [7], and approximately 85% of its total population (13.16 million) resides in rural areas, leading this province to be known as the "powerless province" [8]. The electrification rate in this province stands at around 23%, significantly lower than the average national electrification rate of 72% [9]. As of 2023, the electricity demand is 1650 MW, but only 400–600 MW are supplied, resulting in persistent and regular load shedding (power/grid outages) of 12–18 h/day in main towns, excluding the capital city of Quetta, regardless of the season [10]. The situation in rural areas of this province is even worse, with electricity available for only 4 h/day, and significant areas still are beyond the jurisdiction of the national or regional grid systems [11]. Consequently, this province suffers greatly in terms of agricultural, industrial, and trade activities, in addition to civic problems, making it the least-developed province. The National Transmission and Despatch Company (NTDC) has a total electric transmission line of ~27 km in the province, capable of transmitting a maximum of around 600 MW [9]. Therefore, even when the government issues orders to reduce load shedding across the country, it does not provide any relief to consumers in Baluchistan due to the inadequate and insufficient transmission and distribution network.

The energy needs in Baluchistan Province are predominantly fulfilled through biomass energy sources such as firewood, animal dung, and agricultural waste [12]. However, electricity consumption is increasing at a rate of 17% per year [13]. The province has significant potential for solar, wind, geothermal, and micro-hydro power. Around 40% of its land receives solar energy at 6 kWh/m^2 per day, which adds up to a power generation potential of approximately 1.2 million MW [14]. The government of Pakistan, specifically the Ministry of Planning and Development, has recently shown its endorsement for investigating localized and off-grid alternatives to deliver electricity in the remote regions of Baluchistan [15]. Harnessing solar power for off-grid communities in this province would contribute to improvements in healthcare, education, communication, and water supply, leading to overall socio-economic development and well-being. A detailed survey conducted for the adoption of solar power in rural Baluchistan revealed that 89.2% of the rural population is willing to install solar power systems. However, due to their poor financial condition, they have been unable to install these systems and are awaiting support from the government or international donors [16]. In 2016, the provincial government of Baluchistan allocated USD 4.6 million for solar and wind power to attract private sector investment [17]. However, the private sector has shown reluctance to invest due to several factors, including the poor law and order situation, the remote location of the area, inadequate communications and infrastructure, and a low return on investment [18]. Another potential hurdle associated with emerging technologies is that financial institutions and large-scale investors tend to be risk-averse, often requiring realistic techno-economic information and a pilot plan before providing financing.

As of 2022, the global utility-scale solar sector has witnessed significant growth, with approximately 37,000 MW of operating projects and an additional 112,000 MW in development [19]. The 2030 target of achieving an unsubsidized levelized cost of energy (LCOE) of USD 0.02/kWh for utility-scale solar PV projects was set by the Department of Energy (DOE), USA [20]. However, several challenges hinder the widespread adoption of solar power, including inefficient solar panels and LS and high capital and operational expenditures contributing to a higher LCOE in comparison to power tariffs. To promote the adoption of large-scale solar PV systems in areas with favorable solar energy potential, it is crucial to assess the techno-economic metrics based on local conditions and specific components.

In addition to the continuous research focused on reducing costs in solar systems, it is imperative to address concerns regarding inefficient infrastructure and the development of a proper value chain. Furthermore, to attract private sector investment, make well-informed investment decisions, and gain a comprehensive understanding of the potential and challenges involved, it is of utmost importance to possess detailed techno-economic information about scaled solar power plants in specific geographical locations. Providing this information would greatly contribute to instilling confidence in investors and financial institutions.

The reviewed set of recent studies in Table 1 reveals a disparity in the clarity of the literature concerning the techno-economic metrics of scaled solar power plants. These studies demonstrate variations in technical assumptions and cost estimates and exhibit limitations in scope and procedural deficiencies. Furthermore, they overlook multiple parameters essential for determining the LCOE beyond capital expenditure (CAPEX) and operational expenditure (OPEX). The key deficiencies, as summarized below, underscore the importance of evaluating detailed techno-economic metrics in this field:

- Several significant cost-contributing parameters were either overlooked or arbitrarily selected. For example, Niaz et al. (2022) [21] examined the LCOE considering CAPEX and OPEX over ten years but did not include factors such as power generation scale, salvage value, degradation rate, loss factors, or replacement cost. Similarly, Nadaleti et al. (2020) [22] only considered CAPEX and OPEX. Yates et al. (2020) [23] calculated LCOE using a range of CAPEX and OPEX costs but did not address the impact of economies of scale. Ahshan et al. (2022) [24] investigated the LCOE of wind power, primarily focusing on CAPEX and OPEX. Shehabi et al. (2022) [25] used income tax rate, CAPEX, OPEX, balance of system (BOS) cost, equity, and replacement cost to determine LCOE but did not consider salvage value or the impact of economies of scale.
- The lack of a standardized approach for accounting CAPEX is noted. The direct CAPEX should encompass the costs of PV modules, current balancing devices, installation expenses, and contingency costs when calculating the net present value (NPV). However, it is taken generically, considering CAPEX as the cost/unit-power while excluding the other three cost variables, e.g., by Assowe et al. [26], Alessandro et al. [27], Jang et al. [28], and Burdack et al. [29].
- Furthermore, in the USA as of 2021, out of a total of 1125 proposed photovoltaic (PV) projects, 90% were based on single-axis tracking systems as opposed to fixed tilt systems, and mono-crystalline silicon (mono-c-Si) modules accounted for 69% of installations compared to thin-film modules [30]. Additionally, policy measures in the USA, such as extending the exemption from the 15% import duty through 2026, have encouraged the installation of bi-facial modules. This divergence in plant setup and module specifications in large-scale deployment highlights the need for research to understand how these factors impact energy output and cost.

To facilitate informed decision making regarding the implementation of scaled solar power for electrifying off-grid communities, this study provides a comprehensive techno-economic assessment.

Given the existing knowledge gap and the projected growth in renewable electricity demand, the contribution of this study mainly includes:

- Development of a robust framework for scaled solar PV plants that incorporates all relevant technological, financial, and benefit considerations. This approach enabled the accurate determination of techno-economic metrics, allowing for a fair comparison with fossil fuel-based power generation.
- Incorporate the essential macro- and micro-cost and technical parameters (Table 2) that are integral to the analysis and that must be considered when developing a techno-economic analysis model. Neglecting any of these parameters can lead to underestimation or overestimation of techno-economic metrics, rendering them unreliable. While certain parameters may have specific ranges, completely excluding them may result in misleading conclusions.
- Evaluation of economies of scale impact on techno-economic metrics of a scaled renewable power plant.

- By examining these factors, the primary objective of the study is to generate valuable insights into the feasibility and viability of implementing scaled solar power plants in the region.

Table 1. Key parameters of LCOE (non-exhaustive).

Study (Year Published, Ref.)	2022, [21]	2022, [24]	2022, [25]	2020, [22]	2020, [31]	2020, [23]	2016, [32]	2015, [33]
System lifetime	√	√	√	√	√	√	√	×
Degradation rate	×	√	×	×	×	×	×	×
Technical loss factor	×	×	×	×	√	×	×	×
Carbon trading price	×	√	×	×	×	×	×	×
Residual value	×	×	×	×	×	×	×	×
CAPEX	√	√	√	√	√	√	√	√
OPEX	√	√	√	√	√	√	√	√
Discount rate	×	×	×	√	×	√	√	×
Inflation rate	×	×	×	×	×	×	√	×
BOS and installation cost	×	×	×	×	×	×	×	×
Foundation (land preparation)	×	×	×	×	×	×	×	×
Engineering and developer overhead	×	×	×	×	×	×	×	×
Contingency	×	×	×	×	×	×	×	×

Table 2. Parametric framework of techno-economic assessment.

Category	Parameter	Notes
Site selection	Geographical location	DNI and financial factors, e.g., utility tariffs, tax rate, inflation and discount rate, carbon credits, etc., vary with location/country of interest.
Energy generation	Capacity	The economy of scale has an impact on net technical and cost parameters.
Performance and cost	System lifetime	Various renewable power-generation systems have different life spans and replacement costs.
Feasibility	Energy harvesting system	Solar PV, solar-thermal, wind, and biomass have different energy potentials concerning net output power.
System performance	Efficiency	Module type in the case of PV, turbine class in the case of wind, and Biomass's energy-to-power-conversion technique affect the power generated.
System performance	Loss factor	In practice, at the industrial level of power generation systems, specific energy and exergy losses typically exist, e.g., DC/AC losses in PV plants.
Net output power	Degradation rate	The performance of the power generator degrades with time, reducing the net output power.
System cost	CAPEX	The cost of components related to renewable energy system changes due to ongoing increases in global installed capacity and product improvements; more realistic CAPEX based on a well-defined system's design impacts the net cost.
System cost	Lifetime cost	Power generation systems and BOS have different lifetimes; hence during the specific analysis period, these should be accounted for separately.
System cost	OPEX	Logistics, labor wages, and insurance costs, and other soft costs are often country-specific, hence have a pronounced impact in the long run as OPEX mainly affects annual equivalent costs.
Project finance	Equity	The equity with a specific interest rate through a reasonable estimate must be accounted for in the life cycle costing of the renewable power generation to encourage private sector investment.
Net present value	Residual value	The residual value is a significant cost factor, specifically when the analysis period is shorter than plant life.
Net present value	Discount rate / Inflation rate	Discount and inflation rates directly affect the LCOE, neglecting or assuming it leads to unrealistic results.
Net present value	IRR	To attract private sector investment, IRR with a reasonable estimate should be declared, and its effect should be reflected in net cost.

Table 2. *Cont.*

Category	Parameter	Notes
Annualized cost	Carbon credit	The penalty charges avoided from reducing CO_2 emitting to the atmosphere are accounted for in LCC while evaluating the LCOE.
	Replacement cost	If the analysis period is different than the system life, then replacement cost needs to be accounted for in annualized cost.
Capital cost	Foundation	Renewable power plant requires a considerable land area for installation, which has a significant impact on CAPEX if leased or purchased. The CAPEX must account for land purchase and preparation costs where applicable.
	Engineering and developer overhead	Overhead costs, e.g., computer programming, communication, and encoding, are specifically crucial in setting central control systems in scale enactment.
	Contingency	In LCOE estimation over the 25–35 years period of the power generation system, the cost associated with uncertain and unpredictable risk should be considered.
	BOS and installation cost	BOS equipment and installation cost is a significant direct cost component, and it should be accounted for separately from the CAPEX.

2. Methods and Materials

2.1. Geographical Location

Several factors impact the site selection for a scaled solar power plant, such as the country's economic condition, commitment to the green energy transition, resource constraints, and renewable energy targets [34]. Additional considerations include wind speed, direct normal irradiance (DNI), water resources, transportation, and existing infrastructure such as industrial zones. Yang et al. [35] recommend a cumulative irradiation value of >2000 kWh/m^2/year while cautioning against exposure to less than 1600 kWh/m^2/year of DNI. High wind speeds can have adverse effects on solar PV plant performance, leading to increased thermal losses and structural instability [34]. Given that around 83% of the population in District Chagai, Balochistan, in Table 3 resides in the off-grid area [36], and considering the region's favorable solar energy potential, this study selects this area as a pilot case (Figure 1).

Table 3. Characteristics of selected solar PV plant site.

Production Site (Pakistan)	Co-Ordinates (°N, °E)	DNI (kWh/m^2/d)	Avg. GHI (kWh/m^2/d)	Avg. Wind Speed (m/s)	Avg. T (°C)
Chagai, Baluchistan	29.3058, 64.6945	5.94	5.93	3.1	28.6

2.2. Metrological Data

The metrological data was obtained from the National Solar Radiation Database at the NREL database [37]. By employing these site-specific data, it is believed that the assessment of solar power will yield more precise results compared to relying on average solar irradiation statistics for a given location.

2.3. Power Generation Pathways

The simulation encompasses a range of pathways derived from the combination of three different module configurations and two module types. Through comprehensive enumeration, a total of six pathways were examined in Figure 2. The findings were obtained through open-access simulation tools, the System Advisor Model (SAM.V22.11.21) [38] and a spreadsheet analyzer.

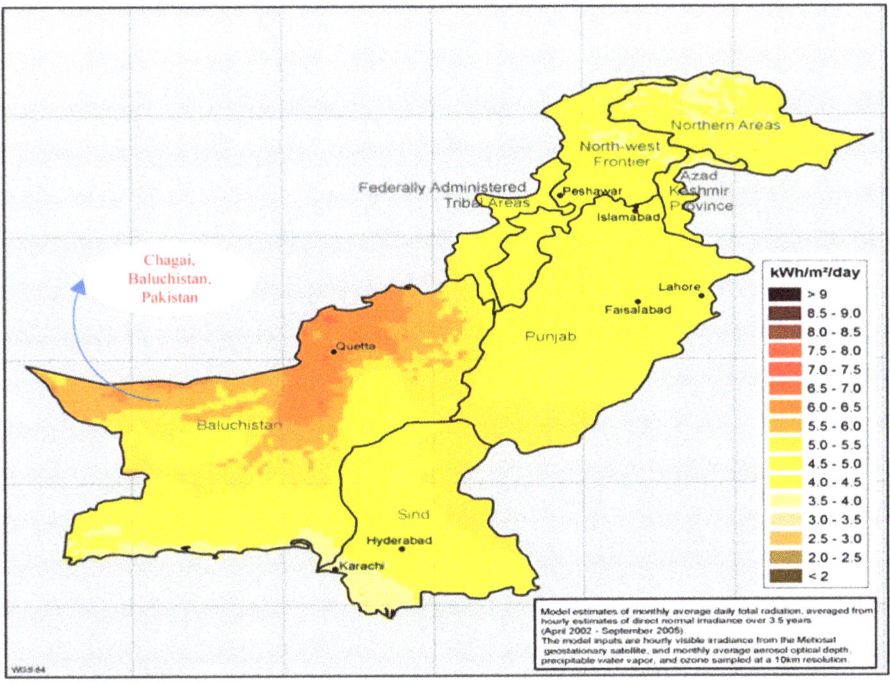

Figure 1. Geographical location of the selected site.

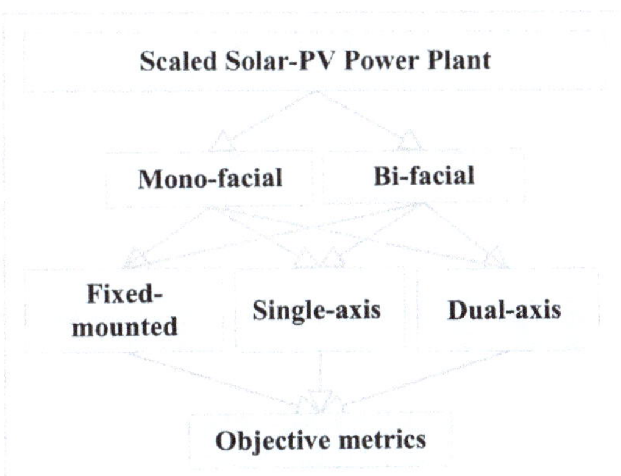

Figure 2. Roadmap of solar PV plant.

2.4. Simulation Algorithm

The algorithm presented in Figure 3 is employed to assess the key objective metrics: The LCOE, capacity factor (CF), total annual energy generated, and energy yield (EY). The technical specifications of the modules, as outlined in Table 4, along with the air-mass modifier polynomial ratio described in Ref. [39], are taken into account to address the effects of the solar spectrum on net power. Additionally, losses resulting from optical lenses, alignment errors, tracker errors, and wind flutter are considered in the loss factor.

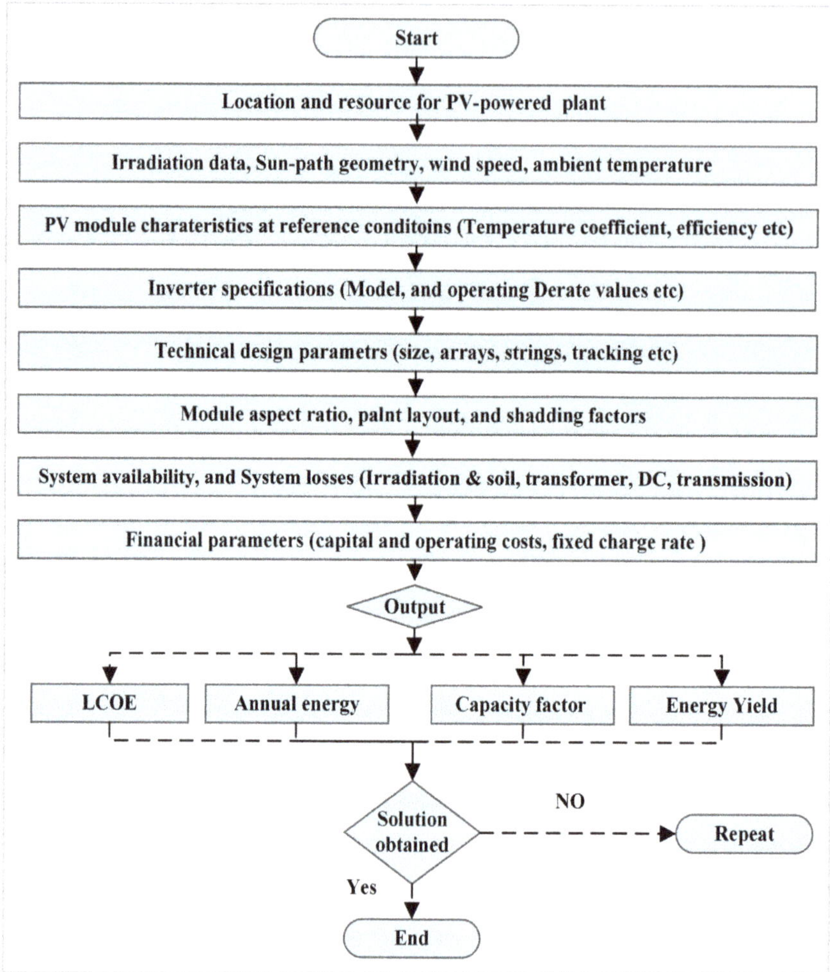

Figure 3. Simulation algorithm of power plant.

Table 4. Specification of solar PV module.

Parameters	Unit	Value	
		Bi-Facial [40]	Mono-Facial [41]
Efficiency	%	21.79	20.88
Power capacity	W_{dc}	671.055	540.696
Performance degradation	%/y	0.45	0.55
Voltage (maximum)	V_{dc}	38.5	31.2
Current (maximum)	A_{dc}	17.4	17.3
Temperature coefficient	W/°C	-0.303	-0.371
Cells	Nos	66	55
Area	m^2	3.080	2.59
Unit mass	kg/m^2	11.092	11.092
Length	m	3.08	2.59

The model incorporates factors such as efficiency, loss factors, thermochemical characteristics, variation in cell efficiency, and the impact of azimuth angle. The performance and cost of the converters, trackers, and voltage optimizers significantly influence the net output of the system. The critical technical characteristics of the modules, along with the application of air-mass modifier polynomial ratios explained in Ref. [39], effectively account for the spectrum effects on net power. The model also considers losses attributed to the visual lens, placement error, tracker error, and wind flap, which are encompassed within the overall loss factors.

The economic model incorporates various input parameters such as module cost, inverter cost, BOS mechanical and electrical costs, installation cost, and non-labor soft costs including approval, procurement, and developer overhead (Table 5). Estimated expenses within the literature typically fall within a specific price range. For instance, predictions for the total installed cost of a solar PV system with single-axis tracking range from 1.3 USD/Wac to 1.14 USD/Wdc for a 100 MW capacity [42,43]. The Solar Energy Industries Association (SEIA) [19] reported that the global average cost of commercial solar PV plants installed in 2021 was 0.77 USD/Wdc for fixed-tilt systems and 0.89 USD/Wdc for single-axis systems. The NREL [44] reports the median cost of 25 different utility-scale solar PV plants as 1.2 USD/Wac and 0.97 USD/Wdc. These costs represent global averages, and the net output is significantly influenced by module scale, location, type, brand, and the presence of clean energy credits. Dedvar et al. [45] examined the scaled impact on the minimum sustainable price (MSP) of solar PV modules in a practical manufacturing plant setting. They observed a progressive decline in MSP with increasing capacity, with reductions of 9%, 8%, and 6% for capacities of 600 MW, 1.2 GW, and 2.4 GW, respectively.

Certain fixed expenditures, such as general administration, vegetation care, and module cleaning, are shared among various plant components, resulting in decreased OPEX with increasing plant capacity. Around a 50% decline in OPEX has been noted in thirteen years, i.e., from 35 USD/kWdc/year in 2007 to 17 USD/kWdc/year in 2019 [46]. Similarly, a report from Berkeley Lab [47] highlights a 51.4% reduction in OPEX over the past 12 years. The Trina-Solar modules examined in this study claim a 6.32% reduction in BOS costs when bi-facial 600+ W modules are installed compared to mono-facial modules [48].

Table 5. Economic parameters.

Parameters	Unit	Value	Reference
Installer margin and overhead	USD/W_{dc}	0.05	[49]
WACC	%	6	Typical value
Installation cost	USD/W_{dc}	0.11	[49]
BOP (mechanical)	USD/W_{dc}	0.10	[50]
BOP (electrical)	USD/W_{dc}	0.09	[50]
Sun-tracker (single-axis)	USD/W_{dc}	0.1	[30]
Sun-tracker (dual-axis)	USD/W_{dc}	0.15	[30]
Engineering and developer overhead	USD/W_{dc}	0.08	[49]
PII	USD/W_{dc}	0.04	[50]
Contingency	% CAPEX	2	Typical value
Fixed-mounted OPEX	USD/kW_{dc}/y	13	[51]
Single-axis OPEX	USD/kW_{dc}/y	14	[52]
Dual-axis OPEX	USD/kW_{dc}/y	16.26	[52]
Depreciation	%/y	MACRS Standards (Industries)	[53]
DC/DC power optimizer	USD/W_{dc}	0.15	[54]
PV Module Performance degradation	%/y	0.5	[50]
Residual value	% CAPEX	20	Typical value
Inflation	%	2.6	[55]

AC: alternate current; DC: direct current; CAPEX: capital expenditures; OPEX: operating expenditures; MARCS: modified accelerated cost recovery system; BOP: balance of plant; PV: photovoltaic; PII: permitting, inspection, and interconnection; WACC: weighted average cost of capital.

The LCOE serves as an economic metric for comparing renewable power generation systems from various sources. Equation (1) defines the LCOE [39], and in this study, modified and detailed relationships are employed as presented in Equations (2) and (3). These equations are solved using the standard life cycle cost (LCC) concept, as illustrated in Figure 4, to determine the NPV. By adopting this approach, a thorough evaluation of the economic feasibility of the solar power plant system can be achieved.

$$\text{LCOE (USD/kWh)} = \frac{Total\ life\ cycle\ cost\ of\ the\ energy\ generation\ system\ (\$)}{Total\ electricity\ generated\ (kWh)} \tag{1}$$

$$\text{LCOE (USD/kWh)} = \frac{\text{CAPEX} + C_{o\&m} - r_{deg}{}^n - R_{value}}{E_n} \tag{2}$$

where:

CAPEX is the capital expenditure (USD)
$C_{o\&m}$ is the operation and maintenance cost (USD)
r_{deg} is the degradation rate (%)
n is the plant's lifetime
R_{value} is the residual value (USD)
E_n is the electricity generated

$$\text{LCOE (USD/kWh)} = \frac{\text{CAPEX} + C_{ins} + \frac{C_{rep}}{(1+d)^n} + \sum_{i=1}^{n} \frac{C_{o\&m}}{(1+d)^i} - \sum_{i=1}^{n} \frac{(r_{dep} \times r_{tax})^i}{(1+d)^i} - \frac{R_{value}}{(1+d)^n}}{\sum_{i=1}^{n} \frac{E_i \times (1-r_{deg})^i}{(1+d)^i}} \tag{3}$$

where:

C_{ins} is the installation cost (power plant) (USD)
C_{rep} is the replacement cost (USD)
I_i is the Year "i"
kW is the Kilowatt
n is the plant's lifetime
USD is the dollar (United States)
r_{dep} is the depreciation rate (%)
r_{tax} is the federal capital tax on investment

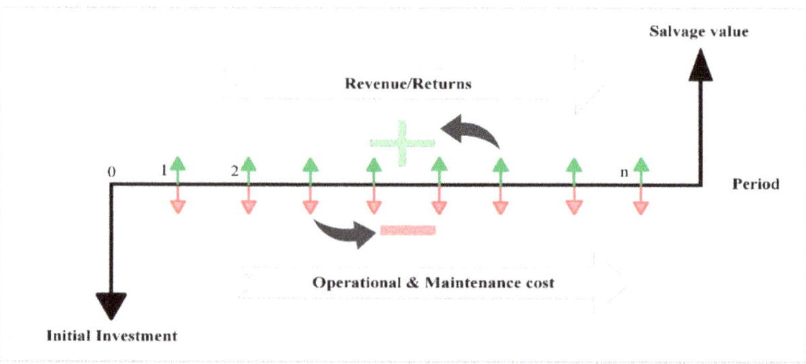

Figure 4. Life cycle cost.

2.5. Assumptions and Exclusions

- The economic model focused on cost analysis without an energy storage system.
- Power transmission and distribution costs were not taken into account.

- The cost of land acquisition or lease was not included and was assumed to be covered by the public development budget.
- Incentives for investment, green power generation, or capacity build-up are not accounted for in the NPV.

3. Results and Discussion

Techno-economic metrics are analyzed across four different generation capacities: 1000 MW, 3000 MW, 5000 MW, and 7000 MW, and six various pathways. The reason for considering different capacities is to evaluate the influence of economies of scale on the overall impact. The chosen range of installed capacity aligns with the requirements of the region's first solar PV plant [56], designed to meet the annual energy demands of an off-grid community. To ensure an equitable comparison across various renewable energy technologies and geographical sites in a global context, the power plant is considered connected directly to the consumer's facility, eliminating the need for an energy storage system. It is worth noting that the irradiation potential varies throughout the year, such that during the winter solstice it is ~44.3% lower compared to the summer solstice in Figure 5. As a result, the monthly energy generation at the selected plant location exhibits significant variability, particularly during the spring and fall equinoxes (Figure 6).

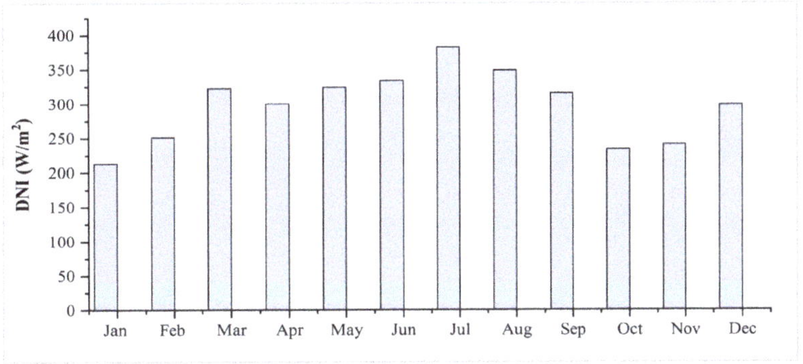

Figure 5. Average direct of normal radiation per month in Pakistan.

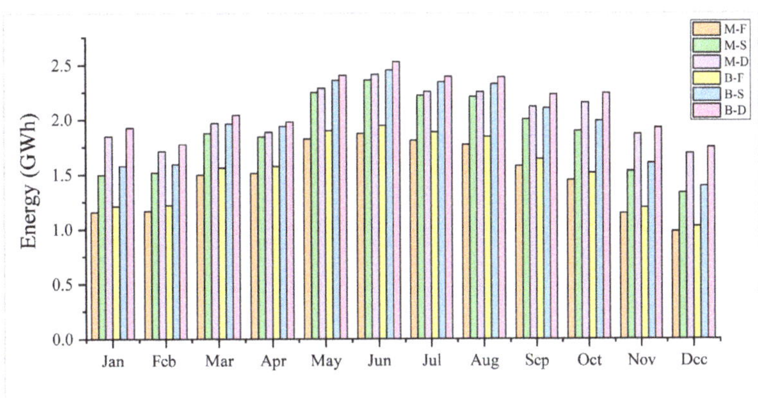

Figure 6. Monthly energy generated from various generation pathways, (M; mono-facial, B; bi-facial, F; fixed-mounted, S; single-axis, D; dual-axis).

Given the variation in the energy received, the size of a power plant required to meet a particular electricity load at this study site is found ~38.4% smaller during the summer solstice compared to the winter solstice. The comparative assessment of current market prices reveals that the cost of bi-facial modules is ~38% higher than mono-facial modules. Additionally, transitioning from a fixed-mounted configuration of the sun-tracking system to a single-axis configuration increases the cost by ~47%, while the cost difference in transitioning from a single-axis to a dual-axis configuration is ~40%. Considering these discrepancies, an evaluation of the energy generation and plant performance in Table 6 is conducted to determine the optimal plant setup using either mono-facial or bi-facial modules for scaled implementation in a standalone position.

Table 6. Plant performance metrics.

Metric	Fixed-Mounted		Single-Axis		Dual-Axis	
	Mono-Facial	**Bi-Facial**	**Mono-Facial**	**Bi-Facial**	**Mono-Facial**	**Bi-Facial**
Annual energy (GWh)	17.76	18.52	22.53	23.64	24.45	25.56
CF (%)	22.9	22.1	28.3	29.5	31.1	32.2
Energy yield (kWh/kW)	1777	1852	2255	2363	2446	2558

The analysis reveals that transitioning from mono-facial to bi-facial modules with a single-axis configuration results in a net power output change of ~5%. Similarly, changing the plant setup from mono-facial to bi-facial modules using the dual-axis configuration leads to a net power output change of ~4.6%. Based on this comparative assessment, the use of bi-facial modules in a single-axis configuration is preferred over the mono-facial configuration due to the higher difference observed in net power output.

The model evaluated the installation of two distinct module types presently accessible on the market, along with three potential mounting structures. Comparable meteorological data and economic and technical input parameters are used to evaluate the influence of module types on BOS costs and LCOE. Six distinct designs are assessed, encompassing two module types and three module orientations, yielding varying optimized LCOE values in Figure 7. The LCOE experiences a decrease of ~12% when shifting from a fixed-mounting to a single-axis configuration, regardless of whether mono-facial or bi-facial modules are used. However, the difference in LCOE when transitioning from a single-axis to a dual-axis configuration is ~2%, which is not considered significant. Considering the higher cost associated with installing dual-axis sun-trackers and the relatively lower increase in energy generation, the LCOE assessment suggests that single-axis tracking is more economically favorable until dual-axis systems become more developed and economically feasible in the future.

Comparing the outcomes of this study with other studies proves challenging due to variations in solar energy potential across different worldwide geographical locations, differences in plant installed capacity, and diverse technical and cost assumptions and limitations. Nonetheless, the results obtained from this study can be compared to recently bid utility-scale regional solar PV projects conducted in areas with similar solar irradiance levels in Table 7. The observed decrease in costs with increasing plant capacity in this study aligns with the awarded prices of the power purchase agreement (PPA) for recent projects. For example, the LCOE for a 2000 MW plant in the UAE is ~55% lower than that of a 1200 MW project. Similarly, the LCOE in Qatar and Oman, which are reported as 1.57 USD/kWh and 1.78 USD/kWh respectively, are also comparable to the findings of this study, with slight variations attributed to differences in solar irradiance due to different geographical locations. Furthermore, the resulting LCOE from this study is comparable to the reported values of 1.67 USD/kWh (PV-battery storage system) and 1.45 USD/kWh (PV-battery storage system-diesel generator) for the geographical location in District Dera Ismail Khan, Pakistan [57]. Similarly, another study by ARENA in Australia [58] reported 1.14 USD/kWh, with the slight difference attributed to ~11% more sunshine hours available for a full load at the selected site of this study as compared to the location in Australia.

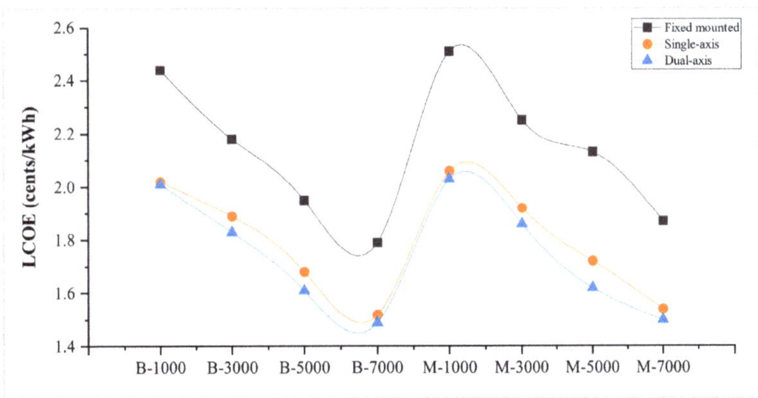

Figure 7. Economy of scale impact of solar PV electricity cost at the selected site (M; mono-facial, B; bi-facial).

Based on a qualitative assessment, the integration of concentrated solar power with conventional plants in Pakistan resulted in a reduced levelized cost of electricity (LCOE) [59]. An evaluation encompassing technical, economic, and environmental considerations advocated for an independent standalone solar PV system in Ref. [60], exhibiting a payback period of 3.125 years and facilitating a substantial reduction of 90,225 tons per annum in CO_2 emissions within the Pakistani context.

Drawing from an inquiry into a hybrid energy system combining wind, PV, and biomass components in Pakistan, an LCOE of 5.744 USD/kWh was ascertained [61]. Moreover, a comparative analysis of the technical and economic dimensions of scaled solar PV installations across five diverse locations, detailed in Ref. [62], identified the Baluchistan Province as the most suitable locale, distinguished by a diminished LCOE of approximately 2.6 USD/kWh. When juxtaposed with the established LCOE of 5.6 USD/kWh attributed to a wind power system as appraised in Ref. [63], the present study's findings—specifically, LCOE values of 2.14 USD/kWh (fixed-mounted PV systems), 1.79 USD/kWh (single-axis tracking systems), and 1.74 USD/kWh (dual-axis tracking systems)—underscore the viability of solar PV plants within the prevailing market dynamics.

Table 7. PPA of recently tendered regional projects.

Country	Plant Capacity (MW)	Awarded PPA (USD/kWh)	Year	Reference
UAE	1200	2.951	2019	[64]
	2000	1.351	2022	[65]
Qatar	800	1.572	2020	[66]
Oman	500	1.781	2019	[67]

PPA: power purchase agreement.

4. Conclusions and Policy Recommendations

The techno-economic analysis conducted on scaled solar PV plants with a power capacity range of 1000–7000 MW has yielded several significant findings. The transition from mono-facial to bi-facial modules, combined with a single-axis configuration, resulted in a noticeable increase in net power output of ~5%. Similarly, the shift from mono-facial to bi-facial modules with a dual-axis configuration led to a net power output increase of around 4.6%. The findings indicate that a power plant utilizing bi-facial modules with a single-axis configuration offers greater feasibility compared to one employing mono-facial modules in a fixed-mount arrangement. In terms of cost factors, the BOS cost and the OPEX, primarily attributed to cleaning and vegetation management, emerged as the most significant after considering the CAPEX. The selected site demonstrated substantial power generation potential, making it well-suited for large-scale commercial solar PV plants. The average LCOE was determined to be 2.14 USD/kWh for fixed-mounted plants, 1.79 USD/kWh for single-axis plants, and 1.74 USD/kWh for dual-axis plants across the examined power generation capacity range. The anticipated economies of scale, driven by the expanding global market for solar PV plants and renewable energy, are expected to further contribute to overall cost reductions. Given the lower cost of green power, as evidenced by the LCOE, particularly in Pakistan and specifically in the Balochistan Province, the region's extensive land area, and high solar energy yield (with 7–13 h of sunshine per day) position it as a compelling leader in the renewable power sector. The comprehensive techno-economic framework developed in this study has yielded a meticulously crafted solar PV system design that is scalable for implementation in off-grid rural settings. This design is a product of rigorous exploration of pertinent techno-economic variables. In forthcoming research endeavors, this framework is positioned for augmentation to encompass the assessment of additional sustainable power systems, including wind and biomass alternatives. This envisioned extension bears the promise of providing invaluable insights to inform the energy system design within off-grid regions.

To foster investment decisions and formulate effective policies to promote off-grid electrification and the development of renewable energy value chains, the following policy recommendations are proposed:

- The active involvement of the private sector is crucial in fostering a resilient renewable energy value chain, especially in regions with limited existing energy infrastructure. It would be helpful to mobilize financial resources from non-budgetary sources and to provide the technical and managerial expertise required for building scaled solar PV plants in off-grid areas.
- The formulation of specialized policies and regulatory frameworks related to solar power generation is required. These initiatives are essential to attract and facilitate investors and to overcome the challenges associated with the renewable energy value chain.
- The BOS cost is the major contributing factor in the LCOE; therefore, solar PV module manufacturers should focus on improving module density within a string to reduce expenses for accessories such as electric cables and racks. Additionally, the production of large-sized and high-power solar PV modules is recommended to decrease the construction time and lower the overall cost of the system.
- Although the LCOE using bi-facial modules is ~7–10% lower, a careful comparison of sun-tracker costs is necessary before deciding. A dual-axis tracking system generates around 19.87% more power than a fixed-mounted system, but it comes with a net installed cost difference of ~15.45%. Therefore, when dealing with limited space, a cost-benefit analysis of installing sun trackers and types of modules is critical.

Author Contributions: Conceptualization, M.S. and P.K.; methodology, M.S.; software, M.S.; validation, P.K. and S.K.; formal analysis, M.S.; investigation, P.K.; resources, M.S.; data curation, M.S. and P.K.; writing—original draft preparation, M.S.; writing—review and editing, M.S. and P.K.; supervision, P.K. and S.K. All authors have read and agreed to the published version of the manuscript.

Funding: This research was funded by the Graduate Scholarship Programme, Chulalongkorn University for ASEAN or Non-ASEAN countries, and the APC was partly funded by Energy Research Institute, Chulalongkorn University.

Institutional Review Board Statement: Not applicable.

Informed Consent Statement: Not applicable.

Data Availability Statement: Data are contained within the article.

Conflicts of Interest: The authors declare no conflict of interest. The funders had no role in the design of the study; in the collection, analyses, or interpretation of data; in the writing of the manuscript; or in the decision to publish the results.

Abbreviations

BOS	Balance of system
CCUS	Carbon capture utilization and storage
CAPEX	Capital expenditures
CF	Capacity factor
DNI	Direct normal irradiance
FCEV	Fuel cell electric vehicle
GHG	Greenhouse gas
GHI	Global horizontal irradiance
IPCC	Intergovernmental Panel on climate change
IRENA	International renewable energy agency
LCOE	Levelized cost of energy
LCC	Life cycle cost
MOE	Ministry of Energy United States of America
MOU	Memorandum of understanding
MT	Million tones
NPV	Net present value
NREL	National Renewable Energy Laboratory
OPEX	Operational expenditures
PV	Photovoltaic
PPA	Power purchase agreement
R&D	Research and development
RES	Renewable energy systems
SAM	System Advisor Model
TWh	Terawatt hours
USD	United States dollar
WACC	Weighted average cost of capital

References

1. Levenda, A.M.; Behrsin, I.; Disano, F. Renewable energy for whom? A global systematic review of the environmental justice implications of renewable energy technologies. *Energy Res. Soc. Sci.* **2016**, *71*, 101837. [CrossRef]
2. Zebra, E.I.C.; van der Windt, H.J.; Nhumaio, G.; Faaij, A.P. A review of hybrid renewable energy systems in mini-grids for off-grid electrification in developing countries. *Renew. Sustain. Energy Rev.* **2021**, *144*, 111036. [CrossRef]
3. Ali, A.; Imtiaz, M. Effects of Pakistan's energy crisis on farm households. *Util. Policy* **2019**, *59*, 100930. [CrossRef]
4. World Bank. Poverty & Equity Brief South Asia Pakistan. Available online: https://databankfiles.worldbank.org/public/ddpext_download/poverty/987B9C90-CB9F-4D93-AE8C-750588BF00QA/current/Global_POVEQ_PAK.pdf (accessed on 1 June 2023).
5. THE NATION. Available online: https://www.nation.com.pk/05-Aug-2022/pakistan-s-electricity-shortfall-jumps-to-7-641mw (accessed on 13 August 2023).
6. Valasai, G.D.; Uqaili, M.A.; Memon, H.R.; Samoo, S.R.; Mirjat, N.H.; Harijan, K. Overcoming electricity crisis in Pakistan: A review of sustainable electricity options. *Renew. Sustain. Energy Rev.* **2017**, *72*, 734–745. [CrossRef]
7. Kumar, V.; Ali, B.S.; Choudry, E.; Khan, S.; Baig, K.; Durrani, N.U.R.; Durrani, N.U.R., Sr. Quality of Neonatal Care: A Health Facility Assessment in Balochistan Province, Pakistan. *Cureus* **2022**, *14*, e22744. [CrossRef] [PubMed]
8. International the News. Available online: https://www.thenews.com.pk/magazine/instep-today/164605-Powerless-Balochistan/ (accessed on 31 May 2023).
9. Japan International Corporation Agency. Available online: https://www.jica.go.jp/pakistan/english/activities/activity02_11.html (accessed on 29 May 2023).
10. Khan, T.; Waseem, M.; Tahir, M.; Liu, S.; Yu, M. Autonomous hydrogen-based solar-powered energy system for rural electrification in Balochistan, Pakistan: An energy-economic feasibility analysis. *Energy Convers. Manag.* **2022**, *271*, 116284. [CrossRef]
11. Solargis, Solar Resource and Photovoltaic Power Potential of Pakistan: Analysis Based on Validated Model with Reduced Uncertainty. Available online: https://pubdocs.worldbank.org/en/175561587077010849/Solargis-Solar-Resource-Report-Pakistan-WBG-ESMAP.pdf (accessed on 29 May 2023).
12. Tareen, W.U.K.; Dilbar, M.T.; Farhan, M.; Ali Nawaz, M.; Durrani, A.W.; Memon, K.A.; Aamir, M. Present status and potential of biomass energy in pakistan based on existing and future renewable resources. *Sustainability* **2020**, *12*, 249. [CrossRef]

13. Raza, M.A.; Khatri, K.L.; Israr, A.; Haque, M.I.U.; Ahmed, M.; Rafique, K.; Saand, A.S. Energy demand and production forecasting in Pakistan. *Energy Strategy Rev.* **2022**, *39*, 100788. [CrossRef]
14. World Bank Report. Balochistan: Development Issues and Prospects Energy Sector. Available online: https://documents1.worldbank.org/curated/zh/352401468145176136/pdf/ACS22580WP0v500art020Energy0Sector.pdf. (accessed on 19 May 2023).
15. Digital Associated Press of Pakistan. Off-Grid Solutions Be Explored to Provide Electricity in Southern Balochistan: Asad. Available online: https://www.app.com.pk/business/off-grid-solutions-be-explored-to-provide-electricity-in-southern-balochistan-asad/ (accessed on 1 June 2023).
16. Naeem, A.; Shah SM, A.; Marri, M.F. Solar Electrification: A Solution For Socio-Economic Development Of Rural Household: Empirical Evidence From District Sibi, Balochistan, Pakistan. *J. Posit. Sch. Psychol.* **2023**, *7*, 807–821.
17. Anwar, N.U.R.; Waqas, A.M.; Khan, F. Renewable energy technologies in Balochistan: Practice, prospects and challenges. In Proceedings of the 5th International Conference on Energy, Environment & Sustainable Development (EESD) 2018, Jamshoro, Pakistan, 14–16 November 2018; pp. 1–9.
18. Asian Development Bank. Available online: https://www.adb.org/sites/default/files/institutional-document/32216/private-sector-assessment.pdf (accessed on 20 June 2023).
19. Solar Energy Industries Association. Utility-Scale Solar. Available online: https://www.seia.org/initiatives/utility-scale-solar-power (accessed on 19 November 2022).
20. United States Department of Energy. The SunShot 2030 Goals. Available online: https://www.energy.gov/sites/prod/files/2020/09/f79/SunShot%202030%20White%20Paper.pdf (accessed on 20 June 2023).
21. Niaz, H.; Shams, M.H.; Liu, J. Evaluating the Economic Impact of Using Curtailed Renewable Energy Sources for Green Hydrogen Production. In Proceedings of the 2022 IEEE International Symposium on Advanced Control of Industrial Processes (AdCONIP), Vancouver, BC, Canada, 7–9 August 2022; pp. 102–107.
22. Nadaleti, W.C.; Borges dos Santos, G.; Lourenço, V.A. The potential and economic viability of hydrogen production from the use of hydroelectric and wind farms surplus energy in Brazil: A national and pioneering analysis. *Int. J. Hydrogen Energy* **2020**, *45*, 1373–1384. [CrossRef]
23. Yates, J.; Daiyan, R.; Patterson, R.; Egan, R.; Amal, R.; Baille, A.H.; Chang, N.L. Techno-economic Analysis of Hydrogen Electrolysis from Off-Grid Stand-Alone Photovoltaics Incorporating Uncertainty Analysis. *Cell Rep. Phys. Sci.* **2020**, *1*, 100209. [CrossRef]
24. Ahshan, R.; Onen, A.; Al-badi, A.H. Assessment of wind-to-hydrogen (Wind-H2) generation prospects in the Sultanate of Oman. *Renew. Energy* **2022**, *200*, 271–282. [CrossRef]
25. Shehabi, M.; Dally, B. Opportunity and Cost of Green Hydrogen Production in Kuwait: A Preliminary Assessment. *SSRN Electron. J.* **2022**, *1*, 1–26. [CrossRef]
26. Dabar, O.A.; Awaleh, M.O.; Waberi, M.M.; Adan, A.B.I. Wind resource assessment and techno-economic analysis of wind energy and green hydrogen production in the Republic of Djibouti. *Energy Rep.* **2022**, *8*, 8990–8996. [CrossRef]
27. Manzotti, A.; Quattrocchi, E.; Curcio, A.; Kwok, S.C.; Santarelli, M.; Ciucci, F. Membraneless electrolyzers for the production of low-cost, high-purity green hydrogen: A techno-economic analysis. *Energy Convers. Manag.* **2022**, *254*, 115156. [CrossRef]
28. Jang, D.; Kim, J.; Kim, D.; Han, W.B.; Kang, S. Techno-economic analysis and Monte Carlo simulation of green hydrogen production technology through various water electrolysis technologies. *Energy Convers. Manag.* **2022**, *258*, 115499. [CrossRef]
29. Burdack, A.; Duarte-Herrera, L.; López-Jiménez, G.; Polklas, T.; Vasco-Echeverri, O. Techno-economic calculation of green hydrogen production and export from Colombia. *Int. J. Hydrogen Energy* **2022**, *48*, 1685–1700. [CrossRef]
30. Lawrence Berkeley National Laboratory. Empirical Trends in Deployment, Technology, Cost, Performance, PPA Pricing, and Value in the United States. Available online: https://eta-publications.lbl.gov/sites/default/files/utility_scale_solar_2022_edition_slides.pdf (accessed on 3 November 2022).
31. Armijo, J.; Philibert, C. Flexible production of green hydrogen and ammonia from variable solar and wind energy: Case study of Chile and Argentina. *Int. J. Hydrogen Energy* **2020**, *45*, 1541–1558. [CrossRef]
32. Shaner, M.R.; Atwater, H.A.; Lewis, N.S.; McFarland, E.W. A comparative techno-economic analysis of renewable hydrogen production using solar energy. *Energy Environ. Sci.* **2016**, *9*, 2354–2371. [CrossRef]
33. Siyal, S.H.; Mentis, D.; Mörtberg, U.; Samo, S.R.; Howells, M. A preliminary assessment of wind-generated hydrogen production potential to reduce the gasoline fuel used in road transport sector of Sweden. *Int. J. Hydrogen Energy* **2015**, *40*, 6501–6511. [CrossRef]
34. Schlecht, M.; Meyer, R. Site selection and feasibility analysis for concentrating solar power (CSP) systems. In *Concentrating Solar Power Technology*; Woodhead Publishing: Cambridge, UK, 2012; pp. 91–119. [CrossRef]
35. Yang, Y.; Wang, Z.; Xu, E.; Ma, G.; An, Q. Analysis and Optimization of the Start-up Process based on Badaling Solar Power Tower Plant. *Energy Procedia* **2015**, *69*, 1688–1695. [CrossRef]
36. Siddiqi, F. Nation-formation and national movement(s) in Pakistan: A critical estimation of Hroch's stage theory. *Natl. Pap.* **2010**, *38*, 16. [CrossRef]
37. National Renewable Energy Laboratory (NREL). Available online: https://www.nrel.gov/docs/fy04osti/36705.pdf (accessed on 10 November 2022).
38. National Renewable Energy Laboratory. Available online: https://sam.nrel.gov/ (accessed on 12 September 2022).

39. National Renewable Energy Laboratory. SAM Photovoltaic Model Technical Reference. Available online: https://www.nrel.gov/docs/fy15osti/64102.pdf (accessed on 12 September 2022).
40. Global Price Source. Available online: https://www.globalsources.com/Solar-grid/foldable-solar-panel-kit-1191486741p.htm (accessed on 1 November 2022).
41. Global Price Source. Available online: https://www.globalsources.com/Solar-panel/Trina-All-Black-Mono-Solar-Panel-1195174784p.htm (accessed on 1 November 2022).
42. National Renewable Energy Laboratory. U.S. Solar Photovoltaic System and Energy Storage Cost Benchmark: Q1 2020. Available online: https://www.nrel.gov/docs/fy21osti/77324.pdf (accessed on 1 November 2022).
43. National Renewable Energy Laboratory. U.S. Solar Photovoltaic System and Energy Storage Cost Benchmarks: Q1 2021. Available online: https://www.nrel.gov/docs/fy22osti/80694.pdf (accessed on 2 November 2022).
44. National Renewable Energy Laboratory. Solar Industry Update. Available online: https://www.nrel.gov/docs/fy23osti/86215.pdf (accessed on 1 November 2022).
45. Dehghanimadvar, M.; Egan, R.; Chang, N.L. Economic assessment of local solar module assembly in a global market. *Cell Rep. Phys. Sci.* **2022**, *3*, 100747. [CrossRef]
46. Electric Power Institute. Budgeting for Solar PV Plant Operations & Maintenance: Practices. Available online: https://www.osti.gov/servlets/purl/1234935 (accessed on 11 November 2022).
47. Wiser, R.H.; Bolinger, M.; Seel, J.; Benchmarking Utility-Scale PV Operational Expenses and Project Lifetimes: Results from a Survey of U.S. Solar Industry Professionals. Available online: https://emp.lbl.gov/publications/benchmarking-utility-scale-pv (accessed on 11 November 2022).
48. Trina Solar. Available online: https://www.trinasolar.com/en-glb/resources/newsroom/mabos-costs-reduced-63-dnv-gl-report-trina-solar-vertex-210mm-modules%E2%80%99-advantages (accessed on 18 November 2022).
49. National Renewable Energy Laboratory. Available online: https://atb.nrel.gov/electricity/2022/utility-scale_pv#operation_and_maintenance_(o&m)_costs (accessed on 10 November 2022).
50. National Renewable Energy Laboratory. Overview of Field Experience-Degradation Rates & Lifetimes. Available online: https://www.nrel.gov/docs/fy15osti/65040.pdf (accessed on 10 November 2022).
51. International Renewable Energy Agency. Renewable Power Generation Costs in 2017. Available online: https://www.irena.org/-/media/Files/IRENA/Agency/Publication/2018/Jan/IRENA_2017_Power_Costs_2018_summary.pdf (accessed on 10 November 2022).
52. García, H.A.; Duke, A.R.; Flores, H.V. Techno-economic comparison between photovoltaic systems with solar trackers and fixed structure in "El Valle de Sula", Honduras. In Proceedings of the 2020 6th International Conference on Advances in Environment Research, Sapporo, Japan, 26–28 August 2020; Volume 776, pp. 1–12. [CrossRef]
53. United States Department of Energy. Hydrogen Program Plan. Available online: https://www.hydrogen.energy.gov/pdfs/hydrogen-program-plan-2020.pdf (accessed on 4 September 2022).
54. Ghanbari, N.; Mobarrez, M.; Madadi, M.; Bhattacharya, S. Comprehensive Cost Comparison and Analysis of Building-Scale Solar DC and AC Microgrid. In Proceedings of the International Conference on DC Microgrids—ICDCM, Matsue, Japan, 20–23 May 2019.
55. Saudi Central Bank. Available online: https://www.sama.gov.sa/en-US/EconomicReports/pages/inflationreport.aspx (accessed on 10 December 2022).
56. Quid-e-Azam Solar Park. Available online: https://cpec.gov.pk/project-details/10 (accessed on 5 July 2023).
57. Ali, F.; Ahmar, M.; Jiang, Y.; Alahmad, M. A techno-economic assessment of hybrid energy systems in rural Pakistan. *Energy* **2021**, *215*, 119103. [CrossRef]
58. Australian Renewable Energy Agency. Opportunities for Australia from Hydrogen Exports-Australian Renewable Energy Agency. Available online: https://arena.gov.au/knowledge-bank/opportunities-for-australia-from-hydrogen-exports/ (accessed on 5 January 2023).
59. Lohana, K.; Raza, A.; Mirjat, N.H.; Ahmed, S.; Ahmed, S. Techno-Economic Feasibility Analysis of Concentrated Solar Thermal Power Plants as Dispatchable Renewable Energy Resource of Pakistan: A case study of Tharparkar. *IJEEIT Int. J. Electr. Eng. Inf. Technol.* **2021**, *4*, 35–40. [CrossRef]
60. Abas, N. Techno-Economic Feasibility Analysis of 100 MW Solar Photovoltaic Power Plant in Pakistan. *Technol. Econ. Smart Grids Sustain. Energy* **2022**, *7*, 16. [CrossRef]
61. Ahmad, J.; Imran, M.; Khalid, A.; Iqbal, W.; Ashraf, S.R.; Adnan, M.; Ali, S.F.; Khokhar, K.S. Techno economic analysis of a wind-photovoltaic-biomass hybrid renewable energy system for rural electrification: A case study of Kallar Kahar. *Energy* **2018**, *148*, 208–234. [CrossRef]
62. Ahmed, N.; Khan, A.N.; Ahmed, N.; Aslam, A.; Imran, K.; Sajid, M.B.; Waqas, A. Techno-economic potential assessment of mega scale grid-connected PV power plant in five climate zones of Pakistan. *Energy Convers. Manag.* **2021**, *237*, 114097. [CrossRef]
63. Adnan, M.; Ahmad, J.; Farooq, S.; Imran, M. A techno-economic analysis for power generation through wind energy: A case study of Pakistan. *Energy Rep.* **2021**, *7*, 1424–1443. [CrossRef]
64. Power Technology. Sweihan Photovoltaic Independent Power Project, Abu Dhabi. Available online: https://www.power-technology.com/projects/sweihan-photovoltaic-independent-power-project-abu-dhabi/ (accessed on 1 November 2022).

65. Emirates Water and Electricity. Abu Dhabi Power Corporation Announces Lowest Tariff for Solar Power in the World. Available online: https://www.ewec.ae/en/media/press-release/abu-dhabi-power-corporation-announces-lowest-tariff-solar-power-world#:~:text=The%20project%20has%20received%2C%20from,Abu%20Dhabi\T1\textquoteright%20project%20%E2%80%93%20Abu%20Dhabi\T1\textquoterights (accessed on 2 November 2022).
66. PV Magazine. Lowest Shortlisted Bid in Saudi 1.47 GW Tender was \$0.0161/kWh. Available online: https://www.pv-magazine.com/2020/04/03/lowest-shortlisted-bid-in-saudi-1-47-gw-tender-was-0-0161-kwh/ (accessed on 29 October 2022).
67. ACWA Power. Financial Closure Achieved for The Largest Solar Photovoltaic Power Plant in Oman. Available online: https://www.acwapower.com/news/financial-closure-achieved-for-the-largest-solar-photovoltaic-power-plant-in-oman/ (accessed on 3 November 2022).

energies

Article

Cost-Effective Optimization of an Array of Wave Energy Converters in Front of a Vertical Seawall

Senthil Kumar Natarajan and Il Hyoung Cho *

Department of Ocean System Engineering, Jeju National University, Jeju 690-756, Republic of Korea;
senthil@stu.jejunu.ac.kr
* Correspondence: cho0904@jejunu.ac.kr; Tel.: +82-64-754-3482; Fax: +82-64-751-3480

Abstract: The present paper focuses on investigating the cost-effective configuration of an array of wave energy converters (WECs) composed of vertical cylinders situated in front of a vertical seawall in irregular waves. First, the hydrodynamic calculations are performed using a WAMIT commercial code based on linear potential theory, where the influence of the vertical wall is incorporated using the method of image. The viscous damping experienced by the oscillating cylinder is considered through CFD simulations of a free decay test. A variety of parameters, including WEC diameter, number of WECs, and the spacing between them, are considered to determine an economically efficient WEC configuration. The design of the WEC configuration is aided by a cost indicator, defined as the ratio of the total submerged volume of the WEC to overall power capture. The cost-effective configuration of WECs is achieved when WECs are positioned in front of a vertical wall and the distance between them is kept short. It can be explained that the trapped waves formed between adjacent WECs as well as the standing waves in front of a seawall significantly intensify wave fields around WECs and consequently amplify the heave motion of each WEC. A cost-effective design strategy of WEC deployment enhances the wave energy greatly and, consequently, contributes to constructing the wave energy farm.

Keywords: wave energy converter; cost-effective analysis; method of image; linear potential theory; vertical seawall

Citation: Natarajan, S.K.; Cho, I.H. Cost-Effective Optimization of an Array of Wave Energy Converters in Front of a Vertical Seawall. *Energies* **2024**, *17*, 128. https://doi.org/10.3390/en17010128

Academic Editors: Konstantinos Aravossis and Eleni Strantzali

Received: 30 November 2023
Revised: 19 December 2023
Accepted: 22 December 2023
Published: 25 December 2023

1. Introduction

The urgency to address global warming and the resulting climate changes highlights the necessity of reducing greenhouse gas emissions, particularly within the energy sector, where electricity and heat production are the largest contributors to global emissions [1]. Meanwhile, energy demand continues to surge. As a promising solution for curbing greenhouse gas emissions, renewable energy, notably solar and wind power, is projected to contribute 43% of the world's electricity by 2030, a significant increase from the current level of 28% [2]. Notably, wave energy boasts a higher power density compared to solar and wind energy. For instance, at a latitude of 15° N within the Northeast Trades, the average power density is 0.17 kW/m^2 for solar, 0.58 kW/m^2 for wind, and 8.42 kW/m^2 for wave energy [3]. Despite its higher potential, wave energy has not yet achieved commercial viability similar to solar and wind power sources due to its levelized cost of energy still lacking competitiveness with other renewable sources [4]. Wave energy exhibits greater availability compared to solar and wind energy, as it remains available both day and night, and persists throughout the entire year. Hence, achieving cost-effectiveness in harnessing wave energy is essential for establishing it as a financially feasible choice among renewable energy alternatives.

There have been many studies focusing on optimizing wave energy converters (WECs) for maximizing power extraction so that they become economically viable. In [5], an optimization of dimensions and layout of an array of heaving buoy WECs have been carried out. A control method is proposed in [6] for an array of WECs maximizing power and satisfying constraints for optimal energy extraction performance, with full consideration of wave and multi-body interactions, as multiple WECs instead of stand-alone devices could increase energy production by over 15% per device. In [7], a comprehensive assessment is carried out focusing on the geometric optimization of wave energy conversion devices. This review offers a critical analysis of the current state-of-the-art in geometry optimization and outlines its limitations. Similarly, [8] provides a review of the geometry optimization of WECs, aiming to discover enhanced hull shapes that can maximize power generation while minimizing the associated costs. An investigation into the hydrodynamic optimization of a sloped motion point absorber WEC has been conducted in [9] using a time-efficient frequency-domain numerical model. In [10], the authors presented a size optimization of WECs on a floating wind–wave combined power generation platform, which has a significant increase in power generation compared with the single point absorber. The structural optimization of oscillating-array buoys was performed in [11] to improve the wave energy capture efficiency with the simulated models of different spacing, placement modes, and actuating arm lengths of buoys. A review was conducted and presented in [12] for the optimal configuration of wave energy conversions concerning nearshore wave energy potential where WECs' shape optimization may significantly boost performance, and where if it is combined with the PTO control approach may lead to better outcomes. In [13], power take-off optimization has been carried out to maximize energy conversion with a two-body WEC system utilizing relative heave motion to extract power.

One method to increase wave power absorption is by positioning WECs in front of a reflective structure which can amplify the WEC's motion, thereby extracting higher power. The authors of [14] analyzed the hydrodynamic performance of a series of truncated cylinders positioned in front of a vertical wall in the frequency domain. This examination aimed to explore the impact of wave reflections coupled with disturbances caused by the bodies themselves. An experimental evaluation was conducted in [15] to determine the hydrodynamic performance of a WEC system integrated into a breakwater, in comparison with conventional WECs. The WECs positioned in front of the breakwater experience increased heave motion, indicating that the presence of the breakwater enhances the energy conversion performance of the WEC array. In [16], the authors conducted a hydrodynamic investigation into an innovative breakwater featuring parabolic openings designed to harness wave energy. Truncated cylinder-type WECs were positioned in front of these openings, which effectively converge propagating waves toward a focal point. This configuration notably increased the extracted wave power. A theoretical assessment of the hydrodynamic attributes of arrays consisting of vertical axisymmetric floaters of various shapes positioned in front of a vertical breakwater was conducted [17] using an analytical method. The image method was utilized to simulate the breakwater's influence on the array. The study investigated three distinct types of floaters and various array configurations situated in front of the vertical wall. The hydrodynamic coefficients either increased or decreased depending on the distance between the wall and the floater. The investigation delved further in [18] into evaluating an array of cylindrical WECs with a vertical symmetry axis positioned in front of a reflective vertical breakwater. This study explored three distinct array configurations: parallel, perpendicular, and rectangular arrangements, considering various distances from the wall, inter-body spacing, wave heading angles, and mooring stiffness values. Results indicated that the most power-efficient WEC arrangement was the one aligned parallel to the breakwater. Additionally, the deployment of WECs in closer proximity to the breakwater demonstrated higher power efficiency across the majority of wave frequencies compared to WECs positioned farther away. Moreover, the presence of the breakwater positively influenced the system's power absorption, significantly enhancing absorbed power across various arrangements, wave angles, and inter-body spacings. This amplification effect was particularly notable for the parallel, close-to-wall configuration across incoming wave angles. In [19], the hydrodynamic efficiency of a WEC placed in front of a bottom-seated, surface-piercing, vertical orthogonal breakwater in the frequency domain has been analyzed. A theoretical approach, employing the image method, was utilized to simulate how the walls affected the device's power absorption, taking into account the infinite length of the walls' arms. The wave power absorption of WEC five by five arrays and a single array of five WECs positioned in front of a vertical wall were computed in [20,21] using an in-house transient wave-multi-body numerical tool called ITU-WAVE, respectively. This tool employed a marching scheme to solve boundary integral equations for analyzing hydrodynamic radiation and exciting forces. The method of images accounted for the perfect reflection of incident waves from the vertical wall. Numerical findings revealed significantly enhanced performance and wave power absorption of WEC arrays in front of a vertical wall compared to arrays without this vertical wall effect. This heightened efficiency primarily resulted from the presence of standing and nearly trapped waves between the vertical wall and the WEC arrays, along with robust interactions between the WECs themselves.

The present models used for estimating the expenses associated with a wave energy project are often oversimplified, leading to a wide range of economic assessments. This variability in evaluations raises uncertainties for potential investors, thus hindering the progress of wave energy development. Indeed, comprehending the costs associated with wave energy is a pivotal area of research within marine renewable energy. Within this context, the paper [4] provides a comprehensive review of all factors essential for an economic analysis of wave energy. This includes considering numerous elements that are typically overlooked. The study aimed to delineate both direct and indirect costs of a wave farm, encompassing preliminary expenses, construction, operational, and maintenance costs, as well as decommissioning costs, alongside potential revenues. The expense associated with WECs constitutes a substantial portion of the total cost of a wave farm. Similar to other renewable sources like solar photovoltaic (PV) and solar thermal systems, the current capital costs for wave energy surpass those of conventional generation technologies such as gas and coal. Nonetheless, these expenses are anticipated to decline as economies of scale come into play with increased wave farm installations. This trend, coupled with the uncertainty surrounding long-term fuel costs and rising construction expenses for traditional generation technologies, is leading to a narrowing of the significant gaps in electricity costs that were previously evident. Additionally, the operational and maintenance costs are notably high, given the sea environment. Concerning the revenues generated by a wave farm, the primary source of income naturally stems from the sale of the generated energy. Currently, WECs exhibit relatively low performance levels, and enhancing their efficiency will significantly strengthen the economic feasibility of wave energy.

In [22], a comprehensive techno-economic optimization of a floating WEC was conducted using a genetic algorithm considering a wide multi-variate design space. This included considerations of the floater's shape, dimensions, subcomponent configuration, and characteristics. Similarly, a techno-economic assessment of the influence on the sizing of WECs was conducted [23]. The articles [24,25] provided economic evaluations and cost estimations for WECs during their initial developmental phases. To assess the cost-effectiveness of WECs in power generation, Ref. [26] introduces a cost indicator that mirrors the expenses linked to WECs, as they constitute a significant portion of the overall project cost.

Although there have been research works on parametric analysis of configurations of WECs in regular waves, exploring the application of prototype-scale WECs in an array subjected to irregular waves is crucial. Specifically, varying key parameters like WEC diameters, the number of WECs, and their spacing while situating them in front of a vertical seawall would provide valuable insights into the hydrodynamic interaction among the WECs themselves and with the reflective wall at the sea site. Therefore, finding an efficient configuration coupled with effective PTO damping is essential for optimization to maximize power extraction, which should also factor into the economic aspect to make it a commercially viable option. Hence, the objective of this research is to conduct an analysis aimed at identifying a cost-efficient arrangement for an array of point absorber-type WECs installed in front of a vertical seawall in irregular waves. Vertical cylindrical floaters have been selected as the WECs, with varying parameters like diameter, number of WECs, and spacing between them, to determine an economically efficient configuration for harnessing wave energy. To facilitate this process, a cost indicator [26] is utilized, representing the ratio of submerged volume to power capture, which provides insight into the cost associated with extracting a unit of electrical power. The hydrodynamic calculations have been performed using the linear potential theory, and the method of image [17–21] has been employed to account for the influence of the vertical wall. The viscous damping was obtained from a computational fluid dynamics (CFD) simulation of heave-free decay test. Optimal PTO damping at the natural frequency as well as PTO damping that results in maximum power output under irregular waves are considered to maximize power extraction.

2. Methodology

An array of N vertical cylindrical WECs is placed in front of a vertical seawall, which is perfectly reflective throughout the constant water depth h. The diameter and draft of WECs are D and d, respectively. WECs are placed at a distance L_w from a seawall in a parallel arrangement, whereas the distance between WECs is L. WECs are independently oscillating vertically in waves while the other modes of motion are restricted. An array of WECs is exposed to plane incident waves with angular frequency ω and amplitude A propagating into the negative x-axis. Figure 1 depicts the array of WECs placed in front of a vertical seawall.

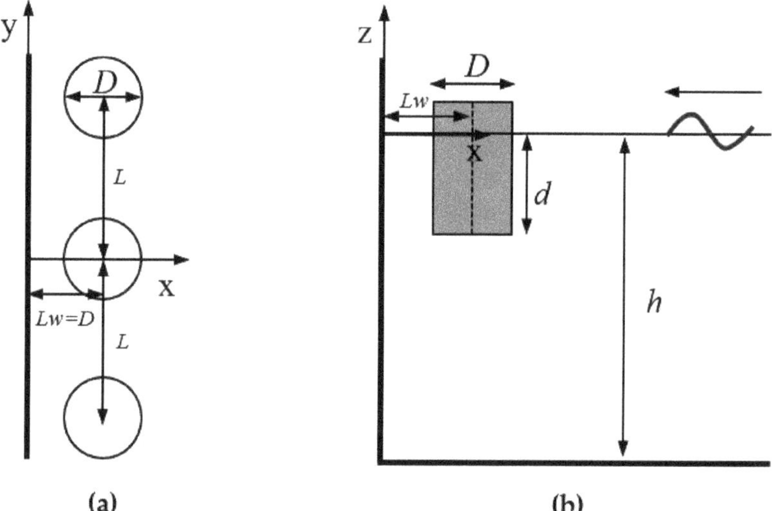

Figure 1. Definition sketch of an array of vertical cylindrical WECs placed in front of a vertical seawall. (**a**) top view (**b**) side view.

2.1. Hydrodynamic Model

The added mass, radiation damping, and wave excitation force on the WECs are computed numerically from a panel-based commercial software WAMIT (Version 7.1), which is widely used in computing wave loads and motions of offshore structures and floating vessels. The linear potential theory has been used in numerical modeling under the assumption of small amplitude, inviscid and incompressible fluid, and irrotational flow. When a floating body oscillates near a rigid lateral boundary like a vertical seawall, the interaction between them must be accounted for. Conventionally, the effect of a rigid wall can be considered by imposing the no-flux boundary condition $\frac{\partial \phi}{\partial n} = 0$ on a rigid wall. However, a more convenient approach is to replace the effect of a seawall with an image body, which is placed symmetrically on the opposite side of the wall, with a prescribed motion to ensure that the boundary condition on the seawall is satisfied [17–21]. Figure 2 shows a top view of an array of WECs positioned in front of a vertical seawall, which is replaced with image bodies.

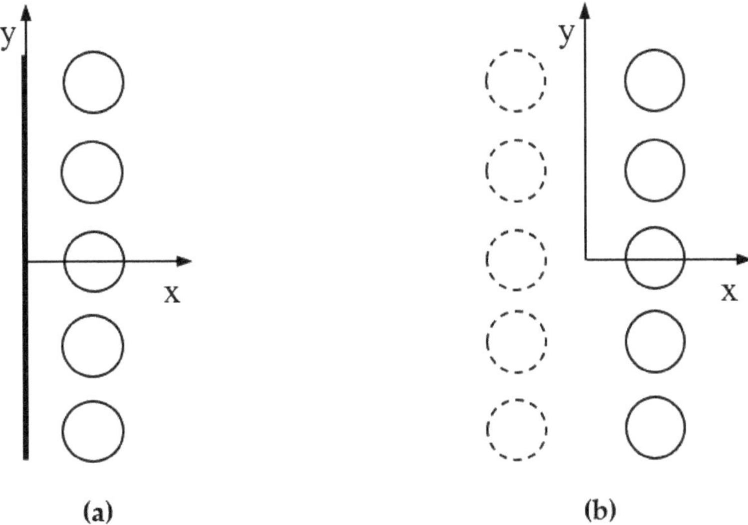

(a)   **(b)**

Figure 2. Top view of an array of (**a**) WECs in front of a vertical seawall and (**b**) WECs and image bodies denoted by a dashed line to replace the vertical seawall.

In this approach, let us consider the WEC p placed in front of the vertical seawall, which is replaced with an image WEC p' symmetrically placed on the opposite side of the seawall. The hydrodynamic parameters of the WEC p with the influence of the vertical seawall can be obtained by combining the hydrodynamic parameters of the forced oscillation of the WEC p and its image p' effect, in the respective mode of motion. For instance, the surge added mass, radiation damping coefficient, and wave excitation force can be derived [17] as $a_{11}^{p} - a_{11}^{p'}$, $b_{11}^{p} - b_{11}^{p'}$ and $f_{1}^{p} - f_{1}^{p'}$ respectively. Similarly, the heave added mass, radiation damping coefficient, and wave excitation force can be obtained by $a_{33}^{p} + a_{33}^{p'}$, $b_{33}^{p} + b_{33}^{p'}$, and $f_{3}^{p} + f_{3}^{p'}$, respectively. a_{ij}^{p}, b_{ij}^{p} and f_{i}^{p} are the hydrodynamic forces of the p floater in i-th direction due to j-th mode of motion, where 1 and 3 denote the surge motion and heave motion respectively.

2.2. Equation of Motion

The p-th WEC is independently oscillating with a vertical mode in incident waves, with the other modes of motion being constrained. The wave power has been extracted with a power take-off (PTO) system, which converts the heave motion of the WEC into electricity. The PTO system is realized by an equivalent linear damping force. The schematic diagram of the WEC in heave motion is shown in Figure 3.

The equation of heave motion can be written as [27–30]

$$(m^{p} + a_{33}^{p})\ddot{z}_{p} + (b_{33}^{p} + b_{vis}^{p} + b_{PTO}^{p})\dot{z}_{p} + c_{33}^{p}z_{p} = f_{3}^{p} \tag{1}$$

where m^{p} is a mass of the p-th WEC, a_{33}^{p}, b_{33}^{p}, f_{3}^{p} are the frequency-dependent heave added mass, radiation damping coefficient, and wave excitation force respectively, c_{33}^{p} is the heave restoring force coefficient, and b_{vis}^p is the heave viscous damping coefficient, which is obtained by

$$b_{vis}^p = \frac{2\kappa^p c_{33}^p}{\omega_N^p} - b_{33}^p(\omega_N^p)$$

(2)

where the undamped heave natural frequency is given by $\omega_N^p = \sqrt{\frac{c_{33}^p}{m^p + a_{33}^p(\omega_N^p)}}$. The damping factor κ^p for the heave mode can be obtained from the heave-free decay test, which can be conducted experimentally or using a CFD simulation. In the present study, the heave-free decay test was conducted in a CFD simulation to obtain the viscous damping coefficient. b_{PTO}^p is the PTO damping coefficient and z_p is the heave motion response of the p-th WEC.

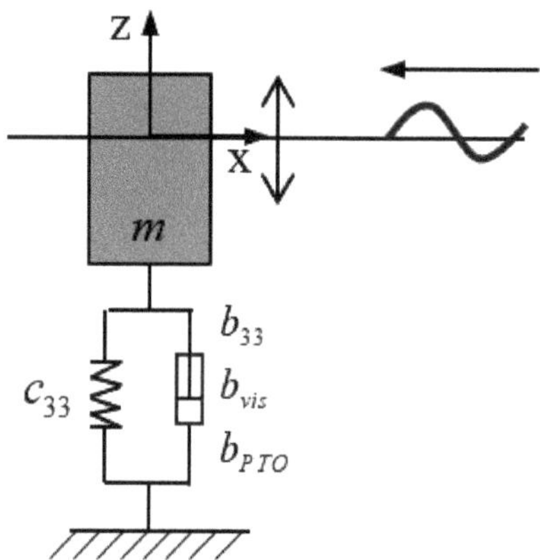

Figure 3. Schematic representation of the WEC in heave motion.

2.3. Extracted Wave Power

Wave power extracted by the WEC depends on the PTO damping and velocity of the WEC. In regular waves, the time-averaged extracted power of p-th WEC per unit wave amplitude is expressed as [27–30]

$$\overline{P}^p(\omega) = \frac{1}{2} b_{PTO}^p \omega^2 |z_p|^2$$

(3)

We can extend the extracted power in regular waves to irregular waves characterized by a significant wave height $H_{1/3}$ and peak period T_P. The JONSWAP spectrum is used for the incident wave spectrum $S_\varsigma(\omega)$, which is obtained by [31]

$$S_\varsigma(\omega) = \beta \frac{H_{1/3}^2 \omega_P^4}{\omega^5} \exp\left[-1.25\left(\frac{\omega}{\omega_P}\right)^{-4}\right] \gamma^{\exp\left[-\frac{(\omega-\omega_P)^2}{2\sigma^2 \omega_P^2}\right]}$$

(4)

$$\text{with } \beta = \frac{0.0624}{0.23 + 0.0336\gamma - 0.185(1.9+\gamma)^{-1}}(1.094 - 0.01915\ln\gamma)$$

where $\omega_P(= \frac{2\pi}{T_P})$ is the peak frequency. The peakedness factor $\gamma = 3.3$, $\sigma = 0.07$ for $\omega < \omega_P$, and $\sigma = 0.09$ for $\omega \geq \omega_P$.

The mean extracted power of the p-th WEC under the irregular waves can be obtained by [27,28,30,32,33]

$$\overline{P}_{irr}^p = \int_0^\infty S_\varsigma(\omega) \overline{P}^p(\omega) d\omega$$

(5)

Thus, the total power of N WECs in an array can be written as

$$P_{Total} = \sum_{p=1}^{N} \overline{P}_{irr}^{p} \tag{6}$$

2.4. PTO Damping

Based on Equation (3), the extracted power will be maximum under the condition of

$$\frac{d\overline{P}^{p}}{db_{PTO}^{p}} = 0 \tag{7}$$

which leads to a derivation of the frequency-dependent optimal condition of

$$\widetilde{b}_{PTO}^{p}(\omega) = \sqrt{\left(b_{33}^{p}\right)^{2} + \left(\frac{c_{33}^{p}}{\omega} - \omega(m^{p} + a_{33}^{p})\right)^{2}} \tag{8}$$

Although applying the variable optimal damping coefficient $\widetilde{b}_{PTO}^{p}$ as a function of wave frequency might yield higher power extraction theoretically, it might not be practical to apply variable PTO damping in the real sea according to incoming wave frequency. Thus, adopting a single best PTO damping is desirable.

Conventionally, the optimal PTO damping at a natural frequency would be the best option as the optimal PTO damping will be $b_{PTO}^{p} = b_{33}^{p}$ at a natural frequency based on Equation (8), which will lead to a maximum power extraction at resonance. Hence the optimal PTO damping at a natural frequency can be a candidate in the maximizing of power extraction.

An alternative approach is to explore a range of PTO damping values, calculate the associated extracted power $(\overline{P}_{irr}^{p})$ for each PTO damping under the irregular waves, and then select the PTO damping that yields the maximum extracted power. In this method, as the PTO damping increases, the heave motion decreases, and these reductions are reflected as the extracted power. However, the extracted power reaches a maximum with the increase in PTO damping, after which it starts to decrease despite further increases in PTO damping. Therefore, the PTO damping that results in maximum power can be chosen as another candidate.

So, in the present study, both the optimal PTO damping at a natural frequency and the PTO damping which yields maximum power extraction are considered and compared to each other.

2.5. Cost Indicator

The cost-effectiveness configuration design of WECs is achieved by reducing the cost of energy production while increasing wave power capture. To demonstrate the cost-effectiveness design, a cost indicator [26], which is defined as a ratio of submerged volume to power capture, is being used. In an array of WECs, it is defined as a ratio of total submerged volume to the total power capture of all the WECs.

$$\text{Cost indicator} = \frac{\text{Total submerged volume } \left(\text{m}^{3}\right)}{\text{Total power capture } \left(\text{kW}\right)} \tag{9}$$

The submerged volume is used to reflect the cost of materials for energy production. Hence the large cost indicator denotes higher fabrication cost of WECs in producing unit power of electricity, which is not desirable in achieving cost-effectiveness. Thus, the analysis for a cost-effective array of WECs focuses on identifying an array of WECs with a smaller cost indicator. Hence, the ranking of cost-effective configurations has been determined based on a scale of lower to higher value of the cost indicator.

3. Numerical Results and Discussion

3.1. Validation

3.1.1. Validation of Hydrodynamic Parameters

The present numerical results of hydrodynamic forces obtained from the commercial software WAMIT are compared with the published analytical results [17] for validation purposes. The WAMIT software supplies frequency-domain solutions based on the low-order panel method for the radiation and diffraction problems under linear potential theory. The hydrodynamic parameters obtained from the WAMIT were modified using the image method to consider the vertical seawall effect.

As a numerical model, an array of five cylindrical WECs is placed in front of a vertical seawall similar to the arrangement shown in Figure 2. The diameter (D) and draft (d) of WECs are 2 m and 1 m. The water depth (h) is identical to the WEC's diameter. The distance (L) between WECs and the distance (L_w) between WECs and a vertical seawall are 8 m and 4 m, respectively. The comparison is shown with the dimensionless added mass $a_{ii}/\frac{\rho D^3}{8}$, radiation damping $b_{ii}/\frac{\omega\rho D^3}{8}$, and wave excitation force $f_i/\frac{\rho g D^2 A}{4}$ where $i = 1, 3$ denote the surge and heave motion mode, which are plotted against $kD/2$ where k is the wave number. Numerical simulations are performed for the cases without the vertical seawall and with the vertical seawall, which is based on the image method. The numerical results of the first WEC in an array denoted as WEC ① are presented here. Figures 4 and 5 show the comparison between the present numerical solutions and analytical results [17] for surge and heave hydrodynamic forces of the WEC ①. The numerical results are in perfect agreement with the analytical results.

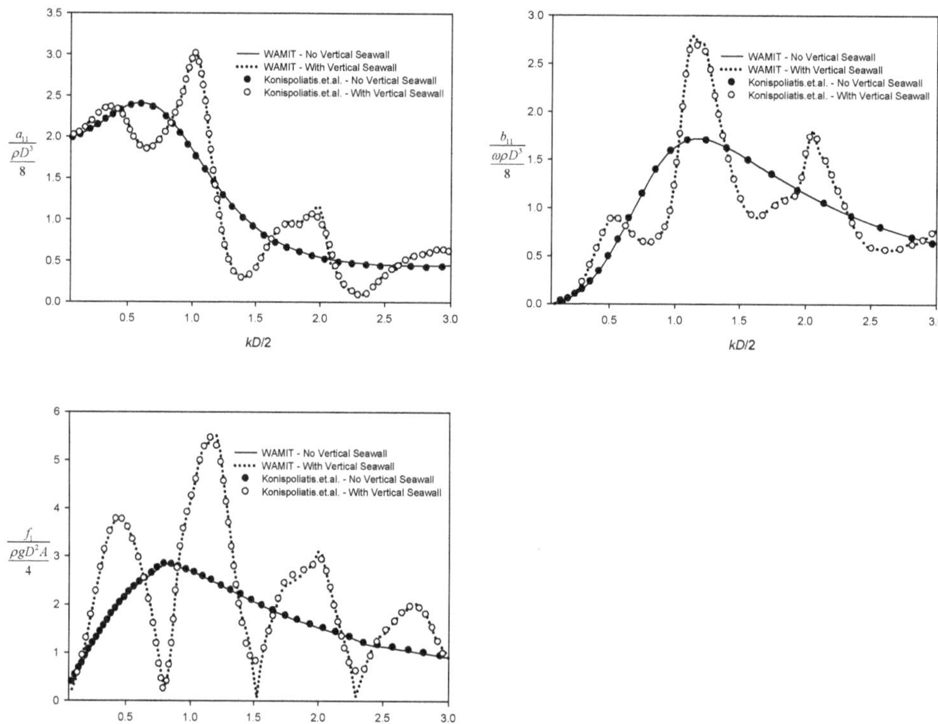

Figure 4. Dimensionless added mass, radiation damping, and wave excitation force of the WEC ① in surge mode [17].

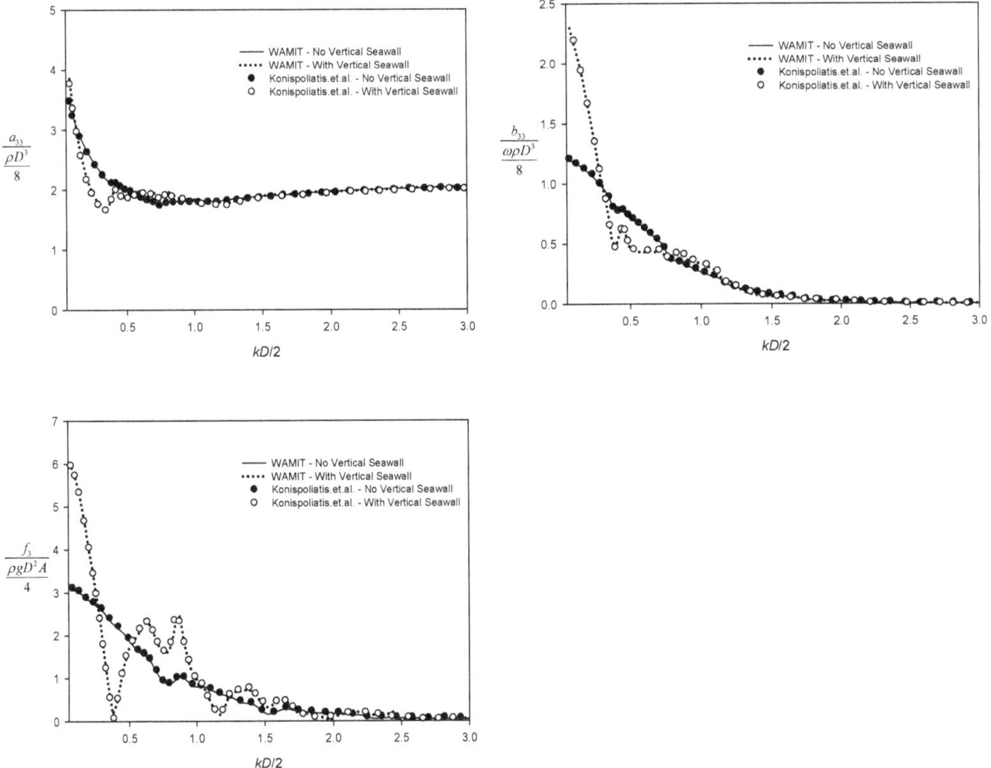

Figure 5. Dimensionless added mass, radiation damping, and wave excitation force of the WEC ① in heave mode [17].

3.1.2. Validation of CFD Simulation

The viscous damping is obtained from a CFD simulation of the free decay test in the heave direction, where the numerical calculation is performed by the commercial CFD code STAR-CCM+. The computational problem is solved by using the three-dimensional continuity, momentum, and K-Omega turbulence model with multiphase interaction in implicit unsteady time-steps. The volume of fluid (VOF) method was adopted to track the free surface in two phases (air and water). Two regions were created, one being the background region and the other surrounding WEC as an overset region with information exchanged through overlapping cells. The domain has been discretized into small cells with a trimmed cell mesher while the surface remesher was selected to create mesh around the WEC and prism layer for handling boundary layer while the domain has been created predominantly with hexahedral elements. The domain's length has been adequately extended to avoid wave reflection, with the outermost and bottom boundaries designated as walls, while the top boundary was a pressure outlet. With the finer time-step, a simulation time of four times the natural period of the WEC has been used to capture peaks of heave decay.

The computational result of the free decay test was compared with an experimental result to validate the numerical model. The model used in the experiment was a vertical cylinder with a diameter of 0.12 m and a draft of 0.25 m, which was placed in a water depth of 0.6 m. The experiment was carried out in a two-dimensional wave flume, located at Jeju National University, which was 20 m long and 0.8 m wide. The model was placed in the middle of a tank with the help of four slack mooring lines which had negligible effect on motion response. The model was initially given a displacement in the heave direction and allowed to oscillate freely. The heave motion was then tracked using image markers on the model, and Python code was used to process the video clips and extract time series data. Figure 6 shows the comparison of the numerical simulation with the experimental observation, which showed good agreement. The validated numerical model was used to simulate the free decay test of WECs with the present study to obtain the viscous damping coefficient of WECs.

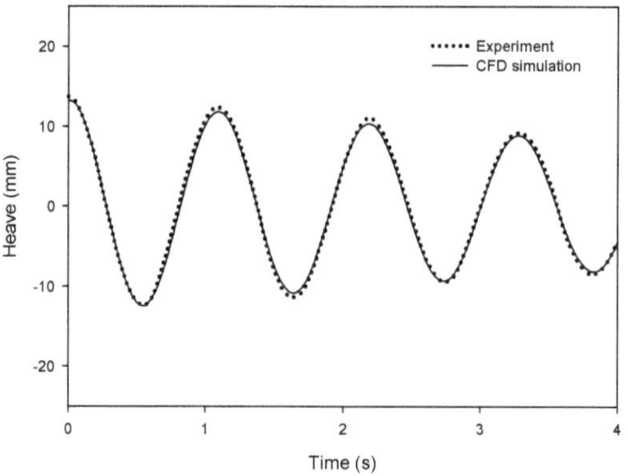

Figure 6. Comparison of experimental results with CFD simulation in the heave free decay test.

3.2. Modeling Parameters

In the calculation, the prototype WECs were considered in irregular wave climates described by a JONSWAP spectrum with a significant wave height ($H_{1/3}$) of 3 m and peak period (T_P) of 5 s. To search for a cost-effective design of WECs, various parameters of WECs such as diameter, number of WECs, and distance between them were considered. The diameters (D) considered were 2 m, 3 m, and 4 m while the drafts (d) of WECs were selected as 5.55 m, 5.25 m, and 5 m, accordingly and mass (m) of 17,872 kg, 38,038 kg and 64,403 kg respectively. These drafts were chosen to tune the heave natural period of WEC to align with the peak period of the wave spectrum such that the heave motion will be maximized at the peak frequency where wave energy is concentrated. The number of WECs under consideration were 1, 3, and 5, which enabled comparison of the performance of a single WEC and multiple WECs in an array. The distance between the WEC and the vertical seawall (L_w) is fixed to be equal to the WEC's diameter (D). The WECs are deployed with a parallel layout to the seawall. The distance between WECs (L) is considered as two- and five-times the diameter of the WEC. These can be representative values of the minimum required distance and a sufficiently distant placement respectively [18,21]. The WECs were placed in a water depth of 10 m. Figure 7 shows a schematic sketch of different configurations of an array of WECs and key parameters with a total of 15 cases.

All these design scenarios were initially tested without a vertical seawall. Subsequently, the WECs were placed in front of a vertical seawall, allowing for a comparison of how the seawall influences the performance of the WEC. Likewise, the comparison of the performance of the different sizes, the number of WECs in an array, and the distance between WECs were analyzed. As explained in Section 2.4, the extracted power was calculated for different conditions like the optimal PTO damping at a natural frequency and the PTO damping that yields maximum power extraction. Both the results were compared and analyzed.

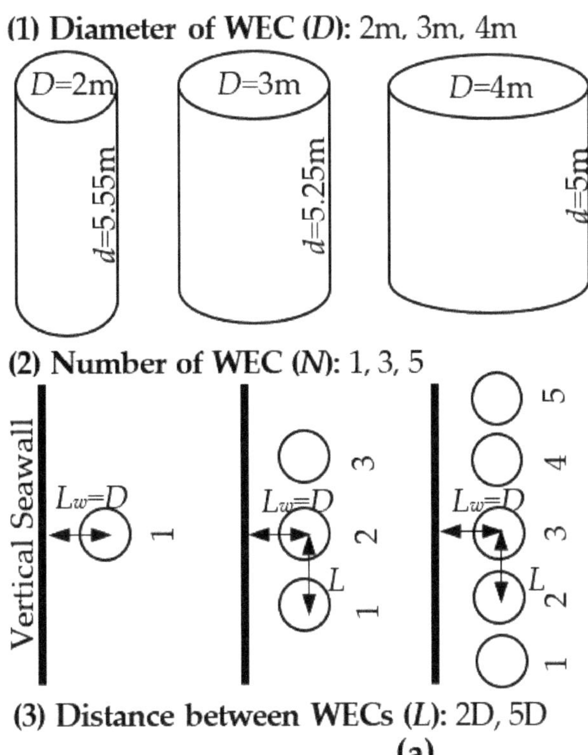

Number of WEC (*N*)	Distance between WEC (*L*)	Diameter of WEC (*D*)		
		2 m	3 m	4 m
1	-	Case 1	Case 6	Case 11
3	2D	Case 2	Case 7	Case 12
	5D	Case 3	Case 8	Case 13
5	2D	Case 4	Case 9	Case 14
	5D	Case 5	Case 10	Case 15

(b)

Figure 7. (**a**) Schematic sketch of different configurations of an array of WECs with various parameters such as WEC diameter, number of WECs, and spacing between WECs. (**b**) Various parameters in a tabular form with a total of 15 cases.

3.3. Viscous Damping

Figure 8 shows the CFD simulation of the heave-free decay test for different diameters of the WEC. The damping factor κ and damped natural period T_N are indicated inside each plot. The viscous damping coefficient b_{vis} can be obtained using Equation (2), which is proportional to the damping factor. It can be observed from the plots that as the WEC's diameter increased, the viscous damping increased due to the increase of the circumference length of the bottom of a cylinder where the generation of vortices occurs.

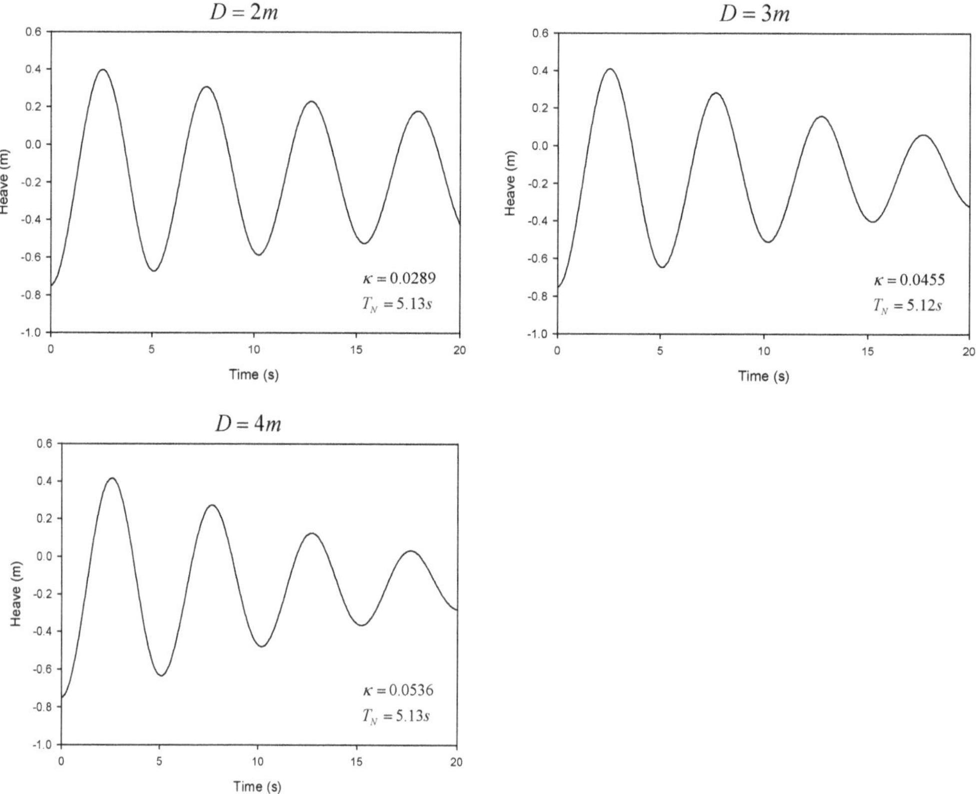

Figure 8. CFD simulation of the heave free decay test with different diameters of the WEC.

3.4. Extracted Power and Cost Indicator

The hydrodynamic parameters, computed from WAMIT with the method of image to incorporate the influence of vertical seawall, were combined with the viscous damping from CFD simulation and PTO damping. These combined parameters were then utilized to compute the heave motion of the WEC, as well as the extracted power from the PTO system and the related cost indicator under irregular wave conditions across different scenarios. A MATLAB code was used to integrate the parameters obtained from WAMIT and CFD calculations, for calculating the heave response of the WEC, along with the extracted power and cost indicators across various scenarios under irregular waves.

Table 1 provides a breakdown of the extracted power of each WEC and the total power of an array, along with the associated PTO damping and cost indicator for WECs positioned without a vertical seawall. The calculations are presented for both the optimal PTO damping at a natural frequency and the PTO damping yielding maximum power. Hereafter, these will be referred to as "optimum" and "maximum", respectively. Meanwhile, Table 2 presents corresponding data for WECs positioned in front of a vertical seawall. It can be observed that the extracted power was significantly higher with the "maximum" than the "optimum". This difference can be attributed to the higher PTO damping associated with the "maximum" resulting in increased power extraction. Likewise, the cost indicator was notably reduced with the "maximum", coinciding with the objective of cost-effectiveness.

Table 1. Extracted power and cost indicator of WECs placed in the open sea, calculated for the optimal PTO damping at a natural frequency and the PTO damping for the maxi mum power (denoted as "optimum" and "maximum" respectively).

D (m)	N	L	Case		$\overline{P}_{irr}$ (kW)		b_{PTO} (kNs/m)		P_{Total} (kW)		Cost Indicator (m³/kW)	
					Opt	Max	Opt	Max	Opt	Max	Opt	Max
2	1	-	1	WEC1	1.01	3.40	0.23	3.00	1.01	3.40	17.23	5.13
	3	2D	2	WEC1	1.01	3.41	0.23	3.00	2.96	10.15	17.66	5.15
				WEC2	0.95	3.32	0.22	3.00				
				WEC3	1.01	3.41	0.23	3.00				
		5D	3	WEC1	1.05	3.50	0.24	3.00	3.18	10.55	16.46	4.96
				WEC2	1.08	3.55	0.24	3.00				
				WEC3	1.05	3.50	0.24	3.00				
	5	2D	4	WEC1	1.06	3.56	0.23	2.90	5.14	17.48	16.98	4.99
				WEC2	1.01	3.46	0.23	3.00				
				WEC3	1.00	3.44	0.23	3.00				
				WEC4	1.01	3.46	0.23	3.00				
				WEC5	1.06	3.56	0.23	2.90				
		5D	5	WEC1	1.03	3.45	0.24	3.00	5.30	17.56	16.46	4.96
				WEC2	1.07	3.53	0.24	3.00				
				WEC3	1.09	3.60	0.24	2.90				
				WEC4	1.07	3.53	0.24	3.00				
				WEC5	1.03	3.45	0.24	3.00				
3	1	-	6	WEC1	2.00	5.82	0.88	8.90	2.00	5.82	18.57	6.37
	3	2D	7	WEC1	2.21	6.18	0.92	8.80	6.62	18.32	16.83	6.08
				WEC2	2.19	5.96	0.96	8.90				
				WEC3	2.21	6.18	0.92	8.80				
		5D	8	WEC1	2.00	5.94	0.86	8.90	6.11	18.24	18.23	6.10
				WEC2	2.11	6.36	0.84	8.80				
				WEC3	2.00	5.94	0.86	8.90				
	5	2D	9	WEC1	2.26	6.31	0.92	8.80	11.81	32.25	15.71	5.75
				WEC2	2.40	6.48	0.96	8.80				
				WEC3	2.49	6.66	0.97	8.80				
				WEC4	2.40	6.48	0.96	8.80				
				WEC5	2.26	6.31	0.92	8.80				
		5D	10	WEC1	2.05	6.01	0.87	8.80	10.30	30.61	18.02	6.06
				WEC2	2.10	6.27	0.84	8.80				
				WEC3	2.00	6.05	0.84	8.90				
				WEC4	2.10	6.27	0.84	8.80				
				WEC5	2.05	6.01	0.87	8.80				
4	1	-	11	WEC1	4.00	9.02	2.45	17.50	4.00	9.02	15.70	6.96
	3	2D	12	WEC1	4.72	10.23	2.57	17.40	14.57	30.75	12.94	6.13
				WEC2	5.13	10.29	2.87	17.50				
				WEC3	4.72	10.23	2.57	17.40				
		5D	13	WEC1	3.94	9.21	2.34	17.50	11.94	28.07	15.79	6.72
				WEC2	4.05	9.64	2.27	17.40				
				WEC3	3.94	9.21	2.34	17.50				
	5	2D	14	WEC1	4.46	9.67	2.60	17.60	25.13	52.55	12.50	5.98
				WEC2	5.31	10.78	2.82	17.40				
				WEC3	5.60	11.67	2.70	17.20				
				WEC4	5.31	10.78	2.82	17.40				
				WEC5	4.46	9.67	2.60	17.60				
		5D	15	WEC1	3.98	9.25	2.35	17.50	19.74	47.00	15.92	6.68
				WEC2	3.95	9.54	2.24	17.50				
				WEC3	3.86	9.42	2.21	17.50				
				WEC4	3.95	9.54	2.24	17.50				
				WEC5	3.98	9.25	2.35	17.50				

Table 2. Extracted power and cost indicator for WECs placed in front of a vertical seawall, calculated for the optimal PTO damping at a natural frequency and the PTO damping for the maximum power (denoted as "optimum" and "maximum" respectively).

D (m)	N	L	Case		$\overline{P}_{irr}$ (kW)		b_{PTO} (kNs/m)		P_{Total} (kW)		Cost Indicator (m³/kW)	
					Opt	Max	Opt	Max	Opt	Max	Opt	Max
2	1	-	1	WEC1	4.50	12.21	0.31	3.00	4.50	12.21	3.87	1.43
	3	2D	2	WEC1	4.91	12.66	0.33	3.00	14.48	37.54	3.61	1.39
				WEC2	4.67	12.20	0.32	3.00				
				WEC3	4.91	12.66	0.33	3.00				
		5D	3	WEC1	4.86	12.94	0.31	2.90	14.97	39.26	3.49	1.33
				WEC2	5.25	13.38	0.33	2.90				
				WEC3	4.86	12.94	0.31	2.90				
	5	2D	4	WEC1	5.31	13.75	0.33	2.90	26.82	67.64	3.25	1.29
				WEC2	5.36	13.40	0.34	3.00				
				WEC3	5.48	13.34	0.35	3.00				
				WEC4	5.36	13.40	0.34	3.00				
				WEC5	5.31	13.75	0.33	2.90				
		5D	5	WEC1	4.73	12.58	0.32	3.00	24.95	65.27	3.49	1.34
				WEC2	5.13	13.20	0.33	3.00				
				WEC3	5.24	13.71	0.32	2.90				
				WEC4	5.13	13.20	0.33	3.00				
				WEC5	4.73	12.58	0.32	3.00				
3	1	-	6	WEC1	9.46	18.78	1.46	8.90	9.46	18.78	3.92	1.98
	3	2D	7	WEC1	12.20	22.26	1.64	8.70	37.01	66.44	3.01	1.68
				WEC2	12.60	21.93	1.78	8.80				
				WEC3	12.20	22.26	1.64	8.70				
		5D	8	WEC1	9.74	19.39	1.46	8.90	30.18	60.61	3.69	1.84
				WEC2	10.69	21.84	1.39	8.80				
				WEC3	9.74	19.39	1.46	8.90				
	5	2D	9	WEC1	12.26	22.81	1.59	8.70	70.22	124.23	2.64	1.49
				WEC2	14.64	25.61	1.74	8.60				
				WEC3	16.42	27.39	1.86	8.60				
				WEC4	14.64	25.61	1.74	8.60				
				WEC5	12.26	22.81	1.59	8.70				
		5D	10	WEC1	10.20	19.86	1.50	8.80	51.07	102.11	3.63	1.82
				WEC2	10.55	21.29	1.42	8.70				
				WEC3	9.56	19.82	1.39	9.00				
				WEC4	10.55	21.29	1.42	8.70				
				WEC5	10.20	19.86	1.50	8.80				
4	1	-	11	WEC1	15.75	24.79	4.19	17.50	15.75	24.79	3.99	2.53
	3	2D	12	WEC1	21.93	32.50	4.54	17.00	72.23	101.43	2.61	1.86
				WEC2	28.37	36.44	5.85	16.50				
				WEC3	21.93	32.50	4.54	17.00				
		5D	13	WEC1	14.84	24.93	3.79	17.60	44.62	76.32	4.22	2.47
				WEC2	14.93	26.46	3.50	17.70				
				WEC3	14.84	24.93	3.79	17.60				
	5	2D	14	WEC1	19.27	28.62	4.65	17.60	124.35	173.59	2.53	1.81
				WEC2	28.16	37.22	5.59	16.80				
				WEC3	29.49	41.90	4.84	16.70				
				WEC4	28.16	37.22	5.59	16.80				
				WEC5	19.27	28.62	4.65	17.60				
		5D	15	WEC1	15.25	25.16	3.88	17.60	72.55	127.30	4.33	2.47
				WEC2	14.28	25.92	3.39	17.80				
				WEC3	13.49	25.15	3.26	17.80				
				WEC4	14.28	25.92	3.39	17.80				
				WEC5	15.25	25.16	3.88	17.60				

These results are further plotted below for a detailed analysis. In Figure 9, the total power output of various configurations of an array is assessed both for WECs without a vertical seawall and those with a vertical seawall. WECs situated in front of the seawall exhibited greater power extraction by the increase of WECs' heave motion due to the formation of standing waves. Also, the extracted power drastically increased with the diameter as the larger WEC possesses the potential to accommodate larger PTO damping, enabling higher power absorption.

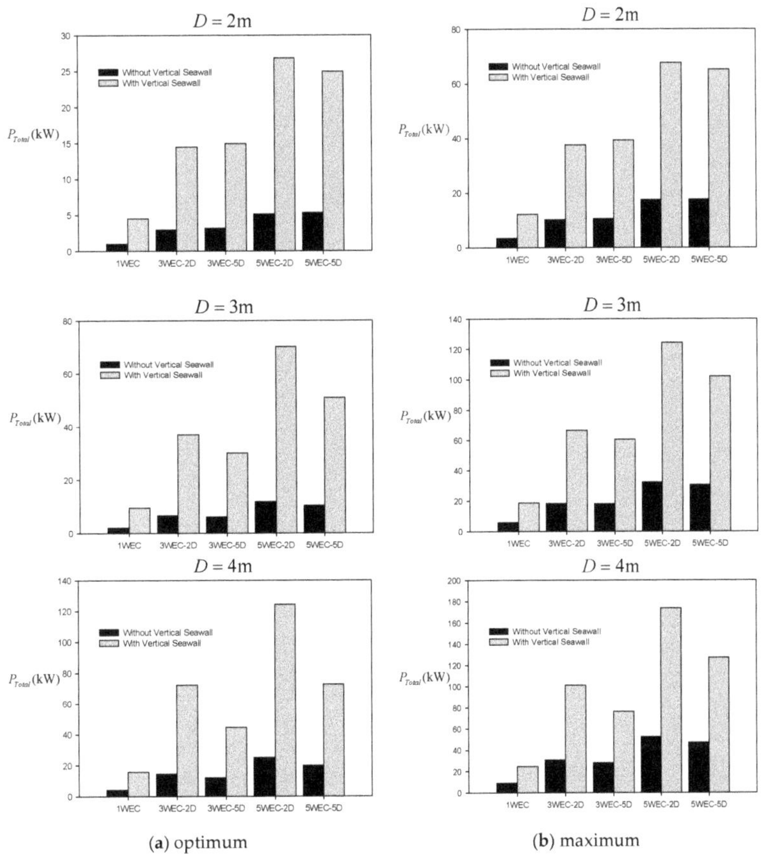

(a) optimum   (b) maximum

Figure 9. Comparison of total power for each configuration of an array of WECs placed in the open sea and in front of a vertical seawall for different diameters of the WEC.

To understand how the number of WEC in an array and the distance between WECs affects the power absorption of each WEC, the "optimum" extracted power of each WEC in an array configuration of single, three, and five WECs with a distance of 2D and 5D between WECs is compared in Figure 10.

The heave motion of the inside-positioned WECs in an array influenced the motion of adjacent WECs. However, the outside-positioned WECs facing the open sea on one side were only affected by the neighboring WEC on the other side. The WECs positioned at the symmetrical placement in an array had the same performance. Therefore, the power extracted from the centered WEC and the outmost WEC in an array were compared, as these serve as representative WECs for the analysis. For the single WEC, the outmost WEC and centered WEC are the same.

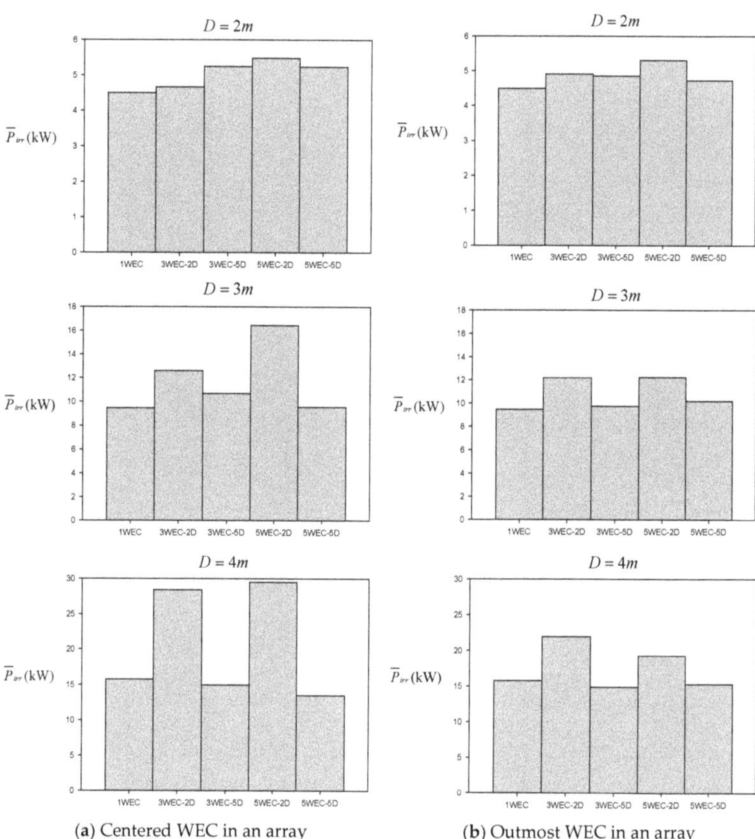

(a) Centered WEC in an array   (b) Outmost WEC in an array

Figure 10. Comparison of wave power extracted from each WEC in the configuration of a single WEC and multiple WECs placed in front of the vertical seawall calculated for the "optimum" PTO.

It is noticed from Figure 10 that when compared to a single WEC, configurations with three and five WECs exhibited an increase in power of each WEC, especially five WECs. This enhancement was a result of the interaction between neighboring WECs due to the presence of trapped waves between them. The trapped waves between WECs, coupled with the standing waves resulting from the reflection of incident waves against the vertical seawall, significantly intensified wave fields and consequently amplified the heave motion of each WEC. Thus, the centered WEC showed higher power absorption because of higher interactions of adjacent WECs than the outmost WEC which is open to sea on one side. Likewise, the intermediate WECs (②, ④) in an array of 5 WECs showed increased power extraction. Therefore, an increase in the number of WECs results in increased interactions among them, contributing to the enhanced power output of each WEC. However, the extracted power would also depend on the distance between WECs, which might constructively or destructively affect the performance. Among all cases of multiple WECs, a separation distance of 2D between WECs exhibited greater power enhancement compared to 5D, except for the 2 m diameter with an array of 3 WECs, where only a marginal difference was observed. Notably, the WECs with larger diameters demonstrated a substantial increase in the power extraction for the closer distance. This power enhancement could be attributed to higher interactions among WECs while keeping a closer distance between them than keeping them farther apart. Hence, these individual power enhancements of each WEC within an array collectively contribute to a higher overall power output. The same observations held when employing the "maximum" power calculation method. Therefore, an increased number of WECs arranged in an array with shorter distances between them would be the optimal configuration for maximizing power extraction.

3.5. Ranking of a Cost-Effective Array of WECs

The cost-effectiveness of various configurations of an array of WECs is assessed based on a cost indicator, which reflects the cost associated with extracting unit power. In Figures 11 and 12, the cost-effective configurations are prioritized according to the cost indicator, with the most favorable scenarios placed at the top of the plots. The WECs placed in front of a vertical seawall have substantially reduced the cost indicator with greater power extraction. This occurred because the vertical seawall increases the heave motion of WEC due to the formation of standing waves, resulting in increased power output for the equivalent submerged volume.

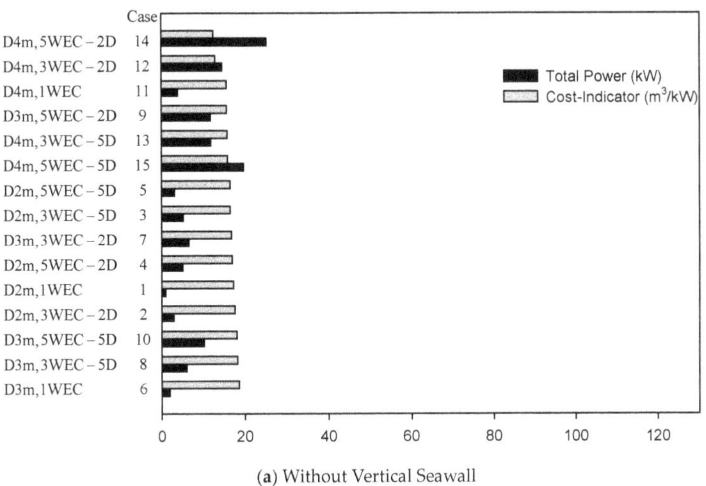

(**a**) Without Vertical Seawall

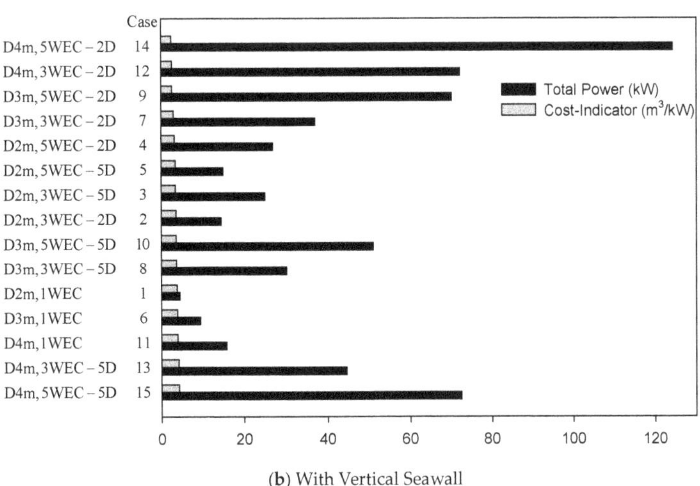

(**b**) With Vertical Seawall

Figure 11. Ranking of the cost-effectiveness of configurations of WECs based on the "optimum" power calculation method.

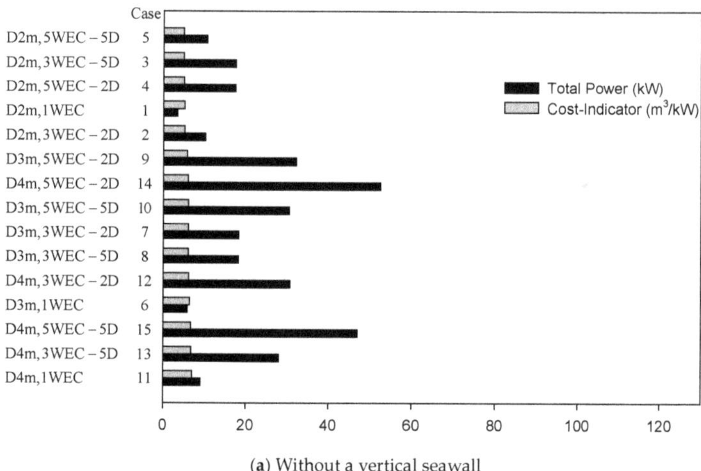

(**a**) Without a vertical seawall

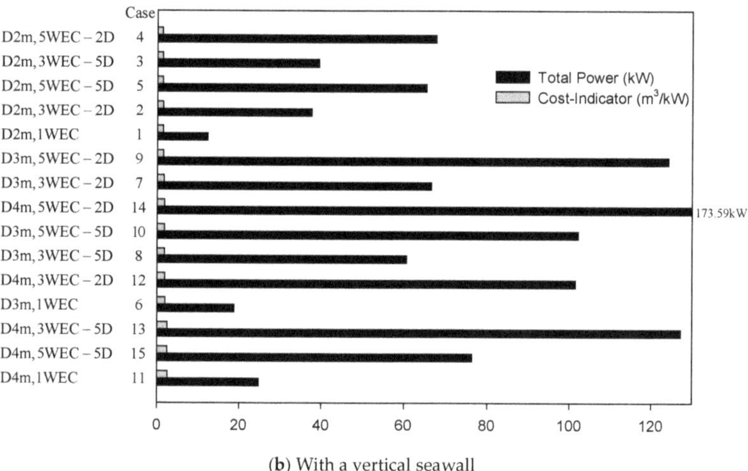

(**b**) With a vertical seawall

Figure 12. Ranking of the cost-effectiveness of configurations of WECs based on the "maximum" power calculation method.

In Figure 11, the trend indicates that the WECs with the sequence of larger-to-smaller diameter, when combined with an increase in the number of WECs in an array and positioned closer together, tended to achieve superior rankings in cost-effectiveness under the "optimum" power calculation method. In contrast, when employing the "maximum" calculation method, the WECs with the decrease of the diameter of WECs tended to attain a superior ranking as shown in Figure 12. When utilizing the "optimum" calculation method, larger diameters tended to exhibit effective PTO damping for maximizing power extraction for an equivalent submerged volume. Conversely, the "maximum" calculation method provided an opportunity for smaller diameters to accommodate efficient damping from a range of PTO damping which yields higher power output for an equivalent submerged volume.

4. Conclusions

An assessment was conducted to determine the cost-effective configuration of an array of WECs positioned in front of a vertical seawall in irregular waves. It involved a parametric study of varying diameters, number of WECs, and distances between the WECs. The WEC oscillates vertically in heave motion while utilizing a linear PTO damping system to harness wave power. The hydrodynamic parameters were numerically obtained using WAMIT with the method of image to incorporate the influence of the vertical seawall. These numerical calculations were validated against previously published analytical results. The viscous damping was obtained from a CFD simulation of the free decay test, which was validated beforehand against the experimental measurement for the cylinder model in a 2D wave tank. The power calculations were performed using both the optimal PTO damping at a natural frequency and the PTO damping that result in maximum power output. The cost-effectiveness was evaluated using a cost indicator, represented as the ratio of the total submerged volume of WECs to the overall power captured which reflects the production cost associated with extracting a unit power.

Based on the parametric analysis, the WECs placed in front of the vertical seawall achieve greater power extraction compared to the WECs placed in the open sea. The formation of standing waves due to total reflection by vertical seawall increases the heave motion of WECs, leading to higher power extraction. When compared to a single WEC, an increase in the number of WECs in an array shows higher power absorption due to interactions among WECs caused by trapped waves between them. The cost-effectiveness of WECs increases when the WECs are placed in front of the seawall, with a larger number of WECs in an array while keeping a shorter distance between them. The larger diameter of WECs excels in cost-effectiveness rankings when considering the optimal PTO damping at a natural frequency, whereas the smaller diameter of WECs exhibits superior performance with the PTO damping for maximum power extraction. These differences are attributed to the methods employed in implementing effective PTO damping, which enables higher power output for an equivalent submerged volume of WEC.

These findings demonstrate that achieving economically efficient wave power extraction is possible by installing multiple WECs in front of a reflecting seawall, even in nearshore shallow water regions. The shorter distance between WECs and the larger number of WECs in an array enables the production of more power. In addition, the nearshore installation of WECs also allows cost-effective power transmission connectivity to the onshore grid and the cost associated with maintenance compared to the offshore installation. Nevertheless, the challenges associated with nearshore installation encompass securing suitable space for WEC installation, implementing PTO systems, and establishing grid connectivity, particularly in densely developed coastal areas, could pose significant hurdles that must be addressed. These multiple WECs can also be installed in front of the offshore wind power platform that has a reflective wall similar to the seawall. This hybrid power system utilizing wind and wave power simultaneously could potentially offer cost-effectiveness by sharing the supported structure, power grid, and connectivity.

Author Contributions: Conceptualization, S.K.N.; Methodology, S.K.N.; software, S.K.N.; validation, S.K.N.; investigation, S.K.N.; writing—original draft preparation, S.K.N.; writing—review and editing, I.H.C.; supervision, I.H.C.; funding acquisition, I.H.C. All authors have read and agreed to the published version of the manuscript.

Funding: This research was supported by the Basic Science Research Program through the National Research Foundation of Korea (NRF) funded by the Ministry of Education (NRF-2022R1I1A3066608).

Data Availability Statement: Data is contained within the article.

Conflicts of Interest: The authors declare no conflicts of interest.

References

1. Hannah Ritchie and Max Roser, Emissions by Sector. Available online: https://ourworldindata.org/emissions-by-sector (accessed on 16 November 2023).
2. IEA. *World Energy Outlook 2022*; IEA: Paris, France, 2022.
3. McCormick, M.E. *Ocean Wave Energy Conversion*; Dover Publication Inc.: New York, NY, USA, 2007; Chapter 1.
4. Astariz, S.; Iglesias, G. The economics of wave energy: A review. *Renew. Sustain. Energy Rev.* **2015**, *45*, 397–408. [CrossRef]

5. Lyu, J.; Abdelkhalik, O.; Gauchia, L. Optimization of dimensions and layout of an array of wave energy converters. *Ocean Eng.* **2019**, *192*, 106543. [CrossRef]

6. Zhong, Q.; Yeung, R.W. On optimal energy-extraction performance of arrays of wave-energy converters, with full consideration of wave and multi-body interactions. *Ocean Eng.* **2022**, *250*, 110863. [CrossRef]

7. Guo, B.; Ringwood, J.V. Geometric optimisation of wave energy conversion devices: A survey. *Appl. Energy* **2021**, *297*, 117100. [CrossRef]

8. Garcia-Teruel, A.; Forehand, D.I.M. A review of geometry optimisation of wave energy converters. *Renew. Sustain. Energy Rev.* **2021**, *139*, 110593. [CrossRef]

9. Rodríguez, C.A.; Rosa-Santos, P.; Taveira-Pinto, F. Hydrodynamic optimization of the geometry of a sloped-motion wave energy converter. *Ocean Eng.* **2020**, *199*, 107046. [CrossRef]

10. Zhang, X.; Li, B.; Hu, Z.; Deng, J.; Xiao, P.; Chen, M. Research on Size Optimization ofWave Energy Converters Based on a Floating Wind-Wave Combined Power Generation Platform. *Energies* **2022**, *15*, 8681. [CrossRef]

11. Sun, P.; Hu, S.; He, H.; Zheng, S.; Chen, H.; Yang, S.; Ji, Z. Structural optimization on the oscillating-array-buoys for energy-capturing enhancement of a novel floating wave energy converter system. *Energy Convers. Manag.* **2021**, *228*, 113693. [CrossRef]

12. Shadmani, A.; Nikoo, M.R.; Al-Raoush, R.I.; Alamdari, N.; Gandomi, A.H. The Optimal Configuration of Wave Energy Conversions Respective to the Nearshore Wave Energy Potential. *Energies* **2022**, *15*, 7734. [CrossRef]

13. Sheng, W.; Lewis, A. Power takeoff optimization for maximizing energy conversion of wave-activated bodies. *IEEE J. Ocean. Eng.* **2016**, *41*, 529–540. [CrossRef]

14. Loukogeorgaki, E.; Chatjigeorgiou, I.K. Hydrodynamic performance of an array of truncated cylinders in front of a vertical wall. *Ocean Eng.* **2019**, *189*, 106407. [CrossRef]

15. Zhao, X.L.; Ning, D.Z.; Liang, D.F. Experimental investigation on hydrodynamic performance of a breakwater-integrated WEC system. *Ocean Eng.* **2019**, *171*, 25–32. [CrossRef]

16. Zhang, C.; Ning, D. Hydrodynamic study of a novel breakwater with parabolic openings for wave energy harvest. *Ocean Eng.* **2019**, *182*, 540–551. [CrossRef]

17. Konispoliatis, D.N.; Mavrakos, S.A.; Katsaounis, G.M. Theoretical evaluation of the hydrodynamic characteristics of arrays of vertical axisymmetric floaters of arbitrary shape in front of a vertical breakwater. *J. Mar. Sci. Eng.* **2020**, *8*, 62. [CrossRef]

18. Konispoliatis, D.N.; Mavrakos, S.A. Wave power absorption by arrays of wave energy converters in front of a vertical breakwater: A theoretical study. *Energies* **2020**, *13*, 1985. [CrossRef]

19. Konispoliatis, D.N.; Mavrakos, S.A. Hydrodynamic efficiency of a wave energy converter in front of an orthogonal breakwater. *J. Mar. Sci. Eng.* **2021**, *9*, 94. [CrossRef]

20. Kara, F. Hydrodynamic performances of wave energy converter arrays in front of a vertical wall. *Ocean Eng.* **2021**, *235*, 109459. [CrossRef]

21. Kara, F. Effects of a vertical wall on wave power absorption with wave energy converters arrays. *Renew. Energy* **2022**, *196*, 812–823. [CrossRef]

22. Sirigu, S.A.; Foglietta, L.; Giorgi, G.; Bonfanti, M.; Cervelli, G.; Bracco, G.; Mattiazzo, G. Techno-Economic optimisation for a wave energy converter via genetic algorithm. *J. Mar. Sci. Eng.* **2020**, *8*, 482. [CrossRef]

23. Tan, J.; Polinder, H.; Laguna, A.J.; Wellens, P.; Miedema, S.A. The influence of sizing of wave energy converters on the techno-economic performance. *J. Mar. Sci. Eng.* **2021**, *9*, 52. [CrossRef]

24. Têtu, A.; Fernandez Chozas, J. A proposed guidance for the economic assessment of wave energy converters at early development stages. *Energies* **2021**, *14*, 4699. [CrossRef]

25. Giglio, E.; Petracca, E.; Paduano, B.; Moscoloni, C.; Giorgi, G.; Sirigu, S.A. Estimating the Cost of Wave Energy Converters at an Early Design Stage: A Bottom-Up Approach. *Sustainability* **2023**, *15*, 6756. [CrossRef]

26. He, Z.; Ning, D.; Gou, Y.; Zhou, Z. Wave energy converter optimization based on differential evolution algorithm. *Energy* **2022**, *246*, 123433. [CrossRef]

27. Babarit, A. Impact of long separating distances on the energy production of two interacting wave energy converters. *Ocean Eng.* **2010**, *37*, 718–729. [CrossRef]

28. Cho, I.H.; Kim, M.H. Hydrodynamic performance evaluation of a wave energy converter with two concentric vertical cylinders by analytic solutions and model tests. *Ocean Eng.* **2017**, *130*, 498–509. [CrossRef]

29. Zhou, B.Z.; Hu, J.J.; Sun, K.; Liu, Y.; Collu, M. Motion response and energy conversion performance of a heaving point absorber wave energy converter. *Front. Energy Res.* **2020**, *8*, 553295. [CrossRef]

30. Natarajan, S.K.; Cho, I. New Strategy on Power Absorption of a Concentric Two-Body Wave Energy Converter. *Energies* **2023**, *16*, 3791. [CrossRef]

31. Goda, Y. Statistical variability of sea state parameters as a function of a wave spectrum. *Coast. Eng. Jpn.* **1988**, *31*, 39–52. [CrossRef]

32. Bhattacharyya, R. *Dynamics of Marine Vehicles*; John Wiley & Sons Incorporated: Hoboken, NJ, USA, 1978; Chapter 6.

33. Newman, J.N. *Marine Hydrodynamics*, 40th Anniversary ed.; The MIT Press: London, UK, 2018; Chapter 6.

energies

Article

A Comprehensive Multicriteria Evaluation Approach for Alternative Marine Fuels

Eleni Strantzali [1,*], Georgios A. Livanos [1] and Konstantinos Aravossis [2]

[1] Department of Naval Architecture, University of West Attica, 96 Agiou Spyridonos Street, 12210 Athens, Greece; glivanos@uniwa.gr
[2] Sector of Industrial Management and Operational Research, School of Mechanical Engineering, National Technical University of Athens, Iroon Polytechniou 9, 15780 Athens, Greece; arvis@mail.ntua.gr
* Correspondence: estra@uniwa.gr; Tel.: +30-210-538-5340

Abstract: In the last decade, shipping decarbonization has accelerated rapidly in response to the regulatory framework. Shifting toward alternative marine fuel options is the subject of extensive study from stakeholders and researchers. This study attempts to propose a decision support model for alternative fuel evaluation. The decision-making process is multidimensional, comprising economic, technical, environmental, and social aspects, and has been carried out with the aid of the outranking multicriteria methodology, Promethee II. The approach is based on a comprehensive list of 11 criteria and 25 sub-criteria, covering all the crucial aspects. The weighting criteria process postulates the viewpoints of six stakeholder categories, including all the stakeholders' preferences: shipowners, fuel suppliers, industry and engine manufacturers, academics, banks and the public. The results demonstrated that although LNG, MGO and HFO are classified in the highest positions, there are renewable options that also appear in high-ranking positions in most categories and especially among academics, banks, the public and in the combined case scenario. The commercially available options of drop-in biofuels, bio and e-LNG, fossil and bio methanol were ranked in these high positions. This approach offers insight into the assessment and selection of alternative marine fuel options, providing an incentive for strategic planning.

Keywords: decarbonization; marine fuels; shipping; multicriteria decision making; Promethee

Citation: Strantzali, E.; Livanos, G.A.; Aravossis, K. A Comprehensive Multicriteria Evaluation Approach for Alternative Marine Fuels. *Energies* **2023**, *16*, 7498. https://doi.org/10.3390/en16227498

Academic Editors: Byong-Hun Jeon and Agustin Valera-Medina

Received: 21 October 2023
Revised: 2 November 2023
Accepted: 7 November 2023
Published: 8 November 2023

1. Introduction

Maritime transportation constitutes a crucial part of the transportation sector and the global economy, with an over 80% share of the volume of commodities transported by sea [1]. In 2018, 2.89% of the global anthropogenic CO_2 emissions was due to emissions from the shipping sector [2]. International shipping emits 70% of global shipping energy emissions and, supposing that it was a state, it would have the sixth or seventh largest CO_2 emissions [3]. The International Maritime Organisation (IMO) has proposed ambitious strategies to decrease GHG emissions from international shipping during this century [2]. Initially, the first strategy, Resolution MEPC.304(72), included initial targets to shrink CO_2 emissions per transport work, with a minimum 40% and 70% until 2030 and 2050, respectively, based on emissions from 2008. Furthermore, GHG emissions have to be reduced by at least 50% by 2050 compared to 2008 [4]. Shipping GHG emissions are going to be increased between 50% and 250% by 2050 (compared to 2008 levels) if no actions are taken [3]. In July 2023, the IMO adopted the IMO GHG Strategy from Ships (Resolution MEPC.377(80)), which includes the use of zero/near-zero GHG emission technologies, fuels and energy sources for a minimum 5–10% of the energy used by international shipping until 2030. GHG emissions from international shipping should become net zero by or around 2050. Furthermore, the current Strategy recalls the 2018 Strategy and might be replaced in the future from a revised IMO GHG Strategy in 2028 [5].

Additionally, the European Commission (EC), in 2021, introduced the 'Fit for 55' legislative package, mandating a reduction of 55% of GHG emissions by 2030. The shipping sector is also affected by this European package, as it will be included in the European emission trading system (EU ETS), applying to all vessels exceeding 5.000 gross tonnage (GT) and covering 100% of intra-European Economic Community (EEC) emissions as well as 50% of extra-EEC emissions. The FuelEU Maritime Proposal set a stepwise limit for reducing the carbon content of the maritime fuel, and the European energy taxation directive (EU ETD) also set a gradually increasing minimum tax for maritime fuels [6–9].

A shift to low- or zero-carbon fuels seems to be mandatory to comply with the above-mentioned IMO and EU targets. The decarbonization of the shipping sector encompasses a range of possible and innovative alternative technical and operational measures. Liquified Natural Gas (LNG), methanol, biofuels, hydrogen, ammonia and electricity are discussed worldwide as promising alternative fuel options. Each marine fuel option has to face its own specific challenges for its adoption. New buildings will accelerate the compliance process.

In response to the legislation requirements, there has been a growing trend in the literature to compare and evaluate the alternative fuel options. Ampah et al. reviewed 583 papers, published between 2000 and 2020 in the field of alternative marine fuels, demonstrating the growth of interest in the field. One of the research gaps, they concluded, is that most of their examined studies only considered the effect on emission reductions from their proposed measures. The types of proposed measures mainly covered technical aspects, such as hull design, power and propulsion system, energy sources and operational optimization [10]. However, fuel option selection requires a multidimensional approach and appropriate tools of evaluation to aid in decision making. Although multicriteria decision-making techniques have been popular over decades in the field of supplier selection, there are only a few research studies employing MCDM methods for alternative fuel selection, considering economic, technical, environmental and social aspects. Furthermore, only a part of these published studies (Deniz and Zincir [11], Hansson et al. [12,13], Mandic et al. [14] and Yang et al. [15]) included the alternative fuels of methanol, hydrogen, ammonia, bio and e-fuels in their assessment. Thus, the development of a comprehensive integrated evaluation framework that aligns with independent shipping stakeholders seems to be a demanding challenge.

The aim of the current paper is to present an in-depth evaluation process of alternative marine fuels by identifying a variety of criteria. This approach assesses 16 alternative fuel options, also taking into consideration the current fossil fuels, HFO (Heavy Fuel Oil) and MGO (Marine Gas Oil) and adopting a set of 25 significant and coherent key parameters. As the fuel selection problem has a multidimensional nature, a multicriteria analysis was assumed for the assessment process, covering the economical, technical, environmental and social aspects of the problem.

The rest of this paper is organized as follows: Section 2 presents the examined alternative marine fuels. Section 3 displays a literature review in the field and a brief presentation of the proposed methodology, including the used multicriteria methodology and the weighting criteria process. Section 4 analyzes the criteria used and determines their values, while in Section 5, the results from the implementation of the evaluation methodology are presented, followed by Section 6 with this research's conclusions.

2. Alternative Marine Fuels

Globally, there are several potential marine fuel options as viable solutions to oil-based fuels to aid the shipping industry in achieving the future emission reduction targets. In this study, the examined marine fuels are divided into three categories: (a) commercially available fuel options, (b) fuels in the demonstration phase and (c) fuels under development. HFO and MGO are used as baseline options, given that they are the current dominant fuels in international shipping. A brief general description of the examined alternative fuels is given below:

Liquified Natural Gas (LNG) is the most prolific and commercially available fuel. The main energy source for LNG is natural gas, composed of methane, liquefied at $-162\,^{\circ}\text{C}$, at atmospheric pressure. As a renewable replacement for LNG, bio-LNG has a much lower carbon footprint than other fossil fuels or biofuels. It is made by processing organic waste, such as animal waste or municipal waste. LNG could also be produced synthetically with the power to gas process. This process includes hydrogen production from water, using a renewable electricity source (wind, solar or other option) or it can be processed into methane by adding non-fossil carbon dioxide obtained from carbon capture. E-LNG is interchangeable with LNG and is able to be utilized in existing infrastructure. Nearly all the current LNG production is from natural gas [16].

Methanol is mainly used to produce chemicals, like formaldehyde, plastics, and acetic acid. It is produced from carbon sources, such as natural gas, coal, biomass, and even CO_2. About 65% of methanol production is currently based on natural gas reformation (grey methanol), while the rest (35%) is largely based on coal gasification (brown methanol) [17]. In this study, grey and brown methanol is referred to as fossil methanol. Blue methanol is produced using blue hydrogen in combination with carbon capture technology. Biomass feedstocks, such as forestry and agricultural waste or biogas from landfill and municipal solid waste, can be used as the raw materials for biomethanol production. Green e-methanol is obtained with hydrogen production from renewable electricity sources or with the carbon capture process. It has the advantage of being liquid in ambient conditions and so there is no need for refrigeration or pressurization for transport and storage. Its bunkering process is similar to HFO, and only minor modifications are necessary to existing infrastructure, being already available in some ports [17].

Biofuels can be made from a variety of feedstocks and can be used as drop-in fuels with minimal alterations to the existing equipment. In some cases, an alternative fuel may not be useable in its 100% pure form and may require 'blending' to produce a drop-in solution [16]. Advanced biofuels are produced from specific feedstocks with no indirect land use change (ILUC). In this study, the examined biofuels are as follows: HTL (hydrothermal liquefaction) fuel oil, pyrolysis fuel oil, HVO and FAME and their respective feedstocks of energy crops, lignocellulosic biomass, oil crops, waste oils and fats. Although crop-based feedstocks, like palm oil and soybean, are widely available, their use in Europe is limited due to the policy of ILUC. Lignocellulosic biomass, such as forestry and agricultural residues as well as woody and grassy energy crops, seems to have greater future potential, whereas waste oils and fats in the maritime sector have to face competition from other transport sectors. Biofuels' sustainability depends mostly on the type of feedstock [16]. Hydrothermal liquefaction (HTL) is direct thermochemical conversion of wet biomass into bio-crude at 300–350 °C and 10–25 MPa. Pyrolysis is a thermochemical conversion of biomass to bio-oil. HVO (Hydrotreated Vegetable Oil) is produced through a hydrotreating process, also called hydroprocessing, in order to remove sulphur, oxygen, and nitrogen. Fatty Acid Methyl Ester (FAME), known as biodiesel, is a prevalent biofuel in the EU. Nowadays, waste, used cooking oils and animal fats are the main feedstocks for FAME, and transesterification is the used chemical process [18].

Hydrogen can be produced from both fossil and renewable sources. Each year, almost 95% of global hydrogen production comes from gas and coal (grey hydrogen) [16]. Green hydrogen is hydrogen through water electrolysis fueled by renewable-based electricity. Hydrogen can also be produced from biomass as a biofuel. The production process refers to the steam reformation of methane (biogas) obtained from the anaerobic digestion of organic waste. The choice of an alternative production pathway for hydrogen is determined to a large extent by the local availability of the energy source [16]. It can be utilized in internal combustion engines or fuel cells. Experience from LNG in shipping could be useful, given the similarity of the known technology of cryogenic conditions. A key barrier to the liquefaction of hydrogen is the low temperature needed, −253 °C. Hydrogen facilities have to increase approximately more than 220-times to reach the current LNG facilities, becoming widespread in world trade [19].

Ammonia constitutes the basic product in chemical industries and especially in the production of fertilizers. Although it is a carbon-free fuel, its application is currently limited.

It has the advantage of being used directly as marine fuel or as a hydrogen carrier (ammonia is converted back to hydrogen for combustion) [20]. Ammonia can be produced through three pathways based on the energy source used: natural gas (fossil grey or blue ammonia), renewable ammonia taking advantage of solar photovoltaics and wind (green ammonia) and from residual biomass and municipal waste (bio-ammonia) [16]. Renewable ammonia is chemically identical to fossil-based ammonia. It can be characterized as a versatile fuel as it can be stored in liquid form at atmospheric pressure at $-33\,^{\circ}C$ or at ambient temperature and at least 8 bar and can be used in internal combustion engines, gas turbines and fuel cells [21]. The conversion of the existing ammonia tankers to ammonia-fueled ships could be applicable in the short term, as the issue of fuel availability from ports ceases to exist. Fossil-based ammonia will perform a transitional role as a short-term solution in decarbonization, whereas renewable NH_3 is predicted to have a dominant role in future markets. Although renewable ammonia is able to displace conventional fuels, its use can increase nitrogen oxide emissions, NO_X, nitrous oxide and N_2O, and an aftertreatment technology is obligatory. Furthermore, ammonia is a hazardous toxic chemical [21].

3. Materials and Methods

3.1. Literature Overview

Currently, the shipping industry has a number of possible low- and zero-carbon marine fuels available to meet IMO 2030 and 2050 emission reduction goals. The selection of each fuel option has its own special characteristics, composing a multicriteria decision-making problem with a finite set of criteria comparing stakeholders' priorities [22]. Thus, the comparative evaluation of fuel options needs a rigorous decision support framework able to incorporate the different preferences of the stakeholder groups. Multicriteria methods can solve problems with conflicting and multiple objectives, expressed by the decision makers and stakeholders.

The topic of making marine fuels greener has been investigated by various researchers. Studies are mainly divided into investigations of the feasibility of alternative technologies and literature reviews mapping the research in respective domains. As a preliminary step, an extensive literature review was implemented. Although there are numerous studies in the field of shipping decarbonization, the number of publications dealing with the multicriteria evaluation of marine fuels is only twelve, as shown in Table 1. Their ranking results are utilized as a base for comparison with the obtained results from this study.

Table 1. Literature review—multicriteria applications.

	Authors	Method	Evaluation Alternatives
1	Ren J. and Lützen M., 2015 [23]	Fuzzy AHP and VIKOR	Low sulphur fuel, Scrubber and LNG
2	Deniz C. and Zincir B., 2016 [11]	AHP	Methanol, Ethanol, LNG and Hydrogen.
3	Ren J. and Liang H., 2017 [24]	Fuzzy logarithmic least squares and fuzzy TOPSIS	Methanol, LNG and Hydrogen
4	Ren J. and Lützen M., 2017 [22]	Dempster-Shafer theory and a trapezoidal fuzzy AHP	LNG, Nuclear and Wind power
5	Hansson J. et al., 2019 [12]	AHP	LNG, LBG, Methanol from NG, Renewable methanol, Hydrogen, HVO and HFO
6	Kim A.R. and Seo Y.-J., 2019 [25]	Fuzzy AHP	Low sulphur fuels, Scrubbers and LNG
7	Hansson J. et al., 2020 [13]	AHP	NG-NH3, Elec-NH3, LNG
8	Luciana (Marcu) T.A. et al., 2021 [26]	AHP	LNG and oil gas
9	Mandic N. et al., 2021 [14]	AHP and SAW	Biofuels, LNG, Hydrogen, LPG, Batteries
10	Carvalho F. et al., 2021 [27]	Qualitative analysis	Alternative fuels' production pathways
11	Moshiul A.M. et al., 2023 [28]	TOPSIS	Criteria assessment
12	Yang Z. et al., 2023 [15]	AHP and q-ROLPBM (q-Rung Orthopair Linguistic Partition Bonferroni mean)	E-fuel, Solar fuel, Biofuel, E-biofuel

An overview of the paper's content is summarized below:

Ren J. and Lützen M., 2015 [23], combined Fuzzy AHP and VIKOR to validate three alternative technologies (low-sulphur fuel, scrubber, and LNG), resulting in LNG as the most viable option for long-term use.

Deniz C. and Zincir B., 2016 [11], qualitatively compared methanol, ethanol, LNG and hydrogen with eleven criteria using the AHP methodology, based on given points from five experts. According to their assessment process, LNG was placed in front of the three examined alternative fuel options, followed by hydrogen and closing with methanol and ethanol.

Ren J. and Liang H., 2017 [24], applied fuzzy logarithmic least squares for the weights' calculation and fuzzy TOPSIS for the assessment of three alternative marine fuels, methanol, LNG and hydrogen, taking into consideration 11 criteria (including environmental, economic, technological and social). This resulted in a similar classification to that of Deniz and Zincir (2016) [11].

Ren J. and Lützen M., 2017 [22], combined Dempster–Shafer theory and a trapezoidal fuzzy AHP for the sustainability assessment of nuclear power, LNG and wind energy as possible energy resources for shipping, prioritizing nuclear power as a sustainable alternative for shipping.

Hansson J. et al., 2019 [12], assessed seven alternate marine fuels (LBG, fossil and renewable methanol, fossil and electric hydrogen, HVO and HFO with scrubbers). The alternative fuels were compared through pairwise comparisons with regard to four main categories of criteria, economic, technical, environmental and social, and 10 sub-criteria, based on the preferences of Swedish stakeholders. LNG and HFO were classified at the highest levels, followed by fossil methanol and biofuels. Meanwhile, the evaluation based on the Swedish government expressed priority in renewable marine fuels, renewable hydrogen and renewable methanol, whereas, in 2020, Hansson J. et al. [13] attempted to evaluate the prospects of ammonia compared to LNG, MGO, hydrogen, HVO, LBG and methanol, including 10 criteria and using AHP. The weights of the criteria were retrieved from shipowners, fuel producers, engine manufacturers and Swedish government authorities. They observed that ammonia has restricted potential for large-scale applications, as issues remain to be solved.

Kim A.R. and Seo Y.-J., 2019 [25], used fuzzy AHP to evaluate three existing alternatives for emission reductions, low-sulphur fuels, scrubbers and LNG-powered vessels for Korean shipping companies.

Luciana (Marcu) T.A. et al., 2021 [26], used 6 criteria and AHP methodology to assess LNG and oil gas.

Mandic N. et al., 2021 [14], proposed the application of AHP and Simple Additive Weighting (SAW) for the alternative marine fuel assessment in coastal shipping. Biofuels, LNG, hydrogen, LPG and batteries were prioritized using 10 criteria covering environmental, technological and economical aspects, and the selected study area was Croatia. Electric propulsion stands out from all the alternatives, and the ranking order is differentiated according to the stakeholder groups.

Carvalho F. et al., 2021 [27], developed a qualitative analysis for ranking 14 fuel production options for the Brazilian maritime trade. The analysis incorporated 9 criteria, including technical, economic and environmental. The drop-in fuels dominated in their results followed by bio-methanol and bio-LNG, whereas green hydrogen and green ammonia were the least-promising alternatives for Brazil.

Moshiul A.M. et al., 2023 [28], used the multicriteria technique, TOPSIS, to assess the most important criteria for the selection of fuel alternatives by prioritizing the preferences of shipowners and shipping companies' management of Singapore firms. The criteria assessment process included 15 factors and 77 subfactors, considering technical aspects, technology status, policies, economic, environmental and socio-political aspects. Their assessment indicated technological aspects, technology status, expenditure, ecosystem impact and health and safety as the most crucial criteria.

Yang Z. et al., 2023 [15], evaluated four alternative low-carbon fuel production pathways (e-fuel, solar fuel, biofuel, e-biofuel), using AHP and the q-ROLPBM operator. The evaluation process was carried out with 13 criteria, including economic, environmental, technical and social, for the United Kingdom. Their research indicated e-fuel and e-biofuel as the most promising production pathways.

The above review concludes that the existing literature for comparison and evaluation of multiple alternative fuels is limited. In the majority of papers, the evaluation includes 2–3 fuel options, and only Hansson J. et al. [12,13] and Mandic N. et al. [14] deal with 5–7 fuels, incorporating a manageable set of 10 criteria, while Carvalho F. et al. [27] and Moshiul A.M. et al. [28] assess fuel production pathways and evaluation criteria, respectively. A broader range of fuel options incorporating a broader range of criteria will provide additional insight in the obtained rankings. In the present study, 16 fuel options are considered and assessed through 25 criteria, which also constitute the novelty of the presented methodological framework. In addition, the multicriteria method Promethee II is applied for the first time to a fuel option evaluation.

3.2. Methods

3.2.1. Multicriteria Evaluation Methodology

Each multicriteria approach has its own advantages and disadvantages, and the choice of the appropriate one depends on the nature of the problem. As can be seen in the literature review, the majority of existing studies are based on the AHP method, where all the criteria and alternatives must be compared by the decision maker/user in a pairwise process, which might be impossible in cases with many criteria.

Outranking methods have been developed rapidly during the last few decades, as they incorporate the characteristic of allowing incomparability between a finite number of alternatives and a conflicting set of criteria [29,30]. Electre (elimination and choice translating reality) and Promethee (Preference Ranking Organization Method for Enrichment Evaluation) are the most commonly used outranking multicriteria techniques [31]. TOPSIS (technique for order preference by similarity to ideal solutions) is another option to Electre methodology, based on the comparison of Euclidean distances of alternatives.

The multicriteria methodology Promethee II was adopted for the assessment process in this research, taking advantage of the pairwise comparison of the alternatives and their final ranking as an output of the process, without the involvement of the decision maker in the process of extracting the results. A multicriteria preference index is formulated for each alternative action X (named "Alternative Fuels"). The importance of each criterion is expressed by a weight. The preference functions, V-type and usual type, for quantitative and qualitative criteria, respectively, were selected for the calculation of the preference index. Furthermore, a preference threshold was considered, whereas the indifference threshold was ignored. The alternative fuels were sorted by a positive or negative flow, $\Phi^+(X)$ and $\Phi^-(X)$, where X is the alternative fuel. The positive flow, "$\Phi^+(X)$", indicates how the alternative X outranks all the others, and the negative flow, "$\Phi^-(X)$", indicates a preference among all other alternatives compared to alternative X. The net outranking flow, $\Phi(X)$, determines an overall score for each alternative [30,31].

3.2.2. Weighting Criteria

Elicitation of weights is always a challenging phase in the decision-making process and it is crucial to reflect all the possible preferences. Simos technique is an indirect weighting methodology, aiding in the expression of preferences, even for stakeholders unfamiliar with decision-making methodologies. The initial Simos approach was extended to face robustness issues, creating the revised Simos approach [32].

In this study, a weight was assigned to each criterion, and the process was carried out according to the revised Simos approach. The criteria weights were categorized in seven different strategies obtained through literature review and interviews/workshops with shipping-related stakeholders. Stakeholders are categorized into six groups associated with their expertise: (i) shipowners, (ii) fuel suppliers, (iii) industry-engine manufacturers, (iv) academics, (v) banks and (vi) public. The seventh category, "combined case" scenario, reflects the combined viewpoints of all stakeholders for the criteria and is estimated as the weighted geometric mean for each criterion. All the participants were informed of the content of the research. A 10-point Likert scale was used to capture each criterion's importance. As can be seen in Table 2, there are some similarities among the stakeholder groups, as the importance given to the following criteria is obvious: "emissions reduction", "regulation", "fuel availability" and the economic criteria, "Capex" and "fuel cost".

Table 2. Criteria weights (%).

Criteria Weights (%)		Shipowners	Fuel Suppliers	Industry—Engine Manufacturer	Academics	Banks	Public	Combined Case
Capex	C1	13.8	4.6	8.9	4.6	17	6.2	8.1
Opex	C2	3.1	1.6	7.1	3.1	7.3	6.2	3.9
Fuel Cost	C3	15.3	13.6	10.7	12.1	8.9	6.2	12.4
Fuel Availability	C4	15.3	16.5	14.4	9.1	13.8	6.2	14.5
Adaptability	C5	10.8	12.1	7.1	10.6	5.7	6.2	8.1
Commercial effects	C6	9.2	3.1	3.5	1.6	4.1	6.2	3.8
Risk assessment	C7	12.3	7.6	5.3	7.6	15.4	15	10.3
Emissions reduction	C8	6.2	10.6	12.6	16.5	12.2	16.8	14.5
Fuel properties	C9	4.7	9.1	12.6	13.6	2.5	6.2	6
Regulation	C10	7.7	15.1	16.2	15.1	10.6	13.3	16.7
Job creation	C11	1.6	6.1	1.6	6.1	2.5	11.5	1.7

4. Criteria

The literature review, in Section 3.1, highlighted that the evaluation criteria are usually grouped into four main aspects: economic, technical, environmental and social. In this study, 25 sub-criteria were derived as the most frequently used indicators and were categorized into 11 main criteria and the 4 above-mentioned groups (Figure 1).

4.1. Economic Indicators

Economic indicators (Table 3) can be broken down into (a) capital cost for propulsion (Capex): this includes the expenditures of propulsion and related system components per installed engine capacity (such engines' cost, fuel tanks, pipelines, gas alarm systems, fuel processors, etc.); (b) operational cost (Opex): this includes crew cost, maintenance and insurance cost but excludes fuel cost [12,13]; (c) fuel cost: the expense of the fuel price is divided into two subcriteria, the current fuel cost (based on fuel prices of 2021) and the potential reduction for future cost, according to the prediction of IMO [9,16,33,34].

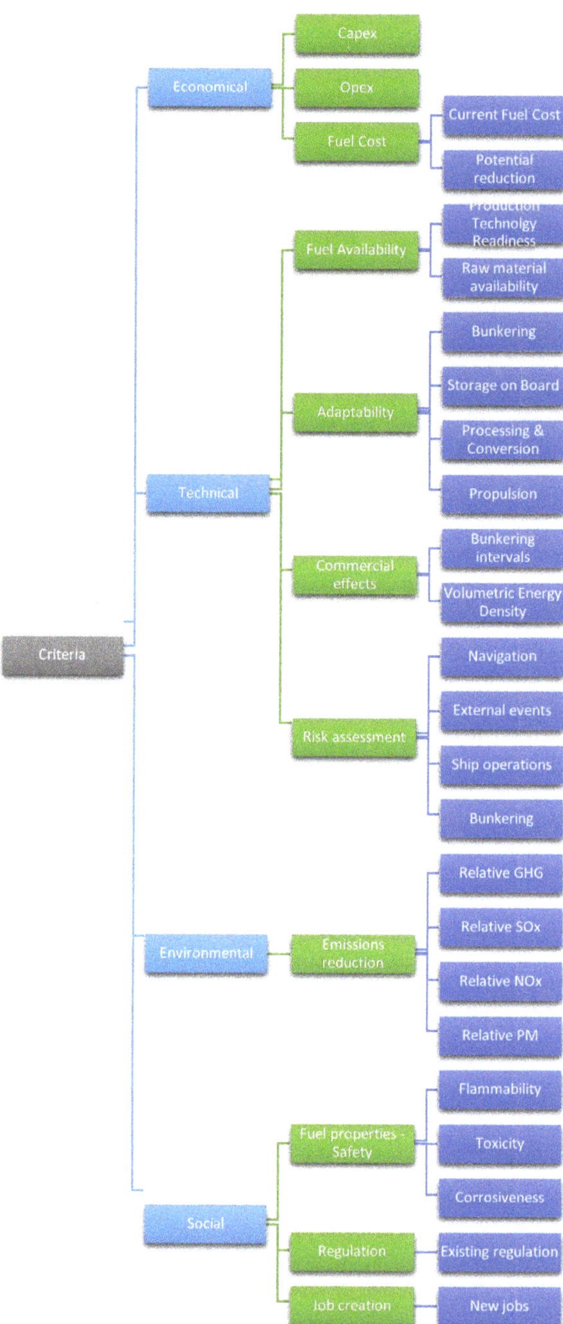

Figure 1. The criteria tree.

Table 3. The values of the economic indicators [9,12,13,16,33,34].

Fuels/Criteria		Capex [1] ($/kW) [2]	Opex [3] ($/MWh) [2]	Fuel Cost [4]	
				Current Fuel cost ($/GJ Fuel) [5]	Potential Cost Reduction
HFO		4800–7300	5	5–12	Low
MGO		4500–7040	5	12–14	Low
Commercially available options					
LNG	LNG	5100–7710	9	7–10	Low
	Bio-LNG			8.5–28.5	Medium
	e-LNG			23–110	High
In demonstration phase					
Methanol	Fossil	4700–7180	6	4–31	Low
	Blue			21–237	High (for CCS)
	Bio			22–35	Medium
	e-methanol			58–463	High
Biofuels (Drop-in)	HTL fuel oil	4500–7040	5	51–98	Medium
	Pyrolysis fuel oil			31–45	
	HVO			24–39	
	FAME			20–35	
Under Development					
Hydrogen	Grey or Blue	6500–12,040	11	11–26	High (for CCS)
	Green			16–33	High
	Bio			20–54	High (for gasification)
Ammonia	Fossil (or blue)	5200–11,400	9–11	16–27	High (for CCS)
	Green			23–27	High (for electrolysers and renewable energy)
	Bio			20–54	High

[1] Capex: includes the cost for onboard infrastructure per engine capacity, [2] in 2015 dollars (2015 is selected as a reference year common for all fuel options based on the data retrieved by the literature review process), [3] Opex: fuel cost in not included, [4] The sub-criteria of "Fuel cost" do not include potential carbon taxes, [5] in 2021 dollars (2021 is selected as a reference year common for all fuel options based on the data retrieved by the literature review process).

4.2. Technical Indicators

This category assesses fuel availability, adaptability of technology, commercial effects of the adoption of the alternative fuel option and their performance in the case of a hazard. More specifically, fuel availability includes the following sub-criteria: (i) production technology readiness, representing the existing level of adequacy of the production technology and the necessary processes, and (ii) raw material availability, meaning the current availability of feedstocks and energy sources [16]. The technology feasibility of the alternative fuels, as regards the onboard procedures of bunkering, storage, processing, conversion and propulsion, is examined through the criterion of "Adaptability" (it is expressed using the TRL score, too). It should be noted that TRL score describes the stage of development of a technology and is measured on a scale from 0 (idea/concept stage) to 9 (full commercial application of technology). In general, alternative fuel systems are more feasible when new building ships, from their application to existing ships [11]. The combination of technological maturity with the growing demand for alternative fuels will have a direct impact on increasing the availability of bunkering infrastructure and operating ships.

The criterion of "Commercial effects" describes the impact of the use of the alternative fuel in a ship's operation and is divided in two sub-criteria: (i) bunkering intervals and (ii) volumetric energy density. Bunkering intervals range from hours to months, depending on the selection of the alternative fuel, and affect the ship route and its bunkering plan [35,36]. Furthermore, energy density should be considered for the different types of fuel, as higher volumetric energy density requires less space for onboard storage of the fuel and, consequently, a higher cargo loading capacity for the ship. LNG is about one-third of the volumetric energy density of diesel, and liquid hydrogen, methanol and ammonia are around 40–50% of LNG, whereas biodiesel is the closest to diesel [16,36–38].

The last indicator for this category is "Risk assessment". "Together in Safety" [39], a non-regulatory shipping industry safety coalition, carried out a risk-ranking process for different hazard scenarios. The examined scenarios included possible events that might occur in the daily operations of a vessel: navigation (loss of maneuverability, motion at sea, etc.), external events (ship collision, ignition), ship operations other than bunkering (crew change, system components etc.), bunkering (misalignment of the bunkering stations, loss of control etc.), and fuel preparation, use and monitoring (loss of control). The "Risk assessment" sub-criteria are measured based on the performance of the alternative fuels in the examined scenarios. The highest score means the best performance. The values for all technical indicators are shown in Table 4.

Table 4. The values of technical indicators [11,16,35–39].

Fuels/Criteria		Fuel Availability			Adaptability [1]			Commercial Effects			Risk Assessment [2]		
		Production Technology (TRL)	Raw Material Availability	Bunkering	Storage on Board	Processing and Conversion	Propulsion	Bunkering Intervals	Volumetric Energy Density (MJ/L)	Navigation	External Events	Ship Operations	Bunkering
HFO		9	Widely	9	9	9	9	Months	39.8	13/13	4/4	6/6	12/12
MGO		9	Widely	9	9	9	9	Months	38.4	13/13	4/4	6/6	12/12
Commercially available options													
LNG	LNG	9	Widely	9	9	9	9	Weeks	20.6	8/13	2/4	1/6	12/12
	Bio-LNG	7–8	Constrained	9	8	9	9						
	e-LNG	6–7	Constrained	9	8	9	9						
In demonstration phase													
Methanol	Fossil	9	Widely	9	7	7	6–7	Weeks	15.7	12/13	2/4	3/6	3/12
	Blue	6	Constrained	9	7	7	6–7						
	Bio	8	Constrained	5.3	7	7	6–7						
	e-	7	Constrained	5.3	7	7	6–7						
Biofuels (Drop-in)	HTL fuel oil	5–6	Widely	9	9	9	9	Months	33	13/13	4/4	6/6	12/12
	Pyrolysis fuel oil	8	Widely	9	9	9	9						
	HVO	9	Widely	9	9	9	9						
	FAME	9	Constrained	9	9	9	9						
Under Development													
Hydrogen	Grey or Blue	9	Widely	3	6	2	5–7	Hours/Days	8.51	7/13	2/4	1/6	12/12
	Green	7–8	Constrained	3	6	2	5–7						
	Bio	6–7	Constrained	3	6	2	5–7						
Ammonia	Fossil (or blue)	9	Widely	3.7	7	4.5	2–7	Weeks	15.7	7/13	1/4	1/6	1/12
	Green	5–6	Constrained	3.7	7	4.5	2–7						
	Bio	5–6	Constrained	3.7	7	4.5	2–7						

[1] Bunkering: it is assumed the average in terms of the technology readiness for equipment, procedures and fuel quality standards. Storage on Board: the maximum value was taken into account among the alternative storage options: structure tank, membrane containment system, IMO type A, B and C tank. Processing and Conversion: In case of ammonia as a fuel, it is obtained the average of the score for internal combustion engines and fuel cells. Propulsion: includes ICE 2-stroke, 4-stroke, main auxiliary boilers and reformers. All the values for the biofuels have been adopted from the values for Bio-diesel. [2] The score for each category and fuel is calculated based on the risk ranking of the examined scenarios. The score for drop-in biofuels has been assumed the same as conventional fuels.

4.3. Environmental Indicators

The main reason to select alternative marine fuels is the emissions' reductions in order to comply with the regulation targets. In this study, a lifecycle perspective (WTW-Well to Wake) was considered for the quantification of GHG emission reduction indicators, including the stages of fuel production, distribution and transport as well as the final consumption from the ship [16,17,40,41]. The percentage values of emission reductions are related to HFO, as a reference case (Table 5).

Table 5. The values of environmental indicators [16,17,40,41].

Fuels/Criteria		Emissions Reduction			
		Relative GHG	Relative SOx	Relative NOx	Relative PM
HFO		0%	0%	0%	0%
MGO		0%	0%	0%	0%
Commercially available options					
LNG	LNG	−15%	−100%	−80%	−100%
	Bio-LNG	−80%			
	e-LNG	−80%			
In demonstration phase					
Methanol	Fossil	+29%	−99%	−60 to −80%	−95%
	Blue	−42% to −60%			
	Bio	−85% to −91%			
	e- methanol	−58% to −94%			
Biofuels (Drop-in)	HTL fuel oil	−80% to −82%	−100% (assuming low sulphur in feedstock)	Uncertain (depends on fuel properties)	~0%
	Pyrolysis fuel oil	−77% to −80%			~0%
	HVO	−53% to −89%	−100%	0% to −20%	Generally reduced
	FAME	−53% to −89%	−99% to −100%	0%	
Under Development					
Hydrogen	Grey or Blue	−22% (blue) +70% (grey)	0% (ICE)—100% (FC)	0% (ICE)—100% (FC)	−100%
	Green	−87%	−100%	−100%	
	Bio	Highly	Moderate [1]	Moderate [1]	
Ammonia	Fossil (or blue)	~−14%	0% (ICE)—100% (FC)	0% (ICE)—100% (FC)	−100%
	Green	~−77%	−100%	−100%	
	Bio	Highly	Moderate [1]	Moderate [1,2]	

[1] It has been considered in line with the respective conventional fuel and depending on the used propulsion system, as at the time of preparing the manuscript there is not available quantitative data for bio-hydrogen and bio-ammonia. [2] Ammonia can be used in modified ICEs or FCs. The combustion of NH_3 produces N_2O and in case of ICE an aftertreatment of removing N_2O is necessary [2].

4.4. Social Indicators

In the literature, the social pillar represents safety factors and economic growth at a local level (Table 6). Safety factors include the fuel properties (such as flammability, toxicity and corrosiveness) and the regulatory compliance, which is one of the most crucial criteria, highly rated in almost all the stakeholders' preferences. The flammability limit is an indicator of the required amount of a fuel to be burnt in the air volumetrically. Indicatively, hydrogen burns easily with the widest flammability limits [11]. The criterion "Regulation" is quantified in a range of 0–5, after studying the existing regulations, standards and guides from numerous organizations: IMO, ISO, Class Society, SGMF, European Committee for Standardization (CEN) and Methanol Institute. Additionally, for the social aspect, the production process of the alternative marine fuels could play a leading role in creating new jobs compared to conventional fuels, generating the indicator "Job creation". Obviously, existing conventional fuel options are not able to create new job opportunities opposed to bio-fuels and e-fuels, which have the highest potential [10,16,28,37,38,40].

Table 6. The values of social indicators [10,11,16,28,37,38,40].

Fuels/Criteria		Fuel Properties—Safety			Regulation [1]	Job Creation [1]
		Flammability (vol%) [2]	Toxicity	Corrosiveness	Existing Regulation	New Jobs
HFO		1–6	Non-toxic	Non-corrosive	5	0
MGO		0.7–5	Non-toxic	Non-corrosive	5	0
Commercially available options						
LNG	LNG	5–15 (Methane)	Non-toxic	Non-corrosive	5	1
	Bio-LNG					5
	e-LNG					4
In demonstration phase						
Methanol	Fossil	6–36	Acutely-toxic	Corrosive	4	1
	Blue			Corrosive		2
	Bio			Corrosive (upon degradation)		5
	e-methanol			Corrosive		4
Biofuels (Drop-in)	HTL fuel oil	0.6–7.5	Non-toxic	Corrosive (upon degradation)	5	1
	Pyrolysis fuel oil					1
	HVO					1
	FAME					1
Under Development						
Hydrogen	Grey or Blue	4–75	Non-toxic	Non-corrosive	3	2
	Green					4
	Bio					5
Ammonia	Fossil (or blue)	15–25	Very toxic	Corrosive	2	2
	Green					4
	Bio					5

[1] The criteria "Existing regulation" and "job creation" are presented in the range of 0–5, where 0 represents the lowest value and 5 the highest. [2] Flammability limits in air (vol%): show the range of vapour concentrations of a certain chemical, over which a flammable mixture gas or vapour in air can be ignited at 25 °C and atmospheric pressure.

5. Results and Discussion

The results generated through the multicriteria evaluation process are displayed in Figure 2. The evaluation process also included HFO and MGO as the current baseline fuel options for the purpose of comparison. The examined biofuels, HTL, pyrolysis fuel oil, HVO and FAME, were grouped together in the evaluation process, called "Drop-in" fuels, as their values are common in almost all the indicators.

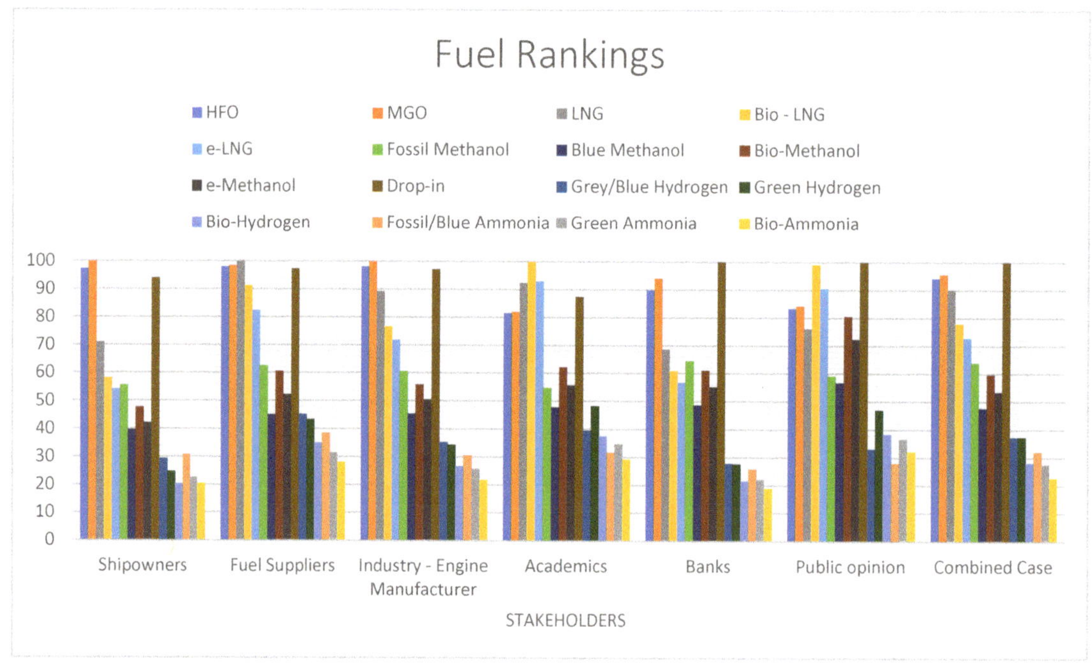

Figure 2. The obtained results for all stakeholders.

The marine fuel rankings present certain similarities among the stakeholder groups. LNG, MGO and HFO are classified in the highest positions in almost all the examined stakeholder categories. The high rating of these fuels is due to their widespread availability, which resulted in their good performance on several criteria, reflecting the current state of the shipping sector, unlike other new marine fuels, which are still in the development phase. When focusing on alternative marine fuels, there are renewable options that also appear in high-ranking positions in the majority of categories and especially among academics, banks, the public and the combined case scenario. In these high positions, commercially available options of drop-in biofuels, bio and e-LNG, and fossil and bio methanol are included. Bio and green ammonia registered the lowest scores in all stakeholder groups due to their high costs in the economic indicators, low adaptability and low performance in risk assessment, as well as the lack of existing regulation. This observation aligns with the conclusions of Hansson et al., 2020 [13], for the use of ammonia. Gradual decarbonization of the current fossil-based ammonia plants with the co-production of renewable hydrogen, replacing a percentage use of natural gas, should be stimulated at an early stage, as well as fostering the development of new production plants. The hydrogen options remain in intermediary positions, with a small lead in conventional hydrogen production.

Figure 3 shows the impact of each criterion in the classification process for the five highest-ranked fuel options for all stakeholder groups, separately. In the shipowners' graph, the lines of MGO and HFO almost overlap, and their highest performance occurred mainly due to their high performance in technical indicators, while Bio-LNG is distinguished for its values in environmental indicators. A similar influence is observed in the fuel suppliers' group and industry-engine manufacturers' group, who prioritized existing regulation, fuel cost, availability and emissions' reductions. On the contrary, the environmental indicators had a key role for the classification of academics and the public. Fossil methanol is the new entry in the banks group because of the relatively good performance in risk assessment, fuel availability and emission reductions.

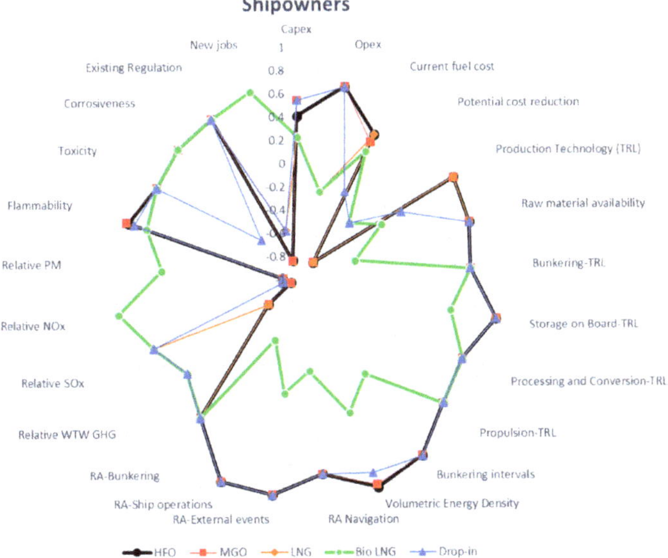

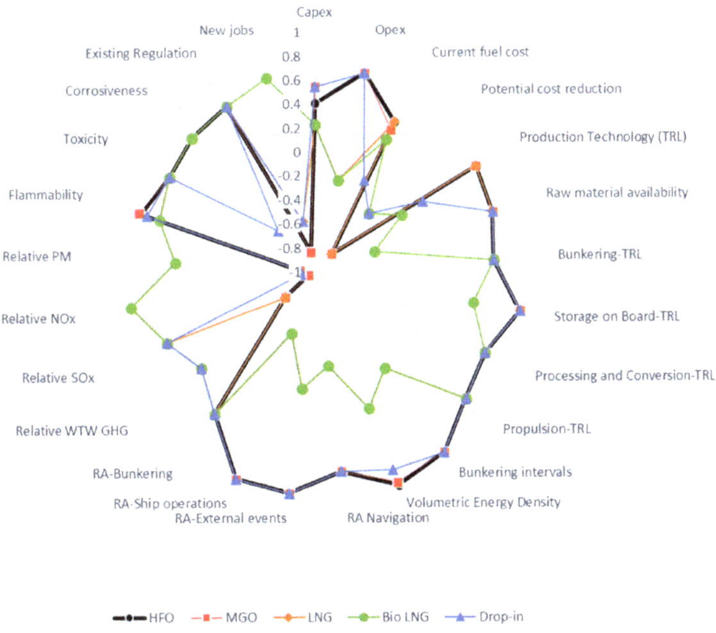

Figure 3. *Cont.*

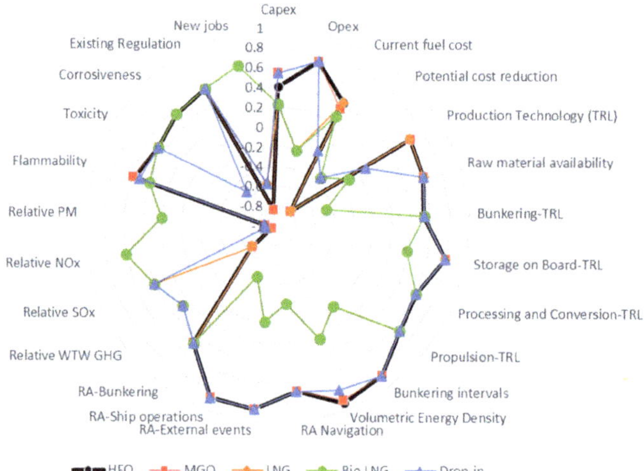

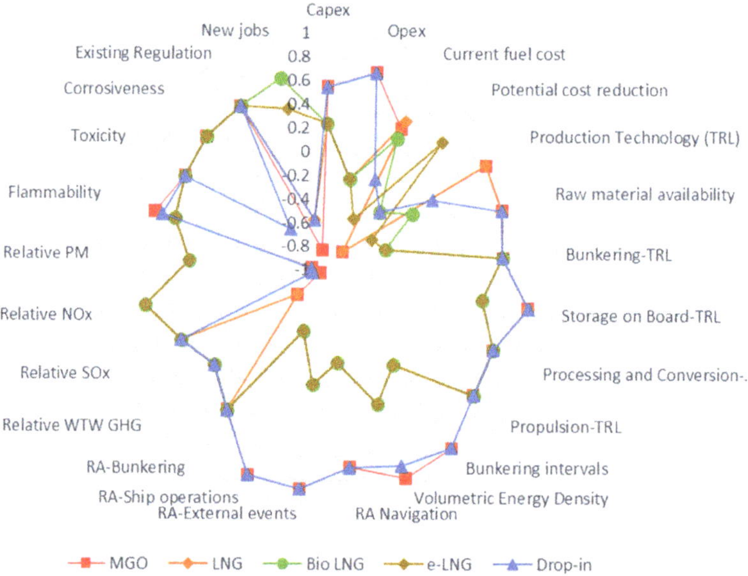

Figure 3. *Cont.*

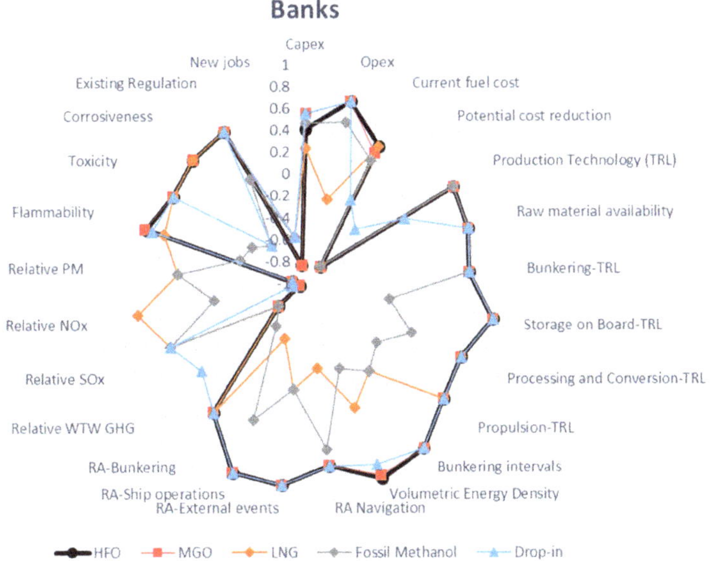

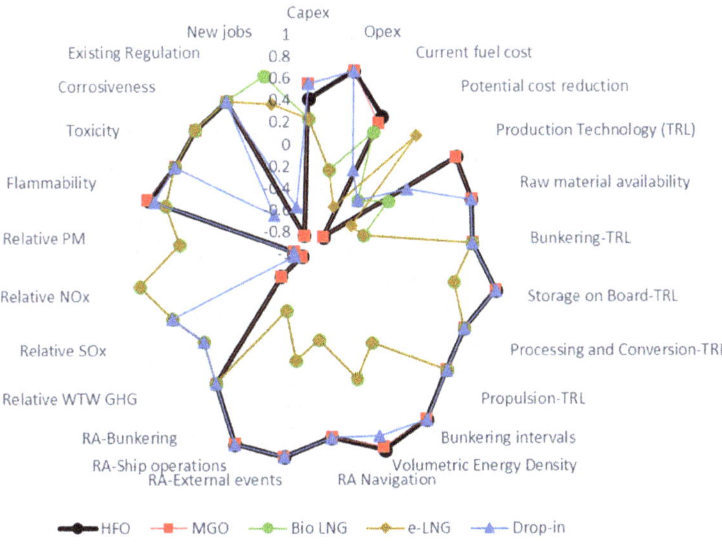

Figure 3. *Cont.*

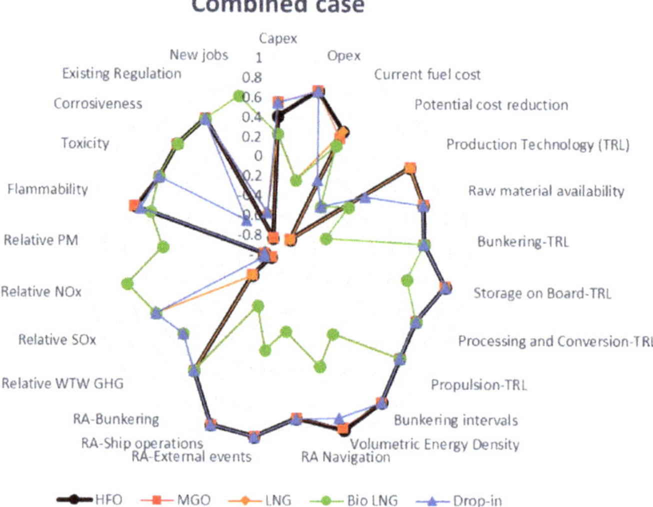

Figure 3. The impact of each criterion for the five highest-ranked fuel options.

Although the existing literature does not include the same alternatives of marine fuels in the multicriteria evaluation, the current classification is in line with the findings of Hansson et el. 2019 and 2020 [12,13]. The common point with the rest of the manuscripts is the dominant role of LNG in the classification. Deviations among the findings of the published studies could also be observed due to the different decision-making methodologies used as well as due to diverse influences from the stakeholders during the weighting process.

The ranking process is obviously influenced by the selection of the set of criteria. The number and range of criteria selected are crucial factors in minimizing the risk of an inaccurate outcome. In this study, the chosen criteria cover a broad range of key aspects, along with the expression of viewpoints by diverse groups of stakeholders. The different views of stakeholders regarding the importance of the criteria also serve as a kind of sensitivity analysis for the obtained results. The importance and performance of the examined fuels might also be differentiated due to new policies or the further development of existing technologies.

Some limitations should be considered in the context of the obtained findings in the current study. The main limitation is that alternative shipping propulsion systems are not integrated in the examined fuel options. Specialization of the engine's types (for instance, ICE or FC), technical parameters of the current and possible future applications of mainstream marine engines and the possible aftertreatment technology could lead to distinct values for certain criteria, such as emission reductions for the case of ammonia. Accordingly, future research might be enriched by specifying the evaluation process for different types of ships (deep-sea, short-sea shipping, coastal shipping).

6. Conclusions

This study developed a holistic evaluation framework, incorporating four sustainability aspects, economic, technical, environmental and social, as well as six stakeholders' views. The variations among the stakeholder group priorities resulted in different classifications of the examined fuel options. The decision-making process through the proposed methodology has the advantage of flexibility and the ability to examine a variety of criteria at the same time as considering the preferences of many decision makers or stakeholders. According to shipowners, engine manufacturers and fuel suppliers, MGO, LNG and

HFO are top ranked, followed by drop-in fuels, while, based on academics, banks and the public, drop-in fuels, bio-LNG, e-LNG and MGO are ranked first, followed by LNG, HFO and MGO. The ranking, which came from a combined case scenario, is a mixture of the above-mentioned outcomes.

The contribution of this research is demonstrated by the multidimensional evaluation of alternative fuel options, incorporating a plethora of criteria for more accurate results, from the perspective of six stakeholder group preferences. The proposed framework may serve as a baseline for decision makers/stakeholders to endorse strategies for existing ships and newbuilds. It is remarkable that the criteria of fuel cost, fuel availability and regulation gain a high priority for the majority of stakeholders. Fuel cost is an uncertain parameter, especially for fuel options that are currently under development, which may heavily influence the outcome.

Further research could focus on the introduction of the factor of carbon tax in the formulation of the fuel cost and the willingness of shipping stakeholders to pay their emissions and up to what amount. For instance, according to the report of IRENA and AEA [21], the cost gap between conventional ammonia and renewable ammonia could be bridged by a carbon tax up to USD 150/ton of CO_2. Furthermore, a thorough forecast of the price and availability of renewable fuels would contribute decisively to the results. Importance should also be given to the social criterion of job creation, as while the creation of new jobs is usually considered, a possible simultaneous reduction in existing jobs has not been examined yet. The feasibility of alternative fuel options is still a long way off, and further research and assessment are required, especially for deep-sea shipping. Although bio-fuels are primarily of interest, competition from the demand of other sectors will influence their applicability in the maritime sector. All the renewable fuel options require support and initiatives for long-term use throughout their supply chain, from the production phase to the selected propulsion system.

Author Contributions: E.S. is the main researcher who studied, implemented and analyzed the presented research and wrote the paper; G.A.L. contributed to alternatives configuration, data selection and the weighting criteria process. K.A. had a general overview of the research. All authors have read and agreed to the published version of the manuscript.

Funding: This research received no external funding.

Data Availability Statement: Data are contained within the article.

Conflicts of Interest: The authors declare no conflict of interest.

Nomenclature

HFO	Heavy Fuel Oil
MGO	Marine Gas Oil
LNG	Liquefied Natural Gas
IMO	International Maritime Organisation
GHG	Greenhouse Gases
PM	Particulate Matter
WTW	Well To Wake
EC	Europoean Commision
ETS	Emission Trading System
GT	Gross Tonnage
ICE	Internal Combustion Engine
FC	Fuel Cells
EEC	European Economic Community
EU ETD	European Energy Taxation Directive
SGMF	Society for Gas as a Marine Fuel
MCDM	Multicriteria decision making

ILUC	Indirect Land Use Change
HVO	Hydrotreated Vegetable Oil
HTL	Hydrothermal liquefaction
FAME	Fatty Acid Methyl Ester
LBG	Liquefied biogas
MeOH	Methanol
H_2	Hydrogen
CCS	Carbon Capture and Storage
AHP	Analytic hierarchy process
VIKOR	Multicriteria Optimization and Compromise Solution
SAW	Simple Additive Weighting
TOPSIS	Technique for Order of Preference by Similarity to Ideal Solution
PROMETHEE	Preference Ranking for Organization Method for Enrichment Evaluation
ELECTRE	ELimination and Choice Expressing REality
TRL	Technology Readiness Level

References

1. UNCTAD. *Review of Maritime Transport 2021*; United Nations Actions Conference on Trade and Development: Geneva, Switzerland, 2021; ISBN 978-92-1-113026-3.
2. Aakko-Saksa, P.T.; Lehtoranta, K.; Kuittinen, N.; Järvinen, A.; Jalkanen, J.P.; Johnson, K.; Jung, H.; Ntziachristos, L.; Gagne, S.; Takahashi, C.; et al. Reduction in greenhouse gas and other emissions from ship engines: Current trends and future options. *Prog. Energy Combust. Sci.* **2023**, *94*, 101055. [CrossRef]
3. IRENA. *A Pathway to Decarbonise the Shipping Sector by 2050*; International Renewable Energy Agency: Abu Dhabi, United Arab Emirates, 2021; ISBN 978-92-9260-330-4.
4. RESOLUTION MEPC.304(72). ANNEX 11. INITIAL IMO STRATEGY ON REDUCTION OF GHG EMISSIONS FROM SHIPS, IMO 2018. Available online: https://wwwcdn.imo.org/localresources/en/KnowledgeCentre/IndexofIMOResolutions/MEPCDocuments/MEPC.304(72).pdf (accessed on 18 October 2023).
5. RESOLUTION MEPC.377(80). ANNEX 1, IMO STRATEGY ON REDUCTION OF GHG EMISSIONS FROM SHIPS, IMO 2023. Available online: https://wwwcdn.imo.org/localresources/en/OurWork/Environment/Documents/annex/MEPC%2080/Annex%2015.pdf (accessed on 18 October 2023).
6. European Commission. Proposal for a REGULATION OF THE EUROPEAN PARLIAMENT AND OF THE COUNCIL on the Use of Renewable and Low-Carbon Fuels in Maritime Transport and Amending Directive 2009/16/EC, European Commission, COM (2021) 562 Final. 2021. Available online: https://eur-lex.europa.eu/legal-content/EN/TXT/?uri=CELEX%3A52021PC0562 (accessed on 18 October 2023).
7. European Commission. Proposal for a DIRECTIVE OF THE EUROPEAN PARLIAMENT AND OF THE COUNCIL Amending Directive 2003/87/EC Establishing a System for Greenhouse Gas Emission Allowance Trading within the Union, Decision (EU) 2015/1814 Concerning the Establishment and Operation of a Market Stability Reserve for the Union Greenhouse Gas Emission Trading Scheme and Regulation (EU) 2015/757, European Commission, COM (2021) 551 Final. 2021. Available online: https://eur-lex.europa.eu/legal-content/EN/TXT/?uri=CELEX%3A52021PC0551 (accessed on 18 October 2023).
8. European Commission. Proposal for a COUNCIL DIRECTIVE Restructuring the Union framework for the Taxation of Energy Products and Electricity (Recast), European Commission, COM (2021) 563 Final. 2021. Available online: https://eur-lex.europa.eu/legal-content/EN/TXT/?uri=CELEX%3A52021PC0563 (accessed on 18 October 2023).
9. Solakivi, T.; Paimander, A.; Ojala, L. Cost competitiveness of alternative maritime fuels in the new regulatory framework. *Transp. Res. Part D Transp. Environ.* **2022**, *113*, 103500. [CrossRef]
10. Ampah, J.D.; Yusuf, A.A.; Afrane, S.; Jin, C. Reviewing two decades of cleaner alternative marine fuels: Towards IMO's decarbonization of the maritime transport sector. *J. Clean. Prod.* **2021**, *320*, 128871. [CrossRef]
11. Deniz, C.; Zincir, B. Environmental and economical assessment of alternative marine fuels. *J. Clean. Prod.* **2016**, *113*, 438–449. [CrossRef]
12. Hansson, J.; Månsson, S.; Brynolf, S.; Grahn, M. Alternative marine fuels: Prospects based on multi-criteria decision analysis involving Swedish stakeholders. *Biomass Bioenergy* **2019**, *126*, 159–173. [CrossRef]
13. Hansson, J.; Brynolf, S.; Fridell, E.; Lehtveer, M. The Potential Role of Ammonia as Marine Fuel—Based on Energy Systems Modeling and Multi-Criteria Decision Analysis. *Sustainability* **2020**, *12*, 3265. [CrossRef]
14. Mandic, N.; Ukic Boljat, H.; Kekez, T.; Luttenberger, L.R. Multicriteria Analysis of Alternative Marine Fuels in Sustainable Coastal Marine Traffic. *Appl. Sci.* **2021**, *11*, 2600. [CrossRef]
15. Yang, Z.; Ahmad, S.; Bernardi, A.; Shang, W.; Xuan, J.; Xu, B. Evaluating alternative low carbon fuel technologies using a stakeholder participation-based q-rung orthopair linguistic multi-criteria framework. *Appl. Energy* **2023**, *332*, 120492. [CrossRef]
16. Green Voyage for 2050. Alternative Fuels and Energy Carriers for Maritime Shipping. IMO Workshop March 2021. Available online: https://greenvoyage2050.imo.org/workshop-packages/ (accessed on 18 October 2023).

17. IRENA and Methanol Institute. *Innovation Outlook: Renewable Methanol*; International Renewable Energy Agency and Methanol Institute: Abu Dhabi, United Arab Emirates, 2021; ISBN 978-92-9260-320-5.
18. Hurtig, O.; Buffi, M.; Scarlat, N.; Motola, V.; Georgakaki, A.; Letout, S.; Mountraki, A.; Joanny, G. *Clean Energy Technology Observatory: Advanced Biofuels in the European Union—2022 Status Report on Technology Development, Trends, Value Chains and Markets*; Publications Office of the European Union: Luxembourg, 2022. [CrossRef]
19. IRENA. *Global Hydrogen Trade to Meet the 1.5 °C Climate Goal: Part II—Technology Review of Hydrogen Carriers*; International Renewable Energy Agency: Abu Dhabi, United Arab Emirates, 2022; ISBN 978-92-9260-431-8.
20. Ashrafi, M.; Lister, J.; Gillen, D. Toward a harmonization of sustainability criteria for alternative marine fuels. *Marit. Transp. Res.* **2022**, *3*, 100052. [CrossRef]
21. IRENA; AEA. *Innovation Outlook: Renewable Ammonia*; International Renewable Energy Agency: Abu Dhabi, United Arab Emirates; Ammonia Energy Association: Brooklyn, NY, USA, 2022; ISBN 978-92-9260-423-3.
22. Ren, J.; Lützen, M. Selection of sustainable alternative energy source for shipping: Multicriteria decision making under incomplete information. *Renew. Sustain. Energy Rev.* **2017**, *74*, 1003–1019. [CrossRef]
23. Ren, J.; Lützen, M. Fuzzy multi-criteria decision-making method for technology selection for emissions reduction from shipping under uncertainties. *Transp. Res. Part D* **2015**, *40*, 43–60. [CrossRef]
24. Ren, J.; Liang, H. Measuring the sustainability of marine fuels: A fuzzy group multi-criteria decision making approach. *Transp. Res. Part D* **2017**, *54*, 12–29. [CrossRef]
25. Kim, A.R.; Seo, Y.-J. The reduction of SOx emissions in the shipping industry: The case of Korean companies. *Mar. Policy* **2019**, *100*, 98–106. [CrossRef]
26. Luciana (Marcu), T.A.; Gasparotti, C.; Rusu, E. Green fuels—A new challenge for marine industry. *Energy Rep.* **2021**, *7*, 127–132. [CrossRef]
27. Carvalho, F.; Müller-Casseres, E.; Poggio, M.; Nogueira, T.; Fonte, C.; Wei, H.K.; Portugal-Pereira, J.; Rochedo, P.R.R.; Szklo, A.; Schaeffer, R. Prospects for carbon-neutral maritime fuels production in Brazil. *J. Clean. Prod.* **2021**, *326*, 129385. [CrossRef]
28. Moshiul, A.M.; Mohammad, R.; Hira, F.A. Alternative Fuel Selection Framework toward Decarbonizing Maritime Deep-Sea Shipping. *Sustainability* **2023**, *15*, 5571. [CrossRef]
29. Strantzali, E.; Aravossis, K. Decision making in renewable energy investments: A review. *Renew. Sustain. Energy Rev.* **2016**, *55*, 885–898. [CrossRef]
30. Strantzali, E.; Aravossis, K.; Livanos, G.A.; Nikoloudis, C. A decision support approach for evaluating liquefied natural gas supply options: Implementation on Greek case study. *J. Clean. Prod.* **2019**, *222*, 414–423. [CrossRef]
31. Strantzali, E.; Aravossis, K.; Livanos, G.A. Evaluation of future sustainable electricity generation alternatives: The case of a Greek island. *Renew. Sustain. Energy Rev.* **2017**, *76*, 775–787. [CrossRef]
32. Papathanasiou, J.; Ploskas, N. *Multiple Criteria Decision Aid*; Springer: Berlin/Heidelberg, Germany, 2018; Volume 136. [CrossRef]
33. DNV. Maritime Forecast for 2050. Energy Transition Outlook 2022. Available online: https://www.dnv.com/maritime/publications/maritime-forecast-2023/download-the-report.html?utm_source=google&utm_medium=paid-search&utm_campaign=23Q3_MaritimeForecast_Search_BroadTarget&gclid=EAIaIQobChMI3PzIkIv_gQMVspGDBx3FAw3zEAAYASAAEgI95fD_BwE (accessed on 18 October 2023).
34. IEA. Indicative Shipping Fuel Cost Ranges. Available online: https://www.iea.org/data-and-statistics/charts/indicative-shipping-fuel-cost-ranges (accessed on 18 October 2023).
35. Lloyd's Register; University Maritime Services (UMAS). Techno-Economic Assessment of Zero-Carbon Fuels. March 2020. Available online: https://www.lr.org/en/knowledge/research-reports/techno-economic-assessment-of-zero-carbon-fuels/ (accessed on 18 October 2023).
36. Veracity by DNV; Alternative Fuels Insight Platform. Available online: https://afi.dnv.com/ (accessed on 18 October 2023).
37. DNV GL. Comparison of Alternative Marine Fuels. Report No.: 2019-0567, Rev. 3; Document No.: 11C8I1KZ-1; 2019-07-05. Available online: https://sea-lng.org/wp-content/uploads/2020/04/Alternative-Marine-Fuels-Study_final_report_25.09.19.pdf (accessed on 18 October 2023).
38. ABS. Hydrogen as Marine Fuel. Sustainability Whitepaper. June 2021. Available online: https://maritimecyprus.com/wp-content/uploads/2021/06/ABS-hydrogen-as-marine-fuel.pdf (accessed on 18 October 2023).
39. Together in Safety 2022. Future Fuels Risk Assessment. Available online: https://togetherinsafety.info/wp-content/uploads/2022/06/Future-Fuels-Report.pdf (accessed on 18 October 2023).
40. Law, L.C.; Foscoli, B.; Mastorakos, E.; Evans, S. A Comparison of Alternative Fuels for Shipping in Terms of Lifecycle Energy and Cost. *Energies* **2021**, *14*, 8502. [CrossRef]
41. Snyder, J. How Shipping Can Blend Its Way to Decarbonisation with bioLNG, Article in Press, 2022. Available online: https://www.rivieramm.com/news-content-hub/news-content-hub/how-shipping-can-blend-its-way-to-decarbonisation-with-biolng-73927 (accessed on 29 September 2023).

Article

Environmental-Economic Analysis for Decarbonising Ferries Fleets

Gerasimos Theotokatos *, Panagiotis Karvounis and Georgia Polychronidi

Maritime Safety Research Centre, Department of Naval Architecture, Ocean and Marine Engineering, University of Strathclyde, Glasgow G4 0LZ, UK
* Correspondence: gerasimos.theotokatos@strath.ac.uk

Abstract: Several countries heavily depend on their domestic ferries, the decarbonisation of which are required following the prevailing and forthcoming international and national carbon reduction targets. This study aims to conduct an environmental-economic analysis to identify the impact of three decarbonisation measures, specifically, hybridisation, liquified natural gas (LNG) and methanol use, for two ferries of different size of a developing country fleet. The study is based on several methodological steps including the selection of key performance indicators (KPIs), the pre-processing of acquired data to identify representative operating profiles, the environmental and economic KPIs calculation, as well as the comparative appraisal of the investigated measures. The required investments for decarbonising the whole domestic fleet of a case country are subsequently estimated and discussed. All the three investigated measures have the potential to reduce CO_2 emissions, however, not beyond the IMO 2030 carbon emissions reduction target. This study provides insights to the involved stakeholders for supporting their decisions pertinent to the domestic ferries sector decarbonisation.

Keywords: decarbonisation; methanol; LNG; environmental-economic analysis; marine engines

Citation: Theotokatos, G.; Karvounis, P.; Polychronidi, G. Environmental-Economic Analysis for Decarbonising Ferries Fleets. *Energies* **2023**, *16*, 7466. https://doi.org/10.3390/en16227466

Academic Editors: Konstantinos Aravossis and Eleni Strantzali

Received: 16 September 2023
Revised: 30 October 2023
Accepted: 31 October 2023
Published: 7 November 2023

1. Introduction

Several developing countries strongly depend on maritime transportation for their inter-island connectivity. Domestic ferries play a crucial role in their economic and social development by transporting goods and people between mainland and islands as well as interconnecting islands. However, the operation of these ferries is associated with significant environmental and economic costs, primarily due to their reliance on fossil fuels [1].

Decarbonising the domestic ferry sectors is a crucial step towards achieving these countries' climate and sustainability goals. The International Maritime Organisation (IMO) lists 176 countries as member states and 3 associate members, which have committed to reduce the carbon dioxide (CO_2) emissions per transport work by 40% by 2030, and reach net-zero emissions by 2050, following the Paris Agreement [2]. The domestic ferry sector significantly contributes to the transport related GHG emissions of several countries, hence rendering its decarbonisation efforts of high priority [3]. Worldwide, the shipping industry has been adopting innovative measures to reduce its environmental impact, particularly through decarbonisation practices [4]. The shipping industry is critical for the global trade and commerce, responsible for transporting approximately 80% of the world's goods by volume [5]. However, this industry's growth has also led to increased carbon emissions, thus exhibiting significant environmental impact, including climate change, air pollution, and ocean acidification. To address these issues, several measures have been proposed to promote sustainable practices in the shipping industry. One such measure is the adoption of alternative fuels, such as liquefied natural gas (LNG), methanol, ammonia, or hydrogen, which result in lower emissions compared to traditional marine fuels including heavy fuel oil (HFO) and marine gas oil (MGO) [6].

Hansson et al. [7] studied ammonia as a potential marine fuel demonstrating that the major challenge for its adoption is the higher price per energy content compared to MGO and LNG. Jovanović et al. [8] studied the feasibility of autonomous ships operating with methanol and LNG along with conventional fuels from an environmental perspective, whilst considering the possible emissions effects on global warming, concluding that methanol has significant advantage compared to LNG and MGO. Hovarth et al. [9] demonstrated that renewable based synthetic fuels, such as methanol, are not economically feasible for decarbonising the shipping sector, without the application of emission taxation schemes. The latter is supported by the findings of Trivyza et al. [10] pertinent to the impact of carbon pricing on cruise ships energy systems. Svanberg et al. [11] argued that renewable methanol is a technically viable option to reduce emissions from shipping as it does not introduce major challenges on the fuel supply chains. Korberg et al. [12] studied alternative propulsion systems along with alternative fuels for ferries operation concluding that large ferries can be cost effective with fuels produced by using renewable energy.

Several alternative low and zero-carbon fuels have been proposed for the shipping sector. The use of ammonia, hydrogen, methanol, and biofuels can lead to lower operational carbon footprint, and may be considered carbon neutral when renewable energy is used for their production. Karvounis et al. [13] reported that fossil-based production of hydrogen and ammonia yields significantly higher CO_{2eq} emissions compared to conventional MGO and LNG fuels (as detailed in Table A1). This is attributed to the energy intensive processes required for these fuels production [14,15]. Bio methanol exhibits around 15% less $CO_{2,eq}$ associated with lower fuel production cost; however, its wide adoption is limited by the production location and scalability [16]. Natural gas extraction and processing is accompanied by methane slip and exhibits 25% higher CO_{2eq} emissions compared to MGO [17]. Methanol can be stored under ambient temperature and pressure, and requires less energy compared to LNG and hydrogen, which are stored at cryogenic conditions [18].

Electrification using batteries is accepted as a potential technology for shipping decarbonisation. Hybrid ship power systems integrating both conventional (mechanical) and electrical components (batteries, electric machinery, converters/inverters) can increase the power plant efficiency, reducing the fuel consumption especially in cases with dynamic operations [19]. Previous studies focusing on hybrid power plants for several ship types and employing different battery sizes reported fuel savings in the range 8–17% [20,21]. Law et al. [22] examined several alternative strategies to decarbonise the shipping operations concluding that carbon capture and storage is the most cost-effective pathway, however, no carbon taxation was considered whilst scaling up to fleet was not presented. Percic et al. [23] considered the lifetime emissions and cost of hybrid inland waterway ships, concluding that electrification can reduce both GHG and NOx emissions; however alternative fuels were not investigated. Jang et al. [24] demonstrated that the use of LNG and fuel cells power systems exhibits lower environmental footprint compared to dual fuel gas engines. Kistner et al. [25] argued that the implementation of alternative fuels and fuel cell technologies require extensive investment cost, which cannot be afforded by developing nations' stakeholders. The use of methanol and electrification were identified as potential solutions for short-term decarbonisation of the shipping sector [26], whilst LNG is already employed as low carbon fuel [13,26].

The aim of this study is to conduct an environmental-economic analysis of decarbonising a fleet of domestic ferries, evaluating the costs and benefits of transitioning the sector to low-emission alternatives. This is achieved by: (i) evaluating the environmental and economic indicators of three short- to medium-term solutions with the use of alternative fuels and hybrid power systems for two typical domestic ferries operating in developing countries, considering their entire lifetime; (ii) assessing the investment costs required for the wide implementation of these technologies whilst monetising the carbon emissions considering a reference fleet; (iii) discussing pathways for policymakers and industry stakeholders to facilitate the decarbonisation of the reference domestic ferry fleet.

This study novelty stems from the investigated case study that includes two typical ferries representing the domestic ferries fleet in a developing country as well as the results extrapolations to the whole fleet. The carbon tax as a policy measure is assessed, comparing with the required investment cost. This study provides valuable insights for policymakers and industry stakeholders on the policy and regulatory actions needed to facilitate the decarbonisation of the domestic ferry sector in the short- to long-term.

2. Materials and Methods

The followed methodology consists of five steps as presented in the flowchart shown in Figure 1. Step 1 involves the selection of the key performance indicators (KPIs) for three categories (technical, environmental, and financial). These KPIs focus on representing the potential technical requirements, such as storage volume or battery weight/volume, as well as to determine the environmental impact and associated costs. An existing lifetime economic-environmental model (LTEEM) is customised to facilitate the calculation of the determined KPIs. Step 2 focuses on the data collection for the selected case ships as well as their pre-processing to estimate the model input parameters, which include the case ships particulars, operating profiles, and fuel consumption datasets. Step 3 investigates four case studies (baseline, hybrid power system, LNG use, methanol use). Step 4 involves the assessment of the environmental, financial, and technical KPIs. Finally, step 5 entails the discussion of this study results facilitating the appraisal of the considered cases feasibility. The presented KPIs did not consider the cost of production and transportation of LNG and methanol fuels whereas, the transport (by ship) costs amount 0.74–1.29 EUR/GJ for LNG and 1.8 EUR/MWh for methanol. However, it is anticipated that those costs are embedded in the fuel price. These factors can be considered in future studies that examine the well-to-wake cost [27,28] as presented in Table A2 of the Appendix A.

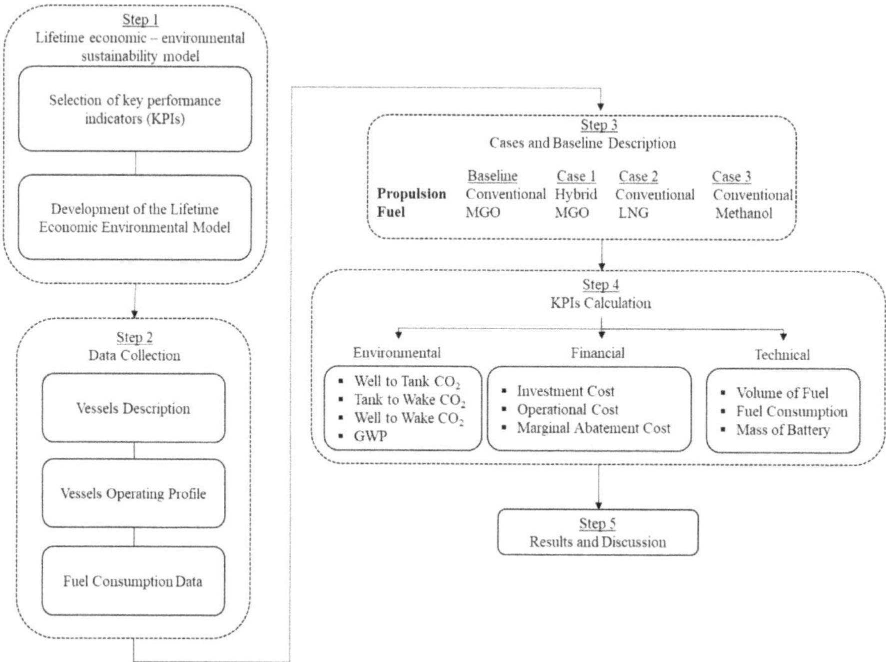

Figure 1. Methodology flowchart.

2.1. Key Performance Indicators

This study employs key performance indicators (KPIs) that are classified in the following groups: environmental, financial, and technical. The environmental KPIs include the CO_2 emissions considering the annual and each voyage timelines, as well as the global warming potential (GWP) that characterises the environmental impact of the considered cases. The CO_2 emissions are considered in a well to tank and tank to wake basis. The financial KPIs include the investment cost (characterising the required capital), the operating expenditure (characterising the operational expenses), and the marginal abatement cost (MAC) that denotes the effectiveness of the emission abatement measures. The technical KPIs include the annual fuel consumption (*FC*), and the fuel required volume, as well as the batteries systems volume and weight, which are required to assess the technical requirements for the investigated cases. The financial KPIs facilitate the appraisal of the potential investment that is essential to accommodate the lower environmental impact power plants.

2.2. Lifetime Economic-Environmental Model

The lifetime economic-environmental model employed in this study is based on Ref. [13]. The model assesses different environmental and economic parameters based on operating profile, employing the typical voyage(s) energy analysis. Since the income streams pertinent to the vessels economic activity are considered the same to the reference ships (with the conventional power plants), they are not used herein. The vessels under consideration can accommodate the alternative fuels storage tanks at free spaces onboard and hence no loss of capital is considered.

The voyage energy analysis is based on the annual fuel consumption, derived from the vessel operating profile, which are estimated based on data received from the ship operators. The determination of the energy required for each voyage is derived by the fuel consumption for each fuel examined by the following equation:

$$E_{trip} = \sum_{f} LHV_f \, FC_i \tag{1}$$

where *LHV* refers to each fuel lower heating value.

The required storage volume for a single voyage is calculated using a storage safety factor (c in Equation (2)) of 20% accounting for the non-used part of the tanks, according to the following equation:

$$V_f = \frac{FC_i}{\rho_f}(1+c) \tag{2}$$

where ρ refers to each fuel density.

The investment cost (*CAPEX*) and annual operational expenditure (*OPEX*) are calculated according to the following equations:

$$CAPEX = P_{ME}\, C_E + AT + C_B \tag{3}$$

$$OPEX = AC_{f_i} + AC_{OM} + AC_O \tag{4}$$

where P_{ME} is the nominal power of the ship main engine; C_E, is the engine cost factor (in EUR/kW); *AT* refers to the NOx after-treatment system cost that is essential equipment for all the examined fuels; AC_f is the annual fuel(s) cost; AC_{OM} denotes the maintenance cost factor (EUR/kWh); AC_O refers to any other annual cost considered, for example, carbon taxation; C_B denotes the cost of batteries and requires systems of the hybrid plant (electric machinery, power electronics, DC/AC converters).

The marginal emission abatement cost that characterises the relative investment needed per abated emissions mass is calculated according to the following equations:

$$MAC_{CAPEX} = \frac{\Delta CAPEX}{\Delta CO_{2i}} \tag{5}$$

$$MAC_{OPEX} = \frac{\Delta OPEX}{\Delta CO_{2i}} \tag{6}$$

where i denotes the case study number, and ΔCO_{2i} denotes the difference of the CO_2 emissions from the baseline case study.

The well to tank and tank to wake carbon emissions are calculated as:

$$EM_{CO_2,i} = M_{CO_2,i} EF_{CO_2,i} \tag{7}$$

where M_{CO_2} refers to the mass of CO_2 and EF_{CO_2} to the CO_2 emission factor, whilst the subscript i corresponds to well to tank or tank to wake emissions.

The global warming potential corresponding to 100 years is calculated by the following equation:

$$GWP_{100y} = M_{CO_2} + 36\, M_{CH_4} + 298\, M_{N_2O} \tag{8}$$

3. Case Studies Description

This study investigates two typical RO-PAX ferries of different sizes, representing the fleet of a developing country. The key characteristics of these ferries (termed Vessel 1 and Vessel 2, henceforth) are listed in Table 1. Vessel 1 length is 97.8 m, whilst Vessel 2 has a length of 50 m. Vessel 1 typical voyage is around 27,000 nm, completing three voyages per week, whereas Vessel 2 typical return voyage is 110 nm, running two voyages per day. The investigated ships main particulars for each propulsion engines of Vessels 1 and 2 are listed in Table 2. The rated power of each generator set installed in Vessels 1 and 2 are 350 kWe and 160 kWe, respectively.

Table 1. Characteristics of the case vessels.

Parameter	Vessel-1	Vessel-2
Type	Ro-pax	Ro-pax
Length/breadth/draught [m]	97.8	50
Typical voyage distance [nm]	27,025	110
GT [t]	5145	2682

Table 2. Main engine characteristics.

Component	Vessel-1	Vessel-2
Type	four-stroke	four-stroke
Fuel	MGO	MGO
Rated Power [kW]	2360	1370
Rated Speed [rpm]	750	850
Cylinders	12	12

Four case studies are investigated for both vessels (1 and 2) as follows. The baseline case study (BL) includes the power plant of the existing ships, which include two main engines (each one drives a propeller via a gear box) and three auxiliary generator sets. Both the ship main engines (ME) and auxiliary engines (AE) use marine gas oil (MGO). Case study.

C1 employs a hybrid propulsion system with installed (retrofitted) batteries to generate electric power partially covering the vessels auxiliary and propulsion power demand. Case study C2 considers the BL layout with the LNG use. The MEs and AEs are converted to dual fuel engines operating with natural gas (90% energy fraction) and pilot diesel (10% energy fraction). Case study C3 considers the BL layout with the use of methanol fuel. The MEs and AEs are converted to dual fuel engines operating with methanol as main fuel (90% energy fraction) and diesel pilot fuel (10% energy fraction). The simplified layouts of the investigated case studies are presented in Figure 2, whereas their main characteristics are reported in Table 3.

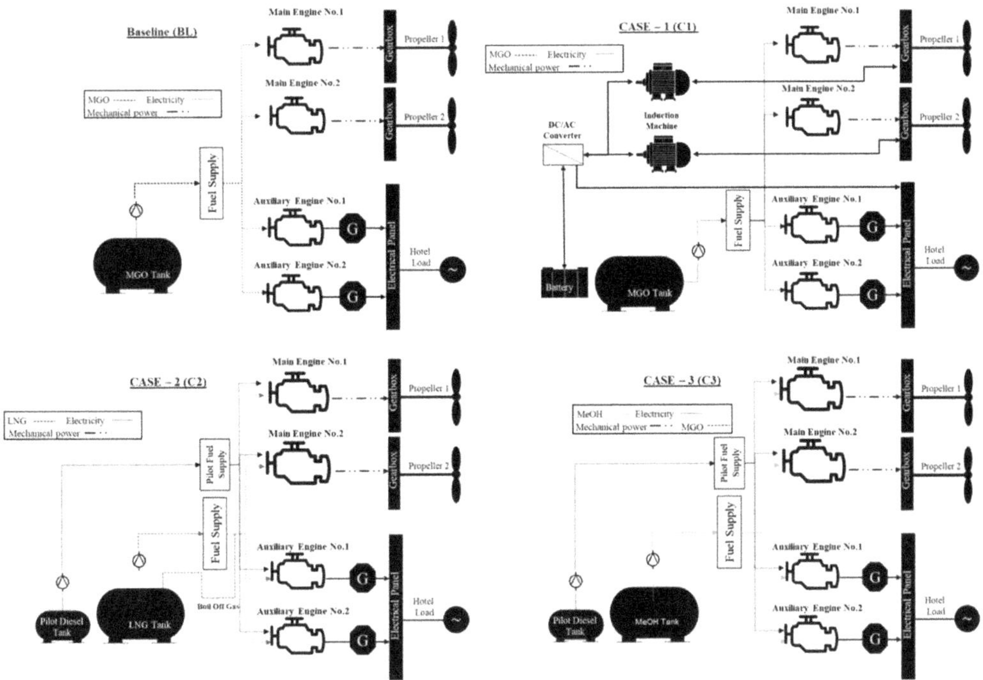

Figure 2. Power plant layouts considered for the four case studies.

Table 3. Main characteristics of the cases studies for each vessel.

Case	Fuels	Main Units	Subsystems
Baseline (BL)	MGO	2 Main diesel engines 2 Auxiliary generator sets	–
CASE—1 (C1)	MGO	2 Main diesel engines 2 Auxiliary generator sets 1 Batteries pack 1 Electric motor/generator	NOx after-treatment unit
CASE—2 (C2)	LNG Pilot diesel	2 Main dual fuel engines 2 Auxiliary dual fuel generator sets	NOx after-treatment unit
CASE—3 (C3)	Methanol Pilot diesel	2 Main dual fuel engines 2 Auxiliary dual fuel generator sets	NOx after-treatment unit

3.1. Input Parameters

For case study C1 (hybrid power system use), the energy storage system consists of a 420 kWh Li-ion battery for Vessel 1 and a 225 kWh Li-ion battery for Vessel 2. These ships power plants include an electric shaft generator, which can be powered by either the battery or by charging the battery through the ship's main engine. The battery sizes were selected by considering batteries capacity of 0.23 kWh per kW of installed power as reported in [26]. According to the same study, hybrid propulsion systems yield an average fuel saving of around 11% with a standard deviation of 3%. In addition to the battery and propulsion system, other components considered in C1 are the DC/AC converter and an electric machine (motor/generator) coupled with the propulsion system gearbox.

Table 4 lists the model input parameters, which include the fuels prices, the emission factors, as well as the cost factors of the marine engines and machinery systems. The emission factor for NG methane slip was adapted from Balcombe et al. [29]. It is worth mentioning that significant progress has been made in recent years to reduce methane slip, with reductions of up to 50% achieved in low-pressure two-stroke gas engines [30]. The cost factors for LNG storage refer to C-type tanks, which are typically employed in maritime applications [31].

Table 4. Model input parameters; adapted from Refs. [8,32–36].

Parameter		Value
Marine Methanol engine cost factor	EUR/kW	780 [1]
Marine LNG engine cost factor [1]	EUR/kW	554
Marine Diesel engine cost factor	EUR/kW	493
Maintenance cost factor	EUR/kWh	0.012
After-treatment unit cost factor	EUR/kW	40
Battery cost factor	EUR/kWh	800
Methanol fuel supply system	M EUR	1.2
MGO CO_2 EF [2]	kg CO_2/kg fuel	3.02
NG CO_2 EF	kg CO_2/kg fuel	2.75
Methanol CO_2 EF	kg CO_2/kg fuel	1.37
MGO CH_4 EF	kg CH_4/kg fuel	0.006
NG CH_4 EF	kg CH_4/kg fuel	0.041
Methanol CH_4 EF	kg CH_4/kg fuel	0
MGO N_2O EF [3]	kg N_2O /kg fuel	1.4×10^{-4}
NG N_2O EF	kg N_2O /kg fuel	0.71×10^{-4}
Methanol N_2O EF	kg N_2O /kg fuel	0.71×10^{-4}
MGO Price [5]	EUR /t	674
LNG Price [4]	EUR /t	1400
Methanol Price [4]	EUR /t	1000
Methanol storage cost	EUR /m^3	3000
LNG storage cost	EUR /m^3	2000

[1] Four stroke gas engine is considered, [2] Provided by industrial sources, [3] Uncertainty regarding the N_2O emission factors is noted, [4] Fuel costs refer to conventional fuel production methods. [5] year average as of 2023 is used for the fuel price of MGO according to [37].

The main properties of the MGO, LNG and methanol fuels are summarised in Table 5. Due to its lower energy content compared to MGO fuel, methanol requires a larger amount of fuel storage to meet the same energy demand. Specifically, the energy content of methanol is less than half of that of MGO fuel [38]. However, LNG would as well require higher storage volume comparing to MGO due to its lower density [39]. The efficiency of the case ships engines when operating with LNG and methanol, is assumed same with the diesel mode, as supported by the data provided in [40].

Table 5. Fuel properties, adapted from [26,41].

Property	MGO	LNG	Methanol
LHV [MJ/kg]	42.7	48.6	20.1
Fuel Density [kg/m^3]	838	428	791
Volumetric Energy Density [MJ/L]	34	22	16
Gross Storage System Size Factor	×1	×2.4	×1.7

The considered ferries fleet characteristics are presented in Table 6. The total gross tonnage of the fleet is 981,500 GT. The examined vessels belong to the category of above

400 GT. These ships can accommodate the batteries and alternative fuels storage tanks at free spaces without loss of payload; hence, no loss of capital is considered.

Table 6. Ferries fleet characteristics.

GT	Number of Vessels
0–100	67
100–399	135
Above 400	160

3.2. Emissions Taxation

Emissions taxation is identified as a potential measure to incentivise the ferries fleet decarbonisation. According to the World Energy Outlook [42], the carbon emissions tax is estimated at 40–50 EUR/t and 100–110 EUR/t for the 2030 and 2040, respectively, for emerging markets and developing countries with net zero targets. Those values are also applied for the energy production sector. Hence, it is assumed that similar values are expected for the shipping industry for the considered developing country.

4. Results

At this section, the derived results are presented and discussed. The following subsections provide the environmental, the financial, and technical KPIs.

4.1. Environmental KPIs

Figure 3 provides the well to tank, tank to wake and their total, annual CO_2 emissions for the four investigated case studies (BL, C1, C2 and C3), for the two vessels. In the case of methanol, fossil (C3-F) and renewable (C3-R) production methods are considered. The former (C3-F) includes the methanol production from natural gas by employing the following processes: steam reforming to produce syngas, methanol synthesis reaction, and methanol purification. The latter (C3-R) considers the use of biomass feedstock and gasification process to produce methanol, whereas the electric energy demand is covered by renewable energy sources. The horizontal lines correspond to 40% Well to Wake CO_2 emissions reduction (compared to the baseline), which aligned with the IMO 2030 targets. For the tank to wake, the presented results demonstrate that the CO_2 emissions can reduce by about 11%, 33% and 8% for the case studies C1, C2 and C3 respectively compared to BL. The methanol use (C3) results in the lowest CO_2 emissions reduction (8%), which is attributed to the methanol lower heating value ratio (compared to the LNG and MGO), leading to higher methanol consumption. However, it is inferred that the three alternative case studies (C1, C2, and C3) cannot achieve the IMO 2030 targets.

Given the well to tank CO_2 emissions for the four cases calculated using the values for the well to tank CO_2 emissions factors listed in Table A1. For BL, the well to tank CO_2 emissions are 864 t CO_2 and 452 t CO_2 for vessels 1 and 2, respectively. Batteries production even when using 15–20% renewable energy mix exhibits significantly lower emission factors [43] Hence, case C1 exhibits better environmental performance (considering the well to tank phase) compared to the other cases. For LNG (case C2), higher well to tank emissions (compared to BL) were estimated, specifically 1161 t CO_2 and 608 t CO_2 for the selected vessels. This is attributed to the increased CO_{2eq} emission factor for the methane slip associated to natural gas extraction. Methanol production using energy from fossil fuels (C3-F) is associated with lower emission factors compared to LNG, and slightly higher compared to MGO. However, the increased methanol consumption yields similar well to tank emissions to the BL case (834 t CO_2 and 437 t CO_2 for vessels 1 and 2, respectively). For methanol produced from biomass feedstock using renewable energy (C3-R), which exhibits potential in developing countries, the well to tank emissions can considerably reduce (709 t CO_2 and 371 t CO_2 for vessels 1 and 2, respectively). The well to tank and corresponds to 26%, 27%, 45%, 27% and 23% of the tank to wake emissions for cases BL,

C1, C2, C3-fossil, and C3-renewables, respectively. Cases C1 and C2 exhibit almost similar well to wake CO_2 emissions (lower by 11% and 10% respectively compared to BL), whereas case 3 exhibits well to wake CO_2 emissions 7% (for fossil based production) and 9% (for biomass based production) lower that the BL and 5% higher than C1.

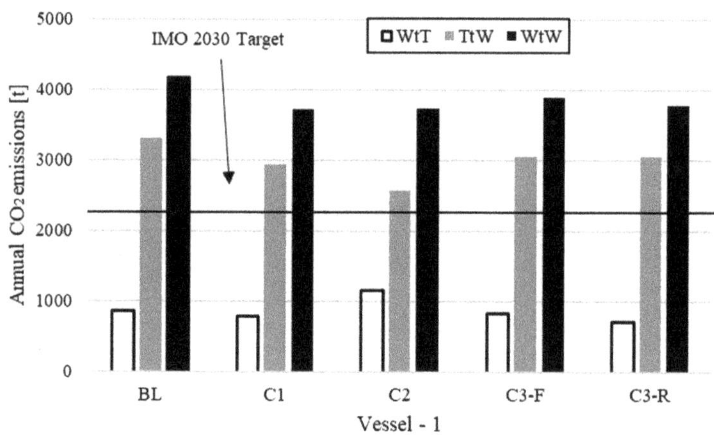

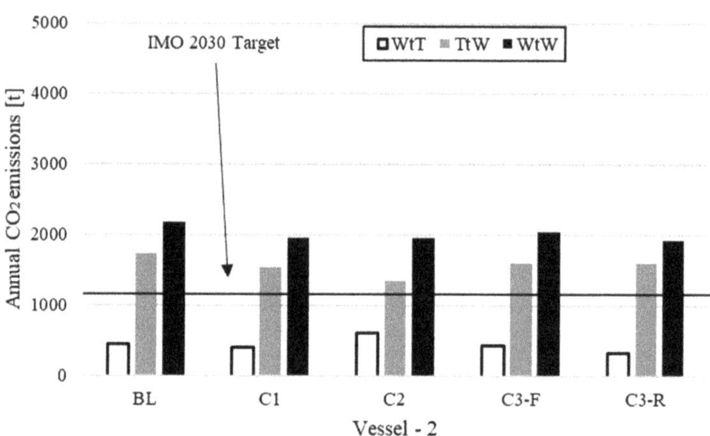

Figure 3. Well to tank, tank to wake and total CO_2 emissions, for vessel-1 (**top**) and vessel-2 (**bottom**) and the considered cases.

Figure 4 illustrates the global warming potential (*GWP*) in CO_2-equivalent emissions of the investigated case studies during the vessels' lifetime. It must be noted that a lifecycle approach considering the fuel production and ship building phases would be more inclusive, hence it is proposed for future studies. However, the lifetime *GWP* is an indicator for the investigated vessels environmental footprint. Case study C3 (methanol use) provides the lowest *GWP*, approximately 22% lower than that of BL, which is attributed to the almost zero N_2O and CH_4 emissions. Case study 2 (LNG use) exhibits 8% higher *GWP* compared to the baseline (BL), due to the significant contribution of N_2O and CH_4 emissions. However, recent advancements in marine gas and dual fuel engines technology have effectively mitigated the methane slip [44,45]. Case study C1 (hybrid system) is also associated with slightly reduced *GWP*, due the lower fuel consumption and corresponding reduction of the CO_2, N_2O and CH_4 emissions.

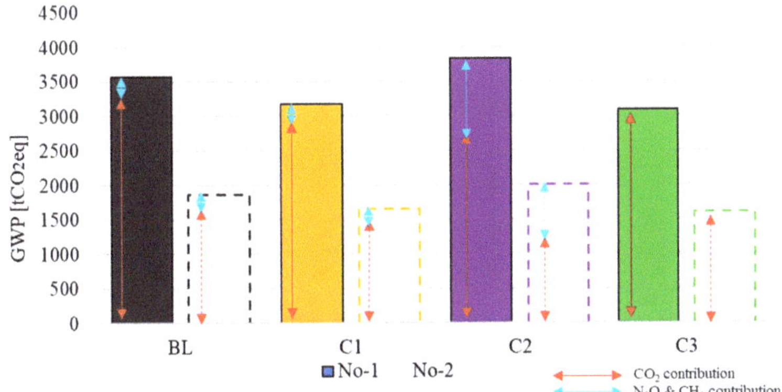

Figure 4. Global warming potential for the operational phase of the examined vessels and different cases (solid lines denote Vessel 1; dashed lines denote Vessel 2).

4.2. Financial KPIs

Figure 5a provides the annual operating expenditure for vessels 1 and 2 (large and small). It is evident that in all case studies the fuel cost amounts more than 95% of the operating costs. For Vessel 1, case studies C2 (LNG use) and C3 (methanol use) correspond to increases of the annual operational expenditure by M EUR 0.52 (42%) and M EUR 1.37 (66%) respectively compared to BL. For Vessel 2, case studies C2 (LNG use) and C3 (methanol use) correspond to increases of the annual operational expenditure by M EUR 0.27 (41%) and M EUR 0.72 (65%) respectively compared to BL. On the contrary, case study C1 (hybrid power plant) reduces the annual operational expenditure by M EUR 0.03 (-4%) and M EUR 0.02 (-4%) for the large and small vessels respectively compared to BL, which is attributed to the considerable fuel savings. Figure 5b provides the investment costs for the four case studies. For the large vessel and cases C1, C2, and C3, the required additional investment costs (compared to the BL investment) amount of M EUR 0.42 (30%), M EUR 0.78 (45%) and M EUR 1.1 (53%) respectively. For the small vessel, the extra investment costs (compared to the BL investment) were found M EUR0.23 (30%), M EUR 0.25 (33%), and M EUR 0.43 (45%) for C1, C2, and C3, respectively. The required investment is greater for the alternative fuel technologies, attributed to the cost required for the retrofitted solutions, storage and feeding systems, safety systems and equipment (Figure 5b). Particularly for methanol use, the higher investment cost is attributed to the considerably higher cost of methanol fuelled marine engines as also indicated by the respective cost factors listed in Table 4.

Table 7 provides the marginal CO_2 emissions abatement costs (MAC) for case studies C1, C2, and to C3 (compared to the BL) considering the required investment cost (MAC_{CAPEX}) and operating cost (MAC_{OPEX}). Considering the investment cost, lower MAC_{CAPEX} denotes more significant contribution of each monetary unit spent for decarbonisation. Hence, for the three case studies, the most significant environmental value for money is attributed to C2 (LNG use), as the CO_2 emissions reduction is higher compared to other case studies. Regarding the carbon benefit based on the operating costs (MAC_{OPEX}), the negative sign of the C1 case denotes that there exist financial benefits along with the carbon emissions reduction, attributed to the fuel consumption reduction, rendering C1 financially most attractive than the others. The overall marginal abatement cost for Vessel-1 and Vessel-2 is calculated as 0.49 M EUR/t CO_2 and 0.84 M EUR/t CO_2 for C1, 6.79 M EUR/t CO_2 and 3.65 M EUR/t CO_2 for C2 and 50.08 M EUR/t CO_2 and 27.19 M EUR/t CO_2 for C3.

Based on these values and the gross tonnage of each ferry, the ratio of cost difference to the GT is calculated and presented in Table 9. The cost difference between the BL case and the cases C1, C2 are used for the examined vessels. The results for case study C1 (hybrid power plant) are employed to identify the trade-off in the whole GT range of the considered fleet (from 100 t to 30,000 t). To address the uncertainty due to the limited number of the investigated ferries (only 2), three trendlines types are considered, namely, linear, exponential, and power.

Table 9. Cost difference to gross tonnage ratio for the two investigated ships considering the hybrid power plan (case study C1) and the LNG use (case study C2).

	Ro-Pax Ferry		**C1**	**C2**	**C1**	**C2**
	Length [m]	**GT**	**ΔCost [M EUR]**	**ΔCost [M EUR]**	$\frac{\Delta Cost}{GT}\left[\frac{EUR}{GT}\right]$	$\frac{\Delta Cost}{GT}\left[\frac{EUR}{GT}\right]$
Vessel 2	50	2682	0.23	0.35	85.76	130.5
Vessel 1	100	5145	0.42	0.96	81.63	186.5

For the LNG use (case study C2), the pertinent investments cost exhibits greater uncertainty. Therefore, the average of the calculated values for the two ferries are employed to subsequently estimate the investment cost for the fleet. The estimated investment costs for the considered fleet for the hybridisation (C1) and LNG use (C2) are presented in Table 10. The hybridisation of the Ro-Pax ferries fleet is estimated to M EUR 28.1, whereas the LNG use in this fleet is more expensive with the pertinent estimated cost amounting at around M EUR 58.

Table 10. Total investment cost for the hybridisation and the use of LNG fuel for the of Ro-Pax ferries fleet.

C1—Ro-Pax hybrid powerplant				
Total IC M EUR	Linear 26.5	Power 30.6	Exponential 27.3	Average 28.1
C2—Ro-Pax LNG fuel use				
Total IC M EUR				Average 58.04

To compare the estimated investment costs, the monetisation of the estimated annual carbon emissions for the considered ferries fleet (amounting to 981,500 GT as reported previously) was carried out by employing a carbon tax based on the ranges discussed in the previous sections. The estimated annual tax is reported in Table 11. Considering a carbon tax of EUR 50 per tonne of CO_2, the annual tax amounts to M EUR 49. It is apparent that the introduction of a carbon tax policy can be used for funding investments for the decarbonisation of the ferries fleet. However, vigilant, and well-planned strategies are needed to avoid disruptions in the ferries sector and ensure that the cost will not pass to the passengers by increasing the fares. To promote decarbonisation initiatives in developing countries, subsidisation or financial support must be sought by national or international authorities. It is recommended that the decarbonisation initiatives are combined with initiatives for the new designs to simultaneously address the safety and cost-effectiveness perspectives. However, this is recommended for future studies.

Table 11. Monetisation of the annual carbon emissions.

Carbon Tax [EUR/t]	**Annual Tax Revenues [M EUR]**
50	49.08
100	98.15

5. Discussion

This study aims to investigate near-term strategies for mitigating carbon emissions within ferry fleets operating in developing countries. It must be noted that the domestic ferries fleets of these countries subject to national regulations (and not IMOs regulations). This study findings based on the estimated lifetime (tank to wake) and lifecycle (well to wake) parameters are summarised as follows.

From a lifetime perspective, hybrid power plants with MGO as fuel, allow for 12% fuel consumption reduction and hence smaller carbon footprint associated with fuel savings.

For deeper decarbonisation, LNG allows for 11% fuel consumption and 23% tank to wake CO_2 emissions reduction. LNG, despite its advantages as a marine fuel, shows increased well-to-tank emissions due to methane slip associated with natural gas extraction. However, it seems to be one the most effective solutions considering the well to wake emissions.

Methanol exhibits the worse financial performance (considering *CAPEX* and *OPEX*) due to its low energy content (that yields increased fuel consumption), which renders its feasibility questionable. From a well to tank perspective, use of locally sourced biomass-based methanol can contribute to the lifecycle CO_2 emissions reduction.

Considering the whole fleet, hybrid power plants would lead to significant savings in fuel costs and a substantial decrease in carbon emissions. However, the reduction in emissions might not be as significant as other alternatives in a well-to-tank perspective. The availability of technology and infrastructure for the hybrid system may influence the feasibility of implementing this solution across the whole fleet.

Contrary, LNG as a fuel for the whole fleet would result in similar fuel consumption reduction, fuel cost savings and reduced emissions combined with established technology and infrastructure, making it a practical choice for fleet-wide adoption especially if advancements in technology continue to mitigate methane slip issues. Fleet-wide adoption of fossil–based methanol would not be financially or environmentally effective, as the increased fuel consumption results to only 7% well to wake CO_2 emissions reduction.

Potential introduction of emission taxation schemes from developing countries is expected to be a key driver towards the adoption of alternative propulsion technologies and fuels. However, associated challenges and measures for not transferring this cost to the end users must be thoroughly investigated in future studies.

This study offers invaluable insights to ferry operators and policymakers of developing counties, to curtail carbon emissions within their fleets. The adoption of short-term measures can facilitate the transition towards decarbonised shipping operations. However, achieving ambitious emissions targets may necessitate the use of synergies and several measures combinations. Furthermore, this study assesses the impact from several measures, contributing towards the enhancement of shipping sustainability.

6. Conclusions

This study examined different short- to medium-term solutions for the Ro-Pax ferry fleet decarbonisation in developing countries. The solutions of power plant hybridisation, LNG fuel use and methanol use were considered for two representative vessels (large and small). A lifetime economic-environmental analysis was carried out to estimate technical, environmental, and economic key performance indicators. The derived results were subsequently employed to comparatively assess these three solutions, whereas the financial impact on the whole fleet was discussed. The study main findings are summarised as follows.

- Hybridised power plants align with a short- to medium-term cost-effective strategy for reducing emissions in ferry operations, as they can yield approximately 11% fuel consumption reduction, leading to proportional emissions reductions.
- The required storage volume for LNG and methanol is expected to increase by 74% and 113% respectively compared to the baseline diesel fuel.

- The hybrid power system is the most cost-effective way to curtail CO_2 emissions, however achieved decarbonisation does not meet the 2030 targets.
- LNG power plants can achieve a 22% reduction in CO_2 emissions, although their *GWP* increases by 8%. Combining LNG use and hybrid power plants can meet the 2030 emission targets.
- The required investments for decarbonising, using LNG, larger and smaller vessels amount to approximately M EUR 0.78 and M EUR 0.25, respectively.
- The use of methanol results in reductions in both CO_2 emissions and *GWP*, but requires substantial investments due to the considerably higher cost of methanol-fuelled marine engines, amounting to M EUR 1.1 and M EUR 0.42 for large and small vessels, respectively.
- From a well to wake perspective the cases C1 and C2 exhibit 11% and 10% lower CO_2 emissions respectively pertinent to BL, whereas C3 exhibits reduction of 7% for fossil-based and 9% for biomass-based production that the BL.
- Considering the RoPax ferries fleet, the total investments required for hybrid propulsion and LNG fuel amount to M EUR 28 and M EUR 58, respectively.
- The introduction of a carbon tax in the range of 50–100 EUR/t CO_2 could be explored as a policy measure to incentivise decarbonisation in this sector. However, financial support for implementing such investments is required to prevent additional costs for end-users.

The limitations of this study are associated with the data uncertainties pertinent to the emission factors and scarcity of data for methanol fuelled marine engines. Given the significance of CO_2 emissions and their impact on the environment, it is crucial to evaluate the overall environmental footprint associated with the use of different fuels. Future studies may employ updated emission factors considering significant developments in marine engine technologies and zero-carbon fuels operations along with the lifecycle assessments.

Author Contributions: Conceptualization, G.T.; methodology, G.T., P.K. and G.P.; software, P.K.; validation, G.T.; formal analysis, P.K.; investigation, G.T., P.K. and G.P.; resources, G.T.; data curation, P.K.; writing—original draft preparation, G.T., P.K. and G.P.; writing—review and editing, G.T., P.K. and G.P.; visualization, P.K.; supervision, G.T.; project administration, G.T. All authors have read and agreed to the published version of the manuscript.

Funding: This research received no external funding.

Data Availability Statement: No available data.

Acknowledgments: The authors also greatly acknowledge the funding from DNV AS and RCCL for the Maritime Safety Research Centre establishment and operation. The opinions expressed herein are those of the authors and should not be construed to reflect the views of Innovate UK, DNV AS, RCCL.

Conflicts of Interest: The authors declare no conflict of interest.

Nomenclature

AC	Annual Cost (EUR)
CAPEX	Capital Expenditure (EUR)
C_i	Cost factor (EUR /kW)
DWT	Dead Weight Tonnage (mt)
FC	Fuel Consumption (t)
GT	Gross Tonnage ($-$)
GWP	Global Warming Potential (t $CO_{2\text{-eq}}$)
MAC	Marginal Abatement Cost (EUR/t CO_2)
OPEX	Operational Expenditure (EUR)
P	Engine Power Output (kW)
V_f	Volume of Fuel (t)

Abbreviation

AT	After Treatment
GHG	Greenhouse Gas
IMO	International Maritime Organisation
KPI	Key Performance Indicator
LNG	Liquified Natural Gas
LTEEM	Lifetime Economic Environmental Model
MGO	Marine Gas Oil

Appendix A

Table A1 lists the characteristics of several fuels including well—to—Tank Emissions factors, shipboard storage conditions, cost factors, and technical maturity. Table A2 provides cost factors associated with the transportation of methanol and LNG.

Table A1. Characteristics of different alternative fuels for the shipping sector [26].

Fuel	Well—to—Tank Emissions Factors				Shipboard Storage Conditions	Cost Factor (EUR/MJ)	Technical Maturity
	CO_2 (g/MJ)	N_2O (g/MJ)	$CO_{2.eq}$ (g/MJ)	NOx (g/MJ)			
Brown NH_3	64.8	4.5×10^{-4}	64.9	4.4×10^{-2}	T: 240–290 K	1.8×10^{-2}	Low
Green NH_3	18.5	4.5×10^{-4}	18.6	4.4×10^{-2}	P: 8–10 bar State: liquid	2.7×10^{-2}	Low
Brown H_2 (liquid)	77.9	2.5×10^{-4}–2.5×10^{-3}	77.9–78.4	3.4×10^{-2}	T: 20 K P: 12.7 bar	1.7×10^{-2}	Low
Green H_2 (liquid)	7.9	4.1×10^{-4}	7.98	3×10^{-2}	State: Cryogenic liquid	4.3×10^{-2}	Low
CH_3OH—NG based	20	2.9×10^{-4}	20	4.6×10^{-2}	T: 293 K	2×10^{-2}	Medium
CH_3OH—biomass based	17	2.2×10^{-4}	17	5.6×10^{-2}	P: 1 bar State: liquid	0.8×10^{-2}	Medium
LNG—Fossil based	26	1.6×10^{-4}	26	6×10^{-2}	T: 134 K P: up to 7 bar State: Cryogenic liquid	2.9×10^{-2}	High
MGO	19.6	5.4×10^{-4}	19.7	23×10^{-2}	T: 293 K P: 1 bar State: liquid	1.9×10^{-2}	High

Table A2. Cost factors for transportation of methanol and LNG.

Fuel	Cost Factor	Transportation Method
Methanol	1.8 EUR /MWh	Ship
	0.16 EUR/t-mile [1]	Truck
	0.071 EUR/t-mile [1]	Rail
LNG	0.74–1.29 EUR/GJ	Ship

[1] Data retrieved by de Fournas et al. [49].

References

1. Fulton, L.; Mejia, A.; Arioli, M.; Dematera, K.; Lah, O. Climate change mitigation pathways for Southeast Asia: CO_2 emissions reduction policies for the energy and transport sectors. *Sustainability* **2017**, *9*, 1160. [CrossRef]
2. International Maritime Organization. IMO GHG Strategy 2023. 2023. Available online: https://www.imo.org/en/OurWork/Environment/Pages/2023-IMO-Strategy-on-Reduction-of-GHG-Emissions-from-Ships.aspx (accessed on 25 August 2023).
3. Liaquat, A.M.; Kalam, M.A.; Masjuki, H.H.; Jayed, M.H. Potential emissions reduction in road transport sector using biofuel in developing countries. *Atmos. Environ.* **2010**, *44*, 3869–3877. [CrossRef]
4. Romano, A.; Yang, Z. Decarbonisation of shipping: A state of the art survey for 2000–2020. *Ocean Coast. Manag.* **2021**, *214*, 105936. [CrossRef]
5. Xia, Q.; Chen, F. Shipping Economics Development: A Review from the Perspective of the Shipping Industry Chain for the Past Four Decades. *J. Shanghai Jiaotong Univ. (Sci.)* **2022**, *27*, 424–436. [CrossRef]
6. Balcombe, P.; Brierley, J.; Lewis, C.; Skatvedt, L.; Speirs, J.; Hawkes, A.; Staffell, I. How to decarbonise international shipping: Options for fuels, technologies and policies. *Energy Convers. Manag.* **2019**, *182*, 72–88. [CrossRef]

7.	Hansson, J.; Brynolf, S.; Fridell, E.; Lehtveer, M. The Potential Role of Ammonia as Marine Fuel—Based on Energy Systems Modeling and Multi-Criteria Decision Analysis. *Sustainability* **2020**, *12*, 3265. [CrossRef]

8.	Maersk Mc-Kinnley Moller Center M. Managing Emissions from Ammonia-Fueled Vessels. 2023. Available online: https://cms.zerocarbonshipping.com/media/uploads/documents/Ammonia-emissions-reduction-position-paper_v4.pdf (accessed on 25 August 2023).

9.	Horvath, S.; Fasihi, M.; Breyer, C. Techno-economic analysis of a decarbonized shipping sector: Technology suggestions for a fleet in 2030 and 2040. *Energy Convers. Manag.* **2018**, *164*, 230–241. [CrossRef]

10.	Trivyza, N.L.; Rentizelas, A.; Theotokatos, G. Impact of carbon pricing on the cruise ship energy systems optimal configuration. *Energy* **2019**, *175*, 952–966. [CrossRef]

11.	Svanberg, M.; Ellis, J.; Lundgren, J.; Landälv, I. Renewable methanol as a fuel for the shipping industry. *Renew. Sustain. Energy Rev.* **2018**, *94*, 1217–1228. [CrossRef]

12.	Korberg, A.D.; Brynolf, S.; Grahn, M.; Skov, I.R. Techno-economic assessment of advanced fuels and propulsion systems in future fossil-free ships. *Renew. Sustain. Energy Rev.* **2021**, *142*, 110861. [CrossRef]

13.	Karvounis, P.; Tsoumpris, C.; Boulougouris, E.; Theotokatos, G. Recent advances in sustainable and safe marine engine operation with alternative fuels. *Front. Mech. Eng.* **2022**, *8*, 994942. [CrossRef]

14.	Armijo, J.; Philibert, C. Flexible production of green hydrogen and ammonia from variable solar and wind energy: Case study of Chile and Argentina. *Int. J. Hydrogen Energy* **2020**, *45*, 1541–1558. [CrossRef]

15.	Sun, S.; Jiang, Q.; Zhao, D.; Cao, T.; Sha, H.; Zhang, C.; Song, H.; Da, Z. Ammonia as hydrogen carrier: Advances in ammonia decomposition catalysts for promising hydrogen production. *Renew. Sustain. Energy Rev.* **2022**, *169*, 112918. [CrossRef]

16.	IEA. Production of Bio-Methanol Technology Brief, s.l. IEA-ETSAP and IRENA Technology Brief. 2013. Available online: https://www.irena.org/-/media/Files/IRENA/Agency/Publication/2013/IRENA-ETSAP-Tech-Brief-I08-Production_of_Bio-methanol.pdf?rev=5ea20e7c84c4472f8eeed8111ff8daf9 (accessed on 12 August 2023).

17.	Alamia, A.; Magnusson, I.; Johnsson, F.; Thunman, H. Well-to-wheel analysis of bio-methane via gasification, in heavy duty engines within the transport sector of the European Union. *Appl. Energy* **2016**, *170*, 445–454. [CrossRef]

18.	McKinlay, J.; Turnock, C.; Hudson, D. Route to zero emission shipping: Hydrogen, ammonia or methanol? *Int. J. Hydrogen Energy* **2021**, *46*, 28282–28297. [CrossRef]

19.	Nguyen, H.P.; Hoang, A.T.; Nizetic, S.; Nguyen, X.P.; Le, A.T.; Luong, C.N.; Chu, V.D.; Pham, V.V. The electric propulsion system as a green solution for management strategy of CO_2 emission in ocean shipping: A comprehensive review. *Int. Trans. Electr. Energy Syst.* **2021**, *31*, e12580. [CrossRef]

20.	Yuan, L.C.W.; Tjahjowidodo, T.; Lee, G.S.G.; Chan, R.; Adnanes, A.K. Equivalent Consumption Minimization Strategy for hybrid all-electric tugboats to optimize fuel savings. *Proc. Am. Control Conf.* **2016**, *2016*, 6803–6808. [CrossRef]

21.	Xie, P.; Tan, S.; Guerrero, J.M.; Vasquez, J.C. MPC-informed ECMS based real-time power management strategy for hybrid electric ship. *Energy Rep.* **2021**, *7*, 126–133. [CrossRef]

22.	Law, L.C.; Foscoli, B.; Mastorakos, E.; Evans, S. A comparison of alternative fuels for shipping in terms of lifecycle energy and cost. *Energies* **2021**, *14*, 8502. [CrossRef]

23.	Perčić, M.; Vladimir, N.; Koričan, M. Electrification of inland waterway ships considering power system lifetime emissions and costs. *Energies* **2021**, *14*, 7046. [CrossRef]

24.	Jang, H.; Jeong, B.; Zhou, P.; Ha, S.; Park, C.; Nam, D.; Rashedi, A. Parametric trend life cycle assessment for hydrogen fuel cell towards cleaner shipping. *J. Clean. Prod.* **2022**, *372*, 133777. [CrossRef]

25.	Kistner, L.; Schubert, F.L.; Minke, C.; Bensmann, A.; Hanke-Rauschenbach, R. Techno-economic and environmental comparison of internal combustion engines and solid oxide fuel cells for ship applications. *J. Power Sources* **2021**, *508*, 230328. [CrossRef]

26.	Karvounis, P.; Dantas, J.L.; Tsoumpris, C.; Theotokatos, G. Ship Power Plant Decarbonisation Using Hybrid Systems and Ammonia Fuel—A Techno-Economic—Environmental Analysis. *J. Mar. Sci. Eng.* **2022**, *10*, 1675. [CrossRef]

27.	Dai, L.; Jing, D.; Hu, H.; Wang, Z. An environmental and techno-economic analysis of transporting LNG via Arctic route. *Transp. Res. Part A Policy Pract.* **2021**, *146*, 56–71. [CrossRef]

28.	Schorn, F.; Breuer, J.L.; Samsun, R.C.; Schnorbus, T.; Heuser, B.; Peters, R.; Stolten, D. Methanol as a renewable energy carrier: An assessment of production and transportation costs for selected global locations. *Adv. Appl. Energy* **2021**, *3*, 100050. [CrossRef]

29.	Balcombe, P.; Heggo, D.A.; Harrison, M. Total methane and CO_2 emissions from liquefied natural gas carrier ships: The first primary measurements. *Environ. Sci. Technol.* **2022**, *56*, 9632–9640. [CrossRef] [PubMed]

30.	Rochussen, J.; Jaeger, N.S.; Penner, H.; Khan, A.; Kirchen, P. Development and Demonstration of Strategies for GHG and Methane Slip Reduction from Dual-Fuel Natural Gas Coastal Vessels. *Fuel* **2023**, *349*, 128433. [CrossRef]

31.	Gore, K.; Rigot-Müller, P.; Coughlan, J. Cost assessment of alternative fuels for maritime transportation in Ireland. *Transp. Res. Part D Transp. Environ.* **2022**, *110*, 103416. [CrossRef]

32.	Livanos, M.; Geertsma, R.D.; Boonen, E.J.; Visser, K.; Negenborn, R.R. Ship energy management for hybrid propulsion and power supply with shore charging. *Control Eng. Pract.* **2017**, *76*, 133–154. [CrossRef]

33.	Livanos, G.A.; Theotokatos, G.; Pagonis, D.N. Techno-economic investigation of alternative propulsion plants for Ferries and RoRo ships. *Energy Convers. Manag.* **2014**, *79*, 640–651. [CrossRef]

34.	Tsoumpris, C.; Theotokatos, G. Performance and reliability monitoring of ship hybrid power plants. *J. ETA Marit. Sci.* **2022**, *10*, 29–38. [CrossRef]

35. Ushakov, S.; Stenersen, D.; Einang, P.M. Methane slip from gas fuelled ships: A comprehensive summary based on measurement data. *J. Mar. Sci. Technol.* **2019**, *24*, 1308–1325. [CrossRef]
36. Gerritse, E.; Harmsen, J. Green Maritime Methanol. 2023 (Report). Available online: https://publications.tno.nl/publication/34640817/zpBGh5/gerritse-2023-green.pdf (accessed on 15 September 2023).
37. Available online: https://www.bunkerindex.com/ (accessed on 1 August 2023).
38. Verhelst, S.; Turner, J.W.; Sileghem, L.; Vancoillie, J. Methanol as a fuel for internal combustion engines. *Prog. Energy Combust. Sci.* **2019**, *70*, 43–88. [CrossRef]
39. Radonja, R.; Bebić, D.; Glujić, D. Methanol and ethanol as alternative fuels for shipping. *Promet-Traffic Transp.* **2019**, *31*, 321–327. [CrossRef]
40. Mrzljak, V.; Poljak, I.; Medica-Viola, V. Dual fuel consumption and efficiency of marine steam generators for the propulsion of LNG carrier. *Appl. Therm. Eng.* **2017**, *119*, 331–346. [CrossRef]
41. Grant, J. Wärtsilä—Update on Future Fuels Developments and W32M Methanol System. In Proceedings of the IMarEST, Glasgow, Scotland, 7 March 2023.
42. IEA. *World Energy Outlook 2022*; IEA: Paris, France, 2022; Available online: https://www.iea.org/reports/world-energy-outlook-2022 (accessed on 25 August 2023).
43. IEA. *Comparative Life-Cycle Greenhouse Gas Emissions of a Mid-Size BEV and ICE Vehicle*; IEA: Paris, France, 2021. Available online: https://www.iea.org/data-and-statistics/charts/comparative-life-cycle-greenhouse-gas-emissions-of-a-mid-size-bev-and-ice-vehicle (accessed on 25 June 2023).
44. Zarrinkolah, M.T.; Hosseini, V. Methane slip reduction of conventional dual-fuel natural gas diesel engine using direct fuel injection management and alternative combustion modes. *Fuel* **2023**, *331*, 125775. [CrossRef]
45. May, I.; Cairns, A.; Zhao, H.; Pedrozo, V.; Wong, H.C.; Whelan, S.; Bennicke, P. *Reduction of Methane Slip Using Premixed Micro Pilot Combustion in a Heavy-Duty Natural Gas-Diesel Engine (No. 2015-01-1798)*; SAE Technical Paper; SAE: Warrendale, PA, USA, 2015.
46. Schinas, O.; Butler, M. Feasibility and commercial considerations of LNG-fueled ships. *Ocean Eng.* **2016**, *122*, 84–96. [CrossRef]
47. Kalikatzarakis, M.; Theotokatos, G.; Coraddu, A.; Sayan, P.; Wong, S.Y. Model based analysis of the boil-off gas management and control for LNG fuelled vessels. *Energy* **2022**, *251*, 123872. [CrossRef]
48. Ammar, N.R. An environmental and economic analysis of methanol fuel for a cellular container ship. *Transp. Res. Part D Transp. Environ.* **2019**, *69*, 66–76. [CrossRef]
49. de Fournas, N.; Wei, M. Techno-economic assessment of renewable methanol from biomass gasification and PEM electrolysis for decarbonization of the maritime sector in California. *Energy Convers. Manag.* **2022**, *257*, 115440. [CrossRef]

Article

Techno-Economic and Environmental Evaluation of a Solar Energy System on a Ro-Ro Vessel for Sustainability

Michail Serris, Paraskevi Petrou, Isidoros Iakovidis and Sotiria Dimitrellou *

Department of Naval Architecture, School of Engineering, University of West Attica, 12243 Athens, Greece; mserris@uniwa.gr (M.S.); na16078@uniwa.gr (P.P.); iiakovidis@uniwa.gr (I.I.)
* Correspondence: sdimitre@uniwa.gr

Abstract: The increased use of fossil fuels in transportation is considered a major cause of environmental pollution and climate change on a global scale. In international shipping, regulations and strict measures have been introduced by the International Maritime Organization to achieve the goal of a 40% reduction in greenhouse gas (GHG) emissions by 2030, with the envisage to reach net-zero GHG emissions close to 2050. Renewable energy sources, such as solar photovoltaic (PV) systems, can be implemented on new-build or existing marine vessels as an effective alternative source for auxiliary power generation, reducing the dependency on fossil fuels and contributing to decarbonization. In the present paper, a sustainable retrofit design using PV panels on an existing Ro-Ro vessel is analyzed for its feasibility. The proposed system is used for energy production during ship cargo operations and takes advantage of the large space area on the upper deck and its continuous exposure to sunlight during its voyage. To investigate the effectiveness of the PV system as an alternative to fossil fuel consumption, an environmental and economic evaluation is performed. According to the results obtained, the solar PV system can provide approximately 88% of the required energy annually for lighting during ship cargo operations, with the corresponding fuel savings and emission reductions, making the investment economically feasible, with a high potential to contribute to environmental sustainability.

Keywords: solar energy; photovoltaic system; fuel savings; ship emissions; environmental sustainability; energy efficiency; techno-economic evaluation; marine transportation

Citation: Serris, M.; Petrou, P.; Iakovidis, I.; Dimitrellou, S. Techno-Economic and Environmental Evaluation of a Solar Energy System on a Ro-Ro Vessel for Sustainability. *Energies* **2023**, *16*, 6523. https://doi.org/10.3390/en16186523

Academic Editors: Konstantinos Aravossis and Eleni Strantzali

Received: 9 August 2023
Revised: 30 August 2023
Accepted: 7 September 2023
Published: 10 September 2023

1. Introduction

Electricity produced from solar irradiation is considered a clean and non-polluting form of energy; it has a minimal environmental impact, and it is a sustainable alternative capable of contributing to the reduction in Greenhouse Gas (GHG) emissions in the atmosphere and the prevention of environmental pollution [1]. In accordance with the recent advances in the field of PV production and the development of efficient energy-storage devices [2], the installation of PV systems emerges as an attractive choice for electrical energy production. It is expected that solar PV technology will have a significant role in the modern world, contributing to a smart and sustainable economy. The construction and operation of PV systems serve the targets and goals for Sustainable Development set by the UN in 2015 [3] well, and are closely related to the strategy of using plastics, metals, and ceramics in a circular economy, enabling the reuse, repair, and recycling of these materials. Additionally, most commercial solar PV panels have an efficiency of 15–20% while the cost of PV panels is between USD 2.60 and 3.20/W [4], making solar energy an attractive option. In general, global primary energy consumption at the end of the 20th century increased by 10 times compared to the beginning of the 20th century [5], while the use of fossil fuels in primary energy increased by 16 times [6]. This clearly points out the necessity of energy transition from fossil-based energy systems to renewable energy sources, such as solar energy. It has been estimated [7] that by 2030, PV systems will contribute to 12–14% to the total electric energy production in the EU energy system, resulting in a notable reduction in GHG emissions.

Maritime transportation has a significant impact on environmental pollution and global warming. The burning of fossil fuels, such as coal, oil, and natural gas, for energy production results in the emission of air pollutants that are responsible for climate change, which could have a long-term effect on biodiversity as well as on human socioeconomic development. Carbon dioxide (CO_2) has been proven to be a major contributor to global warming and unless climate and energy policies are implemented, the average global temperature is projected to be 4.1–4.8 °C higher by the end of the century [8]. In addition, nitrogen oxides can enhance the greenhouse effect and are considered a possible cause of ozone depletion, while sulfur dioxide contributes to acid rain, which can harm sensitive ecosystems. The need to minimize the use of fossil fuels and at the same time to implement a greener and sustainable marine transportation led the shipping community to consider renewable energy sources, such as solar energy, offshore wind energy, hydropower, and biomass, which are environmentally friendly and abundant in the natural environment [4,9,10]. According to the Fourth IMO GHG Study 2020 [11], ship emissions are projected to increase from about 90% in 2018 to 90–130% by 2050 (compared to levels in 2008) for a range of plausible long-term economic and energy scenarios. In this regard, the IMO has adopted mandatory measures to reduce GHG emissions from international shipping through amendments to MARPOL Annex VI Regulations. Existing regulations limit the sulfur oxide (SOx), nitrogen oxide (NOx) and particulate matter (PM) emissions for ships that operate in global waters and Emission Control Areas (ECAs), while new sea territories such as the Mediterranean Sea will be established as an ECA in the forthcoming years. Furthermore, the 2023 IMO GHG Strategy [12] envisages a cutting in carbon intensity (reducing CO_2 emissions per transport work), as an average across international shipping, by at least 40% by 2030 compared to 2008, aiming to reach net-zero GHG emissions by or around 2050. A new ambition was also included in the 2023 IMO GHG Strategy, relating to the uptake of zero or near-zero GHG emission technologies, fuels and/or energy sources that are to represent at least 5%, but striving for 10%, of the energy used by international shipping by 2030. In this context, from 1 January 2023 it is mandatory for ships to calculate their attained Energy Efficiency Existing Ship Index (EEXI) and to report their annual operational Carbon Intensity Indicator (CII).

Although the design of solar energy systems and their utilization on ships have been studied extensively in the last decade [13–17], few studies consider the techno-economic evaluation of the solar system and its influence on the ship's energy efficiency and sustainability. Only recently, researchers and scholars have focused on the application of solar PV systems on ships, investigating the system feasibility from the aspect of investment and the possibility of achieving a reduction in a ship's emissions. As the system feasibility is dependent on the vessel type, its operational profile and navigation routes, most researchers have implemented their techno-economic and environmental evaluations through case studies on a chosen ship type. Salem and Seddiek [18] have examined the effectiveness and challenges of a grid-connected PV solar system utilized for power supply for emergency lighting and navigational equipment, on a research vessel that operates in the Red Sea region. Both the economic and environmental benefits of the proposed PV system have been evaluated, and analytical calculations of the solar array cost and power grid cost based on the system specifications have been performed. Qiu et al. [19] performed a techno-economic evaluation of a PV grid-connected power system on a Pure Car Truck Carrier by proposing a mathematical model for predicting solar radiation along six main navigation routes, and investigated the techno-economic efficiency and environmental performance using single- and multi-criteria evaluation methods. The effectiveness of the application of a solar panel system for obtaining propulsion power on a short-route ferry operating in the Marmara Sea has been examined by means of a life cycle assessment [20]. In this paper, the life cycle environmental impact and the costs and benefits of the system were evaluated through a sensitivity analysis of important and uncertain parameters. Karatuğ and Durmuşoğlu [21] designed a grid-connected solar PV system for a Ro-Ro-type ship navigating between Turkey and Italy and performed its evaluation in terms of fuel consumption, emission reduction and economic profitability. An extensive review on research work concerning stand-alone and hybrid solar energy systems on marine vessels has also been included in their paper. Tercan et al. [22], conducted a technical analysis of an off-grid rooftop PV system for a small tourist boat and investigated the reduction in CO_2 emissions. In addition, a fully electric solar boat, as well as an on-grid PV plant to meet the energy demands for an entire tourist boat fleet, have been analyzed.

In addition to the previous work, some research studies have focused on the energy performance and environmental feasibility of solar hybrid systems using PV panels, diesel engines, fuel cells, and battery storage units. Ling-Chin and Roskilly performed a life cycle assessment study to estimate energy and material consumption, emissions, and the environmental impact of a new-build hybrid system [23], and a retrofit power plant [24], for a Ro-Ro cargo ship, incorporating selected emerging technologies such as lithium-ion batteries, PV systems, and cold ironing. Yuan et al. [25] designed a large-scale solar/diesel hybrid system, with a grid-connected and stand-alone control and battery energy storage unit, for a Pure Car Truck Carrier. In that paper, the energy savings and CO_2 emission reduction were verified through the analysis of the actual ship's experimental data under different PV penetration levels during arrival/departure and normal sailing. A hybrid solar PV/PEM fuel cell/diesel generator power system was simulated in [26], to provide the electric power needed for a cruise ship operating in the Baltic Sea. The fraction of renewable energy was estimated to 13.83%, corresponding to a reduction of 9.84% on GHGs and PM emissions. Yuan et al. [27] explored a stand-alone PV energy system installed onboard an inland river Pure Car Carrier as an auxiliary power source. The GHG emissions and fuel consumption of the case vessel were evaluated using EEDI analysis, while the required data were collected by actual navigation trials. A hybrid solar/wind energy/fuel cell system for an oil tanker [28] was designed and evaluated through an economic and environmental analysis, and the reduced EEDI was calculated. An excessive review on solar, wind and fuel cell energy applications was also included.

The purpose of the present article is to investigate the feasibility and sustainability of implementing a stand-alone solar PV system in a Ro-Ro cargo vessel to supply part of the energy required during the cargo operations (loading/unloading) by performing techno-economic and environmental analysis. The selection of the case study ship was based on its energy load requirements, the deck space available for PV panel installation, and the shipping routes in areas of high solar potential. The Ro-Ro cargo vessel navigates in the broad area of the Mediterranean Sea, Black Sea and on the western shores of Spain and Portugal, and has sufficient deck space for PV panel installation exposed to direct sunlight. The environmental benefits of the proposed PV system are demonstrated through the estimation of CO_2, SOx, NOx and PM emissions, considering the energy provided by the PV system, the emission factors and the fuel savings. To evaluate the economic viability of the investment, important indicators as the Net Present Value, the Internal Return Rate and the payback period are calculated considering the capital cost of the PV system, the operation and maintenance cost and the direct economic benefits from the fuel savings. Although the implementation of PV systems in cargo vessels has been examined in related studies, nevertheless, to the knowledge of the authors, a techno-economic analysis that clearly presents the long-term fuel price assumptions has not been addressed in similar previously published studies. Therefore, this study investigates the economic viability of the PV system using true current fuel prices and representative fuel-price forecast assumptions, and validates the accuracy of the methodology performing a sensitivity analysis with respect to fuel oil price-increase scenarios. The rest of the paper is organized as follows. Firstly, the methodology used for the PV system modeling and the environmental and economic analysis is carried out. Subsequently, a case study referring to the design of the stand-alone PV system on the Ro-Ro vessel is described. Finally, the results of the fuel savings as well as the economic and environmental benefits are evaluated, and the conclusions are presented.

2. Methodology

The methodology used in this study includes three stages: preparation, design, and evaluation. Firstly, to select the type and size of the case vessel for the implementation of the PV system, important information was collected, including ship specifications, deck plans, the operational profile, navigation routes, specific fuel oil consumption, and the required energy for lighting in different operation modes. The sufficient area for the location of the PV panels was determined and a market survey was conducted for the selection of the appropriate PV panels in terms of performance, efficiency, and peak power. Also, the direct normal solar irradiation in the navigation area under study was obtained from global solar radiation databases. Then, the methodology for PV system design was applied, aiming to attain the higher possible PV system capacity to cover the required energy. The maximum energy produced by the PV system and the corresponding fuel savings were calculated. Finally, ship emission reductions and economic indicators were estimated by performing environmental and economic analysis, and sensitivity analysis was used to evaluate the results.

2.1. PV System Modeling

The performance of PV devices is determined by two key parameters [2], namely, the Capacity Factor (*CF*), which is defined as the annual actual *AC* electric energy output (in kWh/year) divided by the annual generated *DC* output peak power rating (in kWp) multiplied by 8760 h/year, according to Equation (1), and the Performance Ratio (*PR*) also called the Quality factor (*Q*), which is defined according to Equation (2).

$$CF = \frac{Actual\_annual\_AC_{output}}{DC\ peak\ power\ rating \cdot 8760} \tag{1}$$

$$PR = \frac{CF}{DCpower_{(peak)} \cdot 8760 \cdot Irr_{(avg)} \cdot 10^{-3}} \tag{2}$$

where $DCpower_{(peak)}$ is the generated output peak power (in kW_p) and $Irr_{(avg)}$ is the average solar irradiation (in W/m^2) in the location.

In general, PV panels are compensated by their high reliability, very low impact to the environment and low maintenance needs. Moreover, their construction is characterized by a low demand for material utilization and a simple manufacturing process. The selection of PV panels to be installed on a ship is based on the energy requirements and the available space, taking into consideration their performance, power rating, and size. It is crucial that the panels are resistant to the marine environment, which is characterized by harsh conditions concerning humidity, salinity, and strong winds. The design of the selected PV system should take into account the construction material of the frame, and the capability of connecting it to a battery, as well as the ease of wiring. However, there are two main problems that need to be addressed: the reverse current flow from the battery to the PV panels, and faulty or partially shaded PV panels. To prevent the back discharge of the battery when the solar panel is in a lower potential, a blocking diode is connected in series to the PV panel or PV string, while to provide an alternative path for current flow in case of a faulty or partially shaded panel, it is recommended to connect a bypass diode in parallel to the PV panel. Performance also depends on degradation, shading, overheating, the presence of impurities on the PV surface and losses in blocking diodes.

A key criterion for selecting a PV panel is the peak power, which represents the maximum electric power produced by the panel under standard control conditions (STC). The efficiency of the PV panel is calculated using Equation (3):

$$n = \frac{P_m}{P_H \cdot S} \tag{3}$$

where P_m is the maximum power of the PV panel (in kW), P_H is the power density of the incident radiation (in kW/m^2) and S is the surface of the incident radiation (in m^2).

The maximum energy produced by the PV system is estimated by Equation (4) [29]:

$$E_{\text{max\_PVsystem}} = B_n \cdot A_{PV} \cdot n_{STC} \cdot AF \cdot SF \cdot DF \cdot T_{coef} \cdot N \tag{4}$$

where B_n is the direct normal solar irradiation (in KWh/m^2), A_{PV} is the PV panel surface area (in m^2), n_{STC} is the efficiency factor of the PV panel in standard temperature conditions (STC), AF is the degradation factor, SF is the shading factor, DF is the factor of energy loss on blocking diodes, T_{coef} is the temperature coefficient and N is the number of PV panels. The temperature coefficient is calculated according to Equation (5):

$$T_{coef} = 1 - \left[(t_{avg} + 30) - 25 \right] \cdot 0.004 \tag{5}$$

where t_{avg} is the average temperature in the location.

To cover the energy requirements and to maximize the system's power output, the PV panels are connected in series (forming a PV string) to increase the output voltage or/and in parallel to increase the output current. The power output of each PV sub-array is determined by the total number of the PV panels and the connection scheme [30].

Hybrid PV systems are environmentally friendly devices and have the capability to generate power using two sources; for instance, they can combine solar or wind energy with the power produced from a generator. This results in a reliable function and a continuous stable power supply [31]. The main purpose of any hybrid PV system installed on a ship is to secure the grid with the supply of constant power. Often, this is hard to achieve due to the prevailing conditions and the capacity limit of the batteries. When the amount of power produced from the PV system is not sufficient, the additional required power is retrieved from the ship's generators. The design must, therefore, include, in addition to the PV panels, a DC-to-AC Inverter, as well as batteries for storage of the excess energy (Figure 1). Solar charge controllers, the main role of which is to charge the batteries and to provide the maximum amount of the required electric power, are also incorporated in the designed installation.

The batteries store much of the energy and can supply the required electric charge to the system whenever there is a drop in the production of electric power from the PV system. The main parameter for choosing the appropriate battery is its autonomy, as well as its main characteristics determined by the nominal voltage and capacity. The autonomy of a battery determines the time period of sustaining the system's energy without the need of recharging, while the nominal voltage determines the type of the system's configuration. The capacity of a battery (in Ah) is calculated according to Equation (6):

$$C = I \cdot n_h \tag{6}$$

where I is the electric current and n_h is the battery charging time (in hours). The maximum electrical energy stored by a battery (in kWh) is calculated by Equation (7):

$$E_B = C \cdot V_B \tag{7}$$

where C is the battery capacity (in Ah) and V_B denotes the nominal voltage.

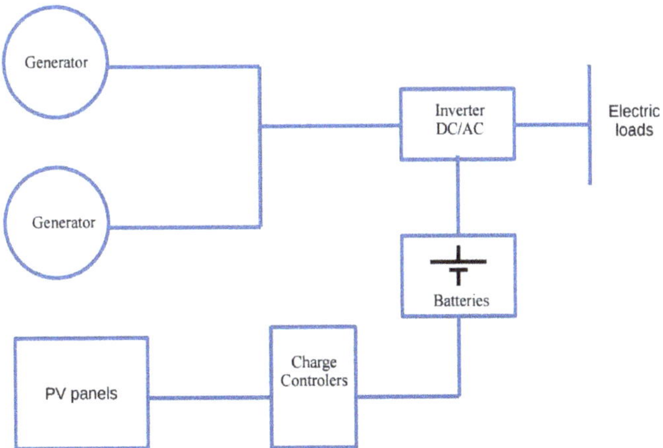

Figure 1. Schematic diagram of a stand-alone solar PV system installed onboard a ship.

The batteries can be connected either in series or in parallel depending on the required power output. The nominal capacity of the system depends on the number of sequential charging and discharging processes of the batteries. The efficiency of a battery is calculated by Equation (8):

$$n_b = \frac{E_{dch}}{E_{ch}} \tag{8}$$

where E_{dch} is the provided energy during discharging and E_{ch} is the supplied energy during charging.

2.2. Environmental Analysis

Emission gases such as CO_2, SOx, NOx, and PM are released through the combustion of the fuels used during the ship operation. These emission gases can be reduced, respectively, to the power generated from the PV system and the fuel savings. A key element for calculating the quantity of emission gases is the emission factors (*EFs*), which can depend on the fuel type, the content of the emission gas in the fuel, the specific fuel oil consumption, and engine characteristics, such as the engine speed and engine production year [32].

The CO_2 fuel-based emission factor (in g/kWh) is calculated as the following [33]:

$$EF_{CO_2} = conversion\ factor \cdot SFOC \tag{9}$$

where *SFOC* is the specific fuel oil consumption (in g/kWh) and the conversion factor denotes the CO_2 content in the fuel type used (in g CO_2 per g of fuel).

SOx and PM fuel-based emission factors (in g/kWh) are calculated according to the following equations (Equations (10)–(12)), based on the sulfur content of the fuel used [33]:

$$EF_{SOx} = SFOC \cdot 2 \cdot 0.97753 \cdot fuel\ sulfur\ fraction \tag{10}$$

$$EF_{PM,HFO} = 1.35 + SFOC \cdot 7 \cdot 0.02247 \cdot (fuel\ sulfur\ fraction - 0.0246) \tag{11}$$

$$EF_{PM,MGO} = 0.23 + SFOC \cdot 7 \cdot 0.02247 \cdot (fuel\ sulfur\ fraction - 0.0024) \tag{12}$$

The NOx emission limit (in g/kWh) is calculated based on the Tier standards according to MARPOL Annex VI Regulations. The different Tiers are based on the ship construction date, and each Tier limits NOx emissions to a specific value that is determined considering the engine's rated speed.

2.3. Economic Analysis

To assess the financial viability of the PV system investment, the Net Present Value (NPV), the Internal Rate of Return (IRR) and the Payback period were estimated. The NPV is the difference between the present value of cash inflows and the present value of cash outflows for a certain period and is used in investment planning to analyze the potential profit or loss of the planned investment. The Net Present Value can be estimated using Equation (13):

$$NPV = \sum_{t=1}^{T} \left(\frac{R_t}{(1+i)^t} - C_0 \right) \tag{13}$$

where C_0 is the initial capital cost, R_t is the total cash flow during the period t, i is the discount rate, and T is the number of time periods.

The discount rate is the rate of return that the investor expects to receive from the particular investment and the cash flows of the specific period. In this study, its value was set considering the respective accepted values in related studies in the literature. The term R_t includes the operation and maintenance cost of the investment as well as the direct economic benefits, and is discounted back to the present value. In this study, the operation and maintenance cost was estimated as a fixed price for the first year of the investment, while an annual increase was considered due to the time value of the currency and the aging of the system. The direct economic benefits, which correspond to the fuel savings that are attained due to the PV system operation, were estimated as a fixed price for the first year according to current fuel prices, while an annual increase was considered for the long-term analysis, combining published oil price forecast assumptions.

The Internal Rate of Return indicates the annual rate of growth that an investment is expected to generate and is used to assess the profitability of the potential investment. In general, when comparing investments, the one with the highest IRR is considered the more desirable. IRR is a discount rate that makes the Net Present Value of all cash flows equal to zero and can be estimated by Equation (14):

$$0 = NPV = \sum_{t=0}^{T} \frac{C_t}{(1+IRR)^t} \tag{14}$$

where C_t is the net cash flow during the period t, and T is the number of time periods.

The Payback period is the period in which the NPV of the investment equals zero and is calculated by dividing the amount of the investment by the annual cash flow. A short payback period indicates a financially viable investment as the cost of the initial investment will be quickly recovered. A long payback period is usually not desirable, as the investment could be risky due to the uncertainties in the long-term predictions.

To assess the accuracy of the proposed methodology and the reliability of the calculation results, a sensitivity analysis was performed. In sensitivity analysis, the values of critical parameters that are most likely to change and affect the system assumptions are modified, and the effect of the values' variation to the results is presented. As the economic evaluation includes long-term assumptions for the cost parameters, sensitivity analysis is also a means to understand how these assumptions affect the derived results.

3. Case Study

3.1. Vessel Information

To investigate the performance and feasibility of the solar PV system, a Roll-on/Roll-off cargo vessel of 170 m length overall, 28.02 m breadth, gross tonnage 36,902 t and carrying capacity 11,010 t DWT was selected as a case study. The vessel has a large space area on the upper deck continuously exposed to sunlight during its voyage, suitable for the installation of solar panels. The ship serves at five routes within the Mediterranean Sea between ports in Spain, France, Italy, the straits of Gibraltar, and the Marmara Sea (Gemlik), two routes on the western shores of Spain and Portugal, and one route from Gemlik to Constanta (Black Sea). It consumes heavy fuel oil (HFO) and marine gas oil (MGO) in a 60–40 ratio and is equipped with scrubbers that limit the SOx emissions of HFO to the allowed levels according to MARPOL Annex VI Regulations.

The ship operates at sea, at port (in/out), at cargo operations (loading/unloading) and at harbor, with different energy requirements at each operating mode. It is equipped with one main engine of 11,010 kW, one shaft generator of 1800 kW that runs at sea mode, two main generators of 1100 kW each that run at other modes, and an emergency generator of 150 kW. In the present case study, the PV system was designed to provide energy for the lighting loads during cargo operations. Lighting loads include machinery space lighting, accommodation lighting, deck lighting, and cargo hold lighting. According to the operational profile of the ship, a cargo operation (loading/unloading) has a duration of 5 h and lighting consumption is equal to 167 kWh. Considering a percentage of simultaneous electrical loads equal to 90%, the required energy for lighting during a cargo operation is equal to 751.5 kWh.

3.2. Solar PV Potential

Solar radiation can be exploited through the application of solar PV systems onboard ships in off-grid or grid-connected operation modes, that utilize deck spaces of high sun exposure and provide a ship with continuous power supply. Especially for ships that operate in shipping routes of high solar potential, solar energy can comprise a greener and more environmentally responsible means for transport, promoting sustainable shipping.

As the Ro-Ro vessel sailing schedule is dependent on the market demand, the sequence and the number of navigation routes is not predetermined for each month. To approximate the solar PV potential for the case study, a specific area was selected that contains representative solar radiation characteristics of the ports where the cargo operations take place, and the PV system was utilized. This area was characterized by the average of solar potentials between the port of Constanta (lower solar potential) and the port of Tanger Med in Straits of Gibraltar (higher solar potential) and included the majority of the ports along the ship's sailing schedule as well as the ports of departure and arrival.

The direct normal solar irradiation for average hour intervals in the area under study is presented in Figure 2, utilizing data obtained from the Global Solar Atlas 2.0 web-based application [34]. Figure 3 presents the in-plane irradiation for the zero tilt angle of the installed PV panels per month, utilizing data from the Photovoltaic Geographical Information System web-based application [35].

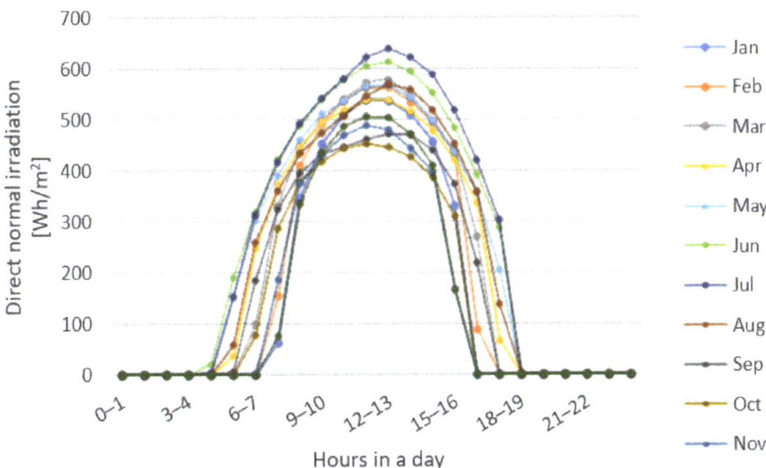

Figure 2. Direct normal irradiation for average hour profiles [34].

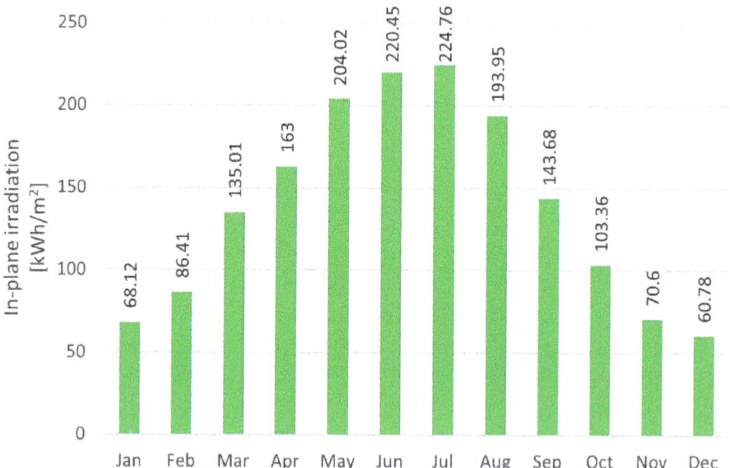

Figure 3. In-plane irradiation for zero tilt angle of the installed PV panels [35].

3.3. Stand-Alone PV System Design

Considering the available space on the decks and to ensure the maximum energy output of the solar system, 450 PV panels of a monocrystalline type, with panel dimensions of 2172 × 1303 × 35 mm and a PV peak power of 605 W, were installed. The panels were connected in 90 strings of five modules each, and mounted in zero tilt angle to avoid shading, with appropriate margins and corridors between the frames. Figure 4 presents the general arrangement of the Ro-Ro vessel and the location of the PV panels on the weather deck (space 1 and 2), the top of bridge deck (space 3) and the upper deck (space 4). Figure 5 demonstrates the stand-alone PV system diagram consisting of 450 PV panels, 45 charge controllers, 1080 solar batteries and 23 DC-AC inverters.

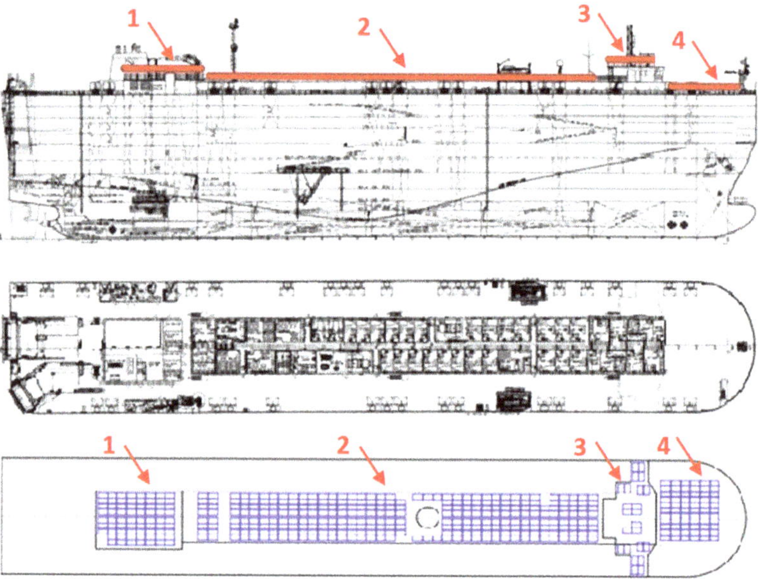

Figure 4. General Arrangement of the Ro-Ro vessel and location of the PV panels.

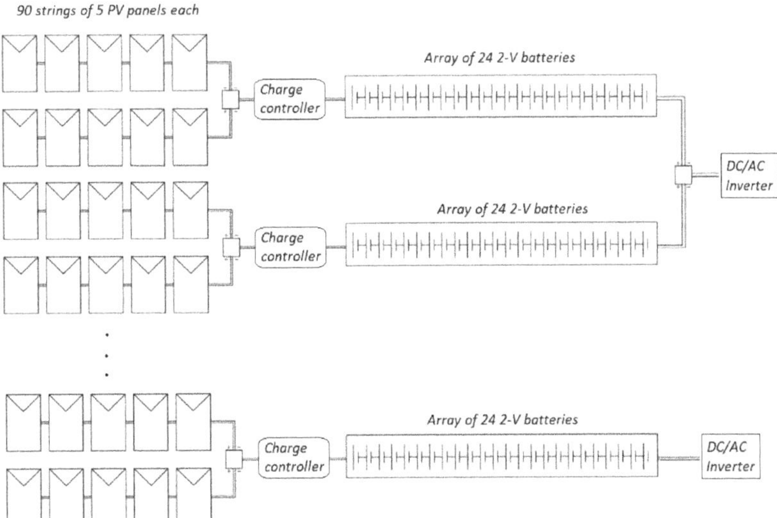

Figure 5. Stand-alone PV system diagram of the Ro-Ro cargo vessel.

To achieve the maximum input power to the batteries by adjusting the voltage and current values of the PV strings, solar charge controllers were installed. Each controller served two strings of five PV panels connected in parallel, resulting in 45 charge controllers of 250 V/100 A in total.

Solar batteries stored the excess energy produced by the PV system and provided stable energy distribution and the avoidance of fluctuations, which could affect and damage the PV system. The total required capacity was calculated to be equal to 40,931 Ah [23,30]. For the PV system, 2-volt batteries of 994 Ah capacity each were connected in a series of 24 to provide the necessary voltage of 48 V. Each array of batteries corresponded to the capacity of 10 PV panels and were connected after the charge controller.

To convert the maximum DC power generated by the PV panels, 23 DC-AC inverters of a continuous output power of 10,000 W (at 25 °C) and peak power of 20,000 W each were installed after the batteries. In particular, the 44 battery arrays were connected to 22 inverters and the 45th was connected to the remaining one. When the solar batteries reached the lowest level (cut-off voltage) which was set as equal to 15%, the two main generators were connected to the inverter inputs (Figure 1), to continue the uninterrupted supply for the loads.

Table 1 presents the maximum energy produced by the PV system according to Equation (4), considering the average temperature in the area under study, the direct normal irradiation for the zero tilt angle, and the technical characteristics of the PV panels (n_{STC} was equal to 21.4% and the factors AF, SF and DF were equal to 0.98, 0.90 and 0.99, respectively) [36].

Table 1. Maximum energy produced by the PV system based on case parameters.

Month	Days	Average Temperature (°C)	Temperature Coefficient *Tcoef*	Direct Normal Irradiation (kWh/m^2)	Maximum Energy Produced by PV System (kWh)
January	31	7.9	0.948	68.12	14,753.29
February	28	8.5	0.946	86.41	18,667.14
March	31	10.9	0.936	135.01	28,870.22
April	30	13.5	0.926	163	34,468.41
May	31	17.0	0.912	204.02	42,490.35
June	30	21.3	0.895	220.45	45,046.27
July	31	23.8	0.885	224.76	45,413.70
August	31	23.9	0.884	193.95	39,170.69
September	30	20.9	0.896	143.68	29,411.75
October	31	17.4	0.910	103.36	21,488.57
November	30	12.1	0.932	70.6	15,019.55
December	31	8.6	0.946	60.78	13,124.74

3.4. Limitations of the PV System

The installation of PV panels onboard a ship at deck areas that are greatly exposed to sunlight with no obstructions can provide a ship with the maximum energy produced by the PV system, especially when it operates in regions of high solar potential. However, there are some potential limitations that can affect the PV system performance and the coverage of energy loads. Firstly, most ships usually have a limited space for the installation of PV panels, to meet the energy demands. Also, environmental conditions such as ambient temperature, the sun's irradiation, humidity, and strong winds, impact the efficiency of the photovoltaic system regarding the conversion of solar energy into electricity. In addition, harsh weather conditions onboard the ship, such as dust, dirt and saltwater, can also affect the efficiency and lifespan of the PV panels, especially if the maintenance of the PV panels is insufficient [37,38]. However, according to trials on the 2400-passenger ferry Blue Star Delos [39], using a thin panel PV technology designed to withstand exposure to dirt and salt, their impact on the performance and the power output of the solar panels was minimal.

4. Results and Discussion

4.1. Fuel Savings

The proposed PV system was used to provide electrical energy for the lighting during the cargo operations (loading/unloading). As the sequence and the number of the vessel's navigation routes was not predetermined, the exact number of cargo operations could not be specified. The shortest and longest duration of the ship journey between two ports were 5.5 h and 4.5 days, respectively. For the calculation of the required energy, the most-demanding scenario with one cargo operation per day was assumed, which resulted in a daily required energy equal to 751.5 kWh. Figure 6 presents the required energy for lighting per month, which was calculated by multiplying the daily required energy by the number of days, and the maximum energy produced by the PV system per month. It was observed that from March to September the PV system provided the full quantity of the required energy, and the excess energy could serve other electrical needs, such as the lighting loads during at the harbor operating mode, that are also supplied by the main generators. During the remaining five months, when the system partially provided the required energy, the main generators were used for full energy coverage.

To calculate the fuel savings (*FS*) per month (in tons) for the operation under study, the following equation was used:

$$FS_i = E_{PV,i} \cdot SFOC_i \cdot 10^{-6} \tag{15}$$

where E_{PV} is the energy provided by the PV system (in kWh), *SFOC* is the specific fuel oil consumption (in g/kWh) obtained from the engine logbook for each month considering mixed fuel of 60% HFO and 40% MGO, and *i* indicates the month.

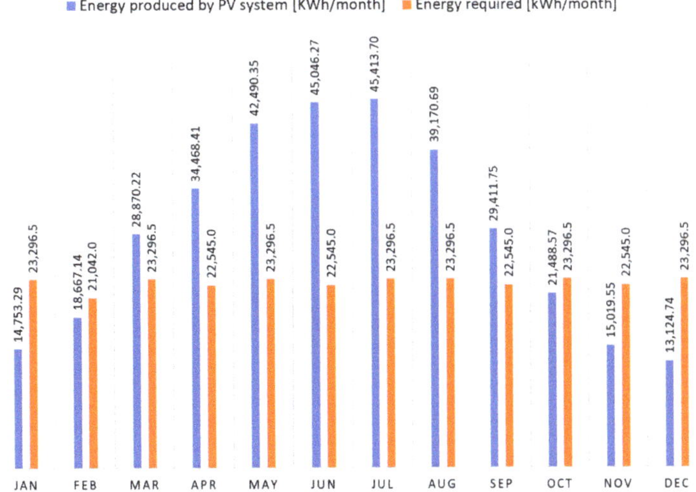

Figure 6. Required energy for lighting and maximum energy produced by the PV system.

Table 2 summarizes the calculations for *FS*, including the number of days per month that the required fuel quantity for lighting was provided by the PV system, fully or partially. According to the calculations, 38.06 t of HFO and 25.37 t of MGO could be saved by the PV system annually; that stands for 88% of the amount of fuel that was consumed for lighting during cargo operations.

Table 2. Fuel savings (in tons) based on case parameters.

Month	SFOC (g/kWh)	Energy Required (kWh)	Fuel Required (tons)	Maximum Energy Produced by PV System (kWh)	Energy Provided by PV System (kWh)	Number of Days	Fuel Savings (tons)
January	300	23,296.50	6.99	14,753.29	14,753.29	19.6	4.43
February	317	21,042.00	6.67	18,667.14	18,667.14	24.8	5.92
March	268	23,296.50	6.24	28,870.22	23,296.50	31	6.24
April	289	22,545.00	6.52	34,468.41	22,545.00	30	6.52
May	254	23,296.50	5.92	42,490.35	23,296.50	31	5.92
June	259	22,545.00	5.84	45,046.27	22,545.00	30	5.84
July	241	23,296.50	5.61	45,413.70	23,296.50	31	5.61
August	214	23,296.50	4.99	39,170.69	23,296.50	31	4.99
September	217	22,545.00	4.89	29,411.75	22,545.00	30	4.89
October	244	23,296.50	5.68	21,488.57	21,488.57	28.5	5.24
November	272	22,545.00	6.13	15,019.55	15,019.55	19.9	4.09
December	286	23,296.50	6.66	13,124.74	13,124.74	17.4	3.75
Sum		274,297.50	72.15	347,924.68	243,874.30		63.43

4.2. Environmental Indicators

To evaluate the environmental performance of the solar system as an alternative to fossil fuel consumption, the reduction in the emissions released to the atmosphere during ship cargo operations was estimated. The quantity of the emissions (in kg) was calculated separately for CO_2, SOx and PM according to Equation (16):

$$Q_{emission} = \sum_{i=1}^{12}(EF_{emission,MGO}\cdot 0.40 + EF_{emission,HFO}\cdot 0.60)E_{PV,i}\cdot 10^{-3} \qquad (16)$$

where E_{PV} is the energy provided by the PV system (in kWh) which also represents the energy saved by the main generators, $EF_{emission,fuel}$ denotes the fuel-based emission factors (in g/kWh), and i indicates the month.

The CO_2 fuel-based *EF*s were estimated according to Equation (9) considering a conversion factor of 3.206 for MGO and 3.114 for HFO [33].

SOx and PM fuel-based *EF*s were calculated in accordance with Equations (10)–(12) considering a fuel sulfur content of 0.1% for MGO and 0.5% for HFO. Actually, the Ro-Ro vessel under the study consumed MGO with a fuel sulfur content of 0.1% and HFO with a high sulfur content of 3.5%. However, the ship is equipped with a scrubber system and the quantity of sulfur that is emitted from the exhausts was reduced to 0.5%, which is also the fuel sulfur limit for ships operating in global waters; therefore, this value was considered for the calculations.

The NOx emission limit (in g/kWh) was calculated by Equation (17), applying the Tier I standard since the ship's construction date is 2010,

$$NO_x\ Limit = 45 \cdot n^{-0.2} \tag{17}$$

where n is the engine's rated speed (in rpm) equal to 900 rpm.

Table 3 summarizes the emission factors and the quantities of the CO_2, SOx, PM, and NOx exhaust gases reduced per month. Adapting the proposed solar PV system, an annual reduction of 199,865 kg of CO_2, 2815 kg of NOx, 421 kg of SOx, and 97 kg of PM can be achieved, indicating that the PV system has a high environmental performance and can contribute to environmental sustainability. Furthermore, using the proposed system to provide energy during cargo operations in port constitutes a cleaner and environmentally friendly solution for reducing air pollution from ships in port areas.

Table 3. Emissions for exhaust gases (in kg) reduced per month.

Month	EF *_CO_2 (g/kWh)		CO_2 (kg)	EF_SOx (g/kWh)		SOx (kg)	EF_PM (g/kWh)		PM * (kg)	NOx (kg)
	MGO *	HFO *		MGO	HFO		MGO	HFO		
January	961.80	934.20	13,945.40	0.59	2.93	29.42	0.16	0.43	4.73	170.32
February	1016.30	987.14	18,644.81	0.62	3.10	39.33	0.16	0.37	5.37	215.50
March	859.21	834.55	19,671.90	0.52	2.62	41.50	0.17	0.52	8.91	268.94
April	926.53	899.95	20,529.05	0.57	2.83	43.31	0.17	0.46	7.71	260.26
May	814.32	790.96	18,644.26	0.50	2.48	39.33	0.17	0.57	9.55	268.94
June	830.35	806.53	18,398.01	0.51	2.53	38.81	0.17	0.55	9.02	260.26
July	772.65	750.47	17,690.03	0.47	2.36	37.32	0.18	0.61	10.13	268.94
August	686.08	666.40	15,708.16	0.42	2.09	33.14	0.18	0.69	11.35	268.94
September	695.70	675.74	15,414.55	0.42	2.12	32.52	0.18	0.68	10.86	260.26
October	782.26	759.82	16,520.31	0.48	2.39	34.85	0.18	0.60	9.22	248.07
November	872.03	847.01	12,872.02	0.53	2.66	27.16	0.17	0.51	5.63	173.39
December	916.92	890.60	11,827.09	0.56	2.80	24.95	0.17	0.47	4.56	151.51
Sum			199,865.60			421.65			97.05	2815.33

* EF: Emission Factor, PM: Particulate Matter, MGO: Marine Gas Oil, HFO: Heavy Fuel Oil.

4.3. Sensitivity Analysis of the Environmental Assessment

In the environmental assessment of the proposed PV system, the quantity of reduced emissions depends highly on fuel savings and, therefore, on the energy provided by the PV system. Critical factors that affect the performance of PV panels and the maximum energy produced are the environmental and weather conditions on board the ship, such as strong winds, cloudy sky, humidity, and dust [37,38]. For the sensitivity analysis, two scenarios of decreased maximum energy produced by the PV system (10% and 20% decrease) were analyzed, and their effect on the emissions are presented in Table 4, relative to the base case.

Table 4. Sensitivity analysis of the environmental results with respect to different PV panel performance (values per year).

	Maximum Energy Produced by PV System (kWh)	Energy Provided by PV System (kWh)	Fuel Savings (tons)	Emissions (in kg) Reduced			
				CO_2	SOx	PM	NOx
Base case	347,924.68	243,874.30	63.43	199,865.60	421.65	97.05	2815.33
Scenario 1 (-10%)	313,132.21	235,568.97	61.09	192,484.62	406.08	94.10	2719.46
Scenario 2 (-20%)	278,339.74	227,063.31	58.69	184,934.50	390.15	91.07	2621.27

The results of the sensitivity analysis show that a decrease of 10–20% in the performance of the PV panels and, consequently, on the maximum energy produced, has a small effect on the energy provided by the PV system to cover the lighting loads. In particular, and according to the analytical calculations, in the case of a 10% decrease, the PV system can provide the full quantity of the required energy from March to September, as in the base case study. For a higher decrease of 20%, the PV system can cover the energy needs from April to September. For the remaining months, the PV system partially provides the required energy; however, the differences in the results between the base study and the two scenarios are small. The robustness of the PV system is due to the high solar potential of the sailing routes of the Ro-Ro vessel, which permits a large amount of solar energy to be converted to electricity for the sunny months (usually spring and summer) that exceeds the required energy in both scenarios. According to the sensitivity analysis, the implementation of the proposed PV system comprises an effective solution for reducing ship emissions, even if the performance of the PV system is decreased due to harsh environmental and/or weather conditions.

4.4. Costs, Benefits and Economic Indicators

Economic analysis is the means to determine if the solar PV system installation is profitable. Costs and benefits as well as the generated cash flows that result from the specific investment were calculated. To assess the economic feasibility of the PV system, the NPV, IRR and payback period were estimated for an investment period of up to 20 years.

Costs to be included in the economic evaluation are the capital cost for the PV system, and the operation and maintenance cost. A market survey was conducted to determine the current price for the purchase and installation of the PV system components. All prices presented in Table 5 were fixed prices for the period of August 2022 from a global supplier, who also provided a 33% discount on charge controllers and DC-AC inverters. The purchase price for the PV panels, as well as the foundation and installation cost of the PV system, are given in EUR/kW. The operation and maintenance cost (O&M) was calculated as a percentage of 0.5% of the capital cost for small-scale PV energy systems [21,40]. This fixed value was set for the first year of the investment, while an annual increase of 1% in the O&M cost due to the time value of the currency and the aging of the system was considered for the investment period.

Table 5. Cost estimation of the designed stand-alone PV system.

Item	Quantity	Price	Discount	Total Cost (EUR)
PV panels	272.25 kW	0.32 EUR /Watt	-	87,120.00
Solar batteries	1080	EUR 260.00	-	280,800.00
Solar charge controllers	45	EUR 1003.00	33%	30,240.45
DC-AC inverters	23	EUR 4290.00	33%	66,108.90
Foundation	272.25 kW	100 EUR /kW	-	27,225.00
Installation	272.25 kW	100 EUR /kW	-	27,225.00
Sum				518,719.35

The benefits of the investment include the direct economic benefits from the fuel savings, while indirect cost benefits such as prior entrance to ports, carbon credits, etc., are neglected. According to an oil market survey, the prices were estimated at USD 1234.5/t for MGO and USD 530.5/t for HFO for the period of August 2022 (exchange rate 1 USD = EUR 1.0066 on 24 August 2022, European Central Bank, Frankfurt, Germany), and these prices were set in the economic analysis for the first year of the investment. To estimate the fuel price increase during the 20-year investment period, various long-term forecast assumptions for the price of crude oil and distillates published by energy information and other organizations [41–44] were considered. For the base case under study, the central oil price scenario assumptions from different sources were combined and an annual increase of 2.47% for the MGO/HFO fuels was assumed.

For the specific investment, the discount rate that provides foresight on profitability was set to 8%, according to the respective accepted values in the literature [19,21]. The NPV and IRR were estimated to be EUR 79,931.21 and 9.76%, respectively. It was also computed that the discounted payback of the system would be achieved in a 16.5-year period, indicating that the proposed PV system could be a long-term profitable investment with financial viability, considering that the estimated operational lifespan of a PV module is about 25–30 years [45,46].

4.5. Sensitivity Analysis for the Economic Assessment

In the base case study, the Ro-Ro vessel consumed 40% MGO fuel and 60% HFO fuel. The vessel navigates in global waters where the fuel sulfur limit is 0.5%, so the use of the MGO and HFO with a scrubber comply with the regulation. However, in December 2022, the MEPC 79 adopted amendments to designate the Mediterranean Sea as an Emission-Control Area for SOx and PM, with the new sulfur limit taking effect from 1 May 2025 [47]. In this context, it is critical to evaluate the proposed PV system for a specific case where the Ro-Ro cargo vessel consumes only MGO fuel. It is possible for this specific case to exist due to more stringent emission regulations, which could designate the whole navigation area of the vessel as ECA, where the limit for fuel sulfur content is 0.1% instead of 0.5% in global waters. In such a case, the Ro-Ro vessel would consume only the MGO fuel that complies with the emission regulations of the IMO. For the "Only MGO" case, the economic analysis was performed, considering the same parameters and fuel cost assumptions as in the base case study. The economic indicators NPV and IRR and the payback period were estimated to be EUR 405,907.15, 16.14% and 9 years, respectively, indicating that installing the proposed stand-alone PV system in such a case could be even more economically profitable for the ship-owners.

Furthermore, one of the most critical factors for the financial viability of the PV system investment is the parameter of the fuel price and its variation during the investment period. Therefore, a sensitivity analysis was conducted according to a Low scenario (1% annual increase on fuels price) and a High scenario (3.5% annual increase on fuels price). The effect of the variation of the fuel prices on the NPV, IRR and payback period are presented in Table 6 relative to the base case of MGO/HFO and the "MGO-Only" case.

Table 6. Sensitivity analysis of the economic indicators with respect to fuel price.

Case	Price Scenario	NPV (EUR)	IRR (%)	Payback Period (Years)
	Low (1%)	12,613.63	8.3	20
MGO/HFO	Central (2.47%)	79,931.21	9.76	16.5
	High (3.5%)	131,450.27	10.73	15
	Low (1%)	303,574.54	14.67	9.8
MGO-Only	Central (2.47%)	405,907.15	16.14	9
	High (3.5%)	484,223.68	17.11	8.6

For the base case of MGO/HFO, the sensitivity analysis indicates that an annual increase of 1% in the oil price (Low scenario) results in an investment with a payback period of 20 years and could lead to the rejection of the PV system project. It is assumed that this long-term scenario corresponds to a sustainable development policy scenario where strong action to reduce carbon emissions is undertaken and the oil demand weakens. The High scenario could be more attractive to ship-owners, with a payback period of 15 years and IRR equal to 10.73%. This long-term scenario responds to a case where current energy policies are put into practice and the global market is characterized by low oil supply and high oil demand. The central scenario (base case scenario) is assumed to be the most realistic scenario to happen long-term (2035 onwards) and corresponds to new energy policies and interventions on emissions reductions as well as a central oil demand.

For the specific case of MGO-Only, all three scenarios can be financially viable and attractive for the ship owners. However, if policies and measures are adopted in the following years for sustainability and further ship emission reductions, the MGO-Only Low scenario seems to have very good prospects, with an NPV, IRR and payback period estimated to be EUR 303,574.54, 14.67% and 9.8 years, respectively.

4.6. Comparison to Previous Studies

A comparison between the results of the economic and environmental analysis between this study and previously published studies on the designed PV systems, on similar sizes and types of vessels [13,15,21,25], is presented in Table 7. The economic indicators and the quantity of the emission reductions depend mainly on the fuel savings according to the specific case study. Fuel savings are related to SFOC and the power generated from the PV panels, which depends on the solar irradiance at the vessel route, the PV panel efficiency, and the actual installation area of the PV panels on the ship. Therefore, a larger available deck area increases the capacity of the PV system and consequently the fuel savings. The values of the economic indicators also depend on the capital cost of the PV system, which is the cash outflow in the analysis. In the case of a grid-connected PV system, the capital cost of solar batteries that comprise a basic component of the stand-alone system is not included.

Table 7. Comparison of economic and environmental results between this study and previously published studies.

Study	Vessel	PV System	Area (m^2)/ No of PVs	NPV (USD)	Payback Period/Discount Rate	Fuel Reduction (tons)	Emissions (in kg) Reduced			
							CO$_2$	NOx	SOx	PM
Qiu et al. [13]	Pure Car Truck Carrier	g-con. *	900/-	165,977.2	7.84/8%	-	163,338	2753	107	70
Yuan et al. [25]	Pure Car Truck Carrier	g-con./ s.a. *	1050/540	-	-	4.02%	8.55%	-	-	-
Yuan et al. [21]	Pure Car Carrier	s.a.	-/135	-	7–20/3.5%	16	28,500	50	630	-
Karatuğ et al. [15]	Ro-Ro vessel	g-con.	2593.5/1274	3,362,397	11.2/8%	73.51	232,393	3942	312	114
This study Base case	Ro-Ro cargo vessel	s.a.	1280/450	79,407.12 **	16.5/8%	63.43	199,865	2815	421	97
This study MGO-Only Low scenario	Ro-Ro cargo vessel	s.a.	1280/450	301,584.09 **	10/8%	63.43	-	-	-	-

* g-con.: grid-connected, s.a.: stand-alone; ** The NPV values have been converted to USD according to the exchange rate used in the study.

The main contribution of this paper, relative to previous studies, is the use of representative fuel price assumptions for three scenarios (Low, Central, High) combining various long-term forecast assumptions published by energy information organizations [32–35]. To strengthen this assertion, we presented the following elements:

Qiu et al. [13] used a price of fuel oil equal to 0.709 USD/L, while no mention was made for future price predictions. Yuan et al. [25] used the historical average price of marine diesel oil from 2014 to 2017, equal to USD 837.96/t, while no mention is made for future price predictions. Yuan et al. [21] used a fuel price equal to RMB 0.426/kWh, and the payback period was estimated according to a different market-price growth rate of fuel to the order of 10–50%, but no information about the sources of these predictions was presented. Karatuǧ et al. [15] performed an analysis in which no detailed fuel prices were mentioned. In other studies of small-size vessels, Tercan et al. [22] simulated a PV system of 15 panels for a small tourist boat, using gasoline prices taken from the Global Petrol Price Index, without mentioning these values, and Wang et al. [20] presented a PV system of 206 PV panels on a short route ferry and used a fuel price equal to 401 USD/t according to Istanbul Bunker Prices, without taking into account future price predictions.

5. Conclusions

The present study focused on the design of a photovoltaic system on a Ro-Ro cargo vessel for electrical energy production for lighting during cargo operations. In this base, the installation of a stand-alone solar PV system that comprised 450 PV panels, solar charge controllers and solar batteries was examined from an environmental and economic point of view. PV panels were installed with a zero-tilt angle, and the technical characteristics n_{STC}, AF, SF and DF were equal to 21.4%, 0.98, 0.90 and 0.99, respectively (Equation (4)). The vessel consumed 40% MGO and 60% HFO and was equipped with a scrubber that limited the SOx emissions of the HFO to the allowed levels. The findings of the study are as follows:

(a) By adopting the proposed PV system, the analysis performed indicates that approximately 88% of the required energy for lighting can be provided by this renewable source of energy. This implies annual fuel savings of 63.43 t and a concomitant reduction in ship direct operating costs.

(b) From March to September the PV system provides the full amount of the required energy and the excess energy can be consumed for other purposes, such as the lighting loads at a harbor operating mode. For the remaining months the main generators can be used for full energy coverage.

(c) According to the environmental analysis results, the annual reduction in exhaust emissions sums up to 199,865 kg of CO_2, 2815 kg of NOx, 421 kg of SOx, and 97 kg of PM, contributing to a cleaner environment in the port areas.

(d) The sensitivity analysis for two scenarios of decreased maximum energy produced by the PV system (10% and 20% decrease) indicates that the PV system can fully cover the lighting loads from March to September and from April to September, respectively, while for the remaining month, the decrease of 10–20% has a small effect on the coverage of the required energy.

(e) The results of the economic analysis of the base case (40% MGO and 60% HFO) indicate that the PV system investment is profitable in the long term, with a Net Present Value of EUR 79,931.21, an Internal Rate of Return of 9.76%, and a payback period of 16.5 years (considering a discount rate of 8% and an annual increase in the fuels price of 2.47% for a 20-year period).

(f) The investment can be more profitable if the ship owners decide to change to Only MGO fuel, considering possible future regulations regarding emissions and ECAs. In such a case, the Net Present Value, the Internal Rate of Return and the payback period are estimated to be EUR 405,907.15, 16.14% and 9 years, respectively.

(g) The sensitivity analysis for two scenarios of different long-term fuel price assumptions (1% and 3.5% increase in fuel price) indicates that if new strategies for a sustainable low carbon economy and stricter regulations on ship emissions are adopted, the MGO-Only case seems to have a promising potential for financial viability.

The findings presented in this paper can lead the field researchers and the shipping community to consider solar energy not only as an effective alternative source for auxiliary power generation that reduces the ship emissions and the dependency on fossil fuels, but also as an economically viable investment. Ship owners and decision makers can assess the economic indicators obtained for the different fuel price scenarios in this study and conclude whether a similarly designed PV system is a profitable investment for a new-build or existing vessel. In addition, it is anticipated that further advancements in renewable energy technology will be implemented in the near future, specifically in photovoltaic systems, in terms of their efficiency and durability. The forthcoming developments are expected to reduce the reliance of marine transportation on fossil fuels, leading to very low or zero emissions and the development of energy-efficient ships.

Supplementary Materials: The following supporting information can be downloaded at: https://www.mdpi.com/article/10.3390/en16186523/s1.

Author Contributions: Conceptualization, M.S.; Methodology, M.S., P.P., I.I. and S.D.; Validation, I.I. and S.D.; Investigation, P.P.; Data curation, P.P.; Writing—original draft, M.S., I.I. and S.D. All authors have read and agreed to the published version of the manuscript.

Funding: This research was supported and funded by the Special Account for Research Grants of the University of West Attica.

Data Availability Statement: Data obtained from the "Global Solar Atlas 2.0", a free, web-based application, is developed and operated by the company Solargis s.r.o. on behalf of the World Bank Group, utilizing Solargis data, with funding provided by the Energy Sector Management Assistance Program (ESMAP). For additional information: https://globalsolaratlas.info (accessed on 1 May 2023). The data presented in this study are available in Supplementary Material upon request.

Conflicts of Interest: The authors declare no conflict of interest.

References

1. Breyer, C.; Bogdanov, D.; Ram, M.; Khalili, S.; Vartiainen, E.; Moser, D.; Medina, E.R.; Masson, G.; Aghahosseini, A.; Mensah, T.N.O.; et al. Reflecting the energy transition from a European perspective and in the global context—Relevance of solar photovoltaics benchmarking two ambitious scenarios. *Prog. Photovolt.* **2022**, *1*, 27. [CrossRef]
2. Choudhary, P.; Srivastav, R.K. Sustainability perspectives—A review for solar photovoltaic trends and growth opportunities. *J. Clean. Prod.* **2019**, *227*, 589–612. [CrossRef]
3. United Nations. *Transforming Our World: The 2030 Agenda for Sustainable Development*; A/Res/70/1; United Nations: New York, NY, USA, 2015.
4. Tay, Z.Y.; Konovessis, D. Sustainable energy propulsion system for sea transport to achieve United Nations sustainable development goals: A review. *Discov. Sustain.* **2023**, *4*, 20. [CrossRef]
5. Ašonja, A.; Vuković, V. The potentials of solar energy in the Republic of Serbia: Current situation, possibilities and barriers. *Appl. Eng. Lett.* **2018**, *3*, 90–97. [CrossRef]
6. Smil, V. Energy in the twentieth century: Resources, Conversions, Costs, Uses, and Consequences. *Annu. Rev. Energy Environ.* **2000**, *25*, 21–51. [CrossRef]
7. Rynska, E. Review of PV Solar Energy Development 2011–2021 in Central European Countries. *Energies* **2022**, *15*, 8307. [CrossRef]
8. Ritchie, H.; Max Roser, M.; Rosado, M. CO_2 and Greenhouse Gas Emissions. 2020. Available online: https://ourworldindata.org/co2-and-greenhouse-gas-emissions (accessed on 6 June 2023).
9. Yin, H.; Lan, H.; Hong, Y.-Y.; Wang, Z.; Cheng, P.; Li, D.; Guo, D. A Comprehensive Review of Shipboard Power Systems with New Energy Sources. *Energies* **2023**, *16*, 2307. [CrossRef]
10. Mallouppas, G.; Yfantis, E.A. Decarbonization in Shipping Industry: A Review of Research, Technology Development, and Innovation Proposals. *J. Mar. Sci. Eng.* **2021**, *9*, 415. [CrossRef]
11. International Maritime Organization. *Fourth IMO Greenhouse Gas Study 2020 Full Report*; IMO: London, UK, 2021.
12. International Maritime Organization. *Resolution MEPC.377(80) (Adopted on 7 July 2023) 2023 IMO Strategy on Reduction of GHG Emissions from Ships*; IMO: London, UK, 2023.
13. Liu, H.; Zhang, Q.; Qi, X.; Han, Y.; Lu, F. Estimation of PV output power in moving and rocking hybrid energy marine ships. *Appl. Energy* **2017**, *204*, 362–372. [CrossRef]
14. Tang, R. Large-scale photovoltaic system on green ship and its MPPT controlling. *Sol. Energy* **2017**, *157*, 614–628. [CrossRef]
15. Lan, H.; Dai, J.; Wen, S.; Hong, Y.-Y.; Yu, D.C.; Bai, Y. Optimal Tilt Angle of Photovoltaic Arrays and Economic Allocation of Energy Storage System on Large Oil Tanker Ship. *Energies* **2015**, *8*, 11515–11530. [CrossRef]
16. Tang, R.; Lin, Q.; Zhou, J.; Zhang, S.; Lai, J.; Li, X.; Dong, Z. Suppression strategy of short-term and long-term environmental disturbances for maritime photovoltaic system. *Appl. Energy* **2020**, *259*, 114183. [CrossRef]

17. Sun, Y.; Yan, X.; Yuan, C.; Tang, X.; Malekian, R.; Guo, C.; Li, Z. The application of hybrid photovoltaic system on the ocean-going ship: Engineering practice and experimental research. *J. Mar. Eng. Technol.* **2018**, *18*, 56–66. [CrossRef]
18. Salem, A.A.; Seddiek, I.S. Techno-Economic Approach to Solar Energy Systems Onboard Marine Vehicles. *Pol. Marit. Res.* **2016**, *23*, 64–71. [CrossRef]
19. Qiu, Y.; Yuan, C.; Tang, J.; Tang, X. Techno-economic analysis of PV systems integrated into ship power grid: A case study. *Energy Convers. Manag.* **2019**, *198*, 111925. [CrossRef]
20. Wang, H.; Oguz, E.; Jeong, B.; Zhou, P. Life cycle and economic assessment of a solar panel array applied to a short route ferry. *J. Clean. Prod.* **2019**, *219*, 471–484. [CrossRef]
21. Karatuğ, C.; Durmuşoğlu, Y. Design of a solar photovoltaic system for a Ro-Ro ship and estimation of performance analysis: A case study. *Sol. Energy* **2020**, *207*, 1259–1268. [CrossRef]
22. Tercan, S.H.; Eid, B.; Heidenreich, M.; Kogler, K.; Akyürek, Ö. Financial and Technical Analyses of Solar Boats as a Means of Sustainable Transportation. *Sustain. Prod. Consum.* **2021**, *25*, 404–412. [CrossRef]
23. Ling-Chin, J.; Roskilly, A.P. Investigating the implications of a new-build hybrid power system for Roll-on/Roll-off cargo ships from a sustainability perspective—A life cycle assessment case study. *Appl. Energy* **2016**, *181*, 416–434. [CrossRef]
24. Ling-Chin, J.; Roskilly, A.P. Investigating a conventional and retrofit power plant on-board a Roll-on/Roll-off cargo ship from a sustainability perspective—A life cycle assessment case study. *Energy Convers. Manag.* **2016**, *117*, 305–318. [CrossRef]
25. Yuan, Y.; Wang, J.; Yan, X.; Li, Q.; Long, T. A design and experimental investigation of a large-scale solar energy/diesel generator powered hybrid ship. *Energy* **2018**, *165*, 965–978. [CrossRef]
26. Ghenai, C.; Bettayeb, M.; Brdjanin, B.; Hamid, A.K. Hybrid solar PV/PEM fuel Cell/Diesel Generator power system for cruise ship: A case study in Stockholm, Sweden. *Case Stud. Therm. Eng.* **2019**, *14*, 100497. [CrossRef]
27. Yuan, C.; Pan, P.; Sun, Y.; Yan, X.; Tang, X. The evaluating on EEDI and fuel consumption of an inland river 800PCC integrated with solar photovoltaic system. *J. Mar. Eng. Technol.* **2021**, *20*, 77–92. [CrossRef]
28. Huang, M.; He, W.; Incecik, A.; Cichon, A.; Krolczyk, G.; Li, Z. Renewable energy storage and sustainable design of hybrid energy powered ships: A case study. *J. Energy Storage* **2021**, *43*, 103266. [CrossRef]
29. Perdios, S. *Photovoltaic Installations*, 3rd ed.; Selka 4M: Athens, Greece, 2011; ISBN 9789608257535.
30. Kobougias, I.; Tatakis, E.; Prousalidis, J. PV Systems Installed in Marine Vessels: Technologies and Specifications. *Adv. Power Electron.* **2013**, *2013*, 831560. [CrossRef]
31. Paulson, M.; Chacko, M. Marine Photovoltaics: A review Of Research and Developments, Challenges and Future Trends. *Int. J. Sci. Technol. Res.* **2019**, *8*, 1479–1488.
32. Lee, J.-U.; Lee, W.-J.; Jeong, E.-S.; Noh, J.-H.; Kim, J.-S.; Lee, J.-W. Algorithm for Monitoring Emissions Based on Actual Speed of Ships Participating in the Korean Vessel Speed Reduction Program. *Energies* **2022**, *15*, 9555. [CrossRef]
33. International Maritime Organization. *Third IMO Greenhouse Gas Study 2014. Executive Summary and Final Report*; IMO: London, UK, 2015.
34. Global Solar Atlas 2.0. Available online: https://globalsolaratlas.info (accessed on 6 June 2023).
35. Photovoltaic Geographical Information System (PVGIS). European Commission. Available online: https://re.jrc.ec.europa.eu/pvg_tools/en/ (accessed on 6 June 2023).
36. Petrou, P. Renewable Energy Sources in Shipping—Photovoltaic System Installation in a Roll-on/Roll-off Cargo Ship. Bachelor's Thesis, University of West Attica, Athens, Greece, 29 November 2022. [CrossRef]
37. Ghazi, S.; Ip, K. The effect of weather conditions on the efficiency of PV panels in the southeast of UK. *Renew. Energy* **2014**, *69*, 50–59. [CrossRef]
38. Dajuma, A.; Yahaya, S.; Touré, S.; Diedhiou, A.; Adamou, R.; Konaré, A.; Sido, M.; Golba, M. Sensitivity of Solar Photovoltaic Panel Efficiency to Weather and Dust over West Africa: Comparative Experimental Study between Niamey (Niger) and Abidjan (Côte d'Ivoire). *Comput. Water Energy Environ. Eng.* **2016**, *5*, 123–147. [CrossRef]
39. Atkinson, G.M. Analysis of marine solar power trials on Blue Star Delos. *J. Mar. Eng. Technol.* **2016**, *15*, 115–123. [CrossRef]
40. Sagani, A.; Mihelis, J.; Dedoussis, V. Techno-economic analysis and life-cycle environmental impacts of small-scale building-integrated PV systems in Greece. *Energy Build.* **2017**, *139*, 277–290. [CrossRef]
41. BEIS 2019 Fossil Fuel Price Assumptions. Department for Energy Security and Net Zero and Department for Business, Energy & Industrial Strategy. Available online: https://www.gov.uk/government/publications/fossil-fuel-price-assumptions-2019 (accessed on 6 June 2023).
42. World Energy Outlook 2019. International Energy Agency (IEA). Available online: https://www.iea.org/reports/world-energy-outlook-2019 (accessed on 6 June 2023).
43. Annual Energy Outlook 2022. U.S. Energy Information Administration (EIA). Available online: https://www.eia.gov/outlooks/aeo/data/browser/#/?id=12-AEO2022&cases=ref2022&sourcekey=0 (accessed on 6 June 2023).
44. DNV Maritime Forecast to 2050, Energy Transition Outlook 2022. DNV. Available online: https://www.dnv.com/maritime/publications/maritime-forecast-2022 (accessed on 6 June 2023).
45. Kim, J.; Rabelo, M.; Padi, S.P.; Yousuf, H.; Cho, E.-C.; Yi, J. A Review of the Degradation of Photovoltaic Modules for Life Expectancy. *Energies* **2021**, *14*, 4278. [CrossRef]

46. Tan, V.; Dias, P.R.; Chang, N.; Deng, R. Estimating the Lifetime of Solar Photovoltaic Modules in Australia. *Sustainability* **2022**, *14*, 5336. [CrossRef]
47. International Maritime Organization. *Resolution MEPC.361(79) (Adopted on 16 December 2022) Amendments to MARPOL Annex VI Concerning a Mediterranean Sea Emission Control Area for Sulphur Oxides and Particulate Matter*; IMO: London, UK, 2022.

Review

Literature Review of Hydrogen Energy Systems and Renewable Energy Sources

Grigorios L. Kyriakopoulos [1,*] and **Konstantinos G. Aravossis** [2,*]

[1] School of Electrical and Computer Engineering, National Technical University of Athens, 9 Heroon Polytechniou Street, 15780 Athens, Greece

[2] School of Mechanical Engineering, Sector of Industrial Management and Operations Research, National Technical University of Athens, 9 Heroon Polytechniou Street, 15780 Athens, Greece

* Correspondence: gregkyr@chemeng.ntua.gr (G.L.K.); arvis@mail.ntua.gr (K.G.A.)

Abstract: The role of hydrogen as a clean energy source is a promising but also a contentious issue. The global energy production is currently characterized by an unprecedented shift to renewable energy sources (RES) and their technologies. However, the local and environmental benefits of such RES-based technologies show a wide variety of technological maturity, with a common mismatch to local RES stocks and actual utilization levels of RES exploitation. In this literature review, the collected documents taken from the Scopus database using relevant keywords have been organized in homogeneous clusters, and are accompanied by the registration of the relevant studies in the form of one figure and one table. In the second part of this review, selected representations of typical hydrogen energy system (HES) installations in realistic in-field applications have been developed. Finally, the main concerns, challenges and future prospects of HES against a multi-parametric level of contributing determinants have been critically approached and creatively discussed. In addition, key aspects and considerations of the HES-RES convergence are concluded.

Keywords: hydrogen energy system (HES); renewable energy sources (RES); clusters; bibliometric analysis; environment; hydrogen efficiency; greenhouse gases (GHGs) emissions; household sector; hydrogen case studies

Citation: Kyriakopoulos, G.L.; Aravossis, K.G. Literature Review of Hydrogen Energy Systems and Renewable Energy Sources. *Energies* **2023**, *16*, 7493. https://doi.org/10.3390/en16227493

Academic Editor: Vladislav A. Sadykov

Received: 2 October 2023
Revised: 1 November 2023
Accepted: 7 November 2023
Published: 8 November 2023

1. Introduction

Globally, the development of economic activities is associated with a great interest in energy-consuming services and the subsequent increase of fuel consumption. However, this fuel demand is commonly followed by high and unregulated carbon dioxide emission, which is the main source generator of the greenhouse gas (GHG) effect. Fossil fuels (mainly in the forms of petroleum, natural gas and coal) are experiencing accelerated consumption and stock depletion, despite meeting today's global energy demand. Such combustion products are the main contributors to environmental problems and uncontrollable climate changes, thus threatening the global environmental safety and sustainability. Among feasible and realistic solutions to these global problems are those which have been proposed by engineers and scientists who agree to replace the existing fossil-fueled energy systems by the hydrogen energy system (HES). Subsequently, hydrogen can be fed to fuel energy-consuming services in order to improve energy security and simultaneously control the GHG effect [1–3]. A plausible hydrogen energy system, which includes its resources, production technologies, storage, fuel-tank, dispensing and utilization, can be primarily analyzed for urban services, but it remains underdeveloped or sparsely studied in alignment with the renewable energy sources (RES). Hydrogen can be characterized as an efficient and clean fuel. Technological advantages are that the hydrogen combustion is neither a GHG producer nor a generator of ozone-layer-depleting chemicals. Moreover, hydrogen combustion does not generate acid rain ingredients or air pollution. It is also noteworthy that hydrogen produced from RES can develop a stable and permanent energy system that is never subject to future changes or modular modifications.

In the following sections of this literature review, a classification of HES-collected studies was developed into five fields (Section 2), followed by a critical overview of the operation characteristics and the technological synergies developed between HES and RES (Section 3). Moreover, the main key aspects, considerations and future prospects of HES and RES convergence are discussed in the Conclusions section (Section 4).

2. Methods and Analysis

Clusters of Hydrogen Energy Systems (HES) and Renewable Energy Sources (RES)

While in the global literature there is a large number of studies focused on HES and RES, the convergence fields of HES and RES have been sparsely investigated or scientifically framed. In response to this research topic, an extensive literature review on both the key phrases HES and RES was undertaken in the first half of the year 2023, in the Scopus database. The historical time of publication dates back to the year 2008 onwards. Then, the collected studies were grouped into the following fields of classification: (a) simulation tests, (b) optimization models and mathematical analyses, (c) experimental pilot industrial and in-field processes including electrolyzer utilities, (d) literature reviews and publication taxonomies of HES, and (e) convergence topics of photovoltaic and hydrogen energy systems. The studies contained in these clusters were grouped and represented in the form of Table 1, following the classifications of (a), (b), (c), (d), and (e). These were collectively represented only in terms of the referencing/citing data collected from a reverse chronological order of publication, without an explanatory text of each one citation listed, due to high conceptual dispersal and heterogeneity of the collected citations. In cases where there arose common fields of classification overlapping among the five determined groups, then the relevant citation was positioned with the classification of the closest conceptual or operation affinity compared to the others. In such a way, all five classifications are presented as follows:

Table 1. Classifications a, b, c, d, and e: HES and RES overview in the classification/cluster of: (a) "simulation tests", (b) "optimization models and mathematical analyses", (c) experimental pilot industrial and in-field processes including the electrolyzer utilities, (d) literature reviews and publication taxonomies of HES, and (e) convergence topics of photovoltaic and hydrogen energy systems.

Classification #	Reference #	Number of Citations in Absolute Numbers or as Percentage of the Total Citations Collected %	Intensity Rations (Per Basis of the Lowest-Documented Field of Cluster #a)
a	Lu et al., 2023; Çiçek, 2023; Ren et al., 2023; Virji et al., 2020; Maghami et al., 2020; Parra et al., 2019; Cho et al., 2018; Rosen and Koohi-Fayegh, 2016; Özden and Tari, 2015; Lanjewar et al., 2014; Contreras and Posso, 2011 [4–14]	11 or 16.7%	2.75
b	Wang et al., 2023; Zhao et al., 2023; Sun et al., 2023; Alanazi et al., 2022; Dong et al., 2022; Schrotenboer et al., 2022; Marocco et al., 2021; Zhang et al., 2021; Wang et al., 2021; Onwe et al., 2020; Manilov, 2019; Zhang et al., 2019a; Zhang et al., 2019b; Alavi et al., 2017; Ren et al., 2017; Lacko et al., 2014; Patricio et al., 2012; Aguado et al., 2009 [15–32]	18 or 27.3%	4.50
c	Acar et al., 2023; Karaca and Dincer, 2023; Lin and Li, 2023; Alex et al., 2022; Ibrahim et al., 2022 Balasubramanian et al., 2021; Endo et al., 2021; Endo et al., 2020a; Endo et al., 2020b; Li et al., 2020; Yilmaz, 2020; Sorgulu and Dincer, 2018; Kalinci et al., 2017; Ewan et al., 2016; Khalid et al., 2016; Maleki et al., 2016; Balabel and Zaky, 2011; Bendaikha et al., 2011; Chao and Shieh, 2011; Kikkinides, 2011; Gabriel García Clúa et al., 2008 [33–53]	21 or 31.8%	5.25

Table 1. *Cont.*

Classification #	Reference #	Number of Citations in Absolute Numbers or as Percentage of the Total Citations Collected %	Intensity Rations (Per Basis of the Lowest-Documented Field of Cluster #a)
d	Cheng et al., 2023; Blanco et al., 2022; Sezgin et al., 2022; Temiz and Dincer, 2022; Yue et al., 2021; Martin et al., 2020; Valente et al., 2019; Valente et al., 2017; Zini and Tartarini, 2012; Rosen, 2012; Noyan, 2011; Li et al., 2011 [54–65]	12 or 18.2%	3.00
e	Yamamoto and Ushifusa, 2022; Deng et al., 2021; Budak and Devrim, 2020; Shiroudi et al., 2013 [66–69]	4 or 6.0%	1.00
	Total references	**66 or 100%**	

According to the literature search outcomes, as presented in Table 1 above, the total number of citations collected for all five clusters is 66, which was further allocated per each cluster in absolute and percentage numbers, as it is shown in column 3, above. This percentage numbering revealed the research focus of a cluster topic compared to the other four. Similarly, the absolute number of citations revealed the least- and most-reported clustering topics, showing that cluster e has the lowest number of citations and cluster c has the most literature studied in the field. Therefore, taking "1" as the number of citations (4 citations) for the least-reported cluster 1, then a calculation of "intensity rations" (column 4 of Table 1) enabled the quantification of the relevant research dynamics for each cluster in terms of the citations collected for each cluster. Indeed, based on the aforesaid clustering of the collected citations, it can be concluded that the order of literature interest and research focus was observed as follows (in descending order):

Clusters: c > b > d > a > e, implying that there are also two subtle peaks in research interests, coupled in the following way:

Cluster c, similar to b, followed by cluster d, similar to a. Taking into consideration the above findings, the most popular fields of HES-RES investigation can be denoted as that of experimental–pilot industrial and in-field processes, including the electrolyzer utilities, together with analyses focused on optimization (including mathematical) models. The second most popular group/pair of fields of HES-RES investigation is that of literature reviews and publication taxonomies of HES, together with the cluster of simulation tests. However, there is little literature interest demonstrated for the convergence topics of PVs and HES. However, the aforementioned argument is not bold and affirmative, since research into "channeling and diffusion" has been reported among the five selected clusters, but the critical point here is that all observable and traceable literature production of HES-RES can be categorized with certainty and taxonomized into one of the aforementioned five clusters proposed.

Based on the descriptive statistics of Table 1 and taking as the basis of comparison the lowest number of references (four references in total were reported in class e), five intensity ratios were calculated (in the right-hand column of Table 1) in order to arrive at some remarks and draw important conclusions that are hidden when first confronting these clusters which developed. Indeed, based on the intensity ratios, it can be further inferred that the clusters a (2.75) and d (3.00) refer predominately to approaches of bridging HES and RES, which, between them, have almost equal intensity scores. However, the most attractive cluster c (5.25), which is that of "experimental, pilot, industrial and in-field processes" is of particular interest among researchers, mainly due to its practical utility and problem-solving orientation compared to the other four clusters. This said, it cannot be ignored that insightful remarks and novel knowledge are also disclosed in the more "theoretically oriented" clusters. Therefore, a synergistic and balanced—rather than an antagonistic or detached/mutually autonomous—approach to both the practical and theoretical clusters is realistic, plausible and recommended.

3. Results and Discussion

3.1. Results of the Bibliometric Analysis

Based on the information collected and classified in this literature review, from Table 1 above, the quantitative information and the relative strength of the five corresponding derived clusters are presented in Figure 1.

Number of references (%) per field of classification

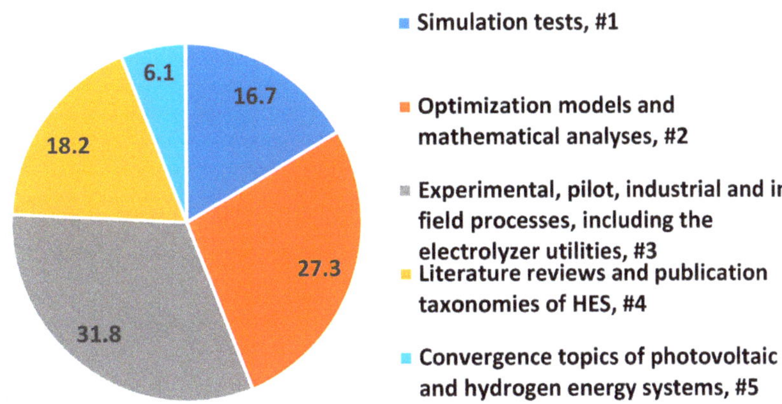

Figure 1. The reference allocation profile for each cluster developed. Source: Authors' own study.

Based on Figure 1, it can be argued that the clusters b and c, "optimization models and mathematical analyses" and "experimental, pilot, industrial and in-field processes" attract the research interest of almost 60% of the total references linking HES and RES. This result is not surprising since the primary role of HES and RES convergence is to establish practical and problem-solving priorities rather than the rather theoretical attribution of the other three clusters a, d, and e.

The research target and the novelty of this literature review is to reveal the research focus on HES publications regarding the contribution of RES technologies. Such studies of HES-powered technology illustrated the increasing importance of hydrogen in contemporary energy power systems, following the adoption of selected technological procedures and those energy-planning criteria for screening the yearly evolution of publications. Such publications investigated different hydrogen power-system applications within the last three decades, as well as the percentage distribution, as shown in Figure 1.

Based on these research studies, it is noteworthy that research into HES-RES contains energy-storage information in terms of storage materials and tanks, as well as an operational/system overview. The key aspects of energy storage in c pointed to an increasing trend in research effort, due to a high level of interest, owing to the rapid development of material engineering, followed by studies on HES for transportation. Less research activity has been focused on applications of power-to-gas and co- and tri-generation, though these are rapidly emerging fields for advantageous research activities in the coming future [58].

Another critical finding of this literature review is the time evolution of research studies regarding the different and multifaceted HES-RES contexts. In particular, the searched findings included issues of cost, efficiency, durability, feasibility and effectiveness of mature technologies or fast-developing HES with scalability caliber. While research on techno-economic and efficiency interest has attracted the most interest in HES, other fields of investigation are those of durability and lifetime infrastructure. This research emphasis has developed especially in the last two decades of research. A challenging topic for future technological orientation is the durability of powered HES, but it still falls behind in reaching the satisfactory level/full scale of diagnostic, prognostic and fault-tolerant control functioning [58].

3.2. Overview of Selected Hydrogen Applications

3.2.1. HES–RES in Practice

When approaching the technologies involved in hydrogen production from RES, besides the apparent environmental advantages of a highly efficient and clean fuel (no GHG contributor, no chemicals causing ozone depletion), it cannot be ignored that the RES technologies (e.g., solar, wind) are leading to permanent and fixed energy systems that cannot later change to include other RES types which we are interested in [70]. In this context, the results of the review developed above are suitable and significant for a wide spectrum of literature-oriented approaches toward RES utility among various HES. Among these, the electrifying transportation sector, which is a promising approach to alleviating climate change issues arising from increased emissions, is noteworthy [71]. In this study, the examined HES can produce hydrogen, using RES, for the transportation sector (in buses). In this case, the electricity demanded for hydrogen production is harvested from the electrolysis of water, covered by RES. Moreover, fuel cells can use hydrogen to power the bus, while an HES exergy analysis referred to a steady-state model of the processes for which exergy efficiencies were calculated for all subsystems. Therefore, those subsystems showing the highest proportion of irreversibility were identified and compared. An exergetic efficiency of 12.74% for the PV panel, 45% for the wind turbine, 67% for the electrolysis, and 40% for the fuel cells was reported [71].

Depending on the production process and the selected (energy planning-based) energy source, the obtained hydrogen can be classified as grey, blue, or green, as shown in Figure 2.

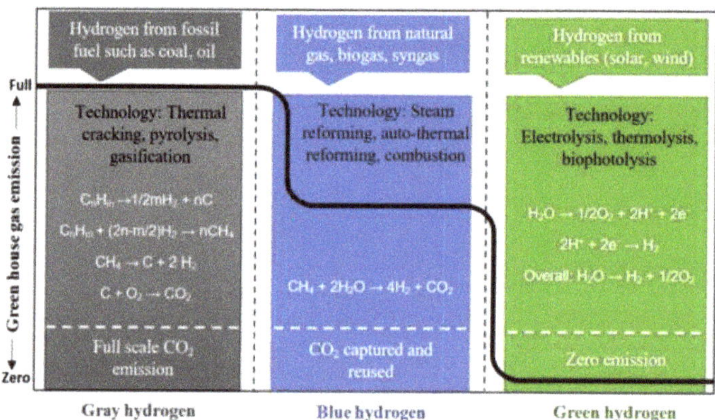

Figure 2. Outline of hydrogen-based feedstocks and technologies. Source: Sarker et al. [72].

As shown in Figure 2, the grey-hydrogen type is hydrogen produced from steam methane reforming and thermal cracking [72]. As is also shown in Figure 2, the main CO_2 production is reported through methane steam reforming, which can be further collected and stored safely in containers in the form of CO_2 vapors. Natural gas, biogas, and syngas can also produce hydrogen that is categorized as blue hydrogen, where the CO_2 gases formed cannot be stored, and will be released into the atmosphere. Blue hydrogen, which is produced from natural gas, unlike grey hydrogen, is capable of significantly reducing CO_2 emissions and simultaneously capturing and reusing carbon. As observed in Figure 2, both grey- and blue-hydrogen production processes generate CO_2 as the by-product, but in the case of green-hydrogen production technologies, zero carbon emissions are also apparent. While solar and wind technologies have been utilized to produce green hydrogen, there are other catalytic-reforming technologies of sound capability for producing green hydrogen.

Among them, biomass gasification and nuclear thermal/chemical pathways can potentially reduce carbon emissions. Because of this, major challenges such as production technology costs, system durability, reliability, infrastructure, and safety are issues of consideration [72]. It is roughly estimated from the life-cycle assessment that hydrogen production through biomass gasification contributes less GHG emission (405–896.61 g CO_2/kg H_2) compared to wind-driven electrolysis (600–970 g CO_2/kg H_2) [72]. However, biomass gasification has not been scaled up so far, and thus it can be expected that its input into global energy production will help to attain its full potential in the near future [72]. Furthermore, higher moisture content, low hydrogen production, and high operating costs are major constraints associated with biomass gasification. On the other hand, the solar- and wind-powered electrolysis techniques were proven to be well-established renewable sources for producing hydrogen through electrolysis [72].

The schematic of the solar–hydrogen hybrid storage system (HESS) for the case study of a house is given in Figure 3. The experimental structure of such an analysis considers a house which is remote from the national power grid with a meteorological solar/wind measurement tower in order to collect data used in the relevant case study. The meteorological measurement system can collect 10 min of data for average solar irradiance, solar duration, wind speed and direction, air temperature and pressure and also relative humidity. The desired characteristics of such an installation are devoted to providing uninterrupted power to houses at any time of the year through the suitable solar-HES.

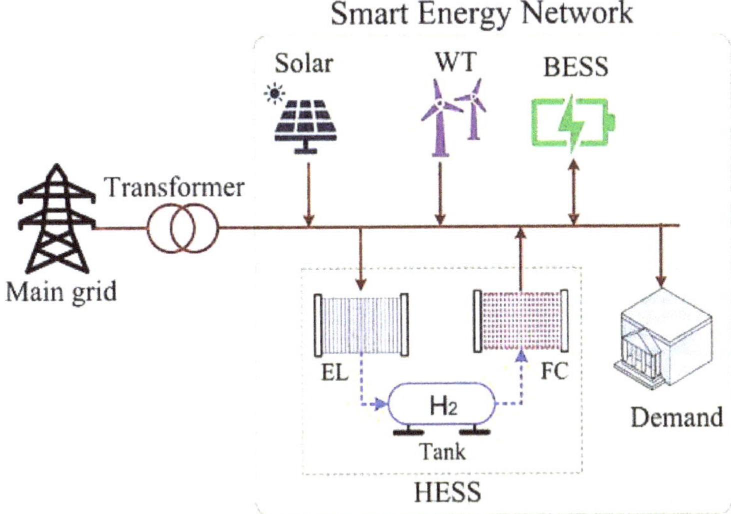

Figure 3. A schematic of a typical domestic solar–hydrogen hybrid energy system. Source: Samende et al. [73].

A diagrammatic structure of a wind–hydrogen energy system, with its elements and some of their characteristics and relationships, are all shown in Figure 3. The projected HESS can transform the energy into a storable product, hydrogen [32]. This transformation process uses water electrolysis and the compression of hydrogen for storing. A technical limitation of the hydrogen tank is its finite capacity. Moreover, the latest transformation process is reforming hydrogen into ready-to-be distributed and sold energy. The pricing of energy selling is related to the energy-storage system through which energy can be sold when prices rise due to a demand increase [32]. Apart from the concurring energy losses due to inefficiencies in the processes of transformation and recovery, there is also the limitation of the cost of the hydrogen equipment required. The extreme fluctuations reported for energy prices depend on whether (or not) the amount to be sold has been pre-committed (the day before). In the case of the price of pre-commitment, this is a higher value, but if the agreed amount is ultimately not supplied, then a penalty is paid. When more than the agreed amount is supplied, then the surplus has a lower price [32].

Besides the aforementioned constraint of finite storage of the hydrogen tank, other operational constraints of different system elements are the existence of a maximum capacity for transforming the energy into hydrogen, as well as a finite power limit, that is, a maximum amount of energy that can be dumped into the grid (and then sold) for each time unit. Another limitation is the inefficiency in the processes of transformation and recovery, since for these processes energy output is less than energy input [32]. Because of this, modelling optimization is needed to determine the proper hydrogen storage capacity for the system's optimization. When selecting from among different tank-size situations, it is noteworthy that storage tank capacity (a) is needed to transform the energy curve into a constant curve through time, and (b) is needed for storage of energy that has been produced above the power limit to dump it into the network again when the production is below the power limit [32].

For an HESS, the simulation design is a key factor, taking into consideration also the constraint of missing data for continuous yearly consumption measurements, whereas measurements are subject to daily and seasonal variations of power consumption in houses. Indeed, energy consumption decreases, since the households are at sleep from sunset and midnight and up until the next morning, where the fuel cell stack is the only available power source in the hybrid system. Regarding seasonal variation, higher energy consumption is reported in winter (the most in-house occupation) than in summer [73]. Regarding the HESS, it is noted that the hydrogen amount in the storage tank decreases from November to February, followed by an increase from March to May, thus shaping repeating hydrogen cycles (second-year cycles) with an arrangement of solar panels and some decades of the cubic meters of storage tanks given. Between June and October there is a full storage tank of hydrogen, showing very little oscillations around the full capacity. Therefore, the decrease and increase in the hydrogen in the storage tank occurs at fewer times of the year, and the hydrogen storage tank stays at almost full capacity for more time of the year. Moreover, changing the volume of the storage tank, therefore, means that the hydrogen cycle in the storage tank changes even though all the other HESS components are kept as they are [73]. The most critical specifications of storage tanks are those of weight, volume, and cost of the cryogenic or high-pressure tanks. Furthermore, the energy efficiency of compressing or liquefying the hydrogen can be also considered. It is also important to signify those pressure tank specifications in future designs which are oriented for concrete/modular applications [70].

From a generalized perspective, the findings of this literature review revealed an imperative need for advances in integrating hydrogen into power systems to be gradually approached in a synergistic way over recent years, ranging from production and storage to re-electrification and safety issues [58]. Therefore, while extensive descriptions of the existing progress can be the research focus of other studies, this literature review reported the need to seek to characterize the current progress in HES-RES integration through novel methods.

3.2.2. HES and Electrolyzer Efficiency

In the relevant literature [49], the general process that researchers follow in order to understand in a comprehensive way the effects of the space between the two electrodes on the rate of hydrogen production, among the different gaps examined, was demonstrated. Loss in mass increase is observed with voltage increase, under the condition of a constant space between the two electrodes. The resulting increase in the percentage mass lost due to the increase in the voltage is attributed to the increase in electrical current. During relevant experimental tests, the pair of electrodes can be exposed to a PV generator with low voltages of 2–5 V. When testing different concentrations of electrolyte solution, then, the percentage water mass lost during a specified time is measurable and the corresponding electrolyzer efficiency is calculated using the mass ratio of the hydrogen in a single molar mass of water [49]. Subsequently, voltage input and the gap between the electrodes play a determining role in the overall performance of the water electrolysis unit. Higher hydrogen production rates are achievable at a closer space between the electrodes, as well as at higher voltage input. Maximum electrolyzer efficiency is realistic for a smaller gap between electrodes, coupled with a specified input voltage value within the designed range [49]. A decrease in space between the pair of electrodes results in lowering the percentage of water mass loss, which increases due to the resulting decrease in electrical resistance between the electrodes, and thus leads to an increase in the electrical current. Hydrogen production rate increase, and thus energy efficiency increase, is due to the decrease in the space between the electrodes (small gaps between the pair of electrodes) being noticeable at higher degrees of input voltage, up to a specified input voltage; then, efficiency decreases by the further increase in the input voltage [49]. Figure 4 illustrates the schematic flow chart for hydrogen production. In the first stage, the electricity is generated from a hybrid renewable source (that is, a combination of wind turbine and solar PV), and then an electrolyzer is used in the second stage. In the electrolyzer, water is split into hydrogen gas as the principal product, while oxygen gas is released and generated as a by-product. Subsequently, hydrogen gas is passed through a compressor for storage purposes. Such compressed hydrogen can be transported in order to be utilized for commercial purposes. In operational terms, an electricity shortfall of hybrid solar and wind electricity is certainly affected by the electrolyzer-produced green hydrogen.

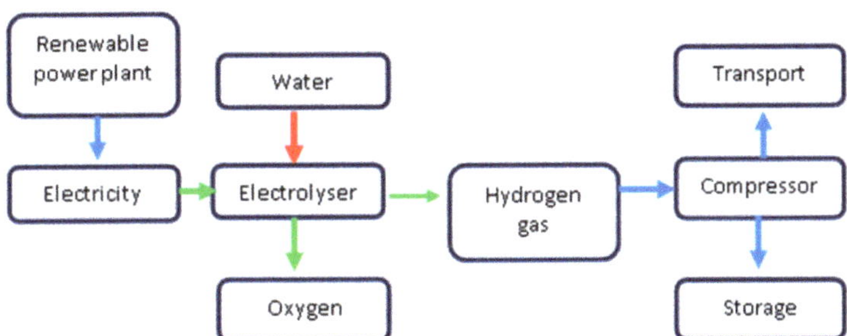

Figure 4. An overview of RES-based hydrogen hybrid system. Source: Sarker et al. [72].

As depicted in Figure 4, above, the RES-based hydrogen hybrid system, which in the literature is referred to also as a hybrid renewable energy system (HRES), has proven a feasible solution to address the issues related to individual energy sources [72]. Usually, a typical hybrid green-energy system uses various RES, such as wind and solar, as shown in Figure 5.

Figure 5. Hydrogen production from hybrid energy-based electrolysis. Source: Sarker et al. [72].

The benefits of HRES rely on multiple RES that are able to supply consistent and uninterruptible energy. Therefore, this energy availability will compensate for the unreliability of single renewable-energy sources and reduce GHG emissions [72]. It is also noteworthy that such systems are mainly located very close to the place of demand, thus lowering the possibility of damage to the transmission wire, and supporting also a prompt access to repair and maintenance when needed. Moreover, the fact that GHG emission is the main contributor to global warming is also concerning; thus, low, or zero, emissions achieved from RES technologies is a promising solution, making of utmost importance the priority for the utilization of those reliable renewable-energy generation systems as the best solutions for research projects worldwide [72].

Given specific configurations of electrolyzers and kW power input, the performance of electrolyzers can be subjected to computer tools that support their management by optimizing their efficiency. Since maximum capacity and an efficiency curve can be supported by the electrolyzer, it is observable that at lower than 25% of nominal power, the system cannot be working or in operation [32]. Generalizing, the higher hydrogen–electricity conversion efficiency is feasible while replacing the internal combustion engine with a fuel cell, but also leads to an equally costly system [72].

3.2.3. Hybrid HES-RES Aspects

Hence, hydrogen storage systems should incorporate more compact, more robust and less-costly fuel cell systems, being installed before wind/hydrogen systems. Additionally, multiple-energy-fed hybrid system solutions—such as wind, solar, and bioenergy—should be selected according to the spatial RES characteristics of each location. Moreover, HES and RES projects should be designed in alignment with "green" incentives, green certificates, and CO_2 tax-free policies. Other economics incentives are the fluctuation in oil and gas prices, the capital monetization of green energy infrastructure and the energy supply security. These factors justify the construction and motivate the development of a full-scale demonstration system (such as the wind/hydrogen plant at Utsira). In this case, such types of demonstrations can plausibly promote public awareness and acceptance, improving RES cost competitiveness, and regulating the new energy-market barriers and the enablers of new technological solutions (in general) and hydrogen technology (in particular) [72].

A typical integrated modular representation of widely used hydrogen-based hybrid systems is depicted in Figure 6.

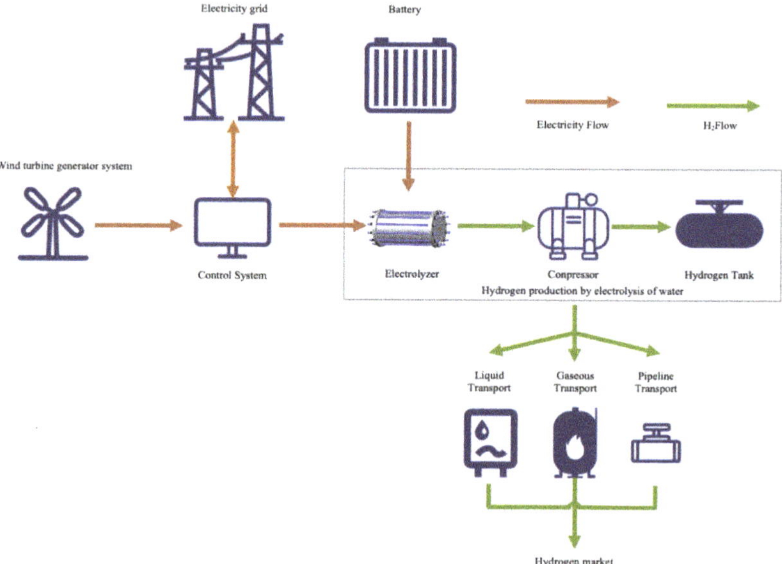

Figure 6. Schematic flow chart of a wind/hydrogen plant. Source: Niu et al. [74].

The system integrates various elements, including renewable energy generation, power storage, hydrogen production, and hydrogen transportation (Figure 5). The electricity generated by a wind turbine can be directly supplied to the power grid or stored in batteries for future use, optimized by the control system. Another fraction of the electricity is utilized in an electrolyzer device, which separates water (H_2O) into hydrogen gas (H_2) and oxygen gas (O_2). The generated hydrogen gas is then compressed and stored in hydrogen tanks for further transportation and utilization [74]. In this process, the electrolyzer, compressor, and hydrogen tanks are essential components for hydrogen production through water electrolysis. Hydrogen gas can be transported to various demand points, being accessible to the hydrogen market through various methods, such as liquid transportation, gas transportation, or pipeline transportation. Liquid transportation is suitable for long-distance transport, while gas and pipeline transportation are commonly used for short-distance distribution. In this context, the term "Electricity Flow" (Figure 6) refers to the movement of electricity from the wind turbine through the control system to the power grid, batteries, and electrolyzer. Similarly, the term "H_2 Flow" (Figure 6) represents the process of hydrogen gas being produced by the electrolyzer, compressed, stored, and ultimately transported to the hydrogen market [74].

Based on the configuration of Figure 6, it can be seen that the PV operating voltage and current determine the power output of the PV array and depend on the insolation on the PV module, the ambient temperature, and the manufacture characteristics of the PV module. Such a successful operation system for the wind turbine unit depends on several climatic and technical factors, such as wind speed and the rated cut-out and cut-in speeds (in m/s) for the wind turbine [26]. From a design perspective, the optimal design for off-grid hybrid renewable systems based on solar and wind energy to continuously meet the load necessitates a consideration of the LPSP (loss of power supply probability) and minimizing of the TLCC (total life-cycle cost), subject to constraints. The optimization algorithm is the most determining factor that can be efficiently used for this type of designed hybrid energy system. In particular, at low LPSP values (0–5%), the combination of a hydrogen and photovoltaic (PV) scheme with weather forecasting data leads to the most cost-effective system, and at LPSP = 10% the combination of wind, PV and a hydrogen scheme is proven to be the most cost-effective hybrid system [26].

In a similar optimization study of a stand-alone renewable H_2-based microgrid, its optimal sizing requires a reliable load demand to satisfy, by means of local renewable energy supported by a hybrid battery/hydrogen storage unit, the minimization of the system costs. A crucial factor is the price reduction, due to the installation and operation of a high number of components. In similar studies, the development and application of a mixed-integer linear programming technique (MILP) optimization framework to an off-grid village in Italy was reported, which is a typical insular case study of the Mediterranean area [21]. To model the seasonal storage, a year-long time horizon was considered necessary for off-grid areas in order to achieve energy independence by relying on local RES. The degradation costs of batteries and H_2-based devices were included in the objective function of the optimization problem, such as the annual cost of the system. Efficiency and investment-cost curves were considered for the electrolyzer and fuel cell components, while the design optimization was also performed while employing a general demand response program (DRP) to assess the sizing effects on the whole performance. It is also noteworthy that there is a reduction in the electricity generation cost, depending on the load amounts that are allowed to participate in the DRP scheme. In this context, a decreased capacity of the battery storage system is associated with the cost reduction in the system [21]. The MILP optimization framework allowed the identification of a cheap system configuration but, due to the larger number of decision variables, intensive computational resources are required for the resolution of the MILP problem. In electricity tariffs, the reliance on local renewables, coupled with batteries and hydrogen, was found to be economically more competitive than the current diesel-based power system. Thus, a levelized cost of energy of around 0.455 EUR/kWh can be computed using the MILP methodology for the renewable hydrogen-battery system. Such a hydrogen storage system is necessary in the optimal hybrid renewable energy system (HRES) configuration, due to its cost-effective long-term capability that can reduce the required battery capacity and lead to a better exploitation of the local RES [21].

3.2.4. HES-RES and Safety Aspects

In HES, hydrogen is an intermediary or secondary form of energy or an energy carrier. Hydrogen complements the primary energy sources, putting them into a convenient form accessible to the desired locations and at times convenient for consumers. Hydrogen can be stored underground in ex-mines, caverns and/or aquifers, which is of utmost importance for large-scale storage cases. The main safety aspects of the HES operation refer to hydrogen re-electrification. Indeed, hydrogen re-electrification refers to electricity generation from hydrogen. Through combustion, hydrogen can be re-electrified. Some combustion engines or turbines can run directly on hydrogen, similarly to internal combustion engines running on gasoline. Nevertheless, comparing the efficiencies of hydrogen and gasoline combustion engines, the former performed less efficiently than the latter, having a thermodynamic efficiency of around 20–25%, since hydrogen has a relatively low volumetric energy density. Moreover, while no CO_2 is released, through hydrogen combustion NOx are emitted. The fuel cell-fed engines can be proven advantageous for hydrogen, as fuel cells convert the hydrogen chemical energy directly into electrical energy, achieving almost 60–80% efficiency, having as a byproduct only water [58]. The wide commercial applicability of fuel cells in various stationary and transportation applications can be stressed [58].

The exact procedure is the following: hydrogen transportation is occurs through pipelines or super tankers, making it directly disposable to energy consumption centers. Subsequently, specific applications regarding the electricity and transportation, and industrial, residential and commercial uses such as a fuel and/or an energy carrier, take place. Regarding the water production, recycling of water and water vapor follows when these effluents are disposed through rain, rivers, lakes and oceans, making up for the water used in the first place to manufacture hydrogen [70]. Similarly, the industrial plant produces oxygen that can be either released into the atmosphere, or shipped or piped to industrial and city centers for use in fuel cells (instead of air) for electricity generation. Rejuvenating the polluted rivers and lakes, or speeding up sewage treatment are similar beneficial non-energy uses of the produced oxygen [70].

Regarding the safety aspects of hydrogen, these involve its toxicity on one hand and its fire hazard properties on the other. In relation to this, regarding the toxicity of hydrogen combustion products, the toxicity increases as the carbon-to-hydrogen ratio increases. For instance, hydrogen and its main combustion product, water or water vapor, are not toxic, but NOx, which can be produced through the flame combustion of hydrogen (also reported in fossil fuel combustion) displays toxic effects [70]. In general, higher specific heat causes a gaseous fuel to be safer, since it slows down the temperature increases for a given heat input. Wider ignition limits, lower ignition energies, and lower ignition temperatures make gaseous fuels less safe, as they increase the limits at which a fire could commence. Higher flame temperature, higher explosion energy, and higher flame emissivity all make gaseous fuels less safe as well, due to the damaging fire consequences [70]. Extra safety concerns and provisions have to be prioritized in the design of in-site plants such as the Utsira plant [75]. However, it is also noteworthy that a serious accident could occur in such a high-profile demonstration project, which would be particularly detrimental to the development of hydrogen as an energy carrier. Another critical issue, similar to the Utsira plant, is that it is possible to contain explosive zones and advanced equipment which regularly receive many unskilled visitors. Because of this, safety measures have to be prioritized, in alignment with the proper training of operators and other personnel and the provision of good working instructions for all parts of the system, together with clear distribution and instructions of responsibility all over the site [75].

3.2.5. HES-RES and Techno-Economic Aspects

As hydrogen can be further used in internal combustion engines that are designed similarly to traditional combustion engines, it can be pointed out that both the mature industry and the vast availability of production infrastructure regarding internal combustion engines are making hydrogen internal combustion engines economically attractive. Moreover, unlike fuel cell vehicles, these types of vehicles do not rely on materials that may limit their large-scale production [63]. These engines support different characteristics, such as traditional gasoline engines that utilize electronic control units for the proper pressure management in the injection and in the hydrogen combustor. In addition, these engines mainly emit nitrogen mixtures, and thus they are not emission free [63].

In economic terms, the increase in pressure in the hydrogen storage from 200 bar to 450 bar, or even 700 bar, would increase the overall energy density of the hydrogen storage, thus making it possible to store more wind energy on the same footprint. However, high-pressure hydrogen storage systems are likely to be more costly than low-pressure systems, both from an investment and an operational point of view. It is noteworthy that investment cost increase is mainly linked to the installment of stronger storage tanks (thicker steel walls and/or use of composite materials), while operational cost increase is mainly linked to the increase in energy consumption for hydrogen compression [75]. It is also likely that extra costs associated with the required safety system concur, being dependent on the type of system installed. because of this, two basic techniques for high-pressure hydrogen gas production are: (a) low-pressure electrolysis, with a long compression stage, or (b) high-pressure electrolysis without compression [75].

While the majority of research studies have been focused on the HES-RES coordination, it is noteworthy that there are also reported RES studies, other than those of solar- and wind-energy infrastructure for hydrogen production. Among these studies, wastewater was characterized as an issue of general concern for environmental sustainability, making the development of a circular low-waste economy a necessity. In this context, lignocellulosic biomass processing (mainly hydrolysis, pyrolysis, and hydrothermal liquefaction) can result in secondary aqueous streams in which there are low quantities of carbon and biomass, making the conventional valorization treatment complicated. In this respect, biodiesel production leads to a glycerol excess on the market, which needs to be valorized [76]. In the relevant literature research, complex feedstock valorizations are considered with respect to real waste streams or synthetic mixtures, demonstrating the outcomes derived from laboratory-scale experiments. Subsequently, aqueous-phase reforming (APR) was a mild-condition process able to convert oxygenated molecules into hydrogen. The catalytic process of APR involves the conversion of water-dissolved oxygenates into a hydrogen-rich gas for biocrude upgrading purposes. The full development and the scalability to industrial level are the main challenges of this type of APR process [76].

In a similar study, hydrothermal liquefaction was proposed as a promising technology for renewable advanced biofuel production [77]. The main constraints in its large-scale applications are the significant carbon loss in the aqueous phase (AP) and the necessity of biocrude upgrading. Therefore, a techno-economic feasibility analysis was followed, in which different lignocellulosic feedstocks, corn stover (CS) and lignin-rich stream (LRS) from cellulosic ethanol production were tested for the evaluation of hydrothermal liquefaction (HTL) coupling with aqueous-phase reforming (APR). Following the carrying out of the mass and energy balances, the equipment design and the capital and operating costs calculation, it was shown that the biofuel minimum selling price (in the case of 0% internal rate of return) was fixed at 1.23 (in the case of LRS) and 1.27 EUR/kg (in the case of CS), respectively [77]. Moreover, the fixed capital investment was devoted to heat exchangers, while electricity and feedstock sustained the highest share of the operating costs. It can be also be pointed out that, in the case of CS, a production of 107% of the required hydrogen was reported for biocrude upgrading, making the APR process particularly profitable. In this context, APR reduced the hydrogen production cost significantly (1.5 EUR/kg), making it a cost-effective and competitive technology compared with conventional electrolysis [77].

3.2.6. HES-RES and Environmental Aspects

As already mentioned above, hydrogen utilization involves oxidation, and the only direct major product from the hydrogen oxidation is water. Small quantities of nitrogen oxides are released when hydrogen is combusted in air, but these effluents can be controlled with careful engine design [63]. The environmental impacts from other phases in the life cycle of a hydrogen system are similar to those for other energy technologies, and may be small or large, depending on the source of the hydrogen [63]. Regarding the environmental concerns and considerations, an HES is prone to cause direct and indirect environmental (especially atmospheric) impacts, at three steps: production, transportation-storage and utilization. At the production step: the leaking of hydrogen and carbon dioxide release in the case of production from fossil fuels; at the transport–storage steps: the leaking of hydrogen; and at the utilization step: the release of water and the leaking of hydrogen. On the basis of experience with technologies associated with the transportation of natural gas and other volatiles, it seems likely that systems of hydrogen production, storage, and transport will involve losses to the atmosphere. Although the average leak rate to be expected in a full-scale hydrogen-driven economy is very uncertain, regarding the relevant literature estimates, around 10% of all hydrogen manufactured is ready to leak into the atmosphere during these steps [64]. Losses during the current commercial transport of hydrogen are substantially greater than this, suggesting a range of 10–20% losses should be expected. If so, and if all current technologies based on oil or gasoline combustion could be replaced by hydrogen fuel cells, then anthropocentric emissions of hydrogen would be in the order of 60–120 Tg/year, or roughly four to eight times the estimates of current anthropocentric hydrogen emissions (15 ± 10 Tg/year) [64].

At this point, it can be seen that human activity can result in approximate duplication or triplication of the scheduled annual production of hydrogen from all sources combined. In the case of replacing all fossil-fuel energy generation with hydrogen fuel cells, then an amount of approximately 60 million tons annually of human-made hydrogen would leak into the atmosphere: this is roughly four times the current amount. Subsequently, such hydrogen leaks might increase water vapor and cool the stratosphere, through retarding the ozone-layer recovery. This excessive release of hydrogen into the atmosphere is also worsening, due to natural sources of hydrogen [64]. Furthermore, hydrogen participates in stratospheric chemical cycles of water and various GHGs, where this substantial increase in its concentration can cause irreversible changes in the stratosphere regarding the imbalance of equilibrium in the concentration of its constituent components. High amounts of hydrogen emissions are also unavoidable from a global fleet of fuel cell vehicles, further impacting on local or regional distribution of water vapor. Water vapor increase can also affect local, regional, and global climatic conditions, mainly due to an increase in relative humidity among areas of widely applied fuel cell technologies, compared to the operation of internal combustion engines. Such an increase in relative humidity can cause shifts in temperature patterns, causing an imbalance in the living conditions of people and ecosystems. On the other hand, hydrogen fuel cells are characterized as a "clean" technology of only water vapor exhaust, GHGs and ozone precursors associated with hydrogen production [64].

3.2.7. Miscellaneous Aspects

Other critical design aspects of hydrogen utility in combined sources of energy generation are the following [33]:

- The selection of the number of fuel cell stacks should consider the price increase of additional fuel cells with the decrease in the price of the smaller storage tank.
- An analysis of the solar–hydrogen hybrid system should consider the effect of the solar irradiance and the ambient air temperature, which are key determinants for calculating the power from the solar panel array. Those reliable and uninterrupted systems of energy are depended on the number of solar panels used in the system.

The number of solar panels used in the system affects the electrolyzer size as well as the storage tank volume.

- The efficiency of the hybrid system is dependent on the nominal electrolyzer power. Proper sizing of the electrolyzer can reduce the non-utilizable energy and therefore increase efficiency. The nominal power of the electrolyzer also determines the number of solar panels and the storage tank volume.
- Modular designers cannot ignore the fact that simulations have to consider short time intervals for constructing the properly sized components in the system; otherwise, there will be inevitable interruptions in the power from time to time with undersized hybrid system components.

Other researchers proposed the following design, aspects of hydrogen utility in combined sources of energy generation [70]:

- Liquid and gaseous hydrogen are valued as the best transportation fuels when compared to liquid fuels such as gasoline, jet fuel and alcohols.
- Hydrogen is a versatile fuel that can be converted to useful thermal-, mechanical- and electrical-energy forms for end-users through a variety of processes, whereas fossil fuels can only be converted through one process, i.e., flame combustion.
- In quantitative terms, hydrogen is 39% more efficient than fossil fuels. Moreover, hydrogen is an energy-conserving fuel that can save primary energy resources.
- In safety terms, hydrogen is safe for use, avoiding fire hazard and toxicity cases.

Moreover, the design aspects cannot ignore the selection of the site for the HES installation. Indeed, an appropriate location has to support the following features: good-to-excellent wind conditions, small but representative load, back-up systems in place, not too remote, a supporting community, and access to service personnel. In general, all equipment should be kept as simple and robust as possible, and redundancy should be considered. Due to the uncertainty in precisely forecasting wind power production and customer power demand, a slightly oversized installation should also be considered. However, there is a tradeoff to be made between plant availability and overall system cost [75].

In general, in urban contexts, HES problems are mainly environmentally-centered. Indeed, in this context the main problems are those of high air-pollution levels in many urban cities and sustainability of the transportation fuels; thus, the need to address control measures using HES has been proposed [78]. The majority of the global transportation vehicle fleets consume the fuels derived from fossil resources. The development of economic activities indicated the increase in transportation services resulting in increased fuel consumption and high emissions, especially the unregulated emission of GHG-induced carbon dioxide. Therefore, the utilization of hydrogen as fuel in vehicle fleets can improve energy security and reduce GHG emission. The feasibility of a hydrogen energy system, which includes its resources, production technologies, storage, fuel transportation, dispensing and utilization, was analyzed for the road transportation sector [78]. Moreover, the methodology of this study focused on the use of hydrogen as a clean and green fuel for road transportation systems, also meeting the criteria of high energy efficiency and zero-carbon-based emissions (CO, CO_2, HC, PM). Subsequently, many countries employed ambitious policies for the development of HES and, with the aid of research organizations, universities and companies, a wider commercialization of hydrogen-fueled vehicles such as internal combustion engines, or fuel cell-based, or hybrid technology should be a reality [78].

Another noteworthy approach of RES and HES is the "Hydrogen Office building (HO)" which presented a wind-hydrogen energy system, located in Fife, in Scotland, prepared to demonstrate the role of hydrogen in reducing the impact of wind intermittency in a grid-tied microgrid. The main components of this system were a wind turbine, alkaline electrolyzer, hydrogen storage and a PEM fuel cell. The building demand was met by the wind turbine, while the fuel cell provided back-up power to the ground floor when wind power was unable to meet the demand. Accurate modelling of wind–hydrogen systems allowed for an effective implementation and operation. However, past research lacks global methodologies for a whole-system scope simulation. Moreover, experimental validation is imperfect, since in most of these models no guidelines are given for parameter estimation, which is essential for repeatability and reproduction [79]. For achieving this, a comprehensive modelling, simulation and performance methodology was employed on wind–hydrogen systems, containing experimental validation and guidelines for parameter estimation. The conducted quantitative analysis also showed how this methodology can contribute to improve the design and performance of such a system in an accurate, reliable, and easily adaptable manner. At its modelling scale, a precise prediction of the HO dynamic behavior was shown, with an error of less than 2% on average. Other modelling parameters for evaluating the HO system were those of average energy production, stand-alone operation and round-trip efficiency [79].

In a similar study, researchers employed a set of updated HES modeling tools (HYDRO-GEMS) for the operation evaluation of the Utsira plant: an autonomous wind/hydrogen energy demonstration system that was located at the island of Utsira in Norway and installed in July 2004. After successive designing and testing optimizations for increasing the wind/hydrogen energy system efficiency, it was observed that specific recommendations and improvements for a full (100%) autonomy can be achieved, following the simulated operation for a specific year, as well as the evaluation performance of several alternative system designs [75].

3.3. Limitations, Challenges and Future Research Prospects of HES-RES Synergies

The shift to a carbon-neutral society requires a substantial transformation of the present global electricity generation and consumption, together with the adoption of other technologies than those of fossil-based technologies for electric power. In response to this need, hydrogen should play a determining role in the scaling-up of a clean, plausible and environmentally friendly RES. The main critique is that its integration into existing power systems remains sparsely studied, while the majority of the existing literature production overlooks potential hydrogen technologies and their application in power systems for hydrogen production, re-electrification and storage [58]. Therefore, in an era where most of the world's energy sources become non-fossil based, hydrogen and electricity can be proven to be two dominant energy carriers for the provision of end-use services. In this context, hydrogen can play a globally decisive role as a future energy carrier regarding energy systems, especially since existing fossil-fuel supplies are becoming scarcer and environmental concerns are increasing. It is also realistic for hydrogen to become an increasingly important chemical-energy carrier and eventually a principal chemical-energy carrier through upgrading the infrastructure needed to support the development of a hydrogen economy [63].

In this type of economy, termed a "hydrogen economy", the two complementary energy carriers, hydrogen and electricity, are coordinated to satisfy most of the requirements of energy consumers. With this prospect, a transition era will bridge the gap between the current fossil-fuel economy and a hydrogen economy, where non-fossil-derived hydrogen can be exploited to extend the lifetime of the global deposits of fossil fuels [63]. In this respect, besides the technological issues of hydrogen as an energy carrier and hydrogen energy systems, the economic–financial dimension has also to be considered in the light of the potential social and political implications of hydrogen energy to be confronted, especially regarding when and where hydrogen is likely to become important player of energy management strategies [63].

Subsequently, hydrogen technologies must be developed to the stage where they can be proven to be economically and commercially viable. This technological and managerial target includes overcoming all existing safety problems regarding the HES infrastructure, probably in a gradual manner. As mentioned before, numerous social and political implications of shifting to a hydrogen era have to be confronted by governments, universities and private industries, and some of them are the following: healthy levels of employment, a reasonably successful rate of economic growth, reasonable stability in prices, a viable balance of payments and an equitable distribution of rising incomes [63].

Financial gains, environmental concerns, safety and knowledge about the new technology seem to be all hot-spot issues in the HES-RES synergies prospected. Moreover, public perceptions of these HES-RES synergies are attributed to the willingness to pay (WtP) in order to participate in projects that foster the use of fuel cell vehicles, despite their current limitations (i.e., reduced range and limited refueling possibilities). Such a WtP reflects the users' expectation of personal financial gains (i.e., reduced running costs), and thus, a financially attractive package is always an essential to be offered to energy end-users and consumers for a feasible and realistic hydrogen promotion among local societies [63]. For other significant stakeholders and those interested in HES, which are the manufacturing industries, it seems that regulations have to be issued for rigorous tests in order to ensure safety and environmental consciousness, to attract public awareness of such technologies, and for the control of air pollution through promoting the wider use of these technologies in the future [63]. Indeed, it has been reported that environmental considerations are found to affect users' WtP for hydrogen technologies, while there are other users—being acquainted with a better knowledge of the technology—who emphasize more the difficulties of a hydrogen future [63].

Regarding the challenges and the prospects of HES-RES synergies, a challenge of utmost importance is the use of wastewater. Indeed, while there are several approaches to wastewater treatment for improving the water-energy nexus with the hydrogen fuel production option, these approaches are limited and confined to the small, pilot, experimental-modeling simulation scale, while there are no fully developed studies on large-scale and in-field applications. Among such small-scale studies, it was argued that the use of fresh water for green-hydrogen fuel production through water electrolysis can exacerbate the challenges of water scarcity [80]. In this context, the utility of non-potable water can effectively design the whole process in a highly secure, reliable and sustainable manner. Therefore, an integration of a solid oxide electrolyzer (SOE) with a water-treatment and recovery process can represent the conceptualization and the techno-economic evaluation of an innovative process of hydrogen energy production. Accordingly, purified wastewater and the waste heat of flue gases from the power plants have been utilized as feedstocks for the electrolysis process. The required thermal energy of the electrolyzer can be supplied under the following two scenarios: first, by the utilization of a preheater based on fossil energies, and second, where the required thermal energy is supplied through a parabolic trough collector (PTC)-based solar farm [80]. The results showed that the integration of fossil fuel-driven power plants with evolving green technologies is plausible for managing the GHG emission crisis and for further reducing the limitations of fossil energies. In parallel, the overall conversion efficiency for the proposed hydrogen production process was calculated as 53.26%. Overall, the proposed system is a competitive and reliable way to support substantial water consumption savings, while further net potential energy savings and carbon emission abatement were also realistic [80].

Regarding the aforementioned term of the "integration" endeavors of energy and environmental impact, it cannot be ignored that energy policy makers and strategic designers also envisage and consider a wider spectrum of optimization features, such as household appliances in the building sector [81–83], the agricultural sector [84], and the business or entrepreneurship sectors [85].

4. Conclusions

During the last three decades of analysis, fossil fuels (mainly petroleum, natural gas and coal) have satisfied the world's energy demand against an unprecedented accelerating pace of stock depletion. Moreover, their combustion products are criticized for global environmental problems such as the GHG effect, ozone layer depletion, acid rain and pollution, threatening the natural environment and causing an imbalance in the natural climate cycles of our planet [86,87]. There is an ongoing debate among engineers, scientists, and governors globally, who agree that the solution to these global problems would be the replacement of, or at least a supplementary role for, HES in the existing fossil fuel systems, since hydrogen is produced from renewable energy (e.g., solar) sources, ensuring a stable and permanent operation [70]. Besides the environmental benefits of HES, the two most prevalent RES for the post-fossil fuel era are those of the solar-HES and the synthetic fossil fuel system. Considering the production costs, the environmental damage and the utilization efficiencies, the solar-HES is the best energy system for ascertaining a sustainable future, being able to replace the fossil fuel system before the end of the 21st century. In this post-petroleum era, the synthetic fossil fuel system is also a promising technology, in which synthetic gasoline and synthetic natural gas can be produced using abundant deposits of coal to ensure the continuation of the existing fossil fuel system [70]. An integrated perception of HES design can consider environmentally friendly modulation like the installment of a solar–hydrogen system for vapor generation. Other design specifications should take into consideration the production, storage, transportation, distribution, utilization, environmental impact and economies-of-scale prospects [70]. Regarding the external costs, which are more common and intensive for fossil fuel utilities, they include the costs of the physical damage done to humans, fauna, flora and the environment due to harmful emissions, oil spills and leaks, and coal strip mining, as well as governmental expenditures for pollution abatement and expenditures for military protection of oil supplies [70].

The research focus of this review was the collection and the classification of the existing studies that jointly address the fields of Hydrogen Energy Systems (HES) and Renewable Energy Sources (RES). The bibliometric analysis of this literature production considered five fields of classification-clusters, revealing firstly the most-reported and the least-reported research areas of interest, secondly the evolution of problem solving approaches, which are primarily experimental, small-scale modelling and simulation studies, followed by problem-solving and large-scale in-field case studies. These large-scale endeavors have been directed toward handling problems of high air-pollution levels in many urban cities, and the sustainability of the transportation fuels could be optimally addressed. The technological dimension of HES and RES convergence and synergies entails the promotion of hybrid systems that ensure stability, storage capacities, operational safety and energy efficiency increase. On the social and the political side, theoretical approaches or legislative measures or policies can be proven to be more flexible and reliable if HES and RES bridge theoretical office designs with practical real world/in-field situations of applicability. On the RES side of utility in HES, the most densely reported studies have been concentrated on the successful collaboration of wind energy (turbines), solar energy (photovoltaics, PVs) and HES. However, the locality characteristics are also the primary precondition of such HES-RES synergy; therefore, among other RES control measures, using HES involves the utility of biomass sources, even at pilot scales of applicability.

In conclusion, we can stress the important economic and environmental consequences that follow the transition of a sustainable hydrogen economy, as a part of a wider energy-solving strategic problem. For this to be achieved, it is important to follow a logical sequence of actions: (a) the design of an optimal hydrogen supply system, (b) the evaluation of its environmental impacts, and (c) an understanding of the effects of various factors that aid the selection and the installation of an optimal hydrogen supply system through a scenario plan and sensitivity analysis. A research limitation of this approach is the inability to directly foresee the actual magnitude of hydrogen emissions associated with a hydrogen fuel cell economy, particularly since today budgets for hydrogen are not fully or well known, while there are also technical constraints, and the future fuel cell industry can be only forecasted. In this case, the evolution and the shift from a fossil-fuel combustion energy planning to the prevalence of hydrogen fuel cells can actually result in unpleasant anthropocentric hydrogen emissions, because fossil fuel combustion is a source of hydrogen itself. On the other hand, researchers are deemed to take into account the climatic effects of HES in the near-future energy plans, especially those based on electrolysis from water, where the simultaneous reduction of fossil-fuel emissions must be also considered [64].

Author Contributions: The authors have contributed equally to this study. All authors have read and agreed to the published version of the manuscript.

Funding: This research received no external funding.

Conflicts of Interest: The authors declare no conflict of interest.

References

1. Ntona, E.; Arabatzis, G.; Kyriakopoulos, G.L. Energy saving: Views and attitudes of students in secondary education. *Renew. Sustain. Energy Rev.* **2015**, *46*, 1–15. [CrossRef]
2. Kyriakopoulos, G.L.; Arabatzis, G. Electrical energy storage systems in electricity generation: Energy policies, innovative technologies, and regulatory regimes. *Renew. Sustain. Energy Rev.* **2016**, *56*, 1044–1067. [CrossRef]
3. Kyriakopoulos, G.L. Energy Communities Overview: Managerial Policies, Economic Aspects, Technologies, and Models. *J. Risk Financ. Manag.* **2022**, *15*, 521. [CrossRef]
4. Lu, Z.; Zhu, Q.; Zhang, W.; Lin, H. Economic operation strategy of integrated hydrogen energy system considering the uncertainty of PV power output. *Energy Rep.* **2023**, *9*, 463–471. [CrossRef]
5. Çiçek, A. Multi-Objective Operation Strategy for a Community with RESs, Fuel Cell EVs and Hydrogen Energy System Considering Demand Response. *Sustain. Energy Technol. Assess.* **2023**, *55*, 102957. [CrossRef]
6. Ren, J.; Xu, D.; Cao, H.; Wei, S.; Dong, L.; Goodsite, M.E. Sustainability decision support framework for the prioritization of hydrogen energy systems. In *Hydrogen Economy: Processes, Supply Chain, Life Cycle Analysis and Energy Transition for Sustainability*, 2nd ed.; Academic Press: Cambridge, MA, USA, 2023; pp. 273–313. [CrossRef]
7. Virji, M.; Randolf, G.; Ewan, M.; Rocheleau, R. Analyses of hydrogen energy system as a grid management tool for the Hawaiian Isles. *Int. J. Hydrogen Energy* **2020**, *45*, 8052–8066. [CrossRef]
8. Maghami, M.R.; Hassani, R.; Gomes, C.; Hizam, H.; Othman, M.L.; Behmanesh, M. Hybrid energy management with respect to a hydrogen energy system and demand response. *Int. J. Hydrogen Energy* **2020**, *45*, 1499–1509. [CrossRef]
9. Parra, D.; Valverde, L.; Pino, F.J.; Patel, M.K. A review on the role, cost and value of hydrogen energy systems for deep decarbonisation. *Renew. Sustain. Energy Rev.* **2019**, *101*, 279–294. [CrossRef]
10. Cho, S.; Won, W.; Han, S.; Kim, S.; Youa, C.; Kim, J. An optimization-based design and analysis of a biomass derived hydrogen energy system. *Comput. Aided Chem. Eng.* **2018**, *44*, 1573–1578. [CrossRef]
11. Rosen, M.A. The prospects for hydrogen as an energy carrier: An overview of hydrogen energy and hydrogen energy systems. In *Hydrogen Production: Prospects and Processes*; Nova Science Publishers, Inc.: Hauppauge, NY, USA, 2012; pp. 1–28.
12. Özden, E.; Tari, I. Numerical investigation of a stand-alone solar hydrogen energy system: Effects of pefc degradation. In Proceedings of the Thermal and Fluids Engineering Summer Conference, New York, NY, USA, 9–12 August 2015; pp. 673–684. [CrossRef]
13. Lanjewar, P.B.; Rao, R.V.; Kale, A.V. A combined graph theory and analytic hierarchy process approach for multicriteria evaluation of hydrogen energy systems. *Int. J. Energy Technol. Policy* **2014**, *10*, 80–96. [CrossRef]
14. Contreras, A.; Posso, F. Technical and financial study of the development in Venezuela of the hydrogen energy system. *Renew. Energy* **2011**, *36*, 3114–3123. [CrossRef]
15. Wang, B.; Sun, L.; Diao, N.; Zhang, L.; Guo, X.; Guerrero, J.M. GaN-based step-down power converter and control strategy for hydrogen energy systems. *Energy Rep.* **2023**, *9*, 252–259. [CrossRef]
16. Zhao, X.; Yao, Y.; Liu, W.; Jain, R.; Zhao, C. A Hydrogen Load Modeling Method for Integrated Hydrogen Energy System Planning. In Proceedings of the 2023 IEEE Power and Energy Society Innovative Smart Grid Technologies Conference, ISGT, Washington, DC, USA, 16–19 January 2023. [CrossRef]
17. Sun, Q.; Wu, Z.; Gu, W.; Liu, P.; Wang, J.; Lu, Y.; Zheng, S.; Zhao, J. Multi-stage Co-planning Model for Power Distribution System and Hydrogen Energy System Under Uncertainties. *J. Mod. Power Syst. Clean Energy* **2023**, *11*, 80–93. [CrossRef]
18. Alanazi, A.; Alanazi, M.; Arabi Nowdeh, S.; Abdelaziz, A.Y.; El-Shahat, A. An optimal sizing framework for autonomous photovoltaic/hydrokinetic/hydrogen energy system considering cost, reliability and forced outage rate using horse herd optimization. *Energy Rep.* **2022**, *8*, 7154–7175. [CrossRef]
19. Dong, W.; Shao, C.; Feng, C.; Zhou, Q.; Bie, Z.; Wang, X. Cooperative Operation of Power and Hydrogen Energy Systems with HFCV Demand Response. *IEEE Trans. Ind. Appl.* **2022**, *58*, 2630–2639. [CrossRef]
20. Schrotenboer, A.H.; Veenstra, A.A.T.; uit het Broek, M.A.J.; Ursavas, E. A Green Hydrogen Energy System: Optimal control strategies for integrated hydrogen storage and power generation with wind energy. *Renew. Sustain. Energy Rev.* **2022**, *168*, 112744. [CrossRef]
21. Marocco, P.; Ferrero, D.; Martelli, E.; Santarelli, M.; Lanzini, A. An MILP approach for the optimal design of renewable battery-hydrogen energy systems for off-grid insular communities. *Energy Convers. Manag.* **2021**, *245*, 114564. [CrossRef]
22. Wang, Z.; Jia, Y.; Yang, Y.; Cai, C.; Chen, Y. Optimal configuration of an off-grid hybrid wind-hydrogen energy system: Comparison of two systems. *Energy Eng. J. Assoc. Energy Eng.* **2021**, *118*, 1641–1658. [CrossRef]
23. Zhang, H.; Yuan, T.; Tan, J. Business model and planning approach for hydrogen energy systems at three application scenarios. *J. Renew. Sustain. Energy* **2021**, *13*, 044101. [CrossRef]

24. Onwe, C.A.; Rodley, D.; Reynolds, S. Modelling and simulation tool for off-grid PV-hydrogen energy system. *Int. J. Sustain. Energy* **2020**, *39*, 1–20. [CrossRef]
25. Manilov, A.I. Hydrogen Energy System Based on Silicon Nanopowders: Prospects and Problems. In Proceedings of the 2019 IEEE 39th International Conference on Electronics and Nanotechnology (ELNANO), Kyiv, Ukraine, 16–18 April 2019; pp. 60–63. [CrossRef]
26. Zhang, W.; Maleki, A.; Rosen, M.A.; Liu, J. Sizing a stand-alone solar-wind-hydrogen energy system using weather forecasting and a hybrid search optimization algorithm. *Energy Convers. Manag.* **2019**, *180*, 609–621. [CrossRef]
27. Zhang, Z.; Zhou, J.; Zong, Z.; Chen, Q.; Zhang, P.; Wu, K. Development and modelling of a novel electricity-hydrogen energy system based on reversible solid oxide cells and power to gas technology. *Int. J. Hydrogen Energy* **2019**, *44*, 28305–28315. [CrossRef]
28. Alavi, O.; Mostafaeipour, A.; Sedaghat, A.; Qolipour, M. Feasibility of a Wind-Hydrogen Energy System Based on Wind Characteristics for Chabahar, Iran. *Energy Harvest. Syst.* **2017**, *4*, 143–163. [CrossRef]
29. Ren, J.; Xu, D.; Cao, H.; Wei, S.; Dong, L.; Goodsite, M.E. Sustainability Decision Support Framework for the Prioritization of Hydrogen Energy Systems. In *Hydrogen Economy: Processes, Supply Chain, Life Cycle Analysis and Energy Transition for Sustainability*; Academic Press: Cambridge, MA, USA, 2017; pp. 225–276. [CrossRef]
30. Lacko, R.; Drobnič, B.; Sekavčnik, M.; Mori, M. Hydrogen energy system with renewables for isolated households: The optimal system design, numerical analysis and experimental evaluation. *Energy Build.* **2014**, *80*, 106–113. [CrossRef]
31. Patrício, R.A.; Sales, A.D.; Sacramento, E.M.; De Lima, L.C.; Veziroglu, T.N. Wind hydrogen energy system and the gradual replacement of natural gas in the State of Ceará-Brazil. *Int. J. Hydrogen Energy* **2012**, *37*, 7355–7364. [CrossRef]
32. Aguado, M.; Ayerbe, E.; Azcárate, C.; Blanco, R.; Garde, R.; Mallor, F.; Rivas, D.M. Economical assessment of a wind-hydrogen energy system using WindHyGen® software. *Int. J. Hydrogen Energy* **2009**, *34*, 2845–2854. [CrossRef]
33. Acar, C.; Erturk, E.; Firtina-Ertis, I. Performance analysis of a stand-alone integrated solar hydrogen energy system for zero energy buildings. *Int. J. Hydrogen Energy* **2023**, *48*, 1664–1684. [CrossRef]
34. Karaca, A.E.; Dincer, I. Development and evaluation of a solar based integrated hydrogen energy system for mobile applications. *Energy Convers. Manag.* **2023**, *280*, 116808. [CrossRef]
35. Lin, X.-M.; Li, J.-F. Applications of In Situ Raman Spectroscopy on Rechargeable Batteries and Hydrogen Energy Systems. *ChemElectroChem* **2023**, *10*, 202201003. [CrossRef]
36. Alex, A.; Petrone, R.; Tala-Ighil, B.; Bozalakov, D.; Vandevelde, L.; Gualous, H. Optimal techno-enviro-economic analysis of a hybrid grid connected tidal-wind-hydrogen energy system. *Int. J. Hydrogen Energy* **2022**, *47*, 36448–36464. [CrossRef]
37. Ibrahim, M.D.; Binofai, F.A.S.; Mohamad, M.O.A. Transition to Low-Carbon Hydrogen Energy System in the UAE: Sector Efficiency and Hydrogen Energy Production Efficiency Analysis. *Energies* **2022**, *15*, 6663. [CrossRef]
38. Balasubramanian, V.; Haque, N.; Bhargava, S.; Madapusi, S.; Parthasarathy, R. Techno-economic evaluation methodology for hydrogen energy systems. In *Bioenergy Resources and Technologies*; Elsevier: Amsterdam, The Netherlands, 2021; pp. 237–260. [CrossRef]
39. Endo, N.; Goshome, K.; Tetsuhiko, M.; Segawa, Y.; Shimoda, E.; Nozu, T. Thermal management and power saving operations for improved energy efficiency within a renewable hydrogen energy system utilizing metal hydride hydrogen storage. *Int. J. Hydrogen Energy* **2021**, *46*, 262–271. [CrossRef]
40. Endo, N.; Segawa, Y.; Goshome, K.; Shimoda, E.; Nozu, T.; Maeda, T. Use of cold start-up operations in the absence of external heat sources for fast fuel cell power and heat generation in a hydrogen energy system utilizing metal hydride tanks. *Int. J. Hydrogen Energy* **2020**, *45*, 32196–32205. [CrossRef]
41. Endo, N.; Shimoda, E.; Goshome, K.; Yamane, T.; Nozu, T.; Maeda, T. Operation of a stationary hydrogen energy system using TiFe-based alloy tanks under various weather conditions. *Int. J. Hydrogen Energy* **2020**, *45*, 207–215. [CrossRef]
42. Li, Z.; Zhang, W.; Zhang, R.; Sun, H. Development of renewable energy multi-energy complementary hydrogen energy system (A Case Study in China): A review. *Energy Explor. Exploit.* **2020**, *38*, 2099–2127. [CrossRef]
43. Yilmaz, C. Life cycle cost assessment of a geothermal power assisted hydrogen energy system. *Geothermics* **2020**, *83*, 101737. [CrossRef]
44. Sorgulu, F.; Dincer, I. A renewable source based hydrogen energy system for residential applications. *Int. J. Hydrogen Energy* **2018**, *43*, 5842–5851. [CrossRef]
45. Kalinci, Y.; Dincer, I.; Hepbasli, A. Energy and exergy analyses of a hybrid hydrogen energy system: A case study for Bozcaada. *Int. J. Hydrogen Energy* **2017**, *42*, 2492–2503. [CrossRef]
46. Ewan, M.; Rocheleau, R.; Swider-Lyons, K.E.; Devlin, P.; Virji, M.; Randolf, G. Development of a hydrogen energy system as a grid frequency management tool. *ECS Trans.* **2016**, *75*, 403–419. [CrossRef]
47. Khalid, F.; Dincer, I.; Rosen, M.A. Analysis and assessment of an integrated hydrogen energy system. *Int. J. Hydrogen Energy* **2016**, *41*, 7960–7967. [CrossRef]
48. Maleki, A.; Pourfayaz, F.; Ahmadi, M.H. Design of a cost-effective wind/photovoltaic/hydrogen energy system for supplying a desalination unit by a heuristic approach. *Sol. Energy* **2016**, *139*, 666–675. [CrossRef]
49. Balabel, A.; Zaky, M.S. Experimental investigation of solar-hydrogen energy system performance. *Int. J. Hydrogen Energy* **2011**, *36*, 4653–4663. [CrossRef]
50. Bendaikha, W.; Larbi, S.; Mahmah, B. Hydrogen energy system analysis for residential applications in the southern region of Algeria. *Int. J. Hydrogen Energy* **2011**, *36*, 8159–8166. [CrossRef]

51. Chao, C.-H.; Shieh, J.-J. Control and management for hydrogen energy systems. Recent Researches in Energy, Environment, Entrepreneurship, Innovation. In Proceedings of the International Conference on Energy, Environment, Entrepreneurship, Innovation, ICEEEI, Canary Islands, Spain, 27–29 May 2011; pp. 27–31.

52. Kikkinides, E.S. Hydrogen-Based Energy Systems: The Storage Challenge. In *Process Systems Engineering: Volume 5: Energy Systems Engineering*; Wiley-VCH Verlag GmbH & Co. KGaA: Weinheim, Germany, 2011; Chapter 3; pp. 85–123. [CrossRef]

53. Gabriel García Clúa, J.; Julián Mantz, R.; De Battista, H. Hybrid control of a photovoltaic-hydrogen energy system. *Int. J. Hydrogen Energy* **2008**, *33*, 3455–3459. [CrossRef]

54. Cheng, L.; Guo, Z.; Xia, G. A Review on Research and Technology Development of Green Hydrogen Energy Systems with Thermal Management and Heat Recovery. *Heat Transf. Eng.* **2023**. [CrossRef]

55. Blanco, H.; Leaver, J.; Dodds, P.E.; Dickinson, R.; García-Gusano, D.; Iribarren, D.; Lind, A.; Wang, C.; Danebergs, J.; Baumann, M. A taxonomy of models for investigating hydrogen energy systems. *Renew. Sustain. Energy Rev.* **2022**, *167*, 112698. [CrossRef]

56. Sezgin, B.; Devrim, Y.; Ozturk, T.; Eroglu, I. Hydrogen energy systems for underwater applications. *Int. J. Hydrogen Energy* **2022**, *47*, 19780–19796. [CrossRef]

57. Temiz, M.; Dincer, I. Development and assessment of an onshore wind and concentrated solar based power, heat, cooling and hydrogen energy system for remote communities. *J. Clean. Prod.* **2022**, *374*, 134067. [CrossRef]

58. Yue, M.; Lambert, H.; Pahon, E.; Roche, R.; Jemei, S.; Hissel, D. Hydrogen energy systems: A critical review of technologies, applications, trends and challenges. *Renew. Sustain. Energy Rev.* **2021**, *146*, 111180. [CrossRef]

59. Martin, A.; Agnoletti, M.-F.; Brangier, E. Users in the design of Hydrogen Energy Systems: A systematic review. *Int. J. Hydrogen Energy* **2020**, *45*, 11889–11900. [CrossRef]

60. Valente, A.; Iribarren, D.; Dufour, J. Cumulative energy demand of hydrogen energy systems. In *Environmental Footprints and Eco-Design of Products and Processes*; Springer: Berlin/Heidelberg, Germany, 2019; pp. 47–75. [CrossRef]

61. Valente, A.; Iribarren, D.; Dufour, J. Life cycle assessment of hydrogen energy systems: A review of methodological choices. *Int. J. Life Cycle Assess.* **2017**, *22*, 346–363. [CrossRef]

62. Zini, G.; Tartarini, P. *Solar Hydrogen Energy Systems: Science and Technology for the Hydrogen Economy*; Springer: Berlin/Heidelberg, Germany, 2012; 184p, ISBN 9788847019980. [CrossRef]

63. Rosen, M.A.; Koohi-Fayegh, S. The prospects for hydrogen as an energy carrier: An overview of hydrogen energy and hydrogen energy systems. *Energy Ecol. Environ.* **2016**, *1*, 10–29. [CrossRef]

64. Noyan, Ö.F. Some approach to possible atmospheric impacts of a hydrogen energy system in the light of the geological past and present-day. *Int. J. Hydrogen Energy* **2011**, *36*, 11216–11228. [CrossRef]

65. Li, Z.; Chang, L.; Gao, D.; Liu, P.; Pistikopoulos, E.N. Hydrogen Energy Systems. In *Process Systems Engineering: Volume 5: Energy Systems Engineering*; Wiley-VCH Verlag GmbH & Co. KGaA: Weinheim, Germany, 2011; Chapter 4, pp. 125–157. [CrossRef]

66. Yamamoto, S.; Ushifusa, Y. Consideration of Charging and Efficiently Using Surplus Photovoltaic power by Hydrogen Energy System. In Proceedings of the 13th Asian Control Conference (ASCC), Jeju, Republic of Korea, 4–7 May 2022; pp. 2298–2303. [CrossRef]

67. Deng, W.; Pei, W.; Zhuang, Y.; Li, N.; Zhang, X.; Yang, Y.; Kong, L. Integration control of renewable energy/hydrogen energy system based on flexible DC interconnection. *J. Phys. Conf. Ser.* **2022**, *1*, 012032. [CrossRef]

68. Budak, Y.; Devrim, Y. Evaluation of hybrid solar-wind-hydrogen energy system based on methanol electrolyzer. *Int. J. Energy Res.* **2020**, *44*, 10222–10237. [CrossRef]

69. Shiroudi, A.; Taklimi, S.R.H.; Mousavifar, S.A.; Taghipour, P. Stand-alone PV-hydrogen energy system in Taleghan-Iran using HOMER software: Optimization and techno-economic analysis. *Environ. Dev. Sustain.* **2013**, *15*, 1389–1402. [CrossRef]

70. Veziroğlu, T.N.; Şahin, S. 21st Century's energy: Hydrogen energy system. *Energy Convers. Manag.* **2008**, *49*, 1820–1831. [CrossRef]

71. Nanaki, E.A.; Koroneos, C.J. Exergetic aspects of hydrogen energy systems-The case study of a fuel cell bus. *Sustainability* **2017**, *9*, 276. [CrossRef]

72. Sarker, A.K.; Azad, A.K.; Rasul, M.G.; Doppalapudi, A.T. Prospect of Green Hydrogen Generation from Hybrid Renewable Energy Sources: A Review. *Energies* **2023**, *16*, 1556. [CrossRef]

73. Samende, C.; Fan, Z.; Cao, J.; Fabián, R.; Baltas, G.N.; Rodriguez, P. Battery and Hydrogen Energy Storage Control in a Smart Energy Network with Flexible Energy Demand Using Deep Reinforcement Learning. *Energies* **2023**, *16*, 6770. [CrossRef]

74. Niu, M.; Li, X.; Sun, C.; Xiu, X.; Wang, Y.; Hu, M.; Dong, H. Operation Optimization of Wind/Battery Storage/Alkaline Electrolyzer System Considering Dynamic Hydrogen Production Efficiency. *Energies* **2023**, *16*, 6132. [CrossRef]

75. Ulleberg, Ø.; Nakken, T.; Eté, A. The wind/hydrogen demonstration system at Utsira in Norway: Evaluation of system performance using operational data and updated hydrogen energy system modeling tools. *Int. J. Hydrogen Energy* **2010**, *35*, 1841–1852. [CrossRef]

76. Zoppi, G.; Pipitone, G.; Pirone, R.; Bensaid, S. Aqueous phase reforming process for the valorization of wastewater streams: Application to different industrial scenarios. *Catal. Today* **2022**, *387*, 224–236. [CrossRef]

77. Tito, E.; Zoppi, G.; Pipitone, G.; Miliotti, E.; Di Fraia, A.; Rizzo, A.-M.; Pirone, R.; Chiaramonti, D.; Bensaid, S. Conceptual design and techno-economic assessment of coupled hydrothermal liquefaction and aqueous phase reforming of lignocellulosic residues. *J. Environ. Chem. Eng.* **2023**, *11*, 109076. [CrossRef]

78. Salvi, B.L.; Subramanian, K.A. Sustainable development of road transportation sector using hydrogen energy system. *Renew. Sustain. Energy Rev.* **2015**, *51*, 1132–1155. [CrossRef]

79. Valverde-Isorna, L.; Ali, D.; Hogg, D.; Abdel-Wahab, M. Modelling the performance of wind-hydrogen energy systems: Case study the Hydrogen Office in Scotland/UK. *Renew. Sustain. Energy Rev.* **2016**, *53*, 1313–1332. [CrossRef]

80. Jing, D.; Mohammed, A.A.; Kadi, A.; Elmirzaev, S.; AL-Khafaji, M.O.; Marefati, M. Wastewater treatment to improve energy and water nexus with hydrogen fuel production option: Techno-economic and process analysis. *Process Saf. Environ. Prot.* **2023**, *172*, 437–450. [CrossRef]

81. Topalis, F.V.; Doulos, L.T. Ambient light sensor integration. In *Handbook of Advanced Lighting Technology*; Springer: Berlin/Heidelberg, Germany, 2017; pp. 607–634. [CrossRef]

82. Anthopoulou, E.; Doulos, L. The effect of the continuous energy efficient upgrading of LED street lighting technology: The case study of Egnatia Odos. In Proceedings of the 2nd Balkan Junior Conference on Lighting, Balkan Light Junior, Plovdiv, Bulgaria, 19–21 September 2019. [CrossRef]

83. Pallis, P.; Braimakis, K.; Roumpedakis, T.C.; Varvagiannis, E.; Karellas, S.; Doulos, L.; Katsaros, M.; Vourliotis, P. Energy and economic performance assessment of efficiency measures in zero-energy office buildings in Greece. *Build. Environ.* **2021**, *206*, 108378. [CrossRef]

84. Arabatzis, G.; Malesios, C. Pro-environmental attitudes of users and non-users of fuelwood in a rural area of Greece. *Renew. Sustain. Energy Rev.* **2013**, *22*, 621–630. [CrossRef]

85. Skordoulis, M.; Kyriakopoulos, G.; Ntanos, S.; Galatsidas, S.; Arabatzis, G.; Chalikias, M.; Kalantonis, P. The Mediating Role of Firm Strategy in the Relationship between Green Entrepreneurship, Green Innovation, and Competitive Advantage: The Case of Medium and Large-Sized Firms in Greece. *Sustainability* **2022**, *14*, 3286. [CrossRef]

86. Kyriakopoulos, G.L.; Sebos, I. Enhancing Climate Neutrality and Resilience through Coordinated Climate Action: Review of the Synergies between Mitigation and Adaptation Actions. *Climate* **2023**, *11*, 105. [CrossRef]

87. Sebos, I.; Progiou, A.G.; Kallinikos, L. Methodological Framework for the Quantification of GHG Emission Reductions from Climate Change Mitigation Actions. *Strateg. Plan. Energy Environ.* **2020**, *39*, 219–242. [CrossRef]

Article

Cluster-Based Approach to Estimate Demand in the Polish Power System Using Commercial Customers' Data

Tomasz Ząbkowski [1,*], Krzysztof Gajowniczek [1], Grzegorz Matejko [2], Jacek Brożyna [3], Grzegorz Mentel [3,4], Małgorzata Charytanowicz [5], Jolanta Jarnicka [5], Anna Olwert [5], Weronika Radziszewska [5] and Jörg Verstraete [6]

[1] Institute of Information Technology, Warsaw University of Life Sciences-SGGW, Nowoursynowska 159, 02-787 Warsaw, Poland; krzysztof_gajowniczek@sggw.edu.pl
[2] Polskie Towarzystwo Cyfrowe, Krakowskie Przedmieście 57/4, 20-076 Lublin, Poland; ge.matejko@gmail.com
[3] Department of Quantitative Methods, The Faculty of Management, Rzeszow University of Technology, Aleja Powstańców Warszawy 10/S, 35-959 Rzeszow, Poland; jacek.brozyna@prz.edu.pl (J.B.); gmentel@prz.edu.pl (G.M.)
[4] INTI International University, Persiaran Perdana BBN, Putra Nilai, Nilai 71800, Malaysia
[5] Systems Research Institute, Polish Academy of Sciences, Newelska 6, 01-447 Warsaw, Poland; malgorzata.charytanowicz@ibspan.waw.pl (M.C.); jolanta.jarnicka@ibspan.waw.pl (J.J.); aolwert@ibspan.waw.pl (A.O.); radzisze@ibspan.waw.pl (W.R.)
[6] Institute of Fluid-Flow Machinery, Polish Academy of Sciences, Fiszera 14, 80-231 Gdańsk, Poland; jorg.verstraete@imp.gda.pl
* Correspondence: tomasz_zabkowski@sggw.edu.pl

Abstract: This paper presents an approach to estimate demand in the Polish Power System (PPS) using the historical electricity usage of 27 thousand commercial customers, observed between 2016 and 2020. The customer data were clustered and samples as well as features were created to build neural network models. The goal of this research is to analyze if the clustering of customers can help to explain demand in the PPS. Additionally, considering that the datasets available for commercial customers are typically much smaller, it was analyzed what a minimal sample size drawn from the clusters would have to be in order to accurately estimate demand in the PPS. The evaluation and experiments were conducted for each year separately; the results proved that, considering adjusted R^2 and mean absolute percentage error, our clustering-based method can deliver a high accuracy in the load estimation.

Keywords: energy usage; commercial customers; clustering; neural networks; demand model; Polish Power System

Citation: Ząbkowski, T.; Gajowniczek, K.; Matejko, G.; Brożyna, J.; Mentel, G.; Charytanowicz, M.; Jarnicka, J.; Olwert, A.; Radziszewska, W.; Verstraete, J. Cluster-Based Approach to Estimate Demand in the Polish Power System Using Commercial Customers' Data. *Energies* **2023**, *16*, 8070. https://doi.org/10.3390/en16248070

Academic Editors: Konstantinos Aravossis and Eleni Strantzali

Received: 5 November 2023
Revised: 7 December 2023
Accepted: 11 December 2023
Published: 14 December 2023

1. Introduction

The basis for ensuring a safe and economically effective operation of each national power system is an appropriate planning of its operation in various time horizons. The priority is to meet the recipients' demand for power and electricity, taking into account the conditions of the grid, the operation of the units and the safety requirements for system operation. When planning the supply side of capacity, it is necessary to ensure a required power surplus over the consumers' demand for power, the so-called power reserves, to be prepared in the event of a failure resulting in a loss of production capacity, as well as for an unexpected increase in power demand by consumers.

The operation of each power system is planned in such a way that no single failure will lead to an overload of network elements or will cause a violation of any other criteria of safe system operation, such as required voltage levels, frequencies, permissible load on network elements, etc. Any violation of the above criteria is associated with emergency events, further increasing the risk of uncontrolled shutdowns of system components, leading to power outages. The dynamics of physical phenomena in the event of emergency shutdowns are very complex, which limits the possibility of reacting to the development of accidents.

Hence, it is very important to properly plan the operation of networks and generation resources with an appropriate safety margin.

In order to guarantee the efficiency of this type of system, the continuous monitoring of electricity demand is necessary. An effective planning of the operation of national power systems is essentially based on the verification of the power balance in a specific planning horizon. Therefore, the question arises whether it is possible to effectively balance the production of electricity with the demand of consumers for this energy, while ensuring the required excess power. The energy usage patterns of the consumers are extremely varied and, as a whole, are affected by known and unknown events, e.g., household usage peaks during football championships and, for the companies, just before Christmas. Understanding the behavior of certain groups in the PPS and foreseeing their changes is a key for effective balancing of the electricity production and demand of consumers, while ensuring the required excess power. The aforementioned excess capacity for the analyzed time horizon is a kind of measure of the future balance situation.

Since electricity consumption behavior may vary between the customers, a cluster-based load curve estimation which describes the variation in load demand from the consumer side on a power source over a period of time, is sometimes considered [1,2]. Clustering enables the discovery of underlying patterns in electricity datasets and serves as the prerequisite for robust modeling, i.e., by first clustering the customers, then modeling the clusters separately, and finally, aggregating the data.

It is considered that the improvement provided with the clustering strategy (compared to the traditional approach on the aggregated level, i.e., on the available population of customers) not only depends on the number of the clusters, but also on the size of the customer base. Therefore, this article presents a cluster-based approach to estimate a demand model of the Polish Power System using neural networks and accounting for the size of the customer base. In particular, energy readings of 27 thousand commercial entities in Poland recorded between 2016 and 2020 are used to deliver the following contributions from the research:

(1) A demonstration of how high frequency customer data can be utilized for the clustering and further, for demand estimation in the Polish Power System;
(2) Confirmation of the minimum requirement in terms of the sample size drawn from the clusters to be able to estimate demand in the system;
(3) The potential implications for the management and policy formulation within the Polish Power System are highlighted. Specifically, by employing a cluster-based approach to estimate demand, our methodology provides a more nuanced understanding of consumer behavior, enabling policymakers and energy managers to align better with strategy of the national power system.

Based on the literature review, as presented in the following section, there are multiple components being analyzed when modeling electricity demand, mainly for residential customers. Often, these works are focused on models and their technical characteristics to solve a problem through estimation or forecasting. In this context, the proposed research fills the gap related to the fact that only a few works use commercial data because the availability of such data for scientific purposes is very limited. Also, the cluster-based approach proposed here is an interesting alternative when modeling electricity demand as only a few works consider clustering as a viable option for robust modeling, mainly due to the fact that a sufficiently large collection of data to enable clustering is hardly available.

This paper is organized as follows. Section 2 describes related works; in Section 3, the dataset used for the analysis of the Polish Power System is characterized. This is followed by a brief methodology outlined in Section 4. Section 5 presents the approach to estimate the demand in the PPS as well as numerical experiments to analyze a minimal sample size to estimate demand with reasonable accuracy. Section 6 describes the insights gained from the analysis and presents conclusions.

2. Literature Review

In the current global situation, national power systems are of great interest to decision-makers in the global economy. There are many reasons for this, but the most important are the willingness to abandon fossil fuels (which are still the main source of electricity), the transition to green energy [3–5], and the possibility of reducing energy consumption, thus also reducing the demand for energy [6]. Moreover, equally important is looking for flexibility in energy markets, aimed primarily at ensuring the stability of power systems [7,8] and the impact of those systems on climate changes [9,10]. So, on the one hand, national power systems face the problem of decarbonization; on the other, they need to maintain a correct functioning of energy systems.

The problems of modern energy and power systems are considerable and complex. The task is basically to define an optimal set of technologies and mechanisms to support the transformation of such systems, while ensuring all interested parties obtain reliable access to the electricity supply. It is therefore a wide field of research for scientists and implementers. The modeling of energy systems is now a primary means for informing, guiding, and supporting decision-makers in this field in their efforts to coordinate the energy transition [11,12]. The main goal of this modeling is the identification of future patterns of energy supply and demand, as well as the development of strategies for the long-term transformation plans of the systems in question [13].

In terms of modeling energy systems, two research approaches can be formally distinguished. The first focuses on models and their technical characteristics, providing information on current methodological trends, challenges, and possible future research directions. These studies are mainly aimed at research dealing with energy modeling. They provide general overviews of different models [11,14–16], compare them in various aspects (scope, capabilities, features, etc.), identify common modeling concepts, outline future research directions or, consequently, identify models appropriate to given needs. At the same time, all analyses in this area relate to specific challenges faced by this type of modeling and refer to various methodological reviews used in the literature.

Thus, in reference [11], for example, the reviewed models were evaluated in terms of their characteristics, like their underlying methodology, analytical approach, time horizon and transformation path analysis, spatial and temporal resolution, licensing and modeling language. Paper [14], in turn, reviewed several existing bottom-up energy system models in order to classify them. In the study [15], the authors also proposed reviews of the changing role of electricity systems modeling but in a strategic manner, focusing on the modeling response to key developments, the move away from monopolies towards liberalized market regimes, and the increasing complexity brought about by policy targets for renewable energy and emissions. Pfenninger S. et al. [16] raised an issue of the crucial factors limiting openness of energy data and models: the lack of practical knowledge as well as personal and institutional inertia.

The second approach relates directly to the comparisons of different concepts developed in the literature. This creates a form of review research: without the need for analyzing the complexity of specific models, but rather through collecting various scenarios of energy and electrical systems, it aims to highlight key trends, differences, similarities, development paths, or the potential risks hidden in them. As a result, this approach considers the use of different analytical methods, the use of different parameters or specific initial or boundary conditions, etc. [17]. Often, it is mentioned in the literature that the models are not very transparent to the users, hence, there is a need to synthesize the available information and transform it into an effective policy. An overview of the results of several studies is presented in Table 1.

Table 1. Overview of the research on power systems modeling.

Authors	Focus
Foley A.M. et al. [15] *	Overview of electricity system modeling techniques and review of proprietary electricity system models.
Gabriel S. et al. [18] *	Estimation of a large-scale mathematical model that computes equilibrium fuel prices and quantities in the U.S. energy sector.
Skinner C.W. [19] *	Development of a new national energy modeling system to provide annual forecasts of energy supply, demand, and prices on a regional basis in the United States and, to a limited extent, in the rest of the world.
Fattahi A. et al. [20] **	Review of nineteen integrated energy system models (ESMs) to: identify the capabilities and shortcomings of current ESMs to adequately analyze the transition towards a low-carbon energy system; assess the performance of the selected models by means of the derived criteria; and discuss some potential solutions to address the ESM gaps.
Yan C. et al. [21] **	Presentation of an integrated evaluation framework to evaluate the possible national multi-energy flow in China in the near future. The framework includes an integrated modeling for a national multi-energy system in China. These key national energy facilities are all modeled in a generalized network flow formulation.
Berntsen P. et al. [22] *	Long-range energy scenarios are used to inform national energy policy decisions. Use of a bottom-up energy system model EXPANSE with modeling to generate alternatives to assess the diversity of the existing ensemble of multi-organization, multi-model Swiss electricity supply scenarios.
Aryanpur V. et al. [23] **	Presentation of national-scale energy systems optimization models, determination of a combination of supply and demand data requirements and socio-economic, environmental, and political issues, can challenge the results of a low-spatial resolution model.
Beaver R. [24] **	Analysis of the structure of energy and economic models.
Mirakyan A. [25] **	The analysis of existing national energy systems, as well as the prediction of potential future scenarios, is usually performed with the aid of an energy system model. The proposed framework can be used to identify and classify different types of uncertainty in context of energy planning in cities or territories.
Baghelai C. et al. [26] *	Characteristics of the uncertainty in the core elements of the US Department of Energy's National Energy Modeling System.
DeCarolis J. et al. [27] **	Energy system optimization models (ESOMs) are widely used to generate insight that informs energy and environmental policy. This paper shows the best practice for energy system optimization modelling and outlines a set of principles and modelling steps to guide ESOM-based analysis.
Pusnik M. et al. [28] **	The main technical, economic, and environmental characteristics of the Slovenian energy system model REES-SLO are described.
Sahoo S. et al. [29] *	An integrated modeling-based approach for regional analysis was proposed. The modeling framework was subdivided into four major blocks: the economic structure, the built environment and industries, renewable energy potentials, and energy infrastructure, including district heating. The results show the added value of regionalized modeling as opposed to relying solely on national energy system models.
Collins S. et al. [30] **	Long-term energy modelling challenges were identified including soft linkages between models of integrated energy systems and models of power systems, as well as an improvement in temporal and technical representation of power systems within models of integrated energy systems.
Gacitua L. at al. [31] **	This publication presents a comprehensive and up-to-date review on expansion planning models and tools, with an emphasis on their application to energy policy analysis. It reviews the most significant policy instruments, with an emphasis on renewable energy integration, the optimization models that have been developed for expansion planning, and existing decision-support tools for energy policy analysis.
Wen X. et al. [32] *	The authors review existing accuracy indicators used for retrospective evaluations of energy models and scenarios.
Chaudry M. et al. [33] *	An integrated energy system model is described. It is used to show the impacts on the environment due to different low carbon options to decarbonize a regional energy system in the context of national targets and constraints.
Hanna R. et al. [34] **	This study explores how different energy systems models and scenarios explicitly represent and assess potential disruptions and discontinuities (socio-economic, political and technological).
Huang K. et al. [35] *	Energy system optimization models (ESOM) to simulate energy and emissions changes under different economic and technological scenarios or prospective policy cases were considered.
Batas Bjelić I. et al. [36] *	In this paper, the achievement of the goals of the EU2030 is modeled by introducing an innovative method of soft-linking EnergyPLAN with the generic optimization program (GenOpt). The result of the optimization loop is an optimal national energy master plan (as a case study, the energy policy in Serbia was used), followed by a sensitivity analysis of the exogenous assumptions and with a focus on the contribution of a smart electricity grid.
Yan C. et al. [37] *	The authors present an analytical method to model the dependent multi-energy capacity outage states and their joint outage probabilities of an integrated energy system for its reliability assessment.
Martinsen T. [38] **	This paper reviews the characteristics of technology learning and discusses its application in energy system modelling in a global–local perspective. Its influence on the national energy system, exemplified by Norway, is investigated using global and national Markal models. The dynamic nature of the learning system boundaries and the coupling between the national energy system and the global development and manufacturing system are elaborated.
Davis M. et al. [39] *	This research presents a framework for developing a scientific tool with a long-range energy alternative planning (LEAP) system for evaluating energy consumption and greenhouse gas (GHG) emission mitigation pathways for a national energy system. The developed framework is applied to create a bottom-up (technology-explicit), data-intensive (over 2 million data points), multi-regional (13 integrated regions) energy model of Canada, one of the world's most energy- and emission-intensive nations.
Lund H. et al. [40] **	The authors analyze diversity of models and their implicit or explicit theoretical backgrounds.

Studies in the field of the first approach are marked in the table with the symbol * and the second approach with **.

Moving from the issues of various modeling concepts of national power systems to issues related to electricity demand, it is worth referring to several studies in this area. The analysis of the demand for this type of energy is the basic element of the stability of the national power systems. Research in this area was carried out by, among others, a team led by Kazemzadeh M. [41], which made attempts to develop a hybrid method for forecasting the annual peak load and total energy demand of Iran's national energy system. For Indonesia, the forecasting of electricity consumption has been recently carried out by McNeil M., Karali N., and Letschert V. [42]. In their research, they considered a novel bottom-up modeling approach to analyze the potential of energy efficiency to reduce the country's electricity demand. The LOADM curve model used in this case combines the total national electricity demand for each end user—as modeled by the bottom up energy analysis system (BUENAS)—with hourly end-user demand profiles. The publication of Ouedraogo N. [43] is an example of this type of analysis for the African continent. The paper developed a scenario-based model to identify and provide a range of electricity needs in Africa and to derive them from the African energy system. The approach was implemented through the application of the scenario methodology developed by Schwartz in the context of the "Long-range Energy Alternative Planning" energy and economic modeling platform. Although most analyses of this type relate essentially to highly industrialized regions, there are studies referring to exceptionally underdeveloped economies. An example is the first multi-purpose, long-term energy planning optimization model adapted to national power systems with a small existing energy infrastructure developed for Uganda by Trotter P., Cooper N., and Wilson P. [44]. Assessment and evaluation of flexible demand in a Danish future energy scenario was the basis of the research by Kwon P.S. and Østergaard P. [45]. They assessed the distant future potential of elastic demand in the energy system.

Of course, there are many more examples of energy system analyses. They are conducted for various applications, taking into account many concepts. The cluster-based approach proposed by the authors of this publication may be an interesting alternative when modeling electricity demand.

It should be emphasized that the data used in the analyzes are unique and real. The analyzed dataset has already been used by the authors in one study [46]. An important contribution here is to clarify the energy demand in the national energy system through commercial customer data. Also, the goal is to draw attention to the so-called minimum sample size taken from clusters necessary to estimate the demand in the considered energy system.

3. Data Characteristics

This study focuses on the Polish Power System, as the data which were obtained are tightly connected to companies operating on Polish territory. The data include profiles of supply and demand from the Polish Power System for the years from 2016 to 2020 and the energy readings of 27,160 commercial entities in Poland recorded for the same period.

The research is based on unique data, but the methods and analysis can be applied to any national power system of any European country.

Each power system is characterized by the volatility of electricity demand due to the fact that the customers' demand varies throughout the days, weeks, and years. It is closely related to the behavior of energy users who cover their energy needs. The changes in the load can be seasonal and recurring, related to the daily activities of people, or to the technological processes of production plants. They can also be irregular, for example as a result of changes in weather conditions, such as temperature or cloudiness. Higher peak loads in the winter months are associated with greater energy needs of the end-users for heating in case they have electric heating or heating pumps. Additionally, the highest peak loads occur on working days, lower ones on non-holiday Saturdays, and the lowest ones on Sundays and public holidays. Figure 1 shows the average weekly volatility of the load in the Polish Power System in 2018 and for each month separately.

Despite significant changes in the load volume from month to month, there are characteristic night valleys with relatively low loads, which remain stable between 22:00 and 6:00. The load volume in working days when comparing months is very similar. There also is a visible reduction in the load on Saturdays and, especially, on Sundays.

An important regularity regarding the load in the Polish Power System is the shift in the evening load peaks due to the change in sunset times. This shift is not only associated with the use of artificial lighting, but also with the fact that people's activity after dusk is moved to houses which triggers the use of various electrical devices. The average daily volatility of the Polish Power System for each month of 2018 is presented in Figure 2.

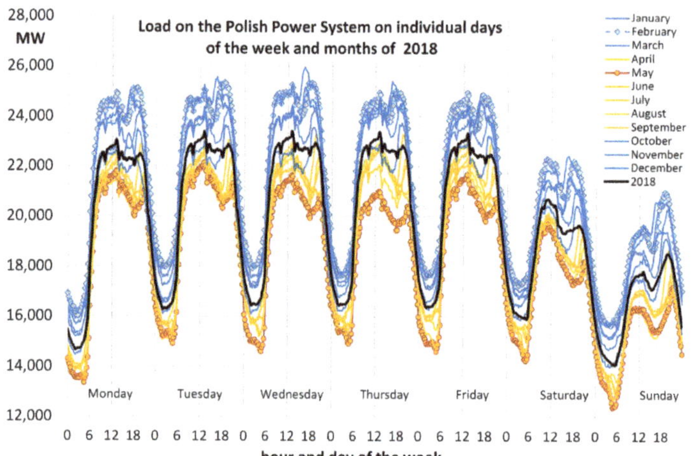

Figure 1. Weekly volatility in the electricity load in the Polish Power System in 2018. Source: [47].

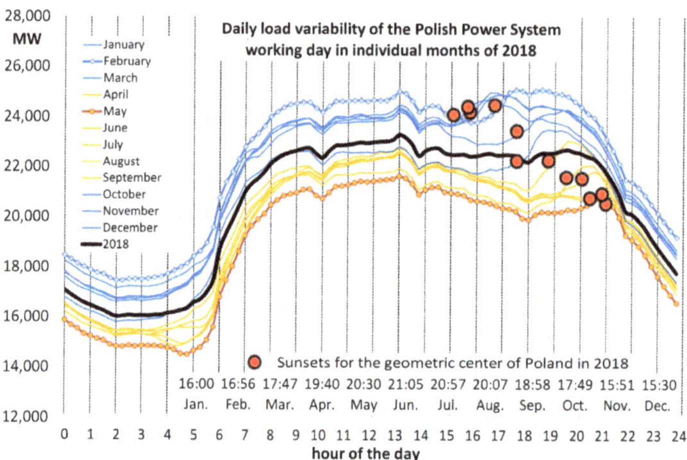

Figure 2. Daily volatility in the Polish Power System for each month of 2018. Source: [47].

The analysis of Figure 2 shows that the evening peak in winter months occurs immediately after sunset. Another important observation is that the load peaks in the summer months occur at noon.

Due to the correlation between the energy demand and the energy price, knowledge about the peak periods and minimum demand observed in the power system is important for managing the costs of electricity supply through switching energy carriers (e.g., coal, gas, etc.) depending on the load level. This enables the standardization of responses from power plants to meet the energy needs of their customers. Those correlations are reflected in the tariff groups and time zones offered by energy companies for making the settlements with customers.

In order to build an accurate model of the electricity demand in the power system, it should be remembered that, in the case of national energy systems, an important issue is related to the availability of disaggregated data on the customer level. Such low level customer data are helpful to analyze the impact of specific customers' groups on the shape of the demand curve, which can be used further for the demand side management (DSM) and demand side response (DSR) programs for the efficient use of the electricity.

This study was prepared based on a historical dataset of 27,160 commercial entities located in the central-eastern part of Poland; the data were obtained from Data Bridge—a company which specializes in gathering data from energy supply companies. This dataset contains hourly data recorded for every customer between 01 January 2016 and 31 December 2020, enriched with calendar data (weekdays, months, and holiday indicators) and meteorological data including temperature and humidity.

Initially, the dataset contained more customers, however, it was necessary to perform data pre-processing to improve the quality, i.e., all the readings whose values were less or equal to zero or those with repeating time stamps were removed. In addition, the customers with less than ten different values in their readings were discarded. The structure of the dataset as well as some basic statistics in terms of the electricity volume are provided in Table 2. Based on Table 2, it can be concluded that between 30% and 50% of the customers are small businesses, i.e., those for whom the average daily demand is less than 10 kWh. Large businesses, i.e., over 150 kWh, represent approximately 10% of all the customer base. Also, the total number of customers for the year 2017 is much smaller compared to other years. This is due to the fact that data were obtained from multiple energy suppliers and their customer base was not stable in 2017 due to market consolidation and migration of the customers between energy suppliers.

Table 2. The structure of the dataset in terms of the electricity volume (in kWh) and the number of entities observed between 2016 and 2020.

Average Daily Usage (in kWh)	Year				
	2016	2017	2018	2019	2020
(0, 2]	9864	1786	4332	5662	4658
(2, 5]	1873	1434	2345	3718	3132
(5, 10]	2127	1923	2770	3824	3929
(10, 25]	2669	3226	4217	5136	5319
(25, 50]	1416	1972	2321	3183	3351
(50, 75]	589	822	955	1299	1764
(75, 100]	378	479	536	740	1088
(100, 150]	384	540	608	808	1225
(150, 200]	153	255	333	464	652
(200, 500]	406	474	752	805	1167
(500, 1000]	124	187	276	335	428
(1000, Inf]	132	207	268	346	447
Total number of entities	20,115	13,305	19,713	26,320	27,160

As shown in Figure 3, a number of weekly and daily cycles is observed on the aggregated load curve. For instance, the daily load curves have different shapes depending on the day (workday, Saturday, Sunday, or holiday). This is visible on the graph: the beginning of May starts on hour 1, which is midnight on the 1st of May 2019. The first and third of May are national holidays in Poland, and in 2019, the first of May was a Wednesday. The second of May is often taken as a bridge-holiday and as such it appears more similar to a non-holiday Saturday. From the fifth of May (hour 96), the normal weekly pattern emerges: 5 weekdays, Saturday and Sunday. During the working days, there are clearly defined peaks in the middle of the days, and smaller peaks in the evenings. Finally, the consumption is significantly lower during the weekend days compared to working days.

Based on Figure 4, it can be concluded that the analyzed data for 27,000 commercial customers constitute approx. 1% of the total volume of the power (in MW) of the Polish Power System and, at the same time, exhibit a load curve similar to the PPS curve. The similarity between both curves is quite high: 0.75 measured with the coefficient of determination (R^2).

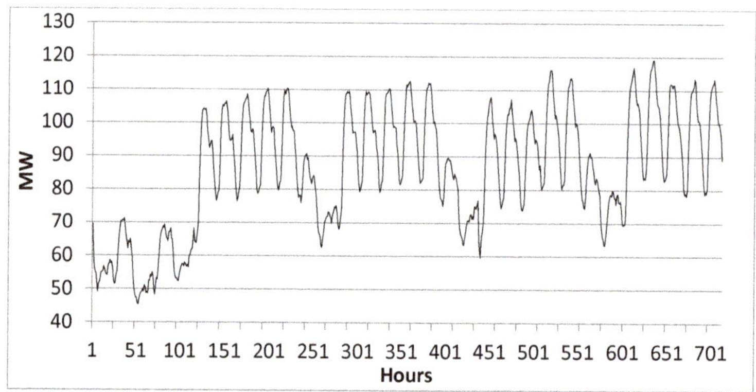

Figure 3. Hourly load data observed between May 1st (0:00) and 31st (23:00), 2019 for the customers.

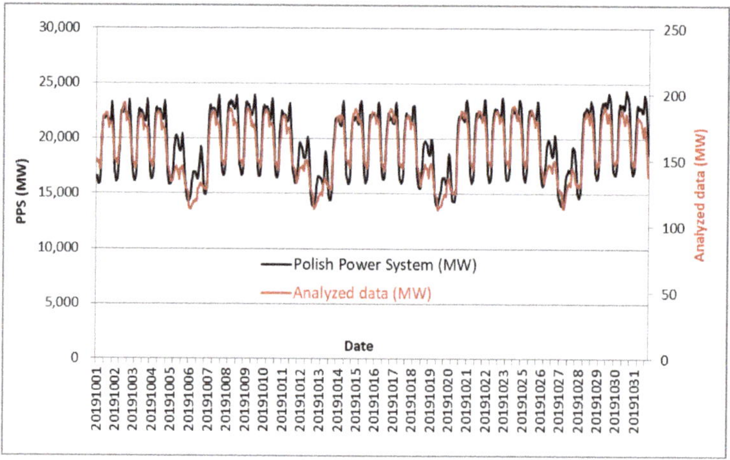

Figure 4. Hourly energy consumption curves in October 2019 for the analyzed data (red pattern) in relation to the Polish Power System (black pattern).

4. Methodology

4.1. Clustering

Since energy consumption behavior might vary among customers, different energy consumption patterns can be grouped using clustering algorithms; these clusters can be used to achieve a better understanding of customer profiles and to perform load modeling. The prerequisite for the clustering of the customers' profiles was proper data preprocessing, i.e., creating, for each customer and each year, matrices containing electricity consumption where the dimensions were month, weekday, and hour. An example of such a matrix for 2016 for one of the customers is presented in Figure 5. Each cell represents the average consumption in each hour for the customer calculated over four or five values, e.g., four Mondays (weekday = 1) in January (month = 1) 2016. It shows an increased consumption with red cells and a lower consumption with green cells.

The data were normalized by row, i.e., the vector with 24 values for each hour, using standardization: (x-mean(x))/std(x); this yielded the matrices with normalized consumption which were used to determine similarity between customers' profiles. Each time, similarity was calculated using the Euclidean distance between two normalized matrices for two customers' profiles, in other words element-wise operations were applied. Next, hierarchical clustering using Ward's method was performed. It considers cluster analysis through an analysis of variance where the minimum variance criterion reduces the total within-cluster variance, instead of using distance measures to create the clusters [48]. The method involves an agglomerative clustering algorithm which starts at the leaves and works its way to the root. During the process, the method looks for groups of leaves that form into branches, the branches into limbs and finally into the root. Ward's method starts out with n clusters of size 1 and continues until all the observations are included in one cluster.

Month	Weekday			Hour										
		1	2	3	4	5	6		19	20	21	22	23	24
1	1	446.64	446.54	443.15	438.09	454.14	396.40		596.40	586.68	579.84	577.60	563.90	554.84
1	2	545.39	540.29	541.49	538.02	453.64	398.88		573.69	564.89	564.19	558.79	546.88	538.99
1	3	533.30	529.58	525.87	524.64	514.49	485.61		552.23	545.83	536.73	538.26	532.75	517.75
1	4	518.81	517.51	513.28	516.33	491.96	374.02		588.19	579.01	575.66	582.15	569.18	557.84
1	5	493.76	491.50	491.11	486.20	479.45	449.25		519.76	517.03	514.87	514.54	502.87	496.46
1	6	487.22	483.25	481.82	483.33	456.97	327.15		393.42	403.56	393.26	388.38	382.94	368.27
1	7	362.86	360.89	360.22	359.65	358.48	360.35	…	429.18	437.80	442.45	447.77	444.17	441.13
2	1	435.02	436.54	441.30	446.25	472.30	461.62		505.72	497.83	493.46	493.13	480.88	471.31
2	2	458.26	451.33	454.31	450.29	446.16	398.08		580.74	578.29	580.03	587.88	587.55	575.01
2	3	567.13	559.43	555.15	545.07	538.89	540.00		587.14	582.79	575.63	575.36	568.12	556.37
2	4	551.74	548.61	548.82	538.94	526.82	484.04		594.95	591.24	587.69	583.66	570.60	557.70
2	5	565.60	561.32	556.41	550.05	537.20	498.13		589.66	581.71	574.81	574.74	561.21	548.59
2	6	537.55	528.24	520.80	516.79	471.73	380.69		375.01	368.82	360.13	353.98	343.59	334.88
2	7	331.63	329.97	330.47	330.94	331.08	335.92		431.06	447.60	449.58	445.78	443.86	434.57
⋮														
11	1	357.20	358.29	356.14	361.01	391.58	422.91		545.25	528.66	526.80	517.83	496.55	482.41
11	2	457.66	456.20	452.27	445.42	442.04	448.64		485.59	477.13	482.65	482.17	480.21	473.79
11	3	473.07	471.14	470.08	467.13	469.12	477.01		522.11	515.78	513.65	518.52	515.25	507.71
11	4	499.32	486.41	472.96	466.53	460.80	471.76		529.24	513.30	503.23	502.92	485.30	476.17
11	5	465.28	460.50	451.67	449.37	451.32	463.73		475.25	468.33	466.87	464.41	458.03	452.71
11	6	451.16	453.59	453.08	452.21	443.83	428.91		382.38	382.72	373.76	363.54	349.98	344.28
11	7	341.41	337.59	338.27	335.65	330.89	333.39	…	357.02	352.71	355.61	361.61	358.54	357.77
12	1	349.89	349.38	347.86	343.82	356.84	396.28		496.55	491.17	483.13	479.61	462.17	458.15
12	2	463.66	459.84	454.71	451.13	455.86	470.11		547.53	535.15	529.82	530.00	512.64	511.98
12	3	518.84	514.70	502.93	500.84	502.62	515.13		564.65	554.05	541.36	540.31	521.71	512.96
12	4	515.46	517.38	515.49	508.87	495.29	500.40		548.99	546.11	545.98	541.57	518.12	502.37
12	5	498.82	499.79	503.15	502.63	501.28	509.50		549.52	540.73	535.17	526.80	507.79	507.07
12	6	508.67	508.43	498.25	478.46	461.39	451.43		375.21	368.61	365.10	361.54	348.45	344.30
12	7	358.28	359.06	355.67	349.61	346.02	356.32		378.19	375.12	376.08	375.50	369.60	369.76

Figure 5. Consumption matrix for one of the customers observed during 2016.

Applications of Ward's method were used to determine the largest number of distinct clusters that have non-overlapping patterns. For this purpose, energy load profiles for working days, Saturdays, and Sundays were plotted for each cluster created in the 2016–2020 data. The clustering into 20 clusters was considered as the one best meeting that goal. Another rationale for selecting 20 clusters as a cut-off was based on the number of observations, i.e., entities, in each of the clusters. It stems from the fact that clusters should contain a sufficient number of observations to create meaningful and actionable groups of customers. Figure 6 presents a visualization of 20 clusters in terms of energy profiles for working days, Saturdays, and Sundays for 2020, while the number of entities in each cluster is provided in Table 3.

As shown in Figure 6, the comparison of clusters and their load profiles allows the identification of significant differences between the profiles which cannot be seen on the aggregated level of the Polish Power System. There are several clusters which show increased consumption during the day, with one or two spikes, as opposed to some other clusters, with low demand during the day but with increased demand during the night. Those profiles are useful for building tariff structures, demand-side management, planning of the distribution system, and for defining critical segments which can impact the power system when balancing the energy market.

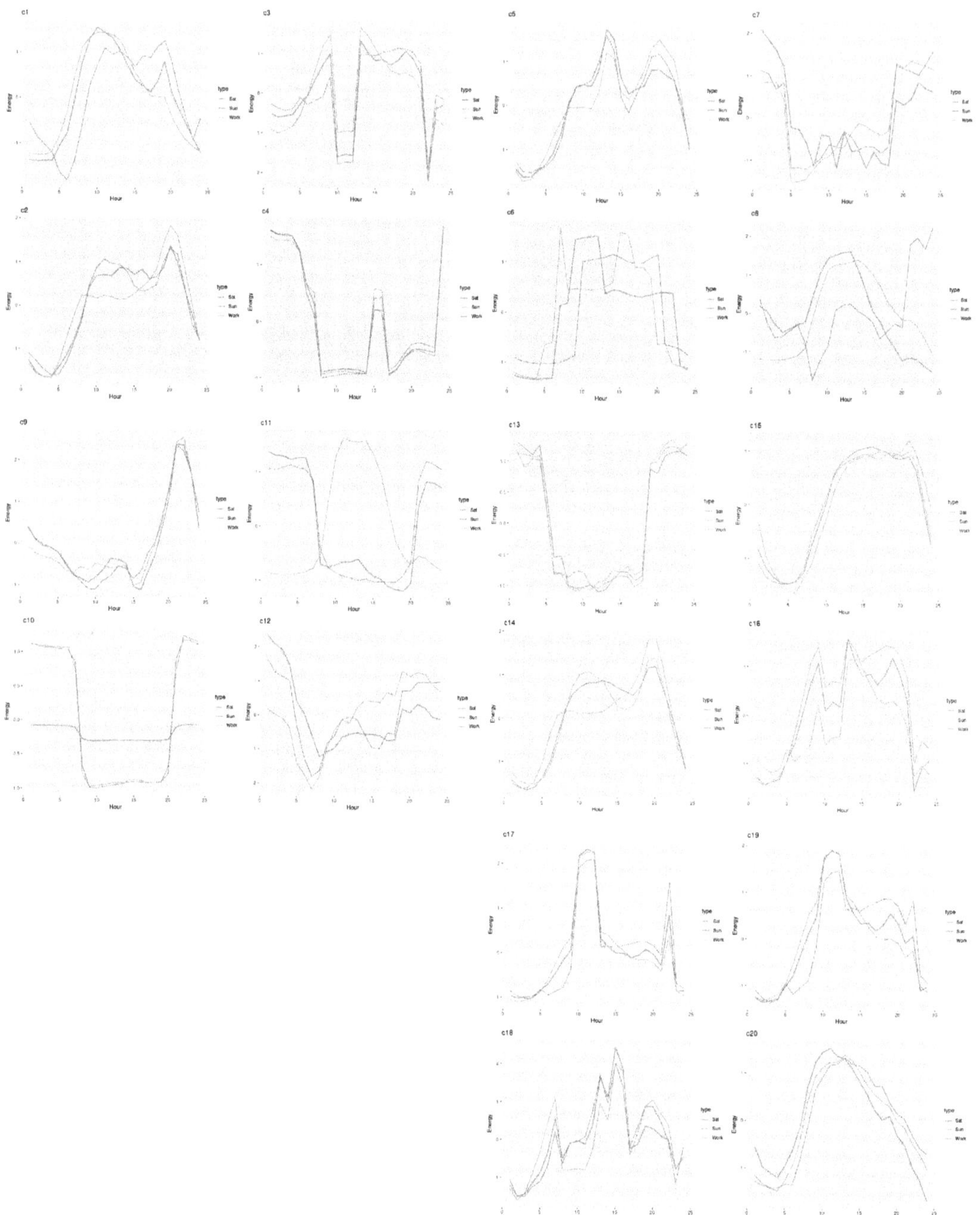

Figure 6. Visualization of 20 clusters (C1–C20) based on 2020 data (profiles are provided separately for working days, Saturdays, and Sundays).

Table 3. The number of entities in each cluster for 2016–2020.

Cluster	Year				
	2016	**2017**	**2018**	**2019**	**2020**
C1	4008	3747	4335	3765	5868
C2	2091	2678	3824	3183	4162
C3	678	858	1647	2367	3673
C4	652	720	1348	2365	2025
C5	488	689	1037	1903	1755
C6	482	647	928	1834	1242
C7	456	492	835	1731	1005
C8	388	453	671	1520	752
C9	275	309	480	1037	641
C10	254	282	454	890	554
C11	197	274	439	664	549
C12	189	259	430	637	524
C13	167	227	374	542	504
C14	146	223	349	424	494
C15	134	222	298	421	477
C16	126	168	271	392	329
C17	96	140	263	375	306
C18	77	123	252	290	297
C19	74	105	139	183	264
C20	60	73	103	153	124

4.2. Neural Networks for Estimation

Artificial neural networks (ANN) were first introduced by Warren McCulloch and Walter Pitts in 1943, who created a computational model for neural networks based on a threshold logic algorithm. The idea of this computational model was inspired by biological nervous systems consisting of a large number of elemental processing units, called neurons, which are organized in input, hidden, and output layers [49,50]. Each neuron in the network is characterized by input weights, an activation function, and a threshold. In the simplest artificial neural networks, neurons are usually connected in a feedforward manner so data processing moves only in one direction, from the input nodes through the hidden layer, to finally reach the output neurons.

A multilayer perceptron (MLP), introduced by Frank Rosenblatt in 1958, is a feedforward artificial neural network model with multiple layers of neurons which are fully connected to the next neurons in each layer. With an adequate learning method and with a sufficient number of neurons in the hidden layers, the MLP networks are able to deliver precise and satisfactory approximation for any type of bounded piecewise continuous functions [51].

The MLP network utilizes a supervised learning backpropagation technique which is widely recommended as the most efficient procedure for the training of neural networks and used in conjunction with gradient descent optimization method [52]. The main issue in the application of neural networks is finding the proper values for the weights between the input and the output layer. Starting with random weights, an input dataset is presented to the network to make initial estimations. During the learning process, the differences between the estimated and the measured values are used to assess the error. Then, the error is propagated back through the whole network to update the weights and to obtain improved results, as we want the algorithm to find those properties of the input data that are most relevant for modeling the target function. More details regarding the MLP architecture and learning algorithms are elaborated on in [50,53,54].

Figure 7 presents an example of a three-layer neural network which consists of an input layer with a set of input neurons, one hidden layer with computation neurons, an output layer, and weights between all the layers.

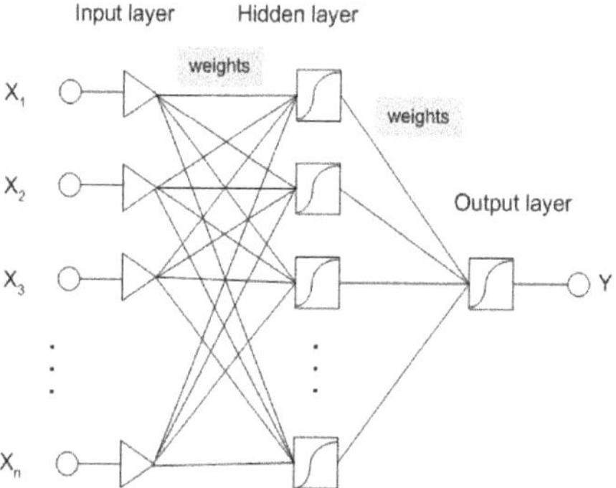

Figure 7. The three layer MLP artificial neural network.

For the purpose of demand estimation, an MLP feedforward artificial neural network was used as this is undoubtedly one of the most commonly used architectures in practical applications of ANN, especially in applications related to estimation and classification [52].

5. Modeling Electricity Demand in the Polish Power System

Energy demand, a critical aspect of energy systems, is the measure of electrical energy required by end-users within a specific timeframe. The knowledge of expected energy demand is essential for ensuring a stable and resilient energy infrastructure. The determinants influencing energy demand are multifaceted. Economic growth plays a pivotal role, as expanding industries and increased commercial activities elevate energy needs. Technological advancements, particularly in energy-efficient appliances and industrial processes, can either mitigate or intensify demand. Furthermore, societal changes, such as shifts in lifestyle and demographic patterns, significantly impact energy consumption. Lastly, climate and weather conditions are also important.

The transition towards renewable energy sources further complicates energy demand dynamics. While renewable integration offers sustainability benefits, the intermittency of sources like solar and wind introduces additional complexities in demand forecasting and grid management. Smart grid technologies, demand response programs, and energy storage solutions emerge as pivotal strategies in addressing these challenges, ensuring a harmonious balance between energy supply and demand.

Understanding energy demand is imperative for policymakers, energy planners, and stakeholders to formulate effective strategies that enhance energy efficiency, reduce environmental impact, and foster a reliable and resilient energy future. Also, it is pivotal for sustainable energy systems, influenced by factors like economic trends, technological shifts, and societal changes.

In the context of the PPS, our research employs clustering techniques on historical electricity usage data from commercial customers. As we delve into modelling electricity demand, this study provides valuable insights into the effectiveness of clustering-based models, showing their potential for accurate load estimation.

5.1. The Approach to Estimate the Demand

The main factors affecting electricity demand on a country-wide level are gross domestic product, energy prices, income, the characteristics of economic urbanization, and climate and seasonal factors. The magnitude of those determinants differs across countries, time periods, and studies, even for the same country; therefore, our goal was to explain the demand through the available high frequency data of commercial customers rather than building a macro-economic model.

Modeling the load in the national power system is an important aspect—not only for the economy, but also for the safety and reliability of the power system operations. The models are necessary for the planning and modernization of the whole system and for creating strategic directions, including market transformation. Moreover, knowledge of the load curve characteristics is necessary to take appropriate actions in balancing the available generation capacity and the demand.

In this context, a data-driven approach, with no theoretical assumptions, was used to estimate a model in which consumers (and specific consumer clusters) create energy demand and thus are responsible for the shape of the load curve. The model that reflects the actual structure of the market through the clusters of customers will help to analyze the demand in the Polish Power System and its fluctuations. For this purpose, various models were tested taking into account up to 25 features, including the results of the clustering. The modeling was not limited to 20 clusters but the whole range of clusters (between 1 and 20 clusters) and their electricity usage were considered in the models. This was to analyze the relation between the number of clusters formulated and the precision of demand estimation.

The following variables were used to create the MLP models:

- Feature 1—day type: working day, Saturday, Sunday, or holiday;
- Feature 2—the time of the day: (1) morning peak: between 7:00 and 13:00 for working days (Monday-Friday), regardless of the month; (2) afternoon peak: between 16:00 and 21:00 during winter months, i.e., between October and March; between 19:00 and 22:00 during summer months, i.e., between April and September; (3) off-peak periods;
- Feature 3—season: (1) summer (May–August); (2) winter (November–February); (3) other (March, April, September, October);
- Feature 4—temperature observed in hourly intervals;
- Feature 5—humidity observed in hourly intervals;
- Features 6—25-aggregated hourly electricity usage within each cluster (between 1 and 20 clusters). Each cluster is the result of the hierarchy of the dendrogram obtained for the hierarchical clustering using Ward's method, as shown in Section 3.

Some basic statistics for the variables which were used to create the MLP models are presented in Table 4. Variables C1 to C20 show aggregated hourly electricity usage within each cluster.

Table 4. Descriptive statistics for the variables which were used to create the models.

Variable	Statistics						
	Min	Q1	Median	Mean	Q3	Max	Sd
C1	114.3	429.7	880.2	5142.5	2681.6	44,699.6	9623.3
C2	15.6	120.6	966.2	4285.8	3022.5	41,144.2	7487.7
C3	68.7	559.6	1524.6	3391.4	3165.6	21,118.1	4498.3
C4	5.8	638.8	1830.7	9542.5	18,570.2	57,701.8	12,640.4
C5	39.1	201.3	1116.0	4267.9	3438.0	53,443.1	6527.9
C6	13.7	47.4	4726.8	4839.8	8503.8	22,143.5	4723.0
C7	8.1	74.5	268.6	862.4	1347.5	5473.9	1036.7
C8	22.5	71.9	220.6	11,332.2	2049.7	93,437.4	25,003.3
C9	38.2	453.9	832.1	1678.2	1478.4	10,971.3	2106.0
C10	32.6	2595.7	4511.4	5409.7	8122.8	34,714.5	4191.2
C11	48.0	1150.4	2707.9	3394.8	5134.5	17,112.1	2935.9
C12	31.5	239.6	1247.6	4482.8	5986.2	22,223.8	6141.6
C13	33.2	162.1	239.4	355.7	447.4	1757.0	299.9
C14	94.9	293.3	1094.2	3966.7	3217.8	29,634.1	6726.4
C15	45.6	323.8	537.4	8910.9	8396.4	58,916.1	14,673.8
C16	16.0	322.0	692.4	1610.8	1568.9	9027.5	2148.5
C17	25.3	104.7	151.9	609.7	579.9	8269.0	1067.2
C18	34.7	196.1	457.3	3869.1	3417.1	28,773.6	6226.6
C19	19.5	177.2	431.4	785.3	1032.9	4021.5	882.0
C20	12.4	341.8	629.9	1221.9	1352.6	7069.4	1423.4
Temperature	−21.8	2.3	9.4	9.2	16.4	34.7	9.05
Humidity	0	66	77.2	82	92	100	17.7

To assess the performance of the models, the mean absolute percentage error (MAPE) and adjusted R^2 were used. MAPE is a measure of prediction accuracy and it expresses the accuracy as a ratio defined by the formula:

$$\text{MAPE} = \frac{1}{n}\sum_{t=1}^{n}\left|\frac{A_t - F_t}{A_t}\right| \times 100\%$$

where A_t is the actual value and F_t is the forecast value.

Adjusted R^2 is a measure which provides better precision by considering the impact of additional independent variables that tend to skew the results of R^2 measurements; it is defined as:

$$R^2_{adj} = 1 - \left[\frac{\left(1 - R^2\right)(n - 1)}{n - k - 1}\right]$$

where n is the number of points in a data sample and k is the number of independent variables in the model, excluding the constant.

All presented simulations were prepared using R software (version 4.1) and nnet (version 7.3-16) library.

For each year, twenty MLP networks were trained; the difference between the networks was in the number of neurons in the input and hidden layers. Each explainable model was built for the hourly demand of the Polish Power System. The models varied from six neurons matching with six input-features (the five features mentioned earlier as well as the aggregated usage treated as a single cluster), up to twenty-five neurons matching with twenty-five input-features (the five aforementioned combined with the separate usage of the twenty clusters). Additionally, another twenty MLP models were created, but this time with random clusters, i.e., clusters of equal size with customers randomly assigned to each of them. The models with random clusters were used for comparison with the models which used the statistically derived clusters as input variables.

Each neural network consisted of one input layer (number of neurons ranging from six up to twenty-five which is in line with increasing number of clusters being considered as the inputs to the network), one hidden layer, and one output layer with one neuron and was trained using the nnet function which implements the Broyden–Fletcher–Goldfarb–Shanno algorithm (BFGS). Importantly, all input features were scaled using standardization. The varying number of features and neurons stems from the fact that, in the simulations, we considered various numbers of clusters and the associated energy volume. Specifically, for the first ANN simulation, there was only one time series for only one cluster, i.e., the entire population. For the second simulation, there were two time series for two clusters dividing the entire population, etc. Eventually, following the dendrogram, there were 20 time series for 20 clusters.

Wanting to follow the golden rule that each subsequent layer has fewer neurons than the previous (pyramid), the hidden layer contained two neurons less than the input layer. Neurons in the hidden layer were activated using sigmoid function, while the neuron in the output layer was activated using linear function. To prevent overfitting, the regularization term (weight decay) that uses as the penalty the sum of squares of the weights was set at its default value 0.0005 The models were built using cross-validation regime and the results were averaged. Due to the fact that analysis deals with feature vector (not a time sequence), these validation samples were randomly selected.

The results of the MLP networks are as follows (as shown in Figures 8 and 9):

○ For both adjusted R^2 and MAPE, there is a clear relation: the higher the number of clusters used, the better the results are;

○ Adjusted R^2 is between 0.96 and 0.99 when the aggregated electricity usage is considered for 20 segments in the models; at the same time, the models with random segments perform worse as adjusted R^2 is much lower, i.e., between 0.89 and 0.96;

○ MAPE is between 1% and 2.5% when the aggregated electricity usage is considered for 20 segments in the models; at the same time the models with random segments perform worse as adjusted MAPE is much higher, i.e., between 2% and 4.5%;

○ The best results are obtained for 2016, 2019, and 2020 which might be due to the fact that our source data contain more data points (as presented in Table 2).

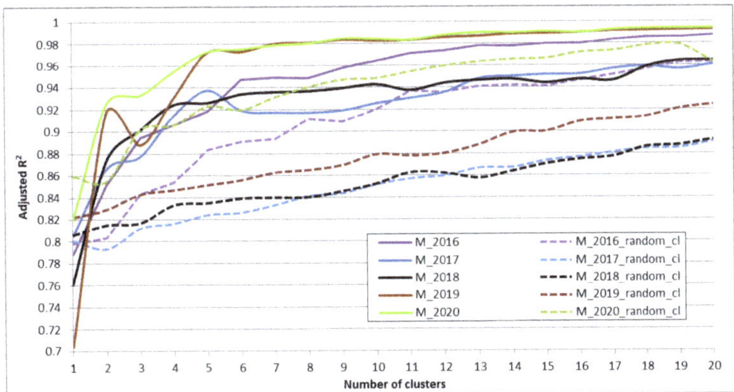

Figure 8. Adjusted R^2 for the models (each year separately) depending on the number of extracted clusters (solid lines) and compared to the models with random clusters (dashed lines).

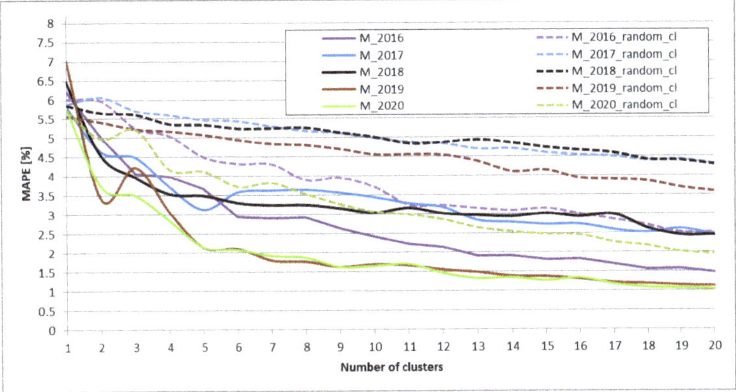

Figure 9. MAPE for the models (each year separately) depending on the number of extracted clusters (solid lines) and compared to the models with the random clusters (dashed lines).

In Figures 10 and 11, a comparison between actual PPS demand and model's estimates is also presented (as an example, January and July 2019 data were considered). These indicate that the models are fitting the PPS curve well and that the results are better when more clusters are considered as the inputs for the models (one cluster vs. ten clusters in the example).

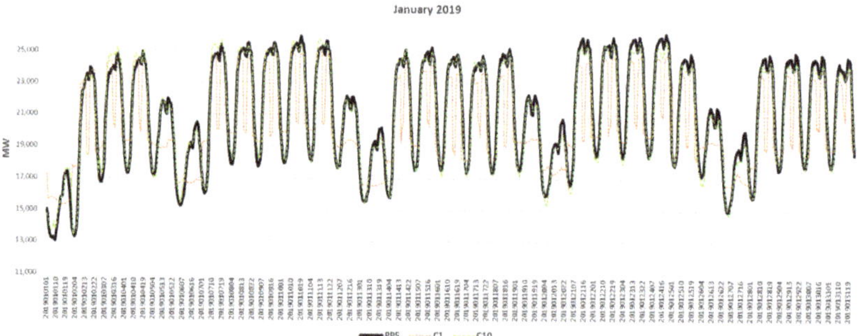

Figure 10. Actual PPS demand (black solid line) for January 2019 compared with the model's estimates; built for one cluster (dashed orange line) and ten clusters (dashed green lines).

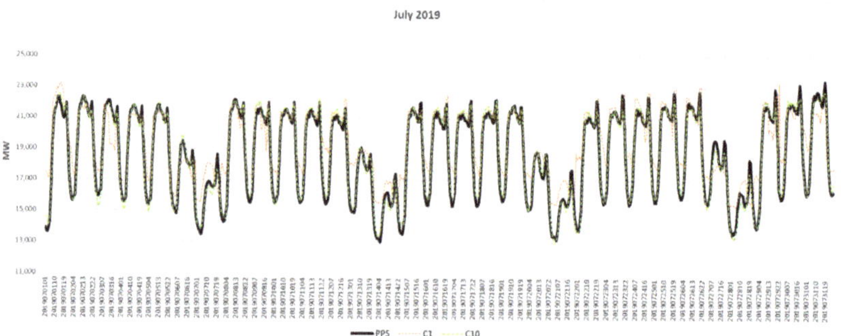

Figure 11. Actual PPS demand (black solid line) for July 2019 compared with the model's estimates; built for one cluster (dashed orange line) and ten clusters (dashed green lines).

5.2. Minimum Sample Size to Estimate Demand

Based on the literature review, the dataset used in this work is considered large as it represents the usage of 27,000 commercial customers. Usually, datasets that are available to scientists are significantly smaller which might impact the results and even make conclusions skewed and biased. Therefore, in this work, it was additionally analyzed what would be a minimal sample size taken from the clusters to estimate demand in the Polish Power System with good precision.

For this purpose, three random samples were prepared from each segment (from 1 to 20) and for each year, having 50%, 10%, and 1% of the data drawn from the original dataset. Additionally, similar samples having 50%, 10%, and 1% of the data drawn from the random clusters were prepared, i.e., segments of equal size with the customers who were randomly assigned to those clusters.

As previously stated, twenty MLP networks were trained for each year and for each sample (50%, 10%, and 1%), starting with a network with six neurons in the hidden layer to one with twenty-five neurons in the hidden layer. Additionally, another twenty MLP networks were created for each year and for each sample (50%, 10%, and 1%), but this time, with random clusters.

The results of the MLP networks are as follows (as shown in Figure 12):

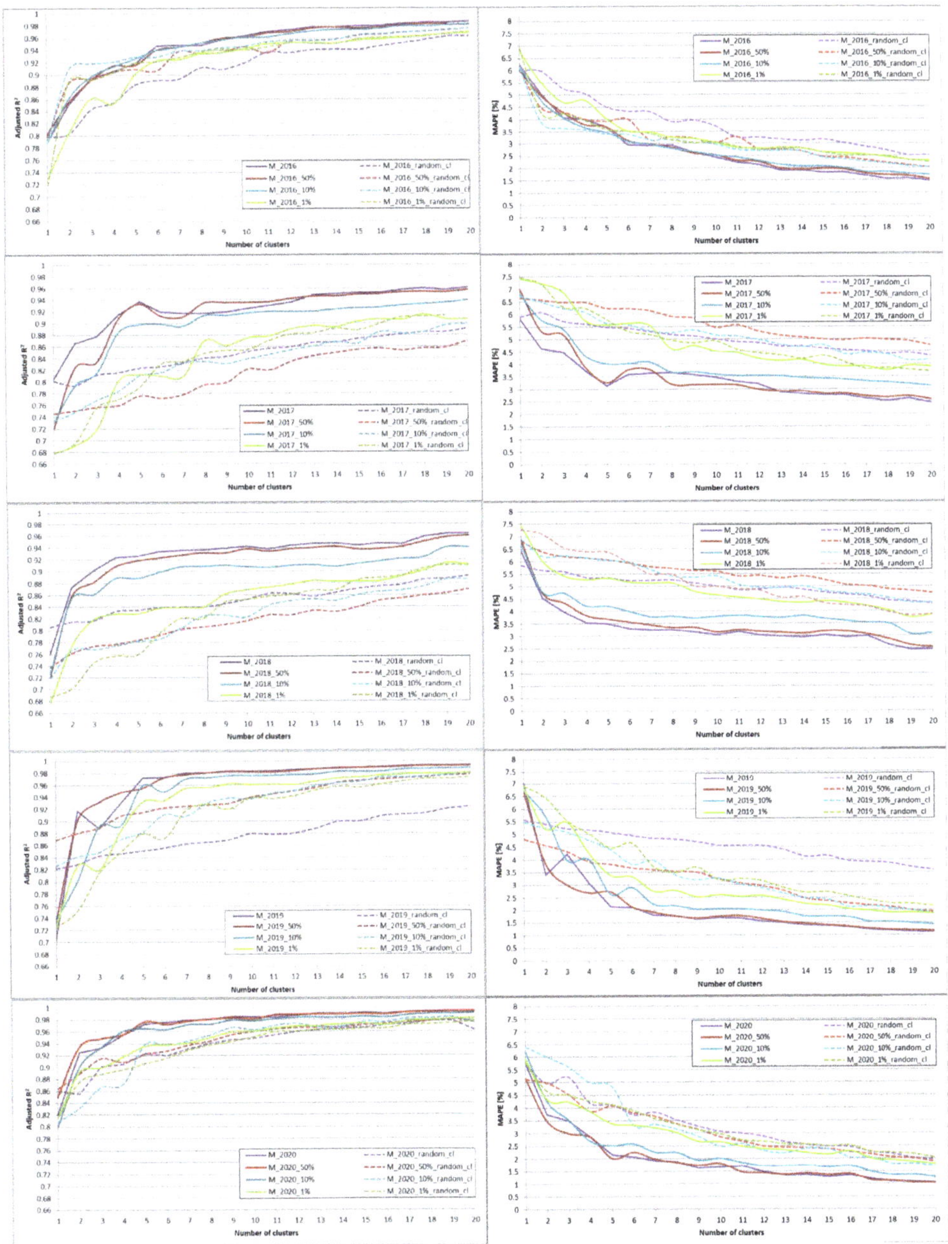

Figure 12. Adjusted R^2 (on the **left**) and MAPE (on the **right**) for the models (each year shown separately) depending on the number of extracted clusters and the sample size (solid lines) compared to the models with random clusters (dashed lines).

- For both measures, adjusted R^2 and MAPE, there is a clear relation: with a bigger sample, better results are achieved;
- When the sample size of 50% is drawn from each of the clusters, then both adjusted R^2 and MAPE are close to the results obtained for the complete dataset;

- When the sample size of 10% is drawn, then some slight deterioration in terms of the adjusted R^2 and MAPE is observed; specifically, R^2 is lower by 0.02 and MAPE is higher by 0.5 p.p. when comparing with the results obtained on the complete dataset; Nevertheless, such a sample still enables the models to be produced with reasonable accuracy;
- When the sample ^{size} of only 1% is drawn, then further deterioration in terms of the adjusted R^2 and MAPE is observed; specifically, R^2 is lower by up to 0.05 and MAPE is higher by 1 p.p. when comparing with the results obtained on the complete dataset;
- The 1% sample is considered too small to build reliable models as the results are close to the results obtained for random clusters;
- For the random clusters and the samples drawn from those clusters, it is observed that adjusted R^2 and MAPE are worse than the results obtained on the complete dataset;
- As previously, the best results are obtained for 2016, 2019, and 2020, which might be due to the fact that more data are available for those years.

6. Conclusions

This study analyzed commercial customers in Poland based on a real dataset with hourly power consumption records of 27,000 businesses spread throughout 2016–2020.

Since electricity consumption behavior may vary between the customers, a cluster-based estimation of the demand curve was considered. Such an approach enables the discovery of underlying patterns in electricity datasets and serves as the prerequisite for robust modeling and estimation, i.e., by first clustering the customers and then modeling the demand in the power system through the profiles associated with the clusters as the input variables.

It was proved that the clustering-based method with MLP models for demand estimation in the Polish Power System decreases the mean absolute percentage error substantially compared to the approach without clusters, while fitting the load curve well, which was confirmed using adjusted R^2.

Through the experiments, it was confirmed that the clustering of customers helps to estimate the demand in the Polish Power System significantly better than on an aggregated level, i.e., using the whole population. Specifically, there is a clear relation: the more clusters are used, the better the results are in terms of adjusted R^2 and MAPE. With 20 clusters, the models deliver MAPE as low as 1% and adjusted R^2 as high as 0.99.

As far as the size of the sample drawn from the clusters is concerned, it is observed that when a sample of size 50% is drawn from each of the clusters, then both adjusted R^2 and MAPE are close to the results obtained on the complete dataset. When a sample size of 10% is considered then a slight deterioration in terms of both these measures is observed; however, such a sample still enables us to build the models with reasonable accuracy. Finally, when a sample of only 1% is drawn, further deterioration in terms of the adjusted R^2 and MAPE is observed and such a sample is considered too small to build reliable models as the results are close to the results obtained for random clusters. This experiment clearly shows that small samples might impact the results and make conclusions skewed or biased, therefore the dataset should be sufficiently large. Finally, by employing a cluster-based approach to estimate demand, our research provides a far-reaching understanding of consumer behavior, enabling policymakers and energy managers to focus their strategies based on distinct customer clusters for proper demand-side management, planning of distribution systems and for defining critical segments which have the biggest impact on the power system. For example, those clusters which contribute to demand peaks should be considered for targeted actions (e.g., incentives) to flatten the peaks, as these pose a problem for the stability of the system.

Some limitations should be considered in the context of the results in the study. Since the research is based on data from Polish market it may impact generalization of the results. Nevertheless, the methods and analysis proposed here can be applied to any national power system of any European country and thus enable generalization of the findings.

There are a couple of promising applications of cluster analysis for managing the load in a power system which could be considered for further research. These are related to tariff design for specific customer groups and demand response programming which helps to flatten the load curve, thus contributing to an increased stability of the system.

Author Contributions: Conceptualization, T.Z. and K.G.; data curation, K.G., A.O. and W.R.; formal analysis, K.G. and J.B.; funding acquisition, T.Z., K.G., G.M. (Grzegorz Matejko) and G.M. (Grzegorz Mentel); investigation, J.J., A.O. and J.V.; methodology, T.Z., K.G. and G.M. (Grzegorz Mentel); project administration, T.Z. and K.G.; resources, J.J.; software, K.G. and J.B.; supervision, T.Z.; validation, G.M. (Grzegorz Matejko), M.C. and W.R.; visualization, K.G. and J.B.; writing—original draft, T.Z. and K.G.; writing—review and editing, G.M. (Grzegorz Matejko), J.B., G.M. (Grzegorz Mentel), M.C., J.J., A.O., W.R. and J.V. All authors have read and agreed to the published version of the manuscript.

Funding: This research was funded by the National Centre for Research and Development, Poland, grant number POIR.01.01.01-00-2023/20-00.

Data Availability Statement: The dataset presented in this study (in anonymized form) is available on request from the corresponding author. It is not available publicly as it belongs to the company that received the funding.

Conflicts of Interest: The authors declare no conflict of interest.

References

1. Wijaya, T.K.; Vasirani, M.; Humeau, S.; Aberer, K. Cluster-based aggregate forecasting for residential electricity demand using smart meter data. In Proceedings of the 2015 IEEE International Conference on Big Data (Big Data), Santa Clara, CA, USA, 29 October–1 November 2015; pp. 879–887. [CrossRef]
2. Laurinec, P.; Lucká, M. Clustering-based forecasting method for individual consumers electricity load using time series representations. *Open Comput. Sci.* **2018**, *8*, 38–50. [CrossRef]
3. IEA. Global Energy Review 2021. Available online: https://www.iea.org/reports/global-energy-review-2021 (accessed on 20 October 2023).
4. IEA. World Energy Outlook 2021. Available online: https://www.iea.org/reports/world-energy-outlook-2021 (accessed on 20 October 2023).
5. IPCC. Summary for policymakers. In *Climate Change 2021: The Physical Science Basis. Contribution of Working Group I to the Sixth Assessment Report of the Intergovernmental Panel on Climate Change*; Cambridge University Press: Cambridge, UK, 2021; Available online: https://www.ipcc.ch/report/ar6/wg1/#SPM (accessed on 20 October 2023).
6. Malinauskaite, J.; Jouhara, H.; Ahmad, L.; Milani, M.; Montorsi, L.; Venturelli, M. Energy efficiency in industry: EU and national policies in Italy and the UK. *Energy* **2019**, *172*, 255–269. [CrossRef]
7. Forouli, A.; Bakirtzis, E.A.; Papazoglou, G.; Oureilidis, K.; Gkountis, V.; Candido, L.; Ferrer, E.D.; Biskas, P. Assessment of Demand Side Flexibility in European Electricity Markets: A Country Level Review. *Energies* **2021**, *14*, 2324. [CrossRef]
8. Heilmann, E.; Klempp, N.; Wetzel, H. Design of regional flexibility markets for electricity: A product classification framework for and application to German pilot projects. *Util. Policy* **2020**, *67*, 101133. [CrossRef]
9. Steinberg, D.C.; Mignone, B.K.; Macknick, J.; Sun, Y.; Eurek, K.; Banger, A.; Livneh, B.; Averyt, K. Decomposing supply-side and demand-side impacts of climate change on the US electricity system through 2050. *Clim. Chang.* **2020**, *158*, 125–139. [CrossRef]
10. Pilli-Sihvola, K.; Aatola, P.; Ollikainen, M.; Tuomenvirta, H. Climate change and electricity consumption—Witnessing increasing or decreasing use and costs? *Energy Policy* **2010**, *38*, 2409–2419. [CrossRef]
11. Lopion, P.; Markewitz, P.; Robinius, M.; Stolten, D. A review of current challenges and trends in energy systems modeling. *Renew. Sustain. Energy Rev.* **2018**, *96*, 156–166. [CrossRef]
12. Jebaraj, S.; Iniyan, S. A review of energy models. *Renew. Sustain. Energy Rev.* **2006**, *10*, 281–311. [CrossRef]
13. Herbst, A.; Toro, F.; Reitze, F.; Jochem, E. Introduction to energy systems modelling. *Swiss J. Econ. Stat.* **2012**, *148*, 111–135. [CrossRef]
14. Prina, M.G.; Manzolini, G.; Moser, D.; Nastasi, B.; Sparber, W. Classification and challenges of bottom-up energy system models—A review. *Renew. Sustain. Energy Rev.* **2020**, *129*, 109917. [CrossRef]
15. Foley, A.M.; Gallachóir, B.P.Ó.; Hur, J.; Baldick, R.; Mc Keogh, E.J. A strategic review of electricity systems models. *Energy* **2010**, *35*, 4522–4530. [CrossRef]
16. Després, J.; Hadjsaid, N.; Criqui, P.; Noirot, I. Modelling the impacts of variable renewable sources on the power sector: Reconsidering the typology of energy modelling tools. *Energy* **2015**, *80*, 486–495. [CrossRef]
17. Pfenninger, S.; Hirth, L.; Schlecht, I.; Schmid, E.; Wiese, F.; Brown, T.; Davis, C.; Gidden, M.; Heinrichs, H.; Heuberger, C.; et al. Opening the black box of energy modelling: Strategies and lessons learned. *Energy Strategy Rev.* **2018**, *19*, 63–71. [CrossRef]
18. Gabriel, S.A.; Kydes, A.S.; Whitman, P. The national energy modeling system: A large-scale energy-economic equilibrium model. *Oper. Res.* **2001**, *49*, 14–25. [CrossRef]

19. Skinner, C.W. National Energy Modeling System. *Gov. Inf. Q.* **1993**, *10*, 41–51. [CrossRef]
20. Fattahi, A.; Sijm, J.; Faaij, A. A systemic approach to analyze integrated energy system modeling tools: A review of national models. *Renew. Sustain. Energy Rev.* **2020**, *133*, 110195. [CrossRef]
21. Yan, C.; Bie, Z. Evaluating National Multi-energy System Based on General Modeling Method. *Energy Procedia* **2019**, *159*, 321–326. [CrossRef]
22. Berntsen, P.B.; Trutnevyte, E. Ensuring diversity of national energy scenarios: Bottom-up energy system model with Modeling to Generate Alternatives. *Energy* **2017**, *126*, 886–898. [CrossRef]
23. Aryanpur, V.; O'Gallachoir, B.; Dai, H.; Chen, W.; Glynn, J. A review of spatial resolution and regionalisation in national-scale energy systems optimisation models. *Energy Strategy Rev.* **2021**, *37*, 100702. [CrossRef]
24. Beaver, R. Structural comparison of the models in EMF 12. *Energy Policy* **1993**, *21*, 238–248. [CrossRef]
25. Mirakyan, A.; De Guio, R. Modelling and uncertainties in integrated energy planning. *Renew. Sustain. Energy Rev.* **2015**, *46*, 62–69. [CrossRef]
26. Baghelai, C.; Moumen, F.; Cohen, M.; Kydes, A.; Harris, C.M. Uncertainty in the National Energy Modeling System. I: Method Development. *J. Energy Eng.* **1995**, *121*, 108–124. [CrossRef]
27. DeCarolis, J.; Daly, H.; Dodds, P.; Keppo, I.; Li, F.; McDowall, W.; Pye, S.; Strachan, N.; Trutnevyte, E.; Usher, W.; et al. Formalizing best practice for energy system optimization modelling. *Appl. Energy* **2017**, *194*, 184–198. [CrossRef]
28. Pusnik, M.; Sucic, B.; Urbancic, A.; Merse, S. Role of the national energy system modelling in the process of the policy development. *Therm. Sci.* **2012**, *16*, 703–715. [CrossRef]
29. Sahoo, S.; van Stralen, J.N.P.; Zuidema, C.; Sijm, J.; Yamu, C.; Faaij, A. Regionalization of a national integrated energy system model: A case study of the northern Netherlands. *Appl. Energy* **2022**, *306*, 118035. [CrossRef]
30. Collins, S.; Deane, J.P.; Poncelet, K.; Panos, E.; Pietzcker, R.C.; Delarue, E.; Pádraig, Ó.; Gallachóir, B. Integrating short term variations of the power system into integrated energy system models: A methodological review. *Renew. Sustain. Energy Rev.* **2017**, *76*, 839–856. [CrossRef]
31. Gacitua, L.; Gallegos, P.; Henriquez-Auba, R.; Lorca, Á.; Negrete-Pincetic, M.; Olivares, D.; Valenzuela, A.; Wenzel, G. A comprehensive review on expansion planning: Models and tools for energy policy analysis. *Renew. Sustain. Energy Rev.* **2018**, *98*, 346–360. [CrossRef]
32. Wen, X.; Jaxa-Rozen, M.; Trutnevyte, E. Accuracy indicators for evaluating retrospective performance of energy system models. *Appl. Energy* **2022**, *325*, 119906. [CrossRef]
33. Chaudry, M.; Jayasuriya, L.; Jenkins, N. Modelling of integrated local energy systems: Low-carbon energy supply strategies for the Oxford-Cambridge arc region. *Energy Policy* **2021**, *157*, 112474. [CrossRef]
34. Hanna, R.; Gross, R. How do energy systems model and scenario studies explicitly represent socio-economic, political and technological disruption and discontinuity? Implications for policy and practitioners. *Energy Policy* **2021**, *149*, 111984. [CrossRef]
35. Huang, K.; Eckelman, M.J. Appending material flows to the National Energy Modeling System (NEMS) for projecting the physical economy of the United States. *J. Ind. Ecol.* **2022**, *26*, 294–308. [CrossRef]
36. Batas Bjelić, I.; Rajaković, N. Simulation-based optimization of sustainable national energy systems. *Energy* **2015**, *91*, 1087–1098. [CrossRef]
37. Yan, C.; Bie, Z.; Liu, S.; Urgun, D.; Singh, C.; Xie, L. A Reliability Model for Integrated Energy System Considering Multi-energy Correlation. *J. Mod. Power Syst. Clean Energy* **2021**, *9*, 811–825. [CrossRef]
38. Martinsen, T. Technology learning in a small open economy-The systems, modelling and exploiting the learning effect. *Energy Policy* **2011**, *39*, 2361–2372. [CrossRef]
39. Davis, M.; Ahiduzzaman, M.; Kumar, A. How to model a complex national energy system? Developing an integrated energy systems framework for long-term energy and emissions analysis. *Int. J. Glob. Warm.* **2019**, *17*, 23–58. [CrossRef]
40. Lund, H.; Arler, F.; Østergaard, P.A.; Hvelplund, F.; Connolly, D.; Mathiesen, B.V.; Karnøe, P. Simulation versus optimisation: Theoretical positions in energy system modelling. *Energies* **2017**, *10*, 840. [CrossRef]
41. Kazemzadeh, M.R.; Amjadian, A.; Amraee, T. A hybrid data mining driven algorithm for long term electric peak load and energy demand forecasting. *Energy* **2020**, *204*, 117948. [CrossRef]
42. McNeil, M.A.; Karali, N.; Letschert, V. Forecasting Indonesia's electricity load through 2030 and peak demand reductions from appliance and lighting efficiency. *Energy Sustain. Dev.* **2019**, *49*, 65–77. [CrossRef]
43. Ouedraogo, N.S. Modeling sustainable long-term electricity supply-demand in Africa. *Appl. Energy* **2017**, *190*, 1047–1067. [CrossRef]
44. Trotter, P.A.; Cooper, N.J.; Wilson, P.R. A multi-criteria, long-term energy planning optimisation model with integrated on-grid and off-grid electrification—The case of Uganda. *Appl. Energy* **2019**, *243*, 288–312. [CrossRef]
45. Kwon, P.S.; Østergaard, P. Assessment and evaluation of flexible demand in a Danish future energy scenario. *Appl. Energy* **2014**, *134*, 309–320. [CrossRef]
46. Ząbkowski, T.; Gajowniczek, K.; Matejko, G.; Brożyna, J.; Mentel, G.; Charytanowicz, M.; Jarnicka, J.; Olwert, A.; Radziszewska, W. Changing Electricity Tariff—An Empirical Analysis Based on Commercial Customers' Data from Poland. *Energies* **2023**, *16*, 6853. [CrossRef]
47. Matejko, G. *Energy Demand Management*; ECCC Foundation: Lublin, Poland, 2021.
48. Ward, J.H., Jr. Hierarchical Grouping to Optimize an Objective Function. *J. Am. Stat. Assoc.* **1963**, *58*, 236–244. [CrossRef]
49. Fausett, L.V. *Fundamentals of Neural Networks: Architectures, Algorithms and Applications*; Pearson: London, UK, 1993.

50. Simon, H. *Neural Networks: A Comprehensive Foundation*; Prentice Hall: Upper Saddle River, NJ, USA, 1999.
51. Funahashi, K.I. On the approximate realization of continuous mappings by neural networks. *Neural Netw.* **1989**, *2*, 183–192. [CrossRef]
52. Werbos, P.J. *The Roots of Backpropagation: From Ordered Derivatives to Neural Networks and Political Forecasting*; John Wiley & Sons: Hoboken, NJ, USA, 1994.
53. Hertz, J.; Krogh, A.; Palmer, R.G. *Introduction to the Theory of Neural Computation*; Westview Press: Hoboken, NJ, USA, 1991.
54. Masters, T. *Neural, Novel and Hybrid Algorithms for Time Series Prediction*; John Wiley & Sons: New York, NY, USA, 1995.

Article

Research on Dynamic Economic Dispatch Optimization Problem Based on Improved Grey Wolf Algorithm

Wenqiang Yang [1], Yihang Zhang [1], Xinxin Zhu [1], Kunyan Li [1] and Zhile Yang [2,*]

1   School of Mechanical and Electrical Engineering, Henan Institute of Science and Technology, Xinxiang 453003, China; yangwqjsj@hist.edu.cn (W.Y.); zyh2587@stu.hist.edu.cn (Y.Z.); zxx2575438269@stu.hist.edu.cn (X.Z.); lky@stu.hist.edu.cn (K.L.)
2   Shenzhen Institutes of Advanced Technology, Chinese Academy of Sciences, Shenzhen 518055, China
*   Correspondence: zl.yang@siat.ac.cn

Abstract: The dynamic economic dispatch (DED) problem is a typical complex constrained optimization problem with non-smooth, nonlinear, and nonconvex characteristics, especially considering practical situations such as valve point effects and transmission losses, and its objective is to minimize the total fuel costs and total carbon emissions of generating units during the dispatch cycle while satisfying a series of equality and inequality constraints. For the challenging DED problem, a model of a dynamic economic dispatch problem considering fuel costs is first established, and then an improved grey wolf optimization algorithm (IGWO) is proposed, in which the exploitation and exploration capability of the original grey wolf optimization algorithm (GWO) is enhanced by initializing the population with a chaotic algorithm and introducing a nonlinear convergence factor to improve weights. Furthermore, a simple and effective constraint-handling method is proposed for the infeasible solutions. The performance of the IGWO is tested with eight benchmark functions selected and compared with other commonly used algorithms. Finally, the IGWO is utilized for three different scales of DED cases, and compared with existing methods in the literature. The results show that the proposed IGWO has a faster convergence rate and better global optimization capabilities, and effectively reduces the fuel costs of the units, thus proving the effectiveness of IGWO.

Keywords: dynamic economic dispatch; improved grey wolf optimization algorithm; constraint-handling methods; chaotic initialization; nonlinear convergence factor

Citation: Yang, W.; Zhang, Y.; Zhu, X.; Li, K.; Yang, Z. Research on Dynamic Economic Dispatch Optimization Problem Based on Improved Grey Wolf Algorithm. *Energies* **2024**, *17*, 1491. https://doi.org/10.3390/en17061491

Academic Editors: Konstantinos Aravossis and Eleni Strantzali

Received: 8 February 2024
Revised: 28 February 2024
Accepted: 8 March 2024
Published: 21 March 2024

1. Introduction

1.1. Power Dispatch Problem

Over the years, with the rapid development of science and technology and the improvement of people's living standard, the consumption of energy has been increasing, and especially the large amount of electricity used will inevitably bring a huge burden to the power grid; thus, in order to optimize the scheduling of the power system and the utilization efficiency of electricity, dynamic economic dispatch has become a popular research topic for meeting the actual situation of electricity consumption at home and abroad [1].

The dynamic economic dispatch (DED) problem was first proposed by Bechert and Kwanty in 1971 as an extension of static economic dispatch (SED) [1]. It mainly refers to dividing a day into several periods and optimizing daily economic dispatches based on daily load forecasts, and takes into account various constraints of thermal power units, which is more consistent with the actual power system operation than static economic dispatch. Economic load allocation (ELD) is a typical optimization problem in power systems and one of the most fundamental optimization tasks in dynamic economic dispatching problems, aiming at a reasonable distribution of power among specific units and minimizing the economic cost while satisfying certain constraints imposed; improving the arrangement of the unit output can result in significant savings [2]. Some simple scheduling problems are generally solved using traditional mathematical methods, such as the prioritization method, dynamic programming method, equal micro-increment rate criterion, and gradient projection method, but modern economic scheduling problems are often much more complex, due to the introduction of network transmission losses and valve point effects of thermal units. As a result, today's scheduling problems are essentially non-convex, nonlinear, high-dimensional, multi-constrained, multi-objective optimization problems, making such problems unable to be satisfactorily solved by traditional mathematical methods [2]. In order to solve the modern economic scheduling problem, some random search algorithms and heuristic swarm intelligence algorithms based on the behavior laws of biological populations are proposed. They can obtain the optimal solution in a reasonable time and avoid falling into the local optimal solution prematurely in the iterative process, and have certain advantages in multi-objective, nonlinear and high-dimensional optimization problems.

Domestic and foreign research on the DED problem has been continuously improving and deepening, and the research directions are also different. At present, domestic current research in power systems mainly conforms to the requirements of low-carbon transition, the construction of clean energy-based power systems has become an important task, and the vigorous development and use of renewable energy has become the main trend of the domestic economic dispatch research. Recently, Yang et al. [3] developed a novel power system dispatching model that incorporates a significant number of plug-in electric vehicle charging and discharging behaviors. They conducted a study on a 10-unit power system with 50,000 plug-in electric vehicles to investigate strategies for mitigating the impact of new energy vehicles on the power grid, ultimately achieving low carbonization. Yang et al. [4] also studied a new hybrid unit commitment problem considering renewable energy generation scenarios and plug-in electric vehicles' charging and discharging management. Due to the vigorous development of renewable energy and the large-scale launch of plug-in electric vehicles, the traditional power system scheduling problem is faced with greater challenges. A series of metaheuristic optimization algorithms are proposed to solve the dilemma, and the effectiveness of these methods for the power scheduling problem is verified by comparison experiments. Foreign scholars, on the other hand, have focused on the aspect of saving energy usage. Liu et al. [5] constructed a hybrid economic emission dispatch (HDEED) mathematical model considering renewable energy generation, which is based on wind-photothermal integrated energy, a moth–flame optimization algorithm was proposed to solve it, and finally three experimental cases were tested to verify the effectiveness of the study. Acharya et al. [6] proposed a multi-objective multiscale optimization scheme for minimizing the dynamic economic load scheduling problem with the valve point effect, and the algorithm retains the ramp constraints on the required rate of the generator units. This method eliminates the discontinuities in the operation of the power system, which leads to a better solution of the dynamic economic load dispatch problem, and makes the generators output the optimal power. Shaheen et al. [7] proposed a manta ray foraging (MRF) optimizer to solve the economic dispatching of the combined heat and power system problem including valve point shocks and wind power. In order to obtain the optimal solution of the EDCS problem, the MRF optimizer with an adaptive penalty function was designed to deal with the constraints of the model efficiently, and the validity of the methodology has been verified through experiments on two kinds of test systems, large and small.

1.2. Intelligent Optimization Algorithm

Swarm intelligent optimization algorithms were first explored for application in power systems in the 1970s and are still widely used in various scheduling problems until now. Intelligent optimization algorithms are search techniques based on biological evolution as well as objective laws of nature, and the more typical ones are the genetic algorithm (GA) [8], evolutionary planning algorithm (EP), simulated annealing algorithm (SA) [9], particle swarm optimization algorithm (PSO) [10], whale optimization algorithm (WOA) [11], ant colony optimization algorithm (ACO) [12], etc. These methods have been proven to be very effective for solving nonlinear ELD problems. Many scholars at home and abroad have carried out a large amount of research on the application of intelligent algorithms for economic scheduling in the past decades, and these algorithms are still being tested and improved continuously. Liu et al. [13] proposed a niche differential evolutionary algorithm (NDE) to solve a large-scale cogeneration economic dispatch problem, which is inspired by the neighborhood concept of the niche approach, and utilized a deterministic congested niche approach and a two-phase selection design of greedy selection, which balanced the algorithm's global and local search capabilities, and thus the algorithm could solve the DED problem more efficiently. A genetic algorithm based on the concept of energy-conserving space and the parallel population technique to solve the DED problem was proposed by Silva et al. [14], who added a new repair strategy based on real value coding, and applied the algorithm to four power systems of different sizes for testing to verify the effectiveness of the improved algorithm. Based on the consideration of fuel prices, Mahdavi et al. [15] proposed a scenario-based model to evaluate the impact of substation expansion on TEP from the perspective of voltage level. Discrete artificial bee colony (DABC)- and quadratic programming (OP)-based methods were used to verify the effectiveness of the model in the actual transmission network, and the aim was to economically determine the optimal number, timing, and location of new transmission lines. Li et al. [16] proposed an optimization algorithm combining a chaotic search based on tent mapping and nonlinear adaptive particle swarm optimization, and established a multi-objective optimization model aiming at determining the operating costs, pollutant emissions, and energy efficiency of the cogeneration system. Simulation results showed that the proposed algorithm applied to the model can effectively improve these objectives. Wang et al. [17] established an economic dispatch model based on regional interconnected multi-microgrid systems, and proposed an improved whale algorithm using five strategies, namely adaptive inertia weight, dynamic spiral search and generalized binary learning, to solve this problem. The improved algorithm was applied to two arithmetic models of the grid-connected operation and non-grid-connected operation for testing, and it was found that the convergence accuracy and speed of the algorithm have been improved and the results obtained were good.

The gray wolf optimization algorithm (GWO) is a novel pack intelligence optimization algorithm proposed in 2014 by Seyedali Mirjalili, an Australian scholar, inspired by the predatory behavior of gray wolf packs [18], which is based on the mechanism of wolf pack collaboration to achieve the purpose of optimization. The gray wolf algorithm has the advantages of a simple structure, few parameters to be adjusted, and easy implementation; it has good performance in terms of its problem-solving accuracy and convergence speed, and is now widely used in various scheduling problems. Ge et al. [19] used an improved GWO to optimize the UAV path-planning problem in an oilfield environment and achieved satisfactory results. Wang et al. [20] used a discrete GWO to solve the stacking problem, which effectively solved this problem and surpassed most of the previously reported metaheuristics; Yuan et al. [21] used GWO to solve the lightning whistle acoustic voice recognition problem, and the accuracy of its recognition results was 2% higher compared to the common recognition methods. Dokur et al. [22] used GWO to solve a short-term wind speed-prediction problem with a multilayer perceptron, and the results showed that the algorithm was more effective than other algorithms. Song et al. [23] used GWO to optimize the shaft straightness error assessment of shaft hole-type parts, and the results confirmed that the algorithm was more accurate in solving this problem.

However, like most intelligent algorithms, the gray wolf optimization algorithm has shortcomings, and the most obvious ones are the balance of exploration and exploitation capability of the algorithm and the tendency to fall into local optimal solutions. In order to solve these problems, scholars have made corresponding improvements. Yan et al. [24] proposed a nonlinear convergence factor combining tangent and logarithmic functions to dynamically adjust the global search ability of the gray wolf algorithm, and also introduced an adaptive position update strategy, which led to a significant improvement in the performance of the algorithm. Mostafa et al. [25] proposed an improved gray wolf algorithm to find an optimal solution for the combined economic and emission dispatch problem, so as to minimize the generation costs and achieve emission reduction. Six mutation operators were applied to the GWO to enhance its performance. A test system that consists of 10 units was simulated, and the results showed the effect of applying the mutation operators to the IGWO. Sahoo et al. [26] proposed a coming-of-age improved version of the conventional grey wolf optimization technique to solve the ELD problem. In the improved GWO, the leadership hierarchy of the grey wolf is ameliorated by taking the random walking behavior of the grey wolfs into consideration. The algorithm aims to modify the existing leaders with the best leaders in order to overcome the drawbacks of the conventional GWO. It was found that the performance of the improved gray wolf algorithm was significantly improved through the test of the unit. Mohamed et al. [27] proposed a hybrid whale–wolf optimization method to accurately solve the economic dispatch problem. The proposed method efficiently integrates the mechanisms of the whale optimization algorithm and gray wolf optimization with crossover and mutation operators. To demonstrate the effectiveness of the proposed method, it was compared with six other optimization methods. Two different test systems (six and ten generating units) were used to evaluate the performance of the proposed method. The experimental results showed that the hybrid whale–wolf optimization method showed better performance in finding the optimal solution to the economic dispatch problem compared to the other methods. Paramguru et al. [28] proposed a new modified grey wolf algorithm to solve the economic dispatch problem. The modification was carried out by the incorporation of the exponential operators into the conventional GWO. The constraints with non-linearities of generating units like ramp rate constraints, effect of valve-point loading, and prohibited operating zones were considered for practical application. Compared with other algorithms, the results show that the proposed algorithm is effective in solving the real ELD problem. This optimization process provides a better capability of exploration and exploitation. According to the above literature, it can be concluded that the improvement of the gray wolf optimization algorithm is mainly based on the following three points: first, introducing strategies such as chaos initialization or reverse learning strategies to enhance the diversity of the initial population. Second, introducing nonlinear convergence factors or designing adaptive parameters to automatically adjust the global and local search ability of the algorithm. Third, changing the weight of the iterative formula or mixing it with other algorithms, and combining the search strategies of different algorithms, can improve the GWO algorithm.

1.3. Constraint-Handling Methods

Since the DED problem has strong constraints, it is necessary to address these constraints to ensure that the generated solution becomes feasible as the number of generations increases [29]. In some past studies, the most commonly used constraint-handling methods include the penalty function method, feasibility-based rule method [30], ε-constraint-handling method [31], repair method, and random ordering method [32]. Although these methods can effectively deal with the constraint problem, they all have obvious drawbacks, and the main shortcomings are as follows: first, some processing methods cannot accurately set the parameters or penalty factors, which will affect the feasibility of the final solution; second, some processing methods cannot solve the constraint problem with a small feasible domain or a complex scale; in addition, some constraint processing methods do not work well with algorithms, and cannot be combined with algorithms to solve multi-objective optimization problems. Due to these shortcomings, many scholars have also studied improvement strategies.

Jin et al. [33] proposed an improved penalty function method to solve multi-objective optimization problems, in which the optimal initial iteration points are obtained via the pseudo-random sequence correlation method, and continuous, non-segmented exponential penalty functions are constructed for the sparse decomposition of high-dimensional vectors.

It was found that the accuracy of the improved penalty function method is nearly twice that of the original method; considering the drawbacks of feasibility rules, Bcw et al. [34] designed an individual-dependent feasibility rule that can enhance the utilization of objective function information in the problem and is combined with a differential evolutionary algorithm to achieve good results in dealing with the constraints of optimization problems. Yang et al. [35] proposed an augmented generalized ε-constraint processing method to solve the multi-objective optimal scheduling model of a cogeneration microgrid, which can better obtain the Pareto optimal solution set of the model, and the experimental results demonstrated that this method can better realize the economic optimization of the model than other constraint methods. Yang et al. [36] proposed an adaptive assignment constraint-processing technique that can decompose the multi-objective optimization problem into several optimization subproblems. Each subproblem has a subentry in a subregion, and the constraint processing method is adaptively assigned to each subregion according to the index, so as to solve the constraint problem better. Liu et al. [37] proposed a double random ordering encryption algorithm based on the form of measurement data, which on the one hand uses randomness to ensure the uncertainty of random sequences and improves the security of measurement data, and on the other hand proposes an approximate recovery strategy based on CGAN to ensure the accuracy of decryption, and finally analyzed the effectiveness of the method through experiments on wind power and photovoltaic datasets.

1.4. The Innovation of This Paper and the Arrangement of the Remaining Content

Considering that there are few studies on solving DED problems by using the grey wolf optimization algorithm, this paper focuses on improving the grey wolf algorithm and using the improved algorithm to optimize the fuel costs of the unit, considering transmission losses and valve point effects. On the one hand, this paper adopts an improved scheme that is more suitable for solving DED problems for the gray wolf optimization algorithm, including a new chaotic mapping for population initialization, a more appropriate improvement of the convergence factors and control parameters, and an improvement of the weight of the position-updating formula. On the other hand, this paper uses a more novel method to deal with constraints, which combines a heuristic repair method and a direct repair method to deal with constraints in the model better.

The rest of this study is organized as follows: first, a mathematical model of a dynamic economic dispatch considering unit fuel costs is given, the original gray wolf algorithm is briefly described, and an improved grey wolf algorithm (IGWO) is introduced according to the initialization and updating formula. Then, a hybrid constraint-processing method combining heuristic repair and direct repair is proposed, in which the feasibility rule method is used to filter out high-quality infeasible solutions, a rough adjustment of the heuristic repair is used to make infeasible solutions closer to feasible solutions, and fine-tuning of the direct repair is used to force infeasible solutions to feasible solutions. Finally, the improved algorithm and constraint processing techniques are applied to the DED problem, the experimental results are observed in comparison with other methods, and the conclusions of the study and the outlook for further research work in the future are given.

2. Dynamic Economic Dispatch Model

The goal of the DED problem is to determine the optimal power generation level for all online units within a specified scheduling time (e.g., 24 h per day) with the minimum power generation fuel cost under various equation and inequality constraints, taking into account the valve point effect (VPE) and network losses. The detailed mathematical expression of the whole problem is as follows.

2.1. Optimization Objective Function

The fuel cost objective function considering the valve point effect can be approximated as a smooth quadratic function with the following expression:

$$minF(P) = \sum_{t=1}^{T}\sum_{i=1}^{N}\left\{a_i + b_i p_{t,i} + c_i p_{t,i}^2 + \left|d_i \sin\left[e_i\left(p_i^{min} - p_{t,i}\right)\right]\right|\right\} \tag{1}$$

In Equation (1), $F(P)$ denotes the total fuel cost of thermal generating units; P is the output power of all online units; T is the whole dispatch cycle of 24 h; N is the total number of system generators; a_i, b_i, c_i, d_i, e_i are the consumption characteristic coefficients of the ith generator; $p_{t,i}$ is the output power of the ith thermal generating unit at moment t; and p_i^{min} is the minimum value of its active output. The part of the formula with the absolute value is the valve point effect.

2.2. Constraints

The dynamic economic dispatch problem contains several equality and inequality constraints, including generation capacity constraints, unit ramp constraints, power balance constraints, and transmission loss constraints.

2.2.1. Capacity Constraints

The generation capacity constraint is an inequality constraint, where the unit's generation capacity must be within an appropriate limit during optimal dispatch, and is expressed as Formula (2).

$$p_i^{min} \leq p_{t,i} \leq p_i^{max} \tag{2}$$

here, p_i^{max} and p_i^{min} are the upper and lower limits of the active output of each unit in the generator set, respectively.

2.2.2. Units Ramp Limits

Due to the large inertia of the thermal power unit, the climb limit constraint is introduced in order to extend the service life of the unit. That is, the output of the unit cannot be greatly adjusted in a short period of time, which is expressed by Formula (3).

$$\begin{cases} p_{t,i} - p_{t-1,i} \leq UR_i \\ p_{t-1,i} - p_{t,i} \leq DR_i \end{cases} \tag{3}$$

In the above equation, UR_i and DR_i denote the maximum allowable rise and fall rate of the ith generation unit, respectively, and represent the generation inertia size of the thermal power unit.

2.2.3. Power Balance Constraint

The power balance constraint is the most important and complex equation constraint in the DED problem. In each time period, the sum of the active output $p_{i,t}$ must be equal to the sum of the total load demand $p_{D,t}$ and the network active loss $p_{L,t}$ of each generator in that time period, and the constraint relationship is expressed by Formula (4).

$$\sum_{t=1}^{T} p_{i,t} - p_{D,t} - p_{L,t} = 0 \tag{4}$$

The mathematical model expression for the network transmission loss $p_{L,t}$ in the above equation is usually simplified as in the following Formula (5).

$$P_{L,t} = \sum_{i=1}^{T}\sum_{j=1}^{N} P_{t,i} B_{ij} P_{t,j} \tag{5}$$

3. Improved Gray Wolf Algorithm

3.1. Description of the Gray Wolf Algorithm

The gray wolf optimization algorithm (GWO) is inspired by the social leadership and hunting behavior of gray wolves in nature. Compared with other metaheuristic algorithms, the GWO algorithm has the advantages of a simple structure, having few control parameters, being easy to improve, and having the ability to achieve a balance between local and global search. The GWO algorithm takes the three leading wolves α, β, and δ with the best adaptation in the population as the best solution and guides the remaining ω wolves in the direction closest to the prey so as to find the global optimal solution. Wolf hunting consists of three main steps: surrounding, hunting, and attacking prey. The number distribution of the social ranks of wolves is shown in Figure 1.

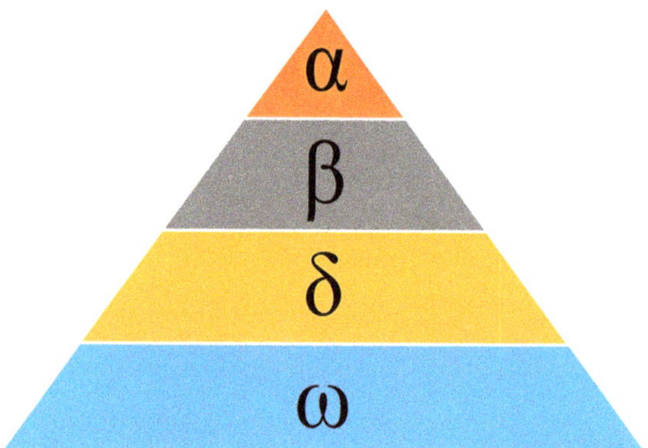

Figure 1. The number distribution of the social ranks of wolves.

3.1.1. Surrounding

In the early stage of algorithm optimization, GWO is mainly expressed as encircling the prey, and its mathematical model is expressed as Formulas (6) and (7).

$$\vec{D} = \left| \vec{C} \times \vec{X_P}(t) - \vec{X}(t) \right| \tag{6}$$

$$\vec{X}(t+1) = \vec{X_P}(t) - \vec{A} \times \vec{D} \tag{7}$$

The two formulas above represent the distance between individual gray wolves and their prey and the position update of gray wolves, respectively, where t is the number of current iterations. $\vec{X_P}$ and $\vec{X}$ are the position vectors of the prey and gray wolves, respectively, and $\vec{A}$ and $\vec{C}$ are the coefficient vectors, which are calculated using Formulas (8) and (9).

$$\vec{A} = 2\vec{a} \times \vec{r_1} - \vec{a} \tag{8}$$

$$\vec{C} = 2 \times \vec{r_2} \tag{9}$$

In the above equations, $\vec{a}$ is the convergence factor, and $\vec{r_1}$ and $\vec{r_2}$ are random numbers with values between 0 and 1, decreasing linearly from 2 to 0 with the number of iterations.

3.1.2. Hunting

In order to mathematically model the hunting behavior of wolves, it is assumed that the three leading wolves have a better ability to identify the location of their prey. Therefore, considering the leading role of the three wolves in searching for the optimal solution, other wolves must follow them. The mathematical model of the hunting behavior of wolves is shown in Equations (10)–(12).

$$\begin{cases} \vec{D}_\alpha = \left| \vec{C}_1 \times \vec{X}_\alpha - \vec{X} \right| \\ \vec{D}_\beta = \left| \vec{C}_2 \times \vec{X}_\beta - \vec{X} \right| \\ \vec{D}_\delta = \left| \vec{C}_3 \times \vec{X}_\delta - \vec{X} \right| \end{cases} \tag{10}$$

$\vec{C}_1$, $\vec{C}_2$, and $\vec{C}_3$ in the above equation are calculated using Equation (9).

$$\begin{cases} \vec{X}_1 = \vec{X}_\alpha - \vec{A}_1 \times \left(\vec{D}_\alpha \right) \\ \vec{X}_2 = \vec{X}_\beta - \vec{A}_2 \times \left(\vec{D}_\beta \right) \\ \vec{X}_3 = \vec{X}_\delta - \vec{A}_3 \times \left(\vec{D}_\delta \right) \end{cases} \tag{11}$$

$\vec{X}_\alpha$, $\vec{X}_\beta$, and $\vec{X}_\delta$ in the above equation represent the best three solutions in the population iterated to moment t; $\vec{A}_1$, $\vec{A}_2$, and $\vec{A}_3$ are calculated using Equation (8); and $\vec{D}_\alpha$, $\vec{D}_\beta$, and $\vec{D}_\delta$ are calculated using Equation (10).

$$\vec{X}(t+1) = \frac{\vec{X}_1 + \vec{X}_2 + \vec{X}_3}{3} \tag{12}$$

To better visualize the surrounding and hunting process of the gray wolf population, the 2D position of the updated gray wolf population is shown in the following Figure 2.

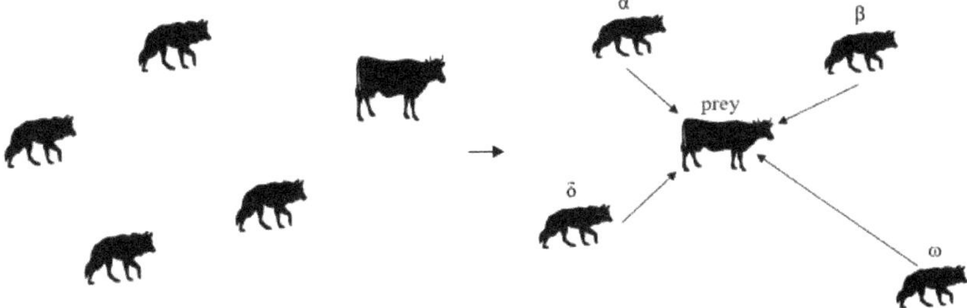

Figure 2. The 2D position maps of the updated gray wolf population.

3.1.3. Attacking Prey

When the prey stops moving, the wolf pack terminates the hunting process and starts attacking the prey. This can be achieved by controlling the value of the linear convergence factor A in the iteration of exploration and development. During the iteration, half of the iteration is used for exploration, and in the case of a smooth transition, the other half is used for exploitation, in which case the wolves change their position to any position between the prey's position and their current position.

In summary, the pseudo-code of the gray wolf optimization algorithm is shown bellow (Algorithm 1). First, the initial population of wolves is randomly generated in the search space, then the position of the wolves is evaluated using the fitness function, and then the following steps are repeated until a predefined number of iterations is reached and stopped. In each iteration, three lead wolves with the optimal solution are identified, and then each wolf updates its position by following the three hunting steps mentioned earlier. The above steps are repeated until the algorithm stops and outputs the optimal position of the prey.

Algorithm 1: The conventional grey wolf optimizer algorithm (GWO)

    Input: N, D, Maxiter
    Output: The global optimum
1: **Begin**
2: Initialize the grey wolf population $X_i (i = 1, 2, \ldots, n)$
3: Initialize a, A, and C
4: Calculate the fitness of each search agent
5: X_α = the best search agent
6: X_β = the second best search agent
7: X_δ = the third best search agent
8: **While** ($t <$ Max number of iterations)
9:      **for** each search agent
10:        Update the position of the current search agent by Equation (12)
11:      **end for**
12:      Update a, A, and C
13:      Calculate the fitness of all search agents
14:      Update X_α, X_β and X_δ
15:      $t = t + 1$
16: **end while**
17: return X_α

3.2. Improvement of the Gray Wolf Algorithm

Although the gray wolf optimization algorithm has the advantages of a simple structure and easy implementation, the algorithm has some obvious shortcomings, such as a lack of diversity in the population, an imbalance between the exploitation and exploration phases, and premature convergence during the iterative process. In order to overcome these shortcomings, this study has made improvements in three aspects. First, the initial population individuals are distributed in a wider solution space through chaotic mapping, thereby improving the diversity of the initial population. Second, by improving the convergence factor of the algorithm model, the convergence accuracy of the algorithm in the later stage can be improved. The third is to change the weight of the iterative formula or combine the optimization ideas of other algorithms to improve the GWO algorithm so that the algorithm can jump out of the local optimum.

3.2.1. Algorithm Population Chaos Initialization

Since the initial population of the original gray wolf algorithm is randomly generated, the coverage of individuals in the solution space is not high, and the diversity of the population cannot be reflected, which affects the optimization search effect of the algorithm. By contrast, the typical features of chaotic mapping include randomness, ergodicity, and regularity, which can ensure the diversity of the population and achieve the purpose of a global search. Therefore, the Bernoulli chaotic mapping method is used to improve the initial population in order to enhance the searching effect of the gray wolf algorithm. Its mathematical model is expressed as Formulas (13) and (14).

$$Z_{k+1} = \begin{cases} \dfrac{Z_k}{(1-\lambda)}, & Z_k \in (0, 1-\lambda] \\ \dfrac{(Z_k - 1 + \lambda)}{\lambda}, & Z_k \in (1-\lambda, 1) \end{cases} \tag{13}$$

$$X_k = X_k^{max} + Z_k \left(X_k^{max} - X_k^{min} \right) \tag{14}$$

In Equation (13), k is the population size, Z_k is the generated chaotic sequence, and the value of λ is a random number between 0 and 1. Then, the chaotic sequence Z_k is combined to further generate the initial position sequence X_k of gray wolf individuals in the search area.

3.2.2. Improvement of the Convergence Factor and Control Parameters

In the mathematical model of the gray wolf algorithm, the coefficient vectors $\vec{A}$, $\vec{C}$ are the key parameters controlling the search range of the wolf pack, where $\vec{A}$ represents the search radius of the wolf pack, which is used to adjust the wolf–prey spacing in stages, and the control parameter $\vec{C}$ also coordinates the global exploration and local exploitation of gray wolf algorithm. In turn, these two parameters are related to the convergence factor a and the random vectors $\vec{r_1}$ and $\vec{r_2}$, so this study improves the control parameters by proposing an exponential convergence factor a updating strategy, which can better fit the actual nonlinear variation process of the convergence factor a. The formula is presented as Formulas (15) and (16).

$$a(l) = 2 - \sqrt{2} \times \left(\left(e^{\frac{l}{maxiter}} - 1 \right)^{\lambda_1} \right)^{\lambda_2} \tag{15}$$

$$\vec{C} = 2 \times r_3 - a \tag{16}$$

In the above equation, l is the number of iterations, $maxiter$ is the maximum number of iterations, λ_1, λ_2 are random numbers between 1 and 6, and r_3 is a random number between 1 and 1.5.

3.2.3. Improvements of the Location Update Formula

In order to better develop the search-seeking capability of the gray wolf algorithm, weigh the different guiding effects of the best three wolves on the position updates of the remaining gray wolf individuals, and prevent falling into premature stagnation at a local scale, a dynamic weight factor b with linear decreasing variation is first introduced, followed by the adaptive scale factors $v1$, $v2$, and $v3$. The method is shown in Formulas (17)–(19).

$$b(l) = b_f - \frac{l}{maxiter} \times \left(b_f - b_s \right) \tag{17}$$

$$\begin{cases} f = |f_\alpha + f_\beta + f_\delta| \\ v_1 = \frac{f_\alpha}{f}, v_2 = \frac{f_\beta}{f}, v_3 = \frac{f_\delta}{f} \quad f > 0 \\ v_1 = v_2 = v_3 = \frac{1}{3} \quad f = 0 \end{cases} \tag{18}$$

$$X(l+1) = b(l) \times r_4 \times (v_1 \times X_1 + v_2 \times X_2 + v_3 \times X_3) \tag{19}$$

In the above equations, b_s takes the value of 0.5 and b_f takes the value of 1 to denote the initial and final values of the weight factor, respectively, f_α, f_β, and f_δ denote the adaptation values of the three wolves, and r_4 is a random number between 0.3 and 1.

To sum up, after applying the above three improvement points to the iterative process of the gray wolf algorithm, the pseudo-code of the improved gray wolf algorithm is shown bellow (Algorithm 2).

Algorithm 2: The improved grey wolf optimizer algorithm (IGWO)

Input: N, D, Maxiter
Output: The global optimum
1: **Begin**
2: Initialize the grey wolf population $X_i(i = 1, 2, \ldots, n)$ with chaotic mapping by Equations (13) and (14).
3: Initialize a, A, and C.
4: Calculate the fitness of each search agent.
5: X_α = the best search agent.
6: X_β = the second best search agent.
7: X_δ = the third best search agent.
8: **While** (t < Max number of iterations)

Algorithm 2: *Cont.*

9:	**for** each search agent
10:	Update the position of the current search agent by Equation (12).
11:	**end for**
12:	Update improved parameters a, A, and C by Equations (15) and (16).
13:	Calculate the fitness of all search agents.
14:	Update X_α, X_β, and X_δ with the improved position update formula by Equation (19).
15:	$t = t + 1$
16:	**end while**
17:	return X_α

---

4. Constraint Handling

Considering that in solving DED problems, some updated candidate solutions are usually infeasible in the early stage of optimization, and it is difficult to meet all constraints, which is not conducive to exploring and developing feasible regions, this paper proposes a hybrid constraint-processing method combining heuristic repair and direct repair to ensure all solutions' feasibility. Different processing methods are used for equality constraints and inequality constraints. The specific constraint processing is as follows.

4.1. Boundary Constraint Handling

The generator units should not only meet the upper and lower limits of capacity constraints, but also meet the ramp constraints in different periods. The critical treatment or the random treatment within the boundary is generally used, and the critical proximity treatment is used for these constraints in this study. To facilitate the processing, a new boundary constraint is formed as shown in Formulas (20) and (21).

$$p_{t,i}^{min} = \begin{cases} p_i^{min} & if\ t = 1 \\ max\left(p_i^{min}, p_{t-1,i} - DR_i\right) & otherwise \end{cases} \tag{20}$$

$$p_{t,i}^{max} = \begin{cases} p_i^{max} & if\ t = 1 \\ max\left(p_i^{max}, p_{t-1,i} + UR_i\right) & otherwise \end{cases} \tag{21}$$

In the above equations, $p_{t,i}^{min}$ and $p_{t,i}^{max}$ denote the lower and upper bounds of the new bound synthesized by the ith unit at the tth moment. Then, any variable $p_{t,i}$ that exceeds its new bound is restricted to its upper and lower boundaries, and the model is expressed as Formula (22).

$$p_{t,i} = \begin{cases} p_i^{min} & if\ p_{t,i} \le p_i^{min} \\ p_i^{max} & if\ p_{t,i} \ge p_i^{max} \end{cases} \tag{22}$$

The repair method mentioned above can guarantee that both ramp constraints and capacity constraints are satisfied at the same time, which is simpler and more efficient than the traditional penalty function method.

4.2. Power Balance Constraint Handling

Considering the network transmission loss, the power balance constraint is the most difficult to repair among all constraints, and it is more difficult to reduce and eliminate violations. Therefore, a hybrid constraint-processing method is proposed, and the whole process is divided into two stages. The first is the rough adjustment stage, which can quickly reduce the degree of violation, and for the infeasible solutions that still cannot solved, the next fine-tuning phase is entered, which helps to speed up the repair speed. The specific implementation steps are as follows.

The first step is to roughly calculate the cost of each unit and form a set E in ascending order to evaluate the real-time efficiency of each unit. The calculation is shown as Equation (23).

$$dif\ f(p_{t,i}) = b_i + 2c_i p_{t,i} \tag{23}$$

In the second step, a cell r is selected according to the efficiency from the set E, and the output power is roughly adjusted using the following Equation (24):

$$p_{t,r} = p_{t,r} - V(t) \tag{24}$$

where $V(t)$ is the constraint violation size of the unit at moment t. If $p_{t,r}$ is on the new boundary, it means that the violation power is not repaired only by a single unit, and thus the r cell is removed from the set M, repeating step 2 until the set M is empty. Otherwise, the next step is executed. After the coarse tuning step, the solution is near the feasible domain.

In the third step, the fine-tuning step proceeds to obtain a feasible solution. Here, the output solution equation of the thermal power unit is rewritten as Formula (25) and the power balance constraint can be transformed into solving the quadratic equation. The unit r is selected according to the efficiency from the unit set U.

$$B_{rr}p_{t,r}^2 + \left(2\sum_{i\in M(Z),\neq r} B_{ri}p_{t,i} - 1\right)p_{t,r} + \left(PD_t + \sum_{i\in M(Z),\neq r}\sum_{j\in M(Z),\neq r} p_{t,i}B_{ij}p_{t,j} - \sum_{i\in M(Z),\neq r} p_{t,i}\right) = 0 \tag{25}$$

In the above equation, let a $= B_{rr}$, b $= 2\sum_{i\in M(Z),\neq r} B_{ri}p_{t,i} - 1$, c $= PD_t + \sum_{i\in M(Z),\neq r}\sum_{j\in M(Z),\neq r} p_{t,i}B_{ij}p_{t,j} - \sum_{i\in M(Z),\neq r} p_{t,i}$. If the roots of the equation exist, then the two roots are $\frac{-b\pm\sqrt{b^2-4ac}}{2a}$.

Two cases are to be discussed here: in case one, if the equation has no solution, the algorithm repeats the third step until the set U is empty. In case two, if there are solutions, the algorithm checks if they satisfy the boundary constraint. If they all satisfy it, $p_{t,r}$ is made equal to any solution. If only one solution satisfies it, $p_{t,r}$ is made equal to the solution that satisfies the constraint.

In the fourth step, Equation (26) is used to adjust the solution further while the feasible solution is not obtained using Equation (25). If it is feasible, the step ends. Otherwise, step 3 is repeated until the set M is empty.

$$p_{t,r} = p_{t,r} - rand(0,1) \times V(t) \tag{26}$$

Steps 1–2 can quickly reduce the amount of violation of the equation constraints associated with the power balance, and steps 3–4 can further reduce or eliminate the amount of constraint violation via fine tuning. If the solution is still infeasible, the overall feasible solution is rigorously screened using the feasibility rule method. In summary, the overall flow chart of the improved gray wolf optimization algorithm for solving the DED problem combining the hybrid constraint processing method can be obtained as shown in Figure 3.

This paper studies day-ahead scheduling and 24 h is a cycle. When continuous scheduling is considered, the generator set is continuous, and the power of the last hour of the previous day should be taken into the ramp constraint of the first hour of the following day. In other words, Formulas (20) and (21) should be modified as Formulas (27) and (28).

$$p_{t,i}^{min} = \begin{cases} max(p_i^{min}, p_{0,i} - DR_i) & if\ t = 1 \\ max(p_i^{min}, p_{t-1,i} - DR_i) & otherwise \end{cases} \tag{27}$$

$$p_{t,i}^{max} = \begin{cases} max(p_i^{max}, p_{0,i} + DR_i) & if\ t = 1 \\ max(p_i^{max}, p_{t-1,i} + UR_i) & otherwise \end{cases} \tag{28}$$

Here, $p_{0,i}$ is the output power of the 24th time interval of the previous day.

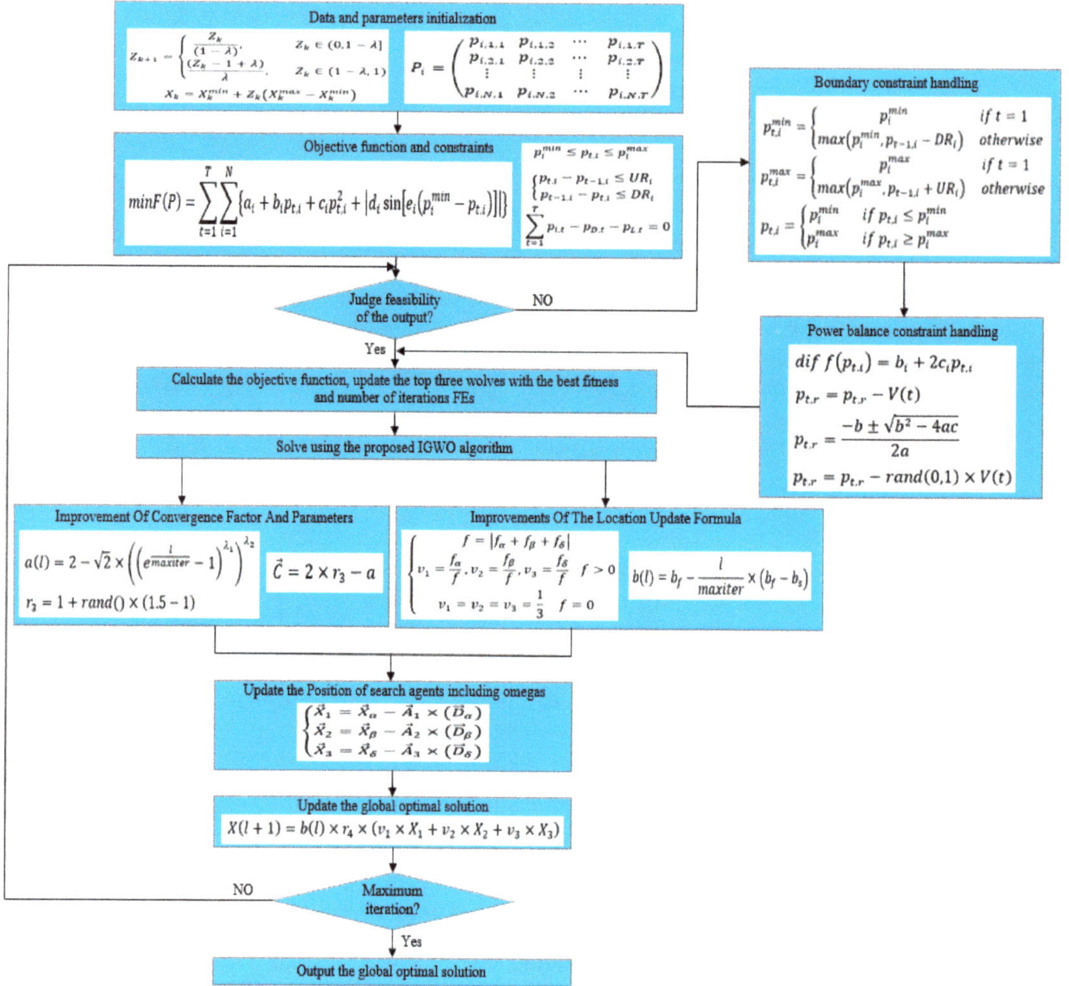

Figure 3. Flowchart of the IGWO for solving the DED problem.

5. Simulation Results and Feasibility Analysis

5.1. Feasibility Analysis of IGWO

To verify the performance of the improved gray wolf algorithm and to determine whether the algorithm is feasible for solving the DED problem, eight benchmark test functions listed in Table 1 were selected for testing, including five unimodal test functions and three multimodal test functions [38]. Several representative intelligent optimization algorithms were then selected for test and comparison, including the original gray wolf algorithm (GWO) [18], the improved gray wolf algorithm based on dimensional learning hunting search strategy (IGWO1) [39], the particle swarm algorithm (PSO) [10], the whale optimization algorithm (WOA) [11], and the grasshopper optimization algorithm (GOA) [40]. The experimental tool used was the computer software MATLAB (https://www.mathworks.com/products/matlab.html).

Table 1. The basic information of eight benchmark functions.

No.	Function	Dimension	Characteristics	Range	Min
F1	Sphere	30, 50, 100	Unimodal	$[-100, 100]$	0
F2	Schwefel	30, 50, 100	Unimodal	$[-10, 10]$	0
F3	Schwefel	30, 50, 100	Unimodal	$[-100, 100]$	0
F4	Rosenbrock	30, 50, 100	Unimodal	$[-30, 30]$	0
F5	Quartic	30, 50, 100	Unimodal	$[-1.28, 1.28]$	0
F6	Rastrigin	30, 50, 100	Multimodal	$[-5.12, 5.12]$	0
F7	Ackley	30, 50, 100	Multimodal	$[-32, 32]$	0
F8	Griewank	30, 50, 100	Multimodal	$[-600, 600]$	0

5.1.1. Parameters Setting

In this experiment, the algorithm was used to test each function in 30, 50, and 100 dimensions and the maximum number of iterations was uniformly set to 2000, and the optimization process of the algorithm was terminated when the evaluation number of the function reaches the maximum number of iterations. In addition, the basic parameters information of each algorithm involved in the comparison test were set as shown in Table 2:

Table 2. Parameter settings of each algorithm.

Algorithms	Parameter Settings
PSO	NP = 30, Vmax = 6, wMax = 0.9, wMin = 0.2, c1 = 2, c2 = 2.
WOA	NP = 30.
GOA	NP = 30, cMax = 1, cMin = 0.00004.
GWO	NP = 30.
IGWO1	NP = 30.
IGWO	NP = 30.

5.1.2. Performance Analysis

The performance of the optimization algorithm should be evaluated in terms of convergence accuracy, convergence speed, robustness, etc. Therefore, the results of 30 independent runs of different algorithms on functions in different dimensions of 30, 50, and 100 were evaluated. The information of the mean, best value, and standard deviation of the results is shown in Tables 3–5, and the comparison graphs of the convergence curves of the six algorithms for the test functions of 30 dimensions are shown in Figures 4–11.

Table 3. The statistics data of 30 runs of the benchmarks of 30 dimension.

NO.	Statistics	PSO	WOA	GOA	GWO	IGWO1	IGWO
F1	Mean	2.82×10^{-15}	2.92×10^{-304}	1.68×10^{-1}	1.36×10^{-121}	1.11×10^{-124}	**0.00**
	Std	9.89×10^{-15}	**0.00**	1.89×10^{-1}	3.76×10^{-121}	2.21×10^{-124}	**0.00**
	Best	3.88×10^{-21}	**0.00**	7.72×10^{-3}	1.52×10^{-126}	4.95×10^{-129}	**0.00**
	Runtime(s)	1.74×10^{-1}	$\mathbf{1.69 \times 10^{-1}}$	7.76×10	2.42×10^{-1}	1.32	3.41×10^{-1}
	Winner	1	0	1	1	1	
F2	Mean	1.19×10^{-7}	1.70×10^{-212}	9.93×10^{-1}	8.32×10^{-71}	4.08×10^{-75}	**0.00**
	Std	2.56×10^{-7}	**0.00**	8.48×10^{-1}	1.23×10^{-70}	1.13×10^{-74}	**0.00**
	Best	2.05×10^{-10}	2.53×10^{-223}	4.13×10^{-1}	7.36×10^{-73}	5.02×10^{-77}	**0.00**
	Runtime(s)	$\mathbf{1.60 \times 10^{-1}}$	1.62×10^{-1}	7.75×10	2.48×10^{-1}	1.33	3.64×10^{-1}
	Winner	1	1	1	1	1	
F3	Mean	1.16	6.93×10^{3}	9.31×10^{2}	4.52×10^{-31}	4.77×10^{-22}	**0.00**
	Std	6.66×10^{-1}	6.00×10^{3}	1.72×10^{2}	2.41×10^{-30}	2.60×10^{-21}	**0.00**
	Best	2.79×10^{-1}	3.57×10	7.00×10^{2}	4.94×10^{-42}	1.43×10^{-28}	**0.00**
	Runtime(s)	4.86×10^{-1}	$\mathbf{4.67 \times 10^{-1}}$	7.76×10	5.73×10^{-1}	1.99	6.81×10^{-1}
	Winner	1	1	1	1	1	
F4	Mean	4.38×10	2.64×10	2.19×10^{2}	2.67×10	$\mathbf{2.16 \times 10}$	2.87×10
	Std	2.82×10	5.31×10^{-1}	1.48×10^{2}	8.26×10^{-1}	4.34×10^{-1}	$\mathbf{2.54 \times 10^{-1}}$

Table 3. *Cont.*

NO.	Statistics	PSO	WOA	GOA	GWO	IGWO1	IGWO
	Best	$\mathbf{1.08 \times 10}$	2.57×10	3.97×10	2.49×10	2.07×10	2.81×10
	Runtime(s)	$\mathbf{1.95 \times 10^{-1}}$	1.99×10^{-1}	7.77×10	2.82×10^{-1}	1.40	3.76×10^{-1}
	Winner	0	-1	1	-1	0	
F5	Mean	3.02×10^{-2}	7.44×10^{-4}	9.55×10^{-3}	4.11×10^{-4}	6.50×10^{-4}	$\mathbf{2.12 \times 10^{-5}}$
	Std	9.44×10^{-3}	8.77×10^{-4}	5.48×10^{-3}	2.22×10^{-4}	3.03×10^{-4}	$\mathbf{1.91 \times 10^{-5}}$
	Best	1.66×10^{-2}	2.52×10^{-5}	4.12×10^{-3}	1.39×10^{-4}	2.59×10^{-4}	$\mathbf{1.99 \times 10^{-7}}$
	Runtime(s)	$\mathbf{3.59 \times 10^{-1}}$	3.63×10^{-1}	7.81×10	4.61×10^{-1}	1.78	5.63×10^{-1}
	Winner	1	1	1	1	1	
F6	Mean	4.22×10	$\mathbf{0.00}$	9.28×10	2.15×10^{-1}	1.41×10	$\mathbf{0.00}$
	Std	9.18	$\mathbf{0.00}$	4.07×10	1.18	4.89	$\mathbf{0.00}$
	Best	2.79×10	$\mathbf{0.00}$	6.72×10	$\mathbf{0.00}$	2.99	$\mathbf{0.00}$
	Runtime(s)	1.86×10^{-1}	$\mathbf{1.66 \times 10^{-1}}$	7.83×10	2.54×10^{-1}	1.39	3.50×10^{-1}
	Winner	1	-1	1	1	1	
F7	Mean	3.88×10^{-8}	$\mathbf{3.52 \times 10^{-15}}$	4.09	9.44×10^{-15}	7.43×10^{-15}	3.88×10^{-15}
	Std	7.75×10^{-8}	$\mathbf{2.59 \times 10^{-15}}$	1.10	2.91×10^{-15}	$\mathbf{6.49 \times 10^{-16}}$	$\mathbf{6.49 \times 10^{-16}}$
	Best	8.85×10^{-12}	$\mathbf{4.44 \times 10^{-16}}$	3.23	7.55×10^{-15}	4.00×10^{-15}	$\mathbf{4.44 \times 10^{-16}}$
	Runtime(s)	1.78×10^{-1}	$\mathbf{1.68 \times 10^{-1}}$	7.84×10	2.52×10^{-1}	1.35	3.51×10^{-1}
	Winner	1	-1	1	1	1	
F8	Mean	7.64×10^{-3}	2.01×10^{-3}	5.09×10^{-1}	1.71×10^{-3}	3.88×10^{-3}	$\mathbf{0.00}$
	Std	7.85×10^{-3}	7.89×10^{-3}	1.12×10^{-1}	6.03×10^{-3}	8.52×10^{-3}	$\mathbf{0.00}$
	Best	$\mathbf{0.00}$	$\mathbf{0.00}$	3.92×10^{-1}	$\mathbf{0.00}$	$\mathbf{0.00}$	$\mathbf{0.00}$
	Runtime(s)	2.02×10^{-1}	$\mathbf{1.95 \times 10^{-1}}$	7.88×10	2.78×10^{-1}	1.42	3.78×10^{-1}
	Winner	1	1	1	1	1	

Table 4. The statistics data of 30 runs of the benchmarks of 50 dimension.

NO.	Statistics	PSO	WOA	GOA	GWO	IGWO1	IGWO
F1	Mean	3.95×10^{-7}	1.54×10^{-295}	6.12×10	1.51×10^{-90}	6.17×10^{-92}	$\mathbf{0.00}$
	Std	6.80×10^{-7}	$\mathbf{0.00}$	1.21×10	7.63×10^{-90}	1.13×10^{-91}	$\mathbf{0.00}$
	Best	9.11×10^{-10}	$\mathbf{0.00}$	4.89×10	1.96×10^{-94}	2.46×10^{-94}	$\mathbf{0.00}$
	Runtime(s)	$\mathbf{1.96 \times 10^{-1}}$	1.98×10^{-1}	1.30×10^2	3.74×10^{-1}	1.53	5.36×10^{-1}
	Winner	1	0	1	1	1	
F2	Mean	5.54×10^{-3}	1.36×10^{-207}	2.42×10	1.31×10^{-53}	1.90×10^{-56}	$\mathbf{0.00}$
	Std	1.19×10^{-2}	$\mathbf{0.00}$	2.80×10	9.31×10^{-54}	2.89×10^{-56}	$\mathbf{0.00}$
	Best	9.86×10^{-5}	2.30×10^{-229}	4.54	2.58×10^{-54}	3.95×10^{-58}	$\mathbf{0.00}$
	Runtime(s)	2.11×10^{-1}	$\mathbf{1.87 \times 10^{-1}}$	1.31×10^2	3.81×10^{-1}	1.55	5.90×10^{-1}
	Winner	1	1	1	1	1	
F3	Mean	1.78×10^2	6.60×10^4	8.30×10^3	6.18×10^{-17}	6.99×10^{-7}	$\mathbf{0.00}$
	Std	5.74×10	2.76×10^4	3.79×10^3	2.63×10^{-16}	1.94×10^{-6}	$\mathbf{0.00}$
	Best	8.16×10	1.57×10^4	4.27×10^3	2.17×10^{-24}	6.55×10^{-12}	$\mathbf{0.00}$
	Runtime(s)	7.49×10^{-1}	$\mathbf{7.22 \times 10^{-1}}$	1.27×10^2	9.69×10^{-1}	2.66	1.13
	Winner	1	1	1	1	1	
F4	Mean	1.08×10^2	4.66×10	1.11×10^4	4.69×10	$\mathbf{4.21 \times 10}$	4.88×10
	Std	4.64×10	3.32×10^{-1}	6.66×10^3	7.87×10^{-1}	3.68×10^{-1}	$\mathbf{2.50 \times 10^{-1}}$
	Best	3.34×10	4.60×10	2.88×10^3	4.56×10	$\mathbf{4.17 \times 10}$	4.81×10
	Runtime(s)	2.35×10^{-1}	$\mathbf{2.23 \times 10^{-1}}$	1.26×10^2	4.10×10^{-1}	1.64	5.67×10^{-1}
	Winner	0	-1	1	-1	0	
F5	Mean	1.59×10^{-1}	8.84×10^{-4}	2.01×10^{-2}	4.86×10^{-4}	1.30×10^{-3}	$\mathbf{2.02 \times 10^{-5}}$
	Std	5.36×10^{-2}	1.25×10^{-3}	7.63×10^{-3}	2.37×10^{-4}	4.94×10^{-4}	$\mathbf{2.11 \times 10^{-5}}$
	Best	8.39×10^{-2}	2.63×10^{-5}	1.11×10^{-2}	1.05×10^{-4}	5.77×10^{-4}	$\mathbf{1.45 \times 10^{-6}}$
	Runtime(s)	5.57×10^{-1}	$\mathbf{5.19 \times 10^{-1}}$	1.25×10^2	7.51×10^{-1}	2.25	8.86×10^{-1}
	Winner	1	1	1	1	1	
F6	Mean	1.15×10^2	$\mathbf{0.00}$	1.52×10^2	$\mathbf{0.00}$	2.76×10	$\mathbf{0.00}$
	Std	2.60×10	$\mathbf{0.00}$	5.56×10	$\mathbf{0.00}$	1.14×10	$\mathbf{0.00}$
	Best	5.77×10	$\mathbf{0.00}$	1.01×10^2	$\mathbf{0.00}$	1.02×10	$\mathbf{0.00}$
	Runtime(s)	2.58×10^{-1}	$\mathbf{1.91 \times 10^{-1}}$	1.26×10^2	3.75×10^{-1}	1.62	5.43×10^{-1}

Table 4. *Cont.*

NO.	Statistics	PSO	WOA	GOA	GWO	IGWO1	IGWO
	Winner	1	-1	1	-1	1	
F7	Mean	2.46×10^{-1}	$\mathbf{4.00 \times 10^{-15}}$	8.29	1.49×10^{-14}	1.37×10^{-14}	$\mathbf{4.00 \times 10^{-15}}$
	Std	5.65×10^{-1}	2.64×10^{-15}	1.71	2.27×10^{-15}	2.07×10^{-15}	0.00
	Best	1.33×10^{-5}	$\mathbf{4.44 \times 10^{-16}}$	5.98	7.55×10^{-15}	7.55×10^{-15}	4.00×10^{-15}
	Runtime(s)	2.43×10^{-1}	$\mathbf{1.92 \times 10^{-1}}$	1.26×10^{2}	3.73×10^{-1}	1.55	5.41×10^{-1}
	Winner	1	-1	1	1	1	
F8	Mean	2.55×10^{-3}	5.87×10^{-3}	1.23	7.17×10^{-4}	1.56×10^{-3}	0.00
	Std	4.91×10^{-3}	1.84×10^{-2}	3.76×10^{-2}	2.73×10^{-3}	4.45×10^{-3}	0.00
	Best	8.91×10^{-11}	0.00	1.17	0.00	0.00	0.00
	Runtime(s)	2.62×10^{-1}	$\mathbf{2.51 \times 10^{-1}}$	1.26×10^{2}	4.12×10^{-1}	1.64	5.76×10^{-1}
	Winner	1	1	1	1	1	

Table 5. The statistics data of 30 runs of the benchmarks of 100 dimension.

NO.	Statistics	PSO	WOA	GOA	GWO	IGWO1	IGWO
F1	Mean	1.72×10^{-1}	1.22×10^{-297}	2.91×10^{3}	8.92×10^{-63}	2.52×10^{-61}	0.00
	Std	2.30×10^{-1}	0.00	7.08×10^{2}	1.65×10^{-62}	4.77×10^{-61}	0.00
	Best	2.15×10^{-2}	0.00	1.88×10^{3}	6.56×10^{-65}	4.19×10^{-64}	0.00
	Runtime(s)	3.13×10^{-1}	$\mathbf{2.59 \times 10^{-1}}$	2.57×10^{2}	6.45×10^{-1}	2.01	1.00
	Winner	1	0	1	1	1	
F2	Mean	2.34	3.07×10^{-207}	7.75×10	9.02×10^{-38}	3.11×10^{-39}	0.00
	Std	1.07	0.00	2.35×10	6.26×10^{-38}	2.30×10^{-39}	0.00
	Best	6.05×10^{-1}	8.18×10^{-222}	5.57×10	2.86×10^{-38}	3.46×10^{-40}	0.00
	Runtime(s)	2.92×10^{-1}	$\mathbf{2.43 \times 10^{-1}}$	2.57×10^{2}	6.62×10^{-1}	2.03	1.05
	Winner	1	1	1	1	1	
F3	Mean	6.11×10^{3}	6.76×10^{-5}	6.84×10^{-4}	5.48×10^{-3}	5.57×10	0.00
	Std	1.60×10^{3}	1.22×10^{5}	2.14×10^{4}	1.96×10^{-2}	5.55×10	0.00
	Best	3.24×10^{3}	3.39×10^{5}	3.86×10^{4}	6.27×10^{-10}	2.42	0.00
	Runtime(s)	1.46	**1.43**	2.60×10^{2}	1.92	4.46	2.27
	Winner	1	1	1	1	1	
F4	Mean	5.76×10^{2}	9.69×10	1.11×10^{6}	9.73×10	$\mathbf{9.34 \times 10}$	9.88×10
	Std	1.57×10^{2}	5.97×10^{-1}	6.30×10^{5}	9.18×10^{-1}	1.46	$\mathbf{2.01 \times 10^{-1}}$
	Best	3.66×10^{2}	9.63×10	7.94×10^{5}	9.49×10	$\mathbf{9.18 \times 10}$	9.81×10
	Runtime(s)	3.28×10^{-1}	$\mathbf{2.92 \times 10^{-1}}$	2.66×10^{2}	7.05×10^{-1}	2.10	1.03
	Winner	1	-1	1	-1	0	
F5	Mean	1.00×10^{3}	1.05×10^{-3}	2.55×10^{-1}	9.83×10^{-4}	3.09×10^{-3}	$\mathbf{2.43 \times 10^{-5}}$
	Std	4.69×10^{2}	1.46×10^{-3}	1.07×10^{-1}	4.61×10^{-4}	8.45×10^{-4}	$\mathbf{2.41 \times 10^{-5}}$
	Best	5.51	6.68×10^{-5}	1.62×10^{-1}	4.44×10^{-4}	1.71×10^{-3}	$\mathbf{3.03 \times 10^{-7}}$
	Runtime(s)	9.73×10^{-1}	$\mathbf{9.13 \times 10^{-1}}$	2.59×10^{2}	1.36	3.48	1.70
	Winner	1	1	1	1	1	
F6	Mean	3.98×10^{2}	0.00	3.11×10^{2}	3.79×10^{-14}	6.13×10	0.00
	Std	6.40×10	0.00	8.40×10	8.08×10^{-14}	3.74×10	0.00
	Best	2.83×10^{2}	0.00	1.91×10^{2}	0.00	6.44	0.00
	Runtime(s)	4.05×10^{-1}	$\mathbf{2.54 \times 10^{-1}}$	2.62×10^{2}	6.79×10^{-1}	2.21	1.02
	Winner	1	-1	1	1	1	
F7	Mean	1.85	$\mathbf{3.29 \times 10^{-15}}$	1.27×10	2.83×10^{-14}	2.70×10^{-14}	4.00×10^{-15}
	Std	4.45×10^{-1}	2.54×10^{-15}	7.32×10^{-1}	4.28×10^{-15}	3.46×10^{-15}	0.00
	Best	3.66×10^{-1}	$\mathbf{4.44 \times 10^{-16}}$	1.21×10	1.82×10^{-14}	2.18×10^{-14}	4.00×10^{-15}
	Runtime(s)	4.20×10^{-1}	$\mathbf{2.58 \times 10^{-1}}$	2.52×10^{2}	6.38×10^{-1}	2.09	1.01
	Winner	1	-1	1	1	1	
F8	Mean	5.43×10^{-3}	0.00	1.70×10	6.04×10^{-4}	2.38×10^{-3}	0.00
	Std	6.75×10^{-3}	0.00	2.21	2.30×10^{-3}	7.09×10^{-3}	0.00
	Best	2.12×10^{-4}	0.00	1.32×10	0.00	0.00	0.00
	Runtime(s)	3.92×10^{-1}	$\mathbf{2.91 \times 10^{-1}}$	2.53×10^{2}	6.81×10^{-1}	2.18	1.06
	Winner	1	-1	1	1	1	

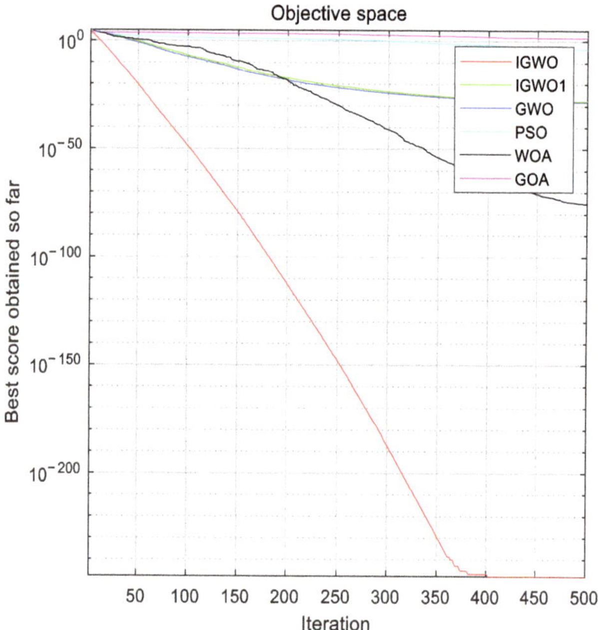

Figure 4. Test result comparison chart of F1.

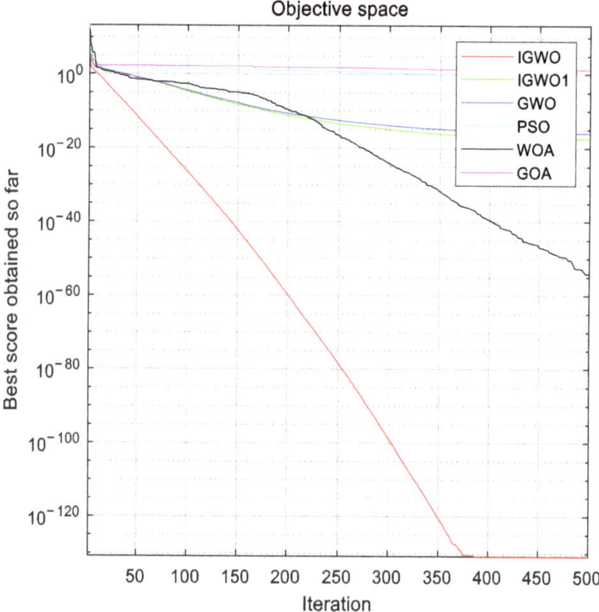

Figure 5. Test result comparison chart of F2.

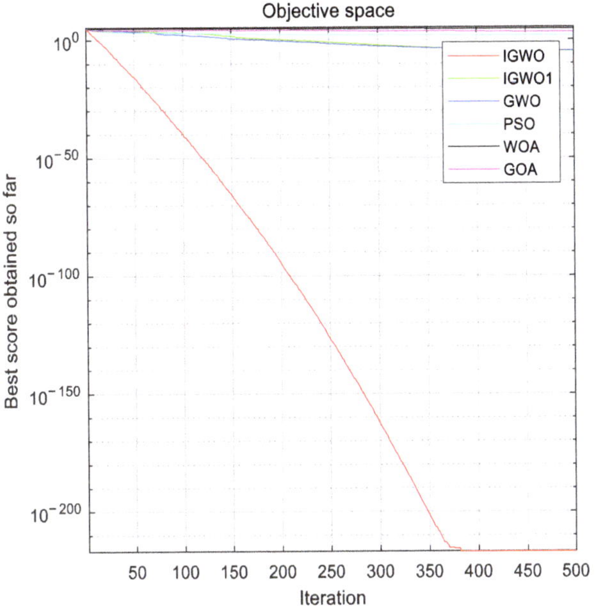

Figure 6. Test result comparison chart of F3.

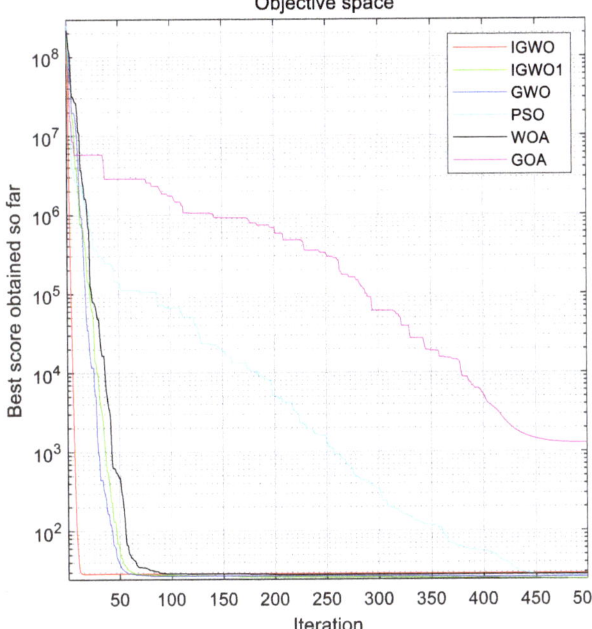

Figure 7. Test result comparison chart of F4.

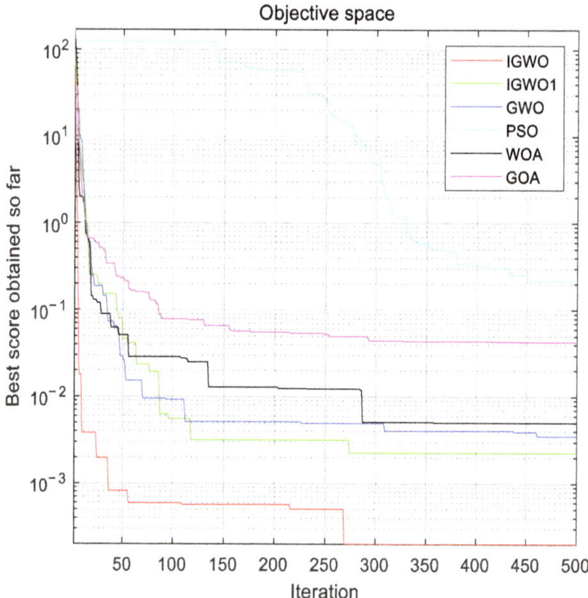

Figure 8. Test result comparison chart of F5.

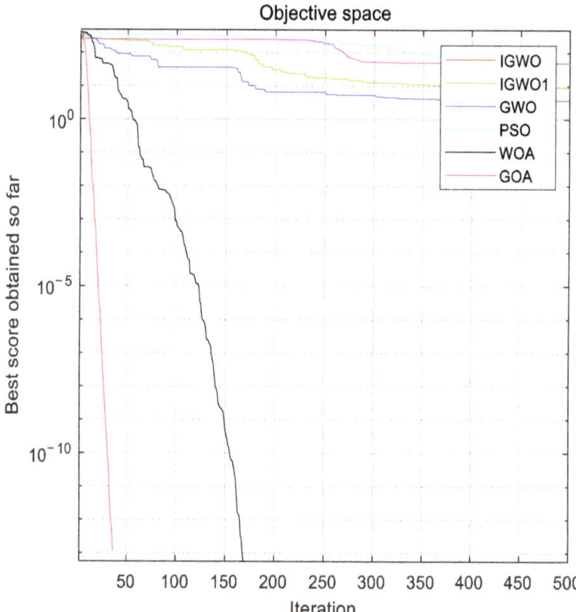

Figure 9. Test result comparison chart of F6.

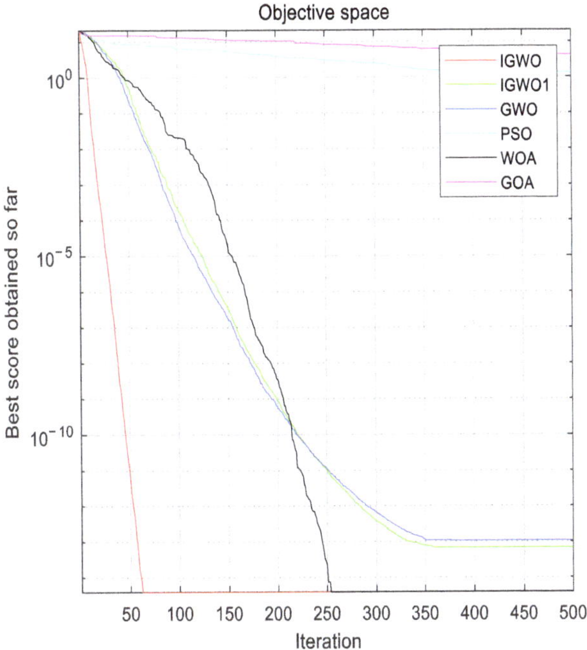

Figure 10. Test result comparison chart of F7.

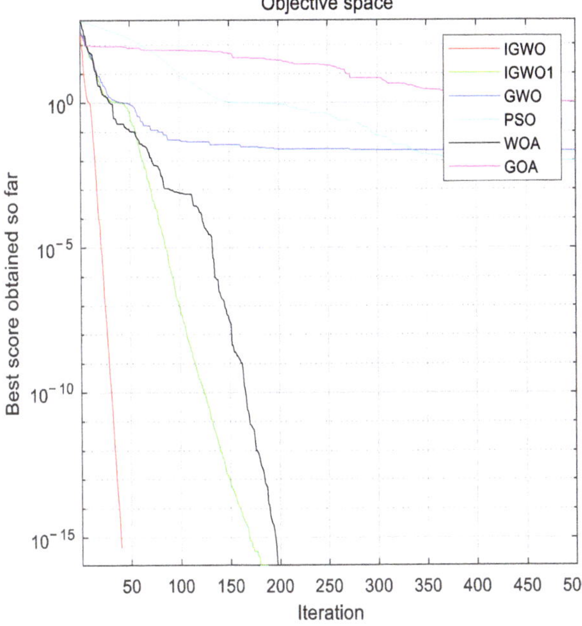

Figure 11. Test result comparison chart of F8.

In Figures 4–11, iterative convergence curves are plotted for eight typical functions representing the full range of features from F1 to F8. It can be seen that the convergence curves of the IGWO show a strong exploration and exploitation capability overall compared to other algorithms. In the F1, F2, F3, F6, and F8 test plots, it can be seen that the convergence curves of the algorithm change more drastically, which indicates that the improved parameters A and C and the updated Formula (19) provide better solutions for the algorithm and speed up the convergence of the algorithm when dealing with single-mode and multi-mode problems. In the F4, F5, and F7 test plots, it can be seen that the convergence curve of the algorithm has an obvious turn in the iterative process, which indicates that the enhanced population diversity of the improved algorithm makes the convergence accuracy of the algorithm show a significant improvement compared with most other algorithms when dealing with single-mode and multi-mode problems, further verifying the detection capability of the IGWO. From the overall test results, it is clear that the first half of the improved algorithm's curve has completed convergence and that the convergence accuracy is closer to the optimal value, illustrating the important role of the proposed algorithm-improvement mechanism in the global search. In addition, compared with other algorithms, the IGWO had a faster convergence speed and maintained a better convergence accuracy, thus verifying the effectiveness of the improved gray wolf algorithm.

In addition, the comparisons of various stability metrics of the six algorithms tested against eight functions in three different dimensions, including the mean, standard deviation, optimal value, and running time, are presented in Tables 3–5, respectively. The meaning of the indicator Winner in the table is the superiority and inferiority of IGWO compared with the test results of other algorithms. Winner 1, 0, and −1 respectively mean that IGWO is superior to, comparable to, or worse than the other algorithms on the whole. First, it can be seen from the table that the performance of the algorithm did not change much as the function dimension increased, and the convergence accuracy of the algorithm improved, which depends on the improvement of the algorithm's exploration capability. Second, the overall test effect of the IGWO algorithm on the unimodal reference function is obviously better than that of other algorithms. Although the effect of the IGWO algorithm on F4 is not as good as that of the WOA and GWO algorithms, the difference of indicators is not large, and the results are still better than other algorithms. In addition, the IGWO algorithm showed good results in testing multimodal benchmark functions, except for the F6 and F7 test results, which were slightly worse than those of the WOA, while the other functional test indicators were better than most algorithms. Finally, compared with the original gray wolf algorithm, the improved gray wolf algorithm has improved all the indicators of the test functions. Only in test F4, the improved gray wolf algorithm was inferior to the original gray wolf algorithm, and the other test function indicators were basically better than the original gray wolf algorithm. Thus, on the whole, IGWO won 98 out of 120 comparisons, thus verifying the effectiveness of the improved gray wolf algorithm.

In summary, the IGWO has excellent performance in testing benchmark problems, especially compared with the original gray wolf algorithm; its exploration and exploitation capability has been greatly enhanced, which avoids the algorithm falling into local optimal solutions to a certain extent, but it needs to be strengthened in solving multimodal benchmark problems.

5.2. Results Analysis of IGWO Test Dynamic Economic Dispatch Model

In order to verify the reliability of the algorithm and constraint treatment proposed in this paper in solving the economic dispatch problem, three cases of different sizes of the single-objective dynamic economic dispatch model were considered, namely Case 1: 5 generating units; Case 2: 10 generating units; and Case 3: 15 generating units [41]. The scheduling period was T = 24 h and the valve point effect and transmission loss constraints were considered for each case, with the corresponding B-factors taken from the studies of Qian and Mohammadi et al. [41,42]. The maximum number of evaluations of the algorithm was D × 10,000, with D representing the dimensionality of the decision variables; in addition, the maximum allowable error at each moment in the process of repairing infeasible solutions was 0.001, and to avoid chance, each case was run 30 times independently. The experiments were simulated in MATLAB software.

Calculation Results and Comparison

To make the comparison results look more intuitive, the optimal solutions of the test results for GWO and IGWO under the three cases are summarized in Tables 6–8, including the feasibility verification of the optimal solutions and the fuel costs corresponding to the optimal solutions, and the optimal solutions for each case tested by other algorithms proposed in the literature are also listed for comparison. In addition, the power output information of the optimal solutions for GWO and IGWO under Case 1 and Case 2 is shown in Tables 9–12, and the stacked histograms of the optimal output results for GWO and IGWO under Case 3 are shown in Figures 12 and 13.

Table 6. Comparison of results with different algorithms for Case I.

Algorithms	Fuel Costs ($)	Power Balance Constraint Violations (MW)
SA [9]	$47.36 K	1.10×10^{-1}
PS [43]	$46.53 K	3.00×10^{-2}
ABC [44]	$44.05 K	2.00×10^{-3}
AIS [45]	$44.39 K	1.00×10^{-3}
PSO-SQP [46]	$43.26 K	1.00×10^{-3}
CMAES [47]	$45.54 K	193.18
DHS [48]	$45.89 K	1.00×10^{-4}
MHS [48]	$45.50 K	1.00×10^{-4}
MBDE [49]	$48.32 K	1.30×10^{-5}
MSL [50]	$48.66 K	2.00×10^{-3}
GWO	$47.15 K	1.26×10^{-5}
IGWO	$43.16 K	2.51×10^{-4}

Table 7. Comparison of results with different algorithms for Case II.

Algorithms	Fuel Costs ($)	Power Balance Constraint Violations (MW)
SPS-DE [29]	$2.47 M	2.00×10^{-3}
DE-SQP [46]	$2.47 M	193.18
PSO-SQP [46]	$2.47 M	184.22
MBDE [49]	$2.60 M	1.30×10^{-5}
HCRO [51]	$2.48 M	1.00
CRO [51]	$2.48 M	1.00×10^{-3}
IBFA [52]	$2.48 M	2.00×10^{-3}
AIS [53]	$2.52 M	1.10×10^{-1}
PSO [53]	$2.57 M	1.00×10^{-3}
EP [53]	$2.59 M	3.00×10^{-2}
GWO	$2.57 M	-2.79×10^{-6}
IGWO	$2.45 M	-2.30×10^{-4}

Table 8. Comparison of results with different algorithms for Case III.

Algorithms	Fuel Costs ($)	Power Balance Constraint Violations (MW)
GWO	$0.69 M	2.80×10^{-5}
IGWO	$0.65 M	3.19×10^{-5}

Table 9. The best results of GWO for Case I.

Hour	P (MW)					PD	V (t)
	P1	P2	P3	P4	P5		
1	12.25	88.02	40.44	49.71	223.59	410	7.88×10^{-5}
2	10.00	104.40	48.22	42.96	233.96	435	-1.26×10^{-4}
3	18.33	96.79	55.55	81.21	228.25	475	-3.37×10^{-5}

Table 9. *Cont.*

Hour	P (MW)					PD	V (t)
	P1	**P2**	**P3**	**P4**	**P5**		
4	14.91	92.12	88.77	107.29	233.06	530	-2.44×10^{-5}
5	11.81	98.35	98.52	122.16	233.93	558	-7.37×10^{-5}
6	31.76	100.32	118.32	129.44	236.05	608	-2.52×10^{-5}
7	47.11	105.57	114.15	127.93	239.60	626	-9.90×10^{-5}
8	66.14	90.01	125.95	142.86	238.03	654	1.28×10^{-4}
9	65.10	100.23	119.99	136.35	278.55	690	-2.09×10^{-5}
10	42.70	102.40	159.99	167.86	241.44	704	5.94×10^{-5}
11	46.83	96.86	136.10	205.46	245.71	720	1.22×10^{-4}
12	58.50	96.96	156.68	211.01	228.32	740	1.12×10^{-4}
13	35.60	110.95	130.49	208.19	229.32	704	9.50×10^{-5}
14	35.13	106.97	122.40	207.76	227.90	690	1.54×10^{-4}
15	11.51	101.11	111.68	207.48	231.49	654	1.62×10^{-5}
16	10.31	78.21	114.00	191.03	193.64	580	8.36×10^{-5}
17	17.54	66.52	114.02	141.06	225.47	558	-3.23×10^{-5}
18	15.18	95.80	111.15	164.71	229.08	608	4.36×10^{-5}
19	17.29	100.83	109.26	213.73	222.15	654	8.82×10^{-5}
20	34.29	103.51	128.47	203.11	245.17	704	-2.01×10^{-6}
21	19.89	101.37	124.88	211.07	232.71	680	2.80×10^{-5}
22	14.80	108.16	101.86	166.69	221.41	605	-1.5×10^{-4}
23	14.34	93.61	72.61	130.51	222.05	527	-1.34×10^{-4}
24	10.54	64.00	42.31	108.10	243.00	463	1.42×10^{-5}

The total fuel cost is $47.15 K.

Table 10. The best results of IGWO for Case I.

Hour	P (MW)					PD	V (t)
	P1	**P2**	**P3**	**P4**	**P5**		
1	11.04	99.13	30.02	124.93	139.85	410	3.24×10^{-4}
2	10.81	98.53	30.00	125.38	189.85	435	1.63×10^{-4}
3	11.29	98.58	32.95	124.97	229.58	475	2.04×10^{-4}
4	15.24	100.57	72.95	125.00	229.59	530	3.13×10^{-4}
5	11.49	98.56	112.58	126.51	229.61	558	3.04×10^{-4}
6	11.19	98.90	112.81	176.51	230.57	608	2.69×10^{-4}
7	11.83	98.77	112.80	209.86	229.62	626	2.09×10^{-4}
8	29.01	98.44	112.77	209.90	229.59	654	3.38×10^{-4}
9	58.85	98.60	112.70	209.88	231.32	690	2.64×10^{-4}
10	75.00	101.70	113.06	209.89	229.79	704	3.22×10^{-4}
11	74.98	99.34	113.35	209.89	229.59	720	3.06×10^{-4}
12	75.00	99.79	113.25	209.86	229.51	740	2.36×10^{-4}
13	75.00	99.34	112.87	210.08	233.56	704	1.64×10^{-4}
14	75.00	99.18	112.91	210.03	229.80	690	3.76×10^{-4}
15	45.98	98.57	112.75	209.99	229.58	654	1.97×10^{-4}
16	15.98	84.89	112.61	209.73	229.47	580	2.52×10^{-4}
17	10.04	68.26	112.50	209.78	227.52	558	3.51×10^{-4}
18	11.35	98.26	113.19	209.85	229.55	608	2.90×10^{-4}
19	10.91	80.57	112.77	209.78	229.54	654	4.14×10^{-4}
20	11.35	98.63	112.80	209.92	229.54	704	2.14×10^{-4}
21	12.89	98.62	112.64	209.83	222.55	680	2.97×10^{-4}
22	11.91	98.97	112.73	209.92	229.54	605	1.35×10^{-4}
23	10.00	103.18	112.49	198.02	179.56	527	8.73×10^{-5}
24	10.05	98.69	112.70	148.55	139.83	463	-6.24×10^{-6}

The total fuel cost is $43.16 K.

Table 11. The best results of GWO for Case II.

Hour	P (MW)										PD	V (t)
	P1	P2	P3	P4	P5	P6	P7	P8	P9	P10		
1	153.18	214.84	91.13	149.57	132.95	97.39	65.73	67.96	56.86	26.71	1036	-8.87×10^{-5}
2	157.15	142.40	131.37	190.03	176.47	86.80	71.41	80.46	47.63	48.95	1110	-5.72×10^{-6}
3	155.42	166.76	147.75	227.63	206.09	95.73	90.22	96.33	53.93	47.24	1258	-2.63×10^{-5}
4	210.48	180.18	193.56	211.73	243.00	103.43	100.29	105.18	50.03	44.91	1406	-2.26×10^{-5}
5	171.86	141.18	270.05	245.39	221.12	115.11	124.95	120.00	59.89	50.40	1480	4.40×10^{-5}
6	214.15	195.01	287.59	245.99	235.30	150.11	130.00	108.24	64.19	46.45	1628	1.91×10^{-5}
7	211.51	199.89	334.27	287.07	228.55	144.05	127.18	112.71	60.13	50.47	1702	1.28×10^{-4}
8	244.84	275.78	329.58	264.17	221.12	158.16	120.59	106.07	73.39	42.01	1776	-8.22×10^{-5}
9	314.48	304.86	323.73	293.00	240.24	148.63	130.00	112.31	78.30	49.89	1924	-1.70×10^{-4}
10	334.91	381.47	322.32	295.97	237.34	159.35	126.58	118.17	76.36	49.80	2022	9.48×10^{-5}
11	369.55	427.22	326.86	300.00	241.98	157.36	130.00	119.30	71.97	50.21	2106	-4.51×10^{-5}
12	408.14	459.12	333.94	287.46	233.37	155.82	124.59	117.62	71.37	52.24	2150	1.14×10^{-4}
13	366.35	395.72	331.29	297.88	237.63	155.24	129.52	116.07	78.76	48.63	2072	-4.81×10^{-5}
14	294.13	360.81	328.01	280.25	243.00	139.51	120.31	118.66	75.72	35.66	1924	-2.86×10^{-5}
15	232.02	300.52	326.20	261.02	230.17	146.59	127.11	106.23	58.88	47.20	1776	4.79×10^{-5}
16	166.10	239.41	292.04	261.42	236.24	131.60	121.90	81.36	38.10	30.74	1554	1.30×10^{-4}
17	170.79	250.74	237.06	245.24	215.07	105.03	125.07	99.60	25.29	47.07	1480	-5.29×10^{-5}
18	202.55	238.83	284.46	266.95	235.64	123.47	120.84	114.72	49.77	40.29	1628	6.16×10^{-5}
19	234.62	231.96	333.60	299.92	242.24	155.95	124.31	108.70	56.75	47.05	1776	-5.84×10^{-5}
20	310.87	311.96	340.00	300.00	243.00	160.00	127.44	120.00	78.90	54.79	1972	-1.54×10^{-5}
21	290.11	303.56	340.00	292.08	235.72	159.51	128.79	119.18	74.42	51.60	1924	-1.05×10^{-4}
22	221.69	237.63	281.14	284.84	215.01	124.87	129.68	92.44	57.28	33.15	1628	3.69×10^{-5}
23	153.64	197.76	248.43	254.99	168.62	108.30	109.86	77.12	35.60	10.45	1332	4.54×10^{-5}
24	150.00	142.77	236.65	205.19	166.19	111.10	84.95	64.88	23.60	24.28	1184	-4.06×10^{-5}

The total fuel cost is $2.57 M.

Table 12. The best results of IGWO for Case II.

Hour	P (MW)										PD	V (t)
	P1	P2	P3	P4	P5	P6	P7	P8	P9	P10		
1	150.00	135.00	92.13	120.16	222.60	122.42	129.54	119.94	20.00	10.00	1036	-8.99×10^{-5}
2	150.00	135.00	167.77	121.16	222.64	122.38	129.67	120.00	20.06	10.00	1110	-2.67×10^{-6}
3	150.01	135.00	185.37	171.16	222.69	149.46	129.70	120.00	49.71	40.00	1258	1.23×10^{-4}
4	150.00	135.00	265.21	219.52	233.40	159.94	129.88	119.93	51.94	43.77	1406	1.04×10^{-5}
5	150.03	135.01	297.50	247.88	223.18	160.00	129.73	120.00	79.99	43.43	1480	-6.04×10^{-5}
6	150.06	135.95	340.00	297.88	243.00	160.00	130.00	120.00	80.00	54.97	1628	-4.35×10^{-5}
7	150.73	215.88	339.88	300.00	243.00	160.00	130.00	120.00	80.00	55.00	1702	3.56×10^{-5}
8	197.50	222.38	340.00	300.00	243.00	160.00	130.00	119.99	80.00	55.00	1776	-4.79×10^{-4}
9	229.48	302.32	340.00	299.99	243.00	160.00	130.00	120.00	80.00	55.00	1924	-1.09×10^{-4}
10	301.03	309.66	340.00	300.00	242.99	159.99	130.00	120.00	80.00	55.00	2022	-7.35×10^{-5}
11	324.58	389.66	340.00	300.00	243.00	160.00	130.00	120.00	80.00	55.00	2106	-3.59×10^{-5}
12	353.15	396.77	340.00	300.00	243.00	160.00	130.00	120.00	80.00	55.00	2150	-9.42×10^{-4}
13	292.77	382.31	340.00	300.00	243.00	160.00	130.00	120.00	80.00	55.00	2072	2.63×10^{-5}
14	225.62	302.31	340.00	300.00	243.00	160.00	130.00	120.00	80.00	55.00	1924	8.51×10^{-6}
15	191.02	222.34	340.00	300.00	242.96	159.99	130.00	119.99	80.00	54.95	1776	-8.83×10^{-5}
16	150.01	142.36	297.95	300.00	241.41	159.98	130.00	119.99	80.00	43.43	1554	-2.17×10^{-4}
17	150.00	135.00	302.08	251.63	242.81	160.00	129.71	119.99	52.10	43.44	1480	6.49×10^{-5}
18	150.19	142.29	339.99	299.96	242.99	159.99	129.88	119.99	80.00	55.00	1628	6.95×10^{-5}
19	207.41	222.29	340.00	300.00	243.00	160.00	130.00	120.00	80.00	55.00	1776	-9.91×10^{-4}
20	255.59	302.14	340.00	300.00	242.99	160.00	130.00	119.99	80.00	55.00	1972	-9.32×10^{-4}
21	226.43	292.46	340.00	300.00	243.00	160.00	130.00	120.00	80.00	55.00	1924	-8.91×10^{-4}
22	150.00	213.14	304.78	299.93	242.86	160.00	130.00	120.00	80.00	43.43	1628	-0.01×10^{-1}
23	150.00	135.00	234.43	250.00	223.54	123.16	129.71	119.99	51.57	13.47	1332	6.82×10^{-5}
24	150.00	135.00	163.63	200.16	223.02	122.73	129.63	119.99	21.63	10.01	1184	3.67×10^{-5}

The total fuel cost is $2.45 M.

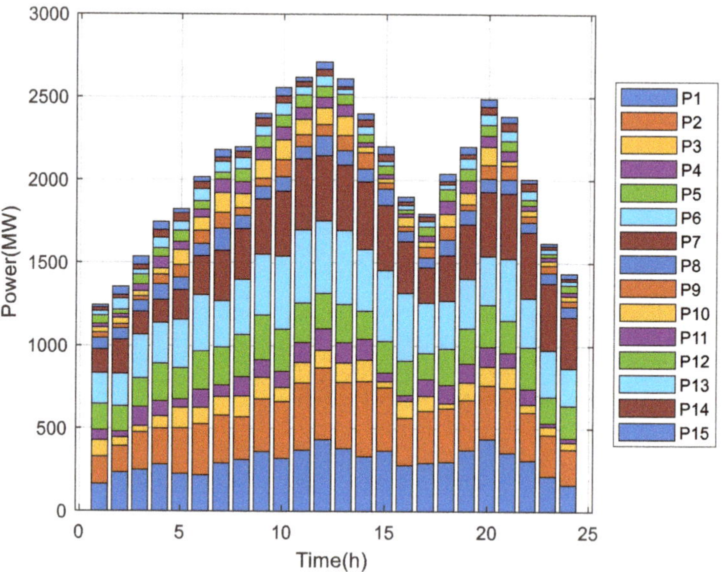

Figure 12. The best results of GWO for Case III.

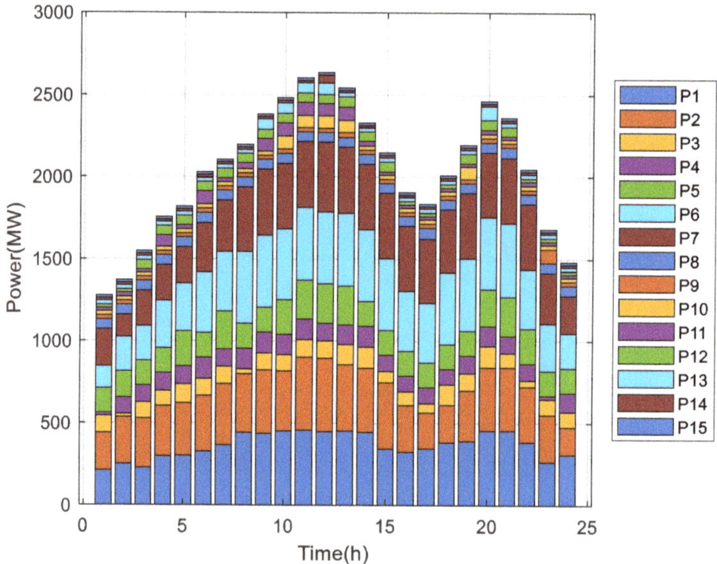

Figure 13. The best results of IGWO for Case III.

Tables 6–8 show the comparison of a series algorithms from the literature with IGWO under three cases, including a comparison with the original GWO algorithm. As can be seen from the table, first, the optimal fuel costs solved by the IGWO are $43.16 K, $2.45 M, and $0.65 M, which are better than the test results of all other algorithms proposed in the literature, which indicates that the innovative method proposed in this paper is effective. On the one hand, the chaotic initialization of the algorithm population makes the algorithm enhance a certain exploration capability in the initial stage, which can help avoid the algorithm falling too quickly into a local optimum; on the other hand, the introduction of the nonlinear convergence factor accelerates the convergence speed of the algorithm; furthermore, the adaptive strategy of the weight of the position update equation also improves the convergence accuracy of the algorithm and further enhances the exploration and exploitation capability of the algorithm. Second, the equation constraint violations of the IGWO are 2.51×10^{-4}, -2.30×10^{-4}, and 3.19×10^{-5}, which are smaller than the constraint violations of most other algorithms, and the constraint violations under each case are smaller than 0.001, which illustrates the feasibility of the coarse and fine tuning, and further verifies that the hybrid constraint-processing method proposed in this paper is effective and thus greatly reduces the constraint violations.

In addition, according to Tables 9–12, comparing the power output information of the IGWO and GWO in the two cases, the GWO and IGWO ran for 18.32 s and 18.82 s on test Case 1, respectively, and 45.68 s and 45.97 s, respectively, on test Case 2. It is found that the optimal solution of the IGWO is significantly better than that of the GWO, which shows that although the constraints are also satisfied, the IGWO has higher exploration capability and convergence accuracy due to the improvement of the population initialization and position update strategy, and also as seen in the stacked histograms of Figures 12 and 13 (Case 3), the GWO and IGWO ran for 83.84 s and 84.24 s, respectively, on test Case 3. It is worth noting here that the disaster of dimension needs to be considered in the case of solving Case 3 (Unit 15). From the results of test Case 3, the operation time of the improved algorithm is shorter, and the obtained fuel cost is also better than in other algorithms, which indicates that the improved algorithm also shows a faster convergence speed and better convergence accuracy for solving high-dimensional problems. In addition, it can be seen from Figures 12 and 13 that the improved algorithm solves the array variance less, which indicates that the improved algorithm has a more stable performance in solving high-dimensional problems, and thus the method proposed in this paper can better reduce the impact of high-dimensional space. In summary, IGWO has an advantage over other algorithms when applied to the DED problem.

6. Conclusions

In order to alleviate the current energy problem and find an effective method to solve the DED problem, an improved gray wolf optimization algorithm (IGWO) and a constraint-processing technique considering the efficiency of the unit are proposed in this paper, and it is proven by the benchmark experimental results that the proposed improving strategy does enhance the population diversity and convergence speed of the gray wolf algorithm, which makes the algorithm avoid falling into a local optimum. The test results of three other DED cases show that the proposed improved algorithm outperforms the original algorithm and other methods proposed in the literature, and that the proposed constraint-processing technique is better able to repair infeasible solutions and thus transform them into high-quality solutions. Additionally, the advantage of IGWO becomes more obvious as the number of units increases, which is of course related to the strong population diversity in the early stage of the algorithm, thus verifying the effectiveness of the proposed method in solving the DED problem in this paper.

There are still several shortcomings in this study, and more factors need to be considered for improvement in solving the DED problem, so future work will start from the following two aspects: (1) In addition to thermal power, the model should also comprehensively consider hydropower, wind power, photovoltaic power, and other power generation types, carry out multi-objective dynamic economic dispatch, and further consider the impact on the environment and ensuring, so as to meet various constraints while ensuring the minimum economic cost and environmental pollution. (2) More work needs to better address the problem of combining constraint processing techniques with swarm intelligence algorithms, by combining the two to solve a case study of the DED problem. Furthermore, how constraint processing techniques affect the performance of intelligent optimization algorithms on a practical problem should be explored.

Author Contributions: W.Y.: Conceptualization, Software, Investigation, Formal analysis, Validation, Visualization, Writing—Original Draft; Y.Z.: Methodology, Software, Data curation, Validation, Visualization, Writing—Original Draft; X.Z.: Writing—Review and Editing; K.L.: Validation, Formal analysis, Investigation; Z.Y.: Funding Acquisition, Resources, Supervision. All authors have read and agreed to the published version of the manuscript.

Funding: This research is financially supported by the National Natural Science Foundation of China (Nos. 52077213 and 62003332), the Youth Innovation Promotion Association CAS (No. 2021358), the Shenzhen Excellent Innovative Talents (No. RCYX20221008093036022), the Scientific and Technological Project of Henan Province (No. 222102110095), and the Higher Learning Key Development Project of Henan Province (No. 22A120007).

Data Availability Statement: The data in this study will be made available upon request from the corresponding author.

Conflicts of Interest: The authors declare no conflicts of interest.

References

1. Niu, Q.; Zhang, H.; Li, K.; Irwin, G.W. An efficient harmony search with new pitch adjustment for dynamic economic dispatch. *Energy* **2014**, *65*, 25–43. [CrossRef]
2. Sinha, N.; Chakrabarti, R.; Chattopadhyay, P.K. Evolutionary programming techniques for economic load dispatch. *Int. J. Emerg. Electr. Power Syst.* **2003**, *7*, 83–94. [CrossRef]
3. Yang, Z.; Li, K.; Guo, Y.; Feng, S.; Niu, Q.; Xue, Y.; Foley, A. A binary symmetric based hybrid meta-heuristic method for solving mixed integer unit commitment problem integrating with significant plug-in electric vehicles. *Energy* **2019**, *170*, 889–905. [CrossRef]
4. Yang, Z.; Li, K.; Niu, Q.; Xue, Y. A novel parallel-series hybrid meta-heuristic method for solving a hybrid unit commitment problem. *Knowl.-Based Syst.* **2017**, *134*, 13–30. [CrossRef]
5. Liu, Z.F.; Li, L.L.; Liu, Y.W.; Liu, J.Q.; Li, H.Y.; Shen, Q. Dynamic economic emission dispatch considering renewable energy generation: A novel multi-objective optimization approach. *Energy* **2021**, *235*, 121407. [CrossRef]
6. Acharya, S.; Ganesan, S.; Kumar, D.V.; Subramanian, S. A multi-objective multi-verse optimization algorithm for dynamic load dispatch problems. *Knowl.-Based Syst.* **2021**, *231*, 107411. [CrossRef]
7. Shaheen, A.M.; Ginidi, A.R.; El-Sehiemy, R.A.; Elattar, E.E. Optimal Economic Power and Heat Dispatch in Cogeneration Systems Including Wind Power. *Energy* **2021**, *225*, 120263. [CrossRef]
8. Holl, J.H. Genetic algorithms. *Sci. Am.* **1992**, *267*, 66–72.
9. Panigrahi, C.K.; Chattopadhyay, P.K.; Chakrabarti, R.N.; Basu, M. Simulated Annealing Technique for dynamic economic dispatch. *Electr. Power Compon. Syst.* **2007**, *34*, 577–586. [CrossRef]
10. Sawyerr, B.A.; Ali, M.M.; Adewumi, A.O. A comparative study of some real-coded genetic algorithms for unconstrained global optimization. *Optim. Methods Softw.* **2011**, *26*, 945–970. [CrossRef]
11. Mirjalili, S.; Lewis, A. The Whale Optimization Algorithm. *Adv. Eng. Softw.* **2016**, *95*, 51–67. [CrossRef]
12. Dorigo, M.; Birattari, M.; Stutzle, T. Ant Colony Optimization. *IEEE Comput. Intell. Mag.* **2006**, *1*, 28–39. [CrossRef]
13. Liu, D.; Hu, Z.; Su, Q.; Liu, M. A niching differential evolution algorithm for the large-scale combined heat and power economic dispatch problem. *Appl. Soft Comput.* **2021**, *113 Pt B*, 108017. [CrossRef]
14. Chávez, J.S.; Calderón, F.; No, S.T.; Centro, C. A Parallel Population Repair Genetic Algorithm for Power Economic Dispatch. *Gen* **2022**, *1*, 1–7.
15. Mahdavi, M.; Kimiyaghalam, A.; Alhelou, H.H.; Javadi, M.S.; Ashouri, A.; Catalao, J.P.S. Transmission expansion planning considering power losses, expansion of substations and uncertainty in fuel price using discrete artificial bee colony algorithm. *IEEE Access* **2021**, *9*, 135983–135995. [CrossRef]
16. Li, M.; Yang, S.; Zhang, M. Power supply system scheduling and clean energy application based on adaptive chaotic particle swarm optimization. *Alex. Eng. J.* **2021**, *61*, 2074–2087. [CrossRef]
17. Wang, Z.; Dou, Z.; Dong, J.; Si, S.; Wang, C.; Liu, L. Optimal Dispatching of Regional Interconnection Multi-Microgrids Based on Multi-Strategy Improved Whale Optimization Algorithm. *IEEJ Trans. Electr. Electron. Eng.* **2022**, *17*, 766–779. [CrossRef]
18. Mirjalili, S.; Mirjalili, S.M.; Lewis, A. Grey Wolf Optimizer. *Adv. Eng. Softw.* **2014**, *69*, 46–61. [CrossRef]
19. Ge, F.; Li, K.; Xu, W. Path Planning of UAV for Oilfield Inspection Based on Improved Grey Wolf Optimization Algorithm. In Proceedings of the 2019 Chinese Control And Decision Conference (CCDC), Nanchang, China, 3–5 June 2019.
20. Wang, P.; Rao, Y.; Luo, Q. An Effective Discrete Grey Wolf Optimization Algorithm for Solving the Packing Problem. *IEEE Access* **2020**, *8*, 115559–115571. [CrossRef]
21. Yuan, J.; Li, C.; Wang, Q.; Han, Y.; Wang, J.; Zeren, Z.; Huang, J.; Feng, J.; Shen, X.; Wang, Y. Lightning Whistler Wave Speech Recognition Based on Grey Wolf Optimization Algorithm. *Atmosphere* **2022**, *13*, 1828. [CrossRef]
22. İnaç, T.; Dokur, E.; Yüzgeç, U. A multi-strategy random weighted gray wolf optimizer-based multi-layer perceptron model for short-term wind speed forecasting. *Neural Comput. Appl.* **2022**, *34*, 14627–14657. [CrossRef]

23. Song, C.; Wang, X.; Liu, Z.; Chen, H. Evaluation of axis straightness error of shaft and hole parts based on improved grey wolf optimization algorithm. *Measurement* **2022**, *188*, 110396. [CrossRef]
24. Yan, C.; Chen, J.; Ma, Y. Grey Wolf Optimization Algorithm with Improved Convergence Factor and Position Update Strategy. In Proceedings of the 2019 11th International Conference on Intelligent Human-Machine Systems and Cybernetics (IHMSC), Hangzhou, China, 24–25 August 2019.
25. Mostafa, E.; Abdel-Nasser, M.; Mahmoud, K. Application of mutation operators to grey wolf optimizer for solving emission-economic dispatch problem. In Proceedings of the International Conference on Innovative Trends in Computer Engineering, Aswan, Egypt, 19–21 February 2018.
26. Sahoo, A.K.; Panigrahi, T.K. A novel modified random walk grey wolf optimization approach for non-smooth and non-convex economic load dispatch. *Int. J. Innov. Comput. Appl.* **2020**, *13*, 59–78. [CrossRef]
27. Mohamed, F.A.; Nasser, M.A.; Mahmoud, K.; Kamel, S. Accurate economic dispatch solution using hybrid whale-wolf optimization method. In Proceedings of the Nineteenth International Middle East Power Systems Conference, Cairo, Egypt, 19–21 December 2017; IEEE: Piscataway, NJ, USA, 2017.
28. Paramguru, J.; Barik, S.K. Modified Grey Wolf Optimization Applied to Non-Convex Economic Load Dispatch in Current Power System Scenario. In Proceedings of the International Conference on Recent Innovations in Electrical, Electronics and Communication Engineering, Bhubaneswar, India, 27–28 July 2018.
29. Shen, X.; Zou, D.; Duan, N.; Zhang, Q. An efficient fitness-based differential evolution algorithm and a constraint handling technique for dynamic economic emission dispatch. *Energy* **2019**, *186*, 115801. [CrossRef]
30. Cimen, M.; Garip, Z.; Boz, A.F. Comparison of metaheuristic optimization algorithms with a new modified deb feasibility constraint handling technique. *Turk. J. Electr. Eng. Comput. Sci.* **2021**, *29*, 3270–3289. [CrossRef]
31. Nikas, A.; Fountoulakis, A.; Forouli, A.; Doukas, H. A robust augmented ε-constraint method (AUGMECON-R) for finding exact solutions of multi-objective linear programming problems. *Oper. Res.* **2020**, *22*, 1291–1332. [CrossRef]
32. Runarsson, T.P.; Xin, Y. Stochastic ranking for constrained evolutionary optimization. *IEEE Trans. Evol. Comput.* **2019**, *4*, 284–294. [CrossRef]
33. Jin, X.; Jiang, B.; Pan, L. Analysis and Simulation of Multipath Separation Based on Improved Penalty Function Method in Time-of-Flight Cameras. In Proceedings of the Symposium on Novel Optoelectronic Detection Technology and Applications, Beijing, China, 3–5 December 2019.
34. Bcw, A.; Yun, F.A.; Hxla, B. Individual-dependent feasibility rule for constrained differential evolution. *Inf. Sci.* **2020**, *506*, 174–195.
35. Yang, X.; Leng, Z.; Xu, S.; Yang, C.; Yang, L.; Liu, K.; Song, Y.; Zhang, L. Multi-objective optimal scheduling for CCHP microgrids considering peak-load reduction by augmented ε-constraint method. *Renew. Energy* **2021**, *172*, 408–423. [CrossRef]
36. Yang, N.; Liu, H.L. Adaptively Allocating Constraint-handling Techniques for Constrained Multi-objective Optimization Problems. *Int. J. Pattern Recognit. Artif. Intell.* **2021**, *35*, 2159032. [CrossRef]
37. Liu, D.; Wang, H.; Liu, Q.; Zhou, Y. A Data Encryption and Approximate Recovery Strategy Based on Double Random Sorting and CGAN. *E3S Web Conf.* **2021**, *256*, 02023. [CrossRef]
38. Roy, P.K.; Bhui, S. A multi-objective hybrid evolutionary algorithm for dynamic economic emission load dispatch. *Int. Trans. Electr. Energy Syst.* **2016**, *26*, 49–78. [CrossRef]
39. Nadimi-Shahraki, M.H.; Taghian, S.; Mirjalili, S. An improved grey wolf optimizer for solving engineering problems. *Expert Syst. Appl.* **2020**, *166*, 113917. [CrossRef]
40. Saremi, S.; Mirjalili, S.; Lewis, A. Grasshopper Optimisation Algorithm: Theory and application. *Adv. Eng. Softw.* **2017**, *105*, 30–47. [CrossRef]
41. Qian, S.; Wu, H.; Xu, G. An improved particle swarm optimization with clone selection principle for dynamic economic emission dispatch. *Soft Comput.* **2020**, *24*, 15249–15271. [CrossRef]
42. Mohammadi-Ivatloo, B.; Rabiee, A.; Ehsan, M. Time-varying acceleration coefficients IPSO for solving dynamic economic dispatch with non-smooth cost function. *Energy Convers. Manag.* **2012**, *56*, 175–183. [CrossRef]
43. Alsumait, J.S.; Qasem, M.; Sykulski, J.K.; Al-Othman, A.K. An improved Pattern Search based algorithm to solve the Dynamic Economic Dispatch problem with valve-point effect. *Energy Convers. Manag.* **2010**, *51*, 2062–2067. [CrossRef]
44. Hemamalini, S.; Simon, S.P. Dynamic economic dispatch using artificial bee colony algorithm for units with valve-point effect. *Int. Trans. Electr. Energy Syst.* **2013**, *21*, 70–81. [CrossRef]
45. Hemamalini, S.; Simon, S.P. Dynamic economic dispatch using artificial immune system for units with valve-point effect. *Int. J. Electr. Power Energy Syst.* **2011**, *33*, 868–874. [CrossRef]
46. Elaiw, A.M.; Xia, X.; Shehata, A.M. Hybrid DE-SQP and hybrid PSO-SQP methods for solving dynamic economic emission dispatch problem with valve-point effects. *Electr. Power Syst. Res.* **2013**, *103*, 192–200. [CrossRef]
47. Manoharan, P.S.; Kannan, P.S.; Baskar, S.; Iruthayarajan, M.W.; Dhananjeyan, V. Covariance matrix adapted evolution strategy algorithm-based solution to dynamic economic dispatch problems. *Eng. Optim.* **2009**, *41*, 635–657. [CrossRef]
48. Li, Z.; Zou, D.; Kong, Z. A harmony search variant and a useful constraint handling method for the dynamic economic emission dispatch problems considering transmission loss. *Eng. Appl. Artif. Intell.* **2019**, *84*, 18–40. [CrossRef]
49. Zhang, H.; Yue, D.; Xie, X.; Hu, S.; Weng, S. Multi-elite guide hybrid differential evolution with simulated annealing technique for dynamic economic emission dispatch. *Appl. Soft Comput.* **2015**, *34*, 312–323. [CrossRef]

50. Hemamalini, S.; Simon, S.P. Dynamic economic dispatch using maclaurin series based Lagrangian method. *Energy Convers. Manag.* **2010**, *51*, 2212–2219. [CrossRef]
51. Elattar, E.E. A hybrid genetic algorithm and bacterial foraging approach for dynamic economic dispatch problem. *Int. J. Electr. Power Energy Syst.* **2015**, *69*, 18–26. [CrossRef]
52. Pandit, N.; Tripathi, A.; Tapaswi, S.; Pandit, M. An improved bacterial foraging algorithm for combined static/dynamic environmental economic dispatch. *Appl. Soft Comput.* **2012**, *12*, 3500–3513. [CrossRef]
53. Basu, M. Artificial immune system for dynamic economic dispatch. *Int. J. Electr. Power Energy Syst.* **2011**, *33*, 131–136. [CrossRef]

 *energies*

Article

Management of Hybrid Wind and Photovoltaic System Electrolyzer for Green Hydrogen Production and Storage in the Presence of a Small Fleet of Hydrogen Vehicles— An Economic Assessment

Anestis G. Anastasiadis [1,2], **Panagiotis Papadimitriou** [1], **Paraskevi Vlachou** [3] **and Georgios A. Vokas** [1,*]

[1] Department of Electrical and Electronics Engineering, University of West Attica, P. Ralli & Thivon 250, 12244 Athens, Greece; a.anastasiadis@uniwa.gr (A.G.A.); ene47337@uniwa.gr (P.P.)
[2] Power Public Corporation (PPC S.A.), Xalkokondyli 22, 10432 Athens, Greece
[3] Department of Mechanical Engineering, University of West Attica, P. Ralli & Thivon 250, 12244 Athens, Greece; vivivlaxou67@gmail.com
* Correspondence: gvokas@uniwa.gr

Abstract: Nowadays, with the need for clean and sustainable energy at its historical peak, new equipment, strategies, and methods have to be developed to reduce environmental pollution. Drastic steps and measures have already been taken on a global scale. Renewable energy sources (RESs) are being installed with a growing rhythm in the power grids. Such installations and operations in power systems must also be economically viable over time to attract more investors, thus creating a cycle where green energy, e.g., green hydrogen production will be both environmentally friendly and economically beneficial. This work presents a management method for assessing wind–solar– hydrogen (H_2) energy systems. To optimize component sizing and calculate the cost of the produced H_2, the basic procedure of the whole management method includes chronological simulations and economic calculations. The proposed system consists of a wind turbine (WT), a photovoltaic (PV) unit, an electrolyzer, a compressor, a storage tank, a fuel cell (FC), and various power converters. The paper presents a case study of green hydrogen production on Sifnos Island in Greece through RES, together with a scenario where hydrogen vehicle consumption and RES production are higher during the summer months. Hydrogen stations represent H_2 demand. The proposed system is connected to the main power grid of the island to cover the load demand if the RES cannot do this. This study also includes a cost analysis due to the high investment costs. The levelized cost of energy (LCOE) and the cost of the produced H_2 are calculated, and some future simulations correlated with the main costs of the components of the proposed system are pointed out. The MATLAB language is used for all simulations.

Keywords: renewable energy sources; wind turbine; photovoltaic; electrolyzer; green hydrogen storage; techno-economic analysis; levelized cost of energy; hydrogen vehicles

Citation: Anastasiadis, A.G.; Papadimitriou, P.; Vlachou, P.; Vokas, G.A. Management of Hybrid Wind and Photovoltaic System Electrolyzer for Green Hydrogen Production and Storage in the Presence of a Small Fleet of Hydrogen Vehicles—An Economic Assessment. *Energies* **2023**, *16*, 7990. https://doi.org/10.3390/en16247990

Academic Editor: Mahmoud Bourouis

Received: 6 November 2023
Revised: 29 November 2023
Accepted: 5 December 2023
Published: 10 December 2023

1. Introduction

When the supply of electricity is generated from non-polluting sources, the electrolytic hydrogen produced via water electrolysis is a clean source of energy. Even though conventional thermal power plants can provide electrical energy at relatively low costs, their impact on the environment is a concern. On the other hand, renewable sources of energy, such as wind and solar energy, can provide the required electricity without negatively impacting the environment. To reduce the emissions of greenhouse gases in the hydrogen production process, RESs are identified as alternatives to fossil fuels in countries that are fully dependent on, and net importers of, fossil fuels. To meet the energy demand, wind and solar energy conversion systems are used. This has been made possible due to the advanced power technologies in the distributed generation systems. To solve this challenge, a hybrid power system (HyPS) is required, which uses storage subsystems and energy management strategies [1–3].

The use of wind energy (WE) for power generation is a promising technology, especially in remote areas such as islands and isolated villages in forests and mountains. Wind farms can harness this abundant, widely distributed renewable energy source without emitting greenhouse gases. There have been tremendous advancements in wind turbine technology over the last decade, with commercial products ranging from a few hundred Watts to 10–15 MW [4,5]. In times of low wind speeds or no wind, energy storage facilities can be integrated into the WT to store the excess electricity generated during off-load periods. Via electrolyzing water, electrical energy can be converted into hydrogen. Additionally, solar energy (SE) is a renewable and green energy source that is environmentally friendly. In achieving sustainable energy solutions, it plays an important role. As a result, solar energy is a very attractive source of electricity due to its massive amount of obtainable energy every day [6,7].

To meet our energy needs, both technologies, concentrated solar power and solar photovoltaics, are constantly being developed. Thus, a large installed capacity of solar energy applications worldwide supports the energy sector and supports the development of the employment market. The world is facing a significant energy crisis, and the depletion of non-renewable energy sources is a significant contributing factor. The growing need for alternative and sustainable energy sources has led to the increased utilization of renewable energy sources like solar and wind power. While renewable energy sources have many advantages over non-renewable sources, their intermittency is one of the most significant challenges [8]. Intermittency means that renewable energy sources cannot provide a constant supply of energy, which limits their ability to replace non-renewable sources entirely. This intermittency issue has led to the development of hybrid renewable energy systems, which combine multiple renewable energy sources to provide a more reliable and consistent energy supply [9].

One potential application of hybrid renewable energy systems is the production of hydrogen through the process of electrolysis. Electrolysis is a process that uses electricity to split water molecules into hydrogen and oxygen. Through the use of renewable energy sources such as wind and solar power, the production of hydrogen can become a sustainable and environmentally friendly process [10]. However, the efficiency of electrolysis is dependent on the quality and consistency of the energy source. This dependence on the energy source's quality and consistency can result in significant inefficiencies in the production of hydrogen. A hybrid wind and photovoltaic system electrolyzer can address the challenges associated with intermittency and provide a more efficient and reliable method of producing hydrogen. The importance of developing an efficient and reliable method of producing hydrogen using RES cannot be overstated. Hydrogen has the potential to be a game changer in the energy industry, as it can be used as a fuel for transportation, heating, and electricity generation. A hybrid wind and photovoltaic system electrolyzer could provide a sustainable and environmentally friendly method of producing hydrogen, which would significantly contribute to reducing the world's dependence on non-renewable energy sources [11,12]. A hybrid wind and photovoltaic system is an RES that combines wind turbines and solar panels to generate electricity. The system is designed to address the challenges associated with intermittency by providing a more reliable and consistent source of renewable energy. The system can be designed to operate in different modes, depending on the availability of wind and solar energy. For example, when the wind is strong and the sun is not shining, the system can rely more on the wind turbine, whereas when the sun is shining and the wind is calm, the system can rely more on the solar panel [13].

Although there are many studies on, and methods of, managing an electrical hybrid system for the production of green hydrogen, usually they use one production unit, for example, WT [14–19]; this study attempted the simultaneous management of wind and solar energy for the production and storage of green hydrogen and meeting the energy needs of consumers but also—attempted for the first time here—the needs of a small hydrogen-powered fleet of cars in a small electrical section of an island. Hydrogen is compressed and stored in a high-pressure container. For grid-level applications, the stored hydrogen is converted into electricity-utilizing fuel cells, effectively addressing the energy demands of the grid. In the context of vehicular applications, hydrogen is directly supplied to vehicles equipped with fuel cells, which subsequently convert the hydrogen into electricity to power the vehicle's propulsion system. The economic assessment of the previously mentioned system, the calculation of the LCOE (levelized cost of energy), and the equivalent cost of hydrogen, in combination with a sensitivity analysis, contribute positively throughout this study. The management of the whole system is carried out per hour, and the data are real, making the results significant for decision making and the management of such systems.

Subsequently, the structure of this study is the following: the following section describes the components of the examinee hybrid system. Simultaneously, we describe the mathematical expressions of the variables that are taking place in the management of the hybrid system. Following is a flow diagram of the proposed management of the hybrid system. Then, the case study is described, and all the data are given in detail. Then, the results are listed to close the paper, with the necessary conclusions and future applications and extensions.

For the input of the data, for the processing based on the proposed management, and for the extraction of the results, we used the MATLAB programming language [20].

2. Description of Sifnos Island

This case study of green hydrogen production management will take place on the island of Sifnos. This is primarily due to the island's substantial wind potential, as well as the intensity of solar radiation. Furthermore, as part of the development of green and environmentally friendly energy in Greece, especially on the islands, there has been, and will continue to be, a growing penetration of renewable energy sources, with the main sources being wind and photovoltaic parks.

In the future, the production of electricity on the island of Sifnos will depend to a significant extent on renewable energy sources combined with energy storage units, as a large portion of the island's thermal units will be phased out.

The island of Sifnos is located in the Cyclades, specifically neighboring the islands of Serifos, Antiparos, and Kimolos, as shown in Figure 1 [21]. It covers just 74 square kilometers and belongs to the Cyclades prefecture, with its capital being Ermoupoli on the island of Syros. Additionally, its coastline extends for approximately 70 km, with a permanent population of approximately 2700 residents. More specifically, the village where renewable energy sources cover the electricity demand is called "Kastro". This village has 118 permanent residents and is situated on the summit of a steep hill near the eastern coast of the island, at an altitude of 80 m.

In general, the road network on Sifnos is quite good and convenient. The official road starts from the port area in the Kamares region and extends to the island's capital, Apollonia. From there, roads lead to the other areas of the island, making transportation easy and quick, with a travel time of less than an hour from the northernmost to the southernmost part, as shown in Figure 2.

The island offers ample space for energy utilization, primarily in the context of renewable energy sources. Specifically, areas like Chersonisos and Kamares have been identified as suitable locations with significant wind potential and good solar data annually. For this reason, some wind and photovoltaic stations have already been installed, as well as hybrid systems. In the future, many investors are expected to undertake investments in renewable energy sources. The municipality's goal is for Sifnos to become the first island with near-exclusive use of electric energy from renewable energy sources and an energy storage system through pumping or other feasible investments.

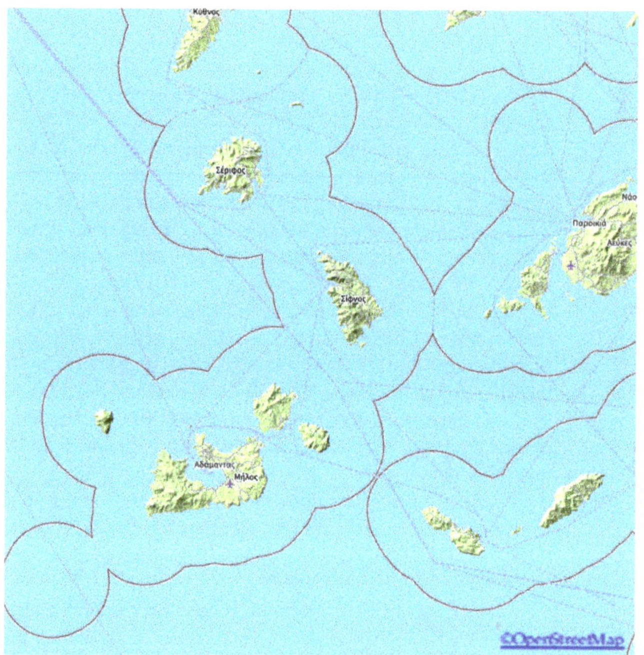

Figure 1. The island of Sifnos on a map of Greek islands (Aegean Sea—Cyclades).

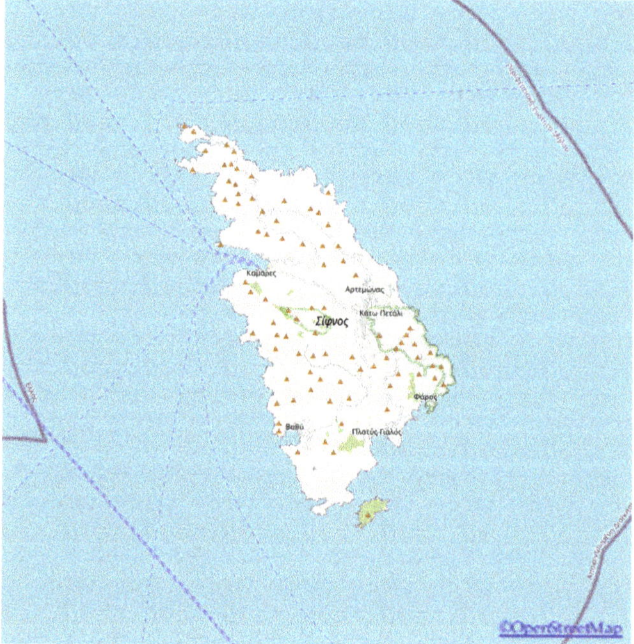

Figure 2. Road map of Sifnos (Triangles – Peaks, Dotted – Ship routes, Purple line - Borders).

3. Component Description of the Proposed Hybrid Power System, Methodology and Mathematic Formulation, and Cost Analysis

3.1. Description of the Proposed Hybrid Power System

The proposed system consists of one WT plant, one PV plant, a hydrogen production and storage facility, a fuel cell (FC), and the local load consumption (electrical load demand and hydrogen vehicles). All these elements are connected to a power grid, which is assumed to be able to cover the need when the RES units do have not sufficient production to do so. A wind turbine with 200 kW power is located at the point of the island with the greatest wind potential. A photovoltaic park with a power of 50 kW is also located on the island. These two renewable energy sources (RESs) are connected to each other and provide energy to the entire island.

The RES production is used to cover the load demand of the island. Any excess energy is used to produce hydrogen. This is achieved using a 64.5 kW electrolysis unit. The hydrogen is then stored in a storage unit with a volume of one cubic meter and a pressure of 700 bar.

This system offers a number of advantages. First, it utilizes renewable energy sources to produce hydrogen, which is a clean and sustainable fuel. Second, it helps to reduce the island's reliance on fossil fuels. Third, it can provide a stable source of energy even when the wind or solar conditions are not ideal.

The system is still in the early stages of development, but it has the potential to be a valuable asset for islands that are looking to reduce their reliance on fossil fuels. The whole system is depicted in Figure 3.

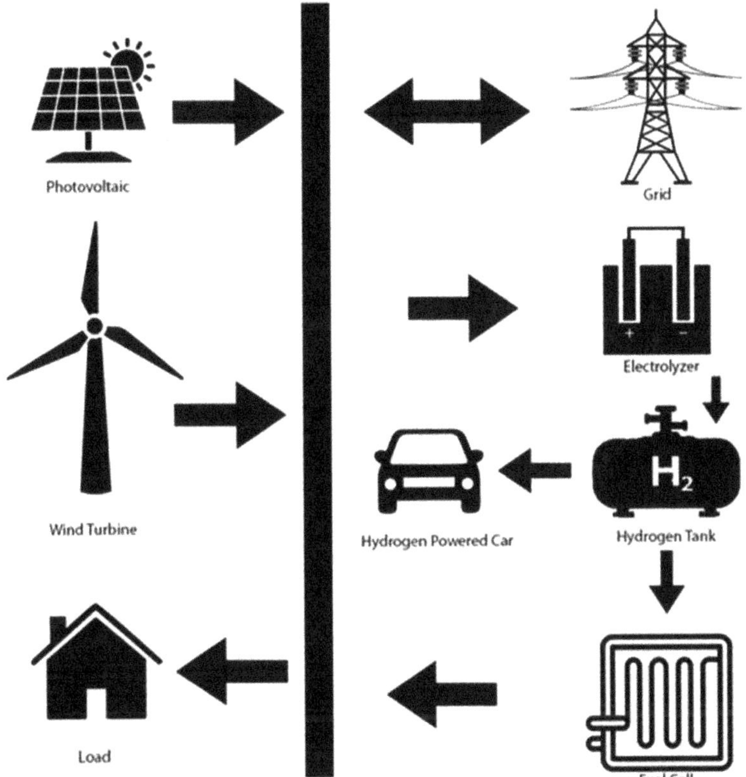

Figure 3. The proposed HyPS.

The advantages of using a hybrid wind and photovoltaic system include increased energy production, improved reliability, reduced environmental impact, and cost savings. The challenges and limitations of using a hybrid wind and photovoltaic system include the complex design, land-use requirements, intermittency, maintenance and repair, and limited scalability [9,13].

3.2. Electrolysis for Hydrogen Production—Electrolyzers and Compressor

Electrolysis is a process that uses electricity to split water molecules into hydrogen and oxygen. The process occurs in an electrolyzer, which consists of two electrodes (an anode and a cathode) separated by a membrane. When an electric current is passed through the water, the hydrogen ions (H^+) are attracted to the cathode, while the oxygen ions (O_2^-) are attracted to the anode. The ions then react at the electrodes to form hydrogen and oxygen gas [10].

There are three main types of electrolyzers [10]: (a) alkaline electrolyzers—the oldest and most mature technology for electrolysis and are often used for industrial-scale hydrogen production [22], (b) polymer electrolyte membrane (PEM) electrolyzers—a newer technology that is gaining popularity due to its high efficiency, rapid response time, and compact size; PEM electrolyzers are often used for small-scale hydrogen production, such as for fuel cell vehicles [23], (c) solid oxide electrolyzers, which operate at high temperatures (800–1000 °C) and can achieve very high efficiencies but are still in the research and development stage [24].

The efficiency of electrolysis is affected by several factors, including the energy input, electrolyte concentration, temperature, and catalysts [25,26].

The model proposed for the electrolyzer and compressor (and their respective equations) in this study is the same as that presented by Greiner, Korpas, and Holen [27]. This is where the electrolyzer and compressor are combined.

According to Equation (1), the electrolyzer power ($P_{electrolyzer}$—kW) is related to the mass flow rate of hydrogen ($M_{electrolyzer,H2}$—kg/h).

$$P_{electrolyzer}\ (t) = SPC_{elelctrolyzer} * M_{electrolyzer,H2}\ (t) \tag{1}$$

According to Equation (2) the H_2 mass storage ($M_{s,H2}$—kg/h) balance is as follows:

$$M_{s,H2}\ (t) = M_{s,H2}\ (t-1) + M_{electrolyzer,H2}\ (t) \tag{2}$$

The H_2 mass storage is limited by the minimum and maximum levels allowed (Equation (3)).

$$M_{s,H2}\ (min) \leq M_{s,H2}\ (t) \leq M_{s,H2}\ (max) \tag{3}$$

Therefore, the specific power consumption of the electrolyzer ($SPC_{electrolyzer,H2}$—kWh/kg) is taken as a summation of the individual power consumptions of the electrolyzer and the compressor.

The production of 1 kg of H_2 at 25 degrees Celsius requires 39.40 kWh divided by the efficiency according to [4]. The power consumption of the compressor is 2.38 kWh/kgH_2 to bring H_2 to 700 bar.

3.3. WT and PV

For the power of the wind turbine (P_{WT}), the wind speed data for every hour of the year are retrieved. Table 1 shows the parameters used to calculate the wind power.

Equations (4)–(7) give the wind power generated each timestep using the parameters of Table 1:

$$P_{WT}\ (t) = 0 \qquad if\ V < V_{ci} \tag{4}$$

$$P_{WT}\ (t) = 1/2 * \rho * A * V(t)^3 * C_p * E_{ffAD} \qquad if\ V_{ci} \leq V < V_r \tag{5}$$

$$P_{WT}(t) = P_r \qquad \text{if } V_r \leq V \leq V_{co} \tag{6}$$

$$P_{WT}(t) = 0 \qquad \text{if } V > V_{co} \tag{7}$$

where V(t) is the airspeed at a given time, and P_r is the nominal power of the wind turbine. Therefore, the wind production for every timestep is calculated.

If the collected wind speed data are related to the height of, e.g., 10 m, but calculations require wind speed at a different height (the height of the turbine blades), then we use the following conversion: $V_2/V_1 = (H_2/H_1)^h$, where V_2 is the wind speed at the height H_2 (which is the height of the turbine blades), and V_1 is the wind speed at the height H_1 (the height of measurement), and h is the power law coefficient, which must be calculated (or it is given for a specific area, e.g., for Sifnos, it is h = 0.2) [28].

Table 1. WT parameters of VESTAS 200 kW [29].

Wind Turbine Parameters	
Swept area of the rotor (m^2)—A	491
Diameter of the turbine (m)—d	25.0
Cut-out turbine power (kW)—P_r	200
AD converter efficiency—Eff$_{AD}$	0.98
Rated speed (m/s)—V_r	13.8
Maximum performance coefficient—C_p	0.59
Cut-in speed (m/s)—V_{ci}	3.8
Air density (kg/m^3)—ρ	1.225
Cut-out speed (m/s)—V_{co}	25
Height of the wind turbine (m)—H	30

3.4. PV, FC, Load Demand, and Hydrogen Vehicle Consumption

The PV production, fuel cell performance, load demand, and hydrogen vehicle consumption are retrieved through an Excel file to be used in the main algorithm (see the below diagrams).

3.5. Cost Analysis and LCOE Calculation [30–32]

The cost analysis of the hybrid system is carried out via the MATLAB program. A 20-year lifespan is assumed for the hybrid system. The major components (WT + PV) will not need to be replaced within the first 20 years of operation and, lastly, we assume an interest rate of 7% and inflation of 2%.

According to Equation (8), the total cost of a component (TC_C) in EUR is the cost in EUR per kilowatt or per kilogram (C) of each component of the hybrid system multiplied by the power rated (PR) of the component. Then, for the calculation of the operational and maintenance cost per component, we use Equation (9).

$$TC_{\_C} = C * PR \tag{8}$$

$$MO_C = (\%C) * PR \tag{9}$$

The maintenance and operational costs (MO_C) are the product of a defined percentage (%) of the C and the PR (%EUR/kW * kW).

Then, we proceed by setting the replacement year for every component of the system. With that in mind, we calculate how many times every component needs replacement. The replacement cost was calculated first in the future, and then we brought it back to the present with Equations (10) and (11), respectively.

$$C_{rep\_future} = C * (1 + inf)^{nrep} \tag{10}$$

$$C_{rep} = C_{rep\_future} * (1 + r)^{nrep} \tag{11}$$

The replacement cost in the future (C_{rep_future}) is the cost per kilowatt or kilogram multiplied by expression one plus inflation (inf) to the power of the lifespan of the component (n_{rep}). Now, the replacement cost at present (C_{rep}) is the replacement cost in the future (C_{rep_future}) multiplied by expression one plus the interest rate (r) to the power of the lifespan (n_{rep}). So, the total replacement cost of a component can be calculated using Equation (12).

$$C_{t,rep} = n\_rep * C_{rep} * PR \tag{12}$$

The total replacement cost of one component ($C_{t,rep}$) is the number of replacements (n_rep) multiplied by the replacement cost at present (C_{rep}) multiplied by the power rating of each component (PR). Therefore, for the calculation of the LCOE, we used Equation (13).

$$LCOE = \frac{\sum TC\_c + \sum_1^n \frac{\text{Maintenance Cost + Replacement Cost}}{(1+r)^n}}{\sum_1^n \frac{\text{Total Energy of(WT + PV)}}{(1+r)^n}} \tag{13}$$

The calculation of the levelized cost of electricity (LCOE) requires the sum of all the initial costs (TC_c) plus all the operational and maintenance costs (maintenance) and all the replacement costs (replacement) divided by expression one plus the interest rate (r) to the power of the years the project takes place (n). Then, we divide all the costs with the fraction of the total electrical energy (total energy of WT and PV) for all the years of the project divided by expression one plus the interest rate (r) and all to the power of the years the project takes place (n).

Two other important equations are (14) and (15). The capital recovery factor (CRF) is the ratio of constant earnings to the present value of receiving those earnings for a given length of time. The real interest rate (RIR) is the interest without inflation, where i is the interest rate and t represents the years.

$$CRF = \frac{i(1 + i)^t}{(1 + i)^t - 1}, \tag{14}$$

$$RIR = ((1 + \text{Nominal Interest Rate})/(1 + \text{Inflation Rate})) - 1 \tag{15}$$

4. The Implemented Management Algorithm of the Proposed Hybrid Power System

The algorithm commences by acquiring the anticipated hydrogen demand for the day from the input. Subsequently, it assesses the current hydrogen inventory within the storage container. If the stored hydrogen quantity is sufficient to meet the demand, the vehicles are promptly served. Conversely, if the stored hydrogen is inadequate, the algorithm prioritizes dispensing the available hydrogen and records the deficit. The algorithm then aggregates the renewable energy sources (RESs) and utilizes them to address the load requirements. If surplus power remains after the load demand is fulfilled, then it is redirected toward hydrogen production. Subsequent to the load assessment, the algorithm evaluates the container's fullness. If the container is at its maximum capacity, then the surplus energy is directed towards the grid. Otherwise, the surplus energy is utilized to replenish the container's hydrogen reserves until its capacity is reached. If, upon reaching its maximum capacity, excess energy persists, then it is transmitted to the grid. In the event that the renewable energy sources are incapable of meeting the load demand, the algorithm verifies the availability of sufficient hydrogen within the container to cover the shortfall. If the requisite hydrogen quantity is present, then a fuel cell is employed to convert the hydrogen into electricity, effectively addressing the load deficit. However, if the necessary hydrogen is unavailable, then the available hydrogen is converted into electricity, and the remaining energy deficit is supplemented via drawing power from the grid. In the absence of any hydrogen reserves, the entire energy requirement is sourced from the grid. A simplified flowchart presentation of the applied algorithm is presented in Figure 4.

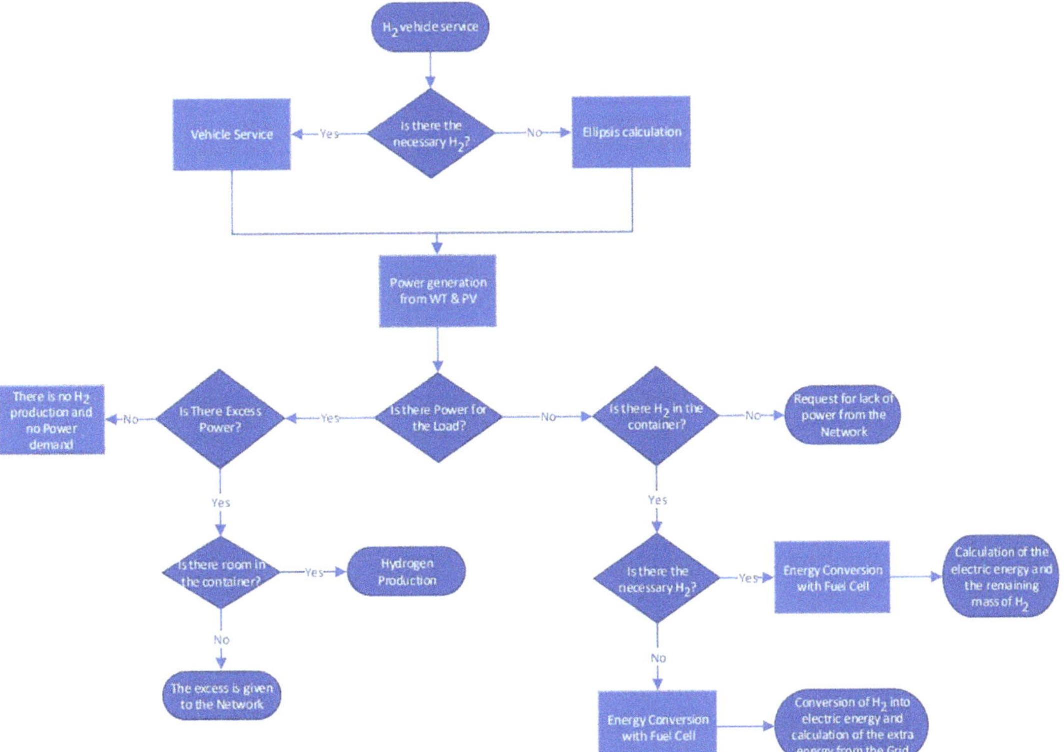

Figure 4. Simplified flow chart of the implemented algorithm.

5. Data and Assumptions

5.1. Assumptions and Input Data

The following subsections present all the assumptions and input data based on the flowchart proposed for the management of the proposed hybrid system. The input data are hourly for one year, as are the results. Characteristic graphical representations are provided on a monthly basis, per hour, or for an average 24 h period in a month. All graphical representations, data, and results come from suitable programs and code in the MATLAB 2022b language.

5.1.1. Assumptions

- On a yearly basis, the hourly data of the wind speed are at 10 m height, and we increased them to 30 m as the hub of the wind turbine VESTAS 200 kW, Table 1 (PVGIS—Year 2020 [33]).
- The installed capacity of the PV is 50 kWp, and the data production comes from [33] (year 2020).
- Load consumption data (charts with the 24 average hours of a month in 30 or 31 days in 24 h using random ±10% fluctuations) [34–36].
- Data for the electrolysis system and the tank are produced from the algorithm.
- Data for the consumption by the hydrogen-powered car Toyota Mirai.

- Economic data for the investment are based on the current prices of the market and in international studies [37,38].

For this study, with the given system on Sifnos Island, the first thing that must be carried out is the obtention of the necessary data.

5.1.2. Input Data

- P_{WT} (wind turbine installed capacity) = 200 kW—VESTAS.
- P_{PV} (photovoltaic installed capacity) = 50 kW.
- Electrolyzer = 236 kVA, 64.5 kWh/kg, 65 kg/24 h, $n_{electrolyzer}$ = 61%, n_{FC} = 50%.
- Max hydrogen mass production = 2.6 kg/h.
- Compressor consumption (2.38 kWh/kg).
- Tank storage H_2 (42 kg, 700 bar, 1 m^3). The hydrogen tank production must remain between 1–42 kg (700 bar, 1 m^3).
- Controller + inverter (200 + 50 + 64.5 + 2.38) kW = 316.88 kW

5.2. Data of the Wind Power Plant

In this study, the WT of Vestas200, with an installed capacity of 200 kW, is used; see Figures 5–7.

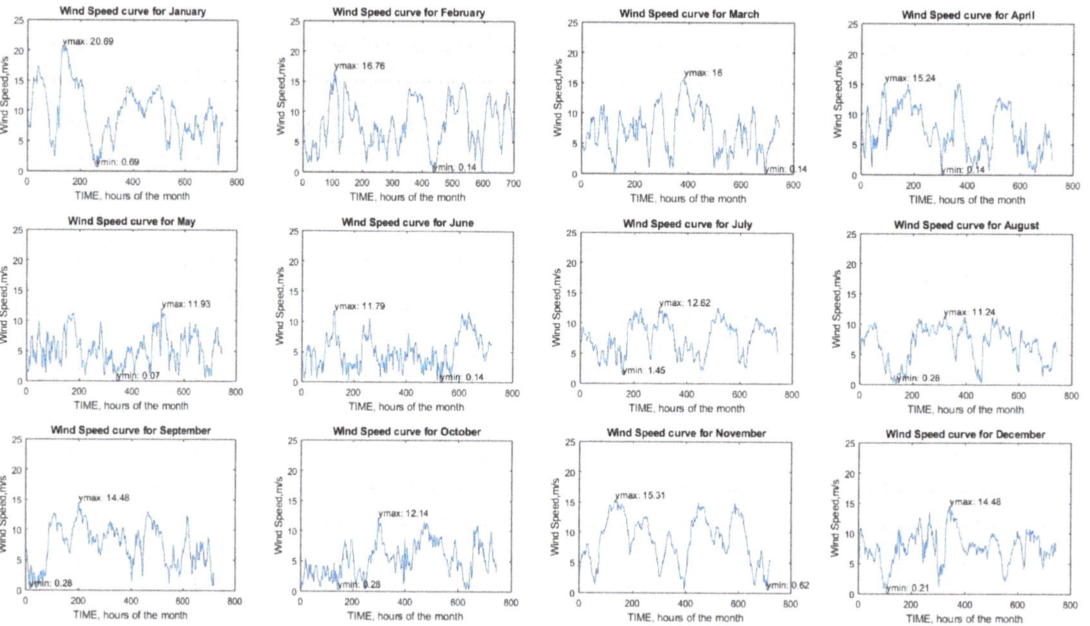

Figure 5. Hourly wind speed curve per month (2020 year).

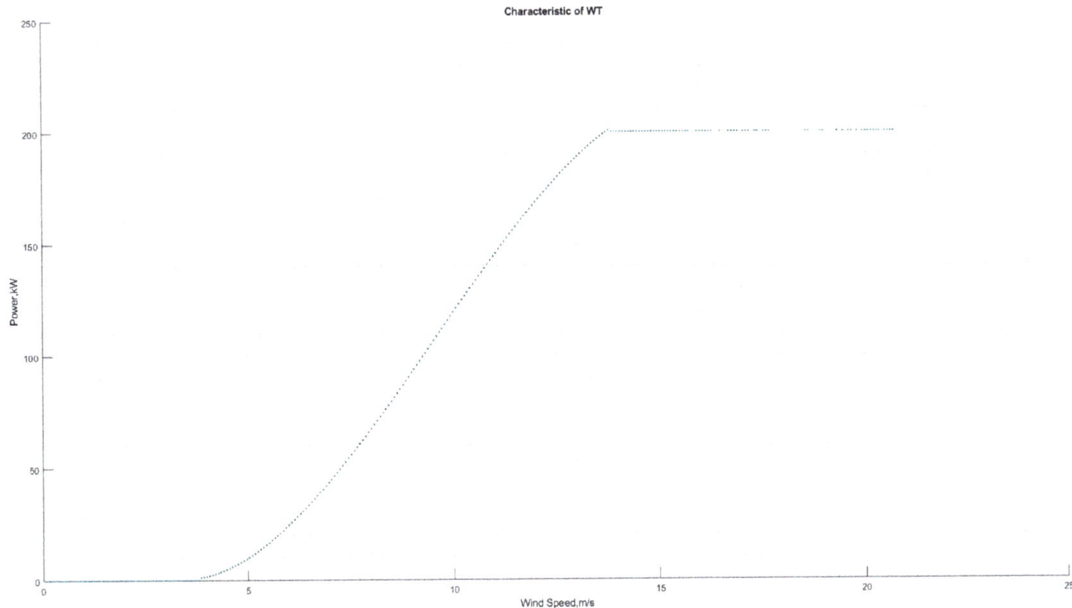

Figure 6. Characteristic wind speed–power curve for WT (VESTAS 200 kW).

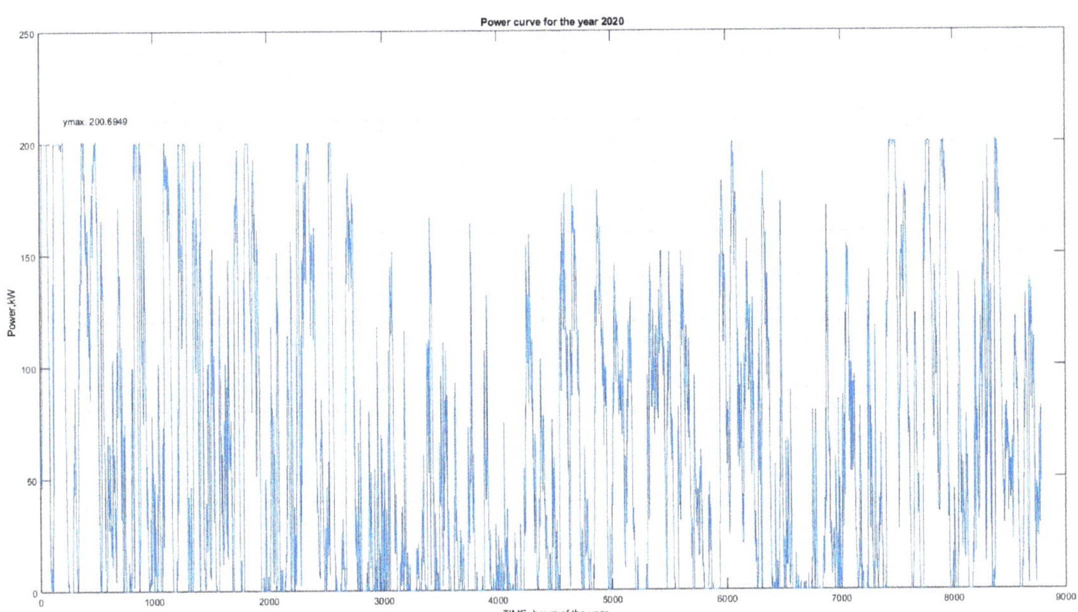

Figure 7. Power curve production for VESTAS200 WT for the whole year (per hour)—estimated capacity factor of WT, CF = 37%.

5.3. Data of the Photovoltaic Power Plant

In this study, the PV with an installed capacity of 50 kW is used; see Figures 8 and 9.

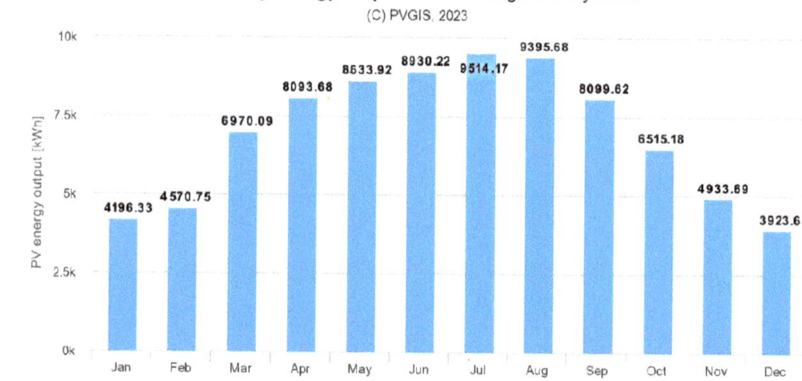

Figure 8. Data for PV system (PVGIS 2020) [33].

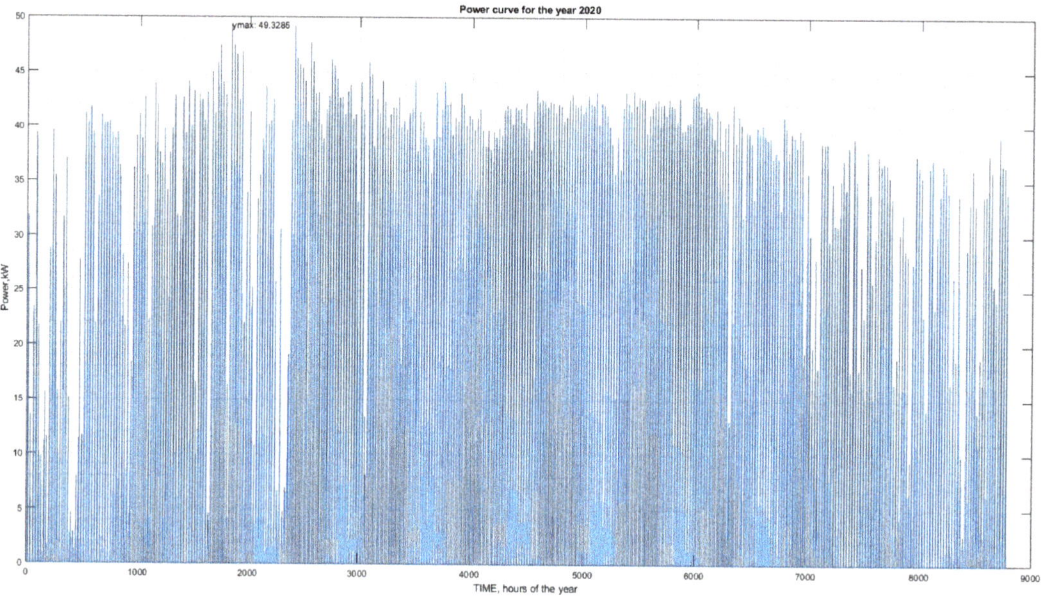

Figure 9. Power production curve for PV system for the whole year (per hour).

5.4. Data of Electrolyzer, Compressor, Storage Tank, and Hydrogen Vehicle [4,39]

All the input data are given in Figure 10 and Table 2.

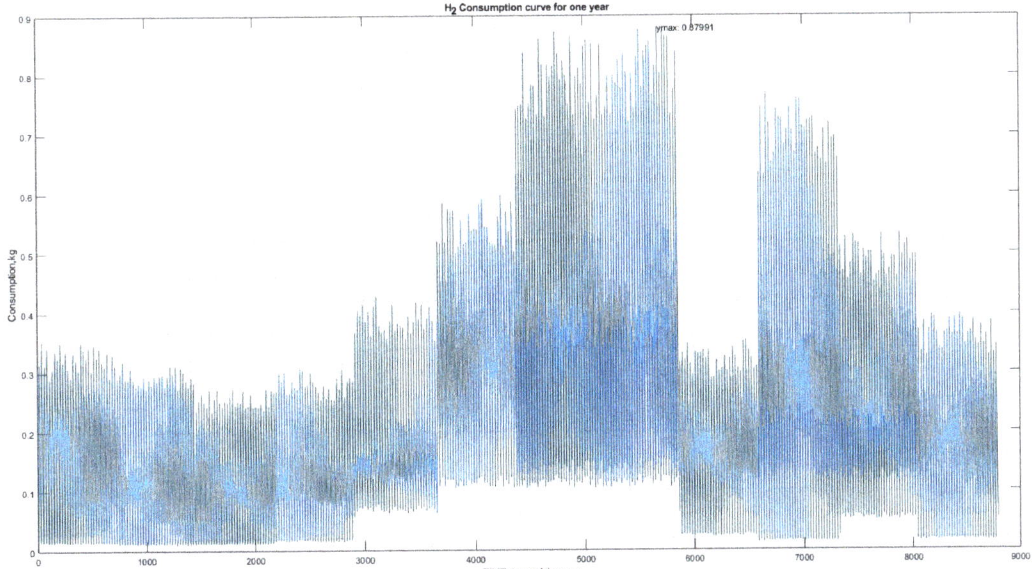

Figure 10. H$_2$ consumption curve for vehicles for the whole year (per hour).

Table 2. Characteristics of H$_2$, electrolyzer specifications, and details of the Toyota Mirai car.

Characteristics of H$_2$	
Energy Density	39.4 kWh/kg
Density at Atmospheric Pressure	0.09 kg/m^3
Density at 350 bar	26.1 kg/m^3
Density at 700 bar	42 kg/m^3
Electrolyzer Specifications	
Electrolyzer	236 kW
Hydrogen Mass Production	2.6 kg/h
Power Consumed per Mass of H$_2$	64.5 kWh/kg
Toyota Mirai Car	
Hydrogen Tank	5.6 kg
Pressure	700 bar
Range	3 km/kWh

5.5. Data of Local Consumption

The input data for local demand are given in Figure 11 [34–36].

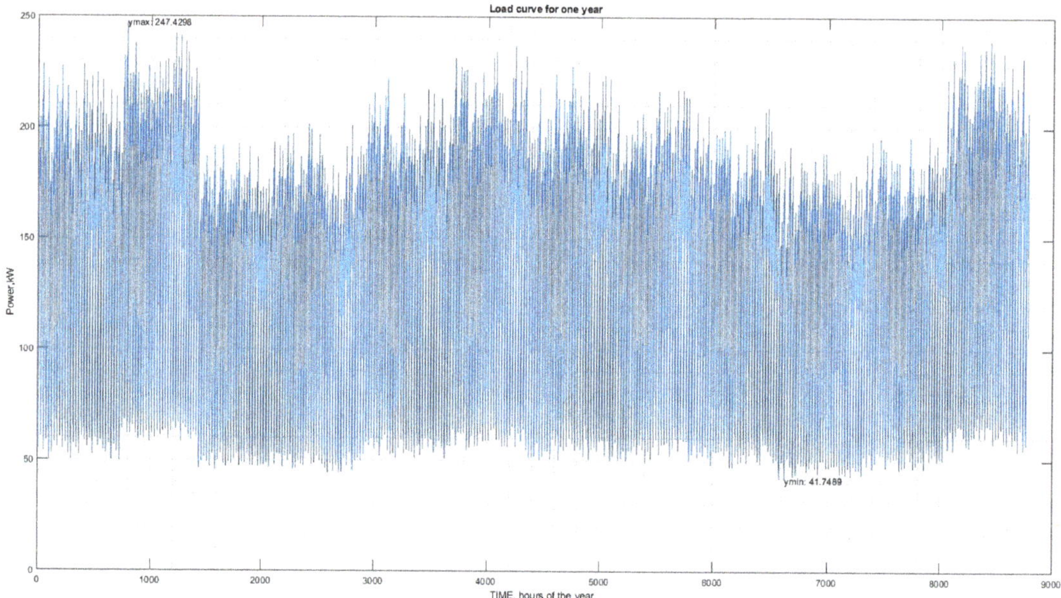

Figure 11. Load demand curve for the whole year (per hour).

5.6. Data for the Economic Assessment

For the economic assessment, using the equations we referred to above, Table 3 presents the initial values that were used to calculate the results shown below, in Table 4.

Table 3. Input data for the cost analysis [8,13,37,38].

Data for Cost Analysis	
Interest Rate (i)	0.07
Inflation Rate (f)	0.02
Project Life for Wind Generator (years)	20
Project Life for PV Generator (years)	20
Project Life for Electrolyzer (years)	10
Project Life for Fuel Cell (years)	20
Project Life for Hydrogen Tank (years)	10
Project Life for Other Items (years)	10
Initial Capital Cost of Wind Generator (EUR/kW)	1400
Initial Capital Cost of PV Generator (EUR/kW)	1200
Initial Capital Cost of Electrolyzer (EUR/kW)	650
Initial Capital Cost of Fuel Cell (EUR/kW)	190
Initial Capital Cost for Hydrogen Tank (EUR/kg)	560
Initial Capital Cost for Other Equipment (EUR/kW)	300
Rated Power of Wind Generator (kW)	200
Rated Power of PV Generator (kW)	50
Rated Power of Electrolyzer (kW)	64.5

Table 3. *Cont.*

Data for Cost Analysis	
Rated Power of Fuel Cell (kW)	190
Rated mass of Hydrogen Tank (kg)	42
Operation and Maintenance Cost for the first year of Wind Generator (EUR/kW)	56
Operation and Maintenance Cost for the first year of PV Generator (EUR/kW)	30
Operation and Maintenance Cost for the First Year of Electrolyzer (EUR/kW)	32.5
Operation and Maintenance Cost for the First Year of Fuel Cell (EUR/kW)	2
Operation and Maintenance Cost for the First Year of Hydrogen Tank (EUR/kg)	5.6
Operation and Maintenance Cost for the First Year for Other Equipment (EUR/kW)	16.5

Table 4. Output results for the cost analysis.

Equipment	Initial Investment (IV) (EUR)	Maintenance Cost in the First Year (EUR)	Annualized Replacement Cost (EUR)	Annualized Total Cost (EUR)
Wind Generator (1)	280,000	4% of IV (11,200)	0	291,200
PV Generator (2)	60,000	2.5% of IV (1500)	0	61,500
Electrolyzer (3)	41,925	5% of IV (2096.25)	51,106.341	95,127.591
Fuel Cell (4)	36,100	1% of IV (361)	0	36,461
Hydrogen Tank (5)	23,520	1% of IV (235.20)	28,670.748	52,425.948
Other Equipment (20%) of (sum = 1 + 2 + 3 + 4 + 5)	94,350	5.5% of IV (5189.25)	115,021.125	214,560.375
Total	535,895	20,581.700	194,789.213	751,274.914

6. Presentation of Results

The overall strategy of this study is to maximize hydrogen production using RES units. The hydrogen tank production must remain between 1 and 42 kg. For this purpose, at least the minimum power required for the hydrogen tank is obtained either from the RES units or the grid if the RES production is not high enough. If there is excess energy from the RES unit's production, then this power is exported to the grid. It is assumed that the grid can cover any needs that arise during the simulation.

It must be noted here that, in every timestep of the simulation, it is checked whether the hydrogen tank production has reached its maximum limit for the period of interest. Then, the hydrogen tank is set not to produce any hydrogen.

The hydrogen that is produced is then used to fuel vehicles. If there is not enough hydrogen in the tank to cover the needs of the vehicles, then this deficit hydrogen is supplied from another station of the power grid. It is again assumed that the power grid can cover those needs.

The following Figures 12–21 and Tables 4 and 5 show the results obtained via the solution of the algorithm.

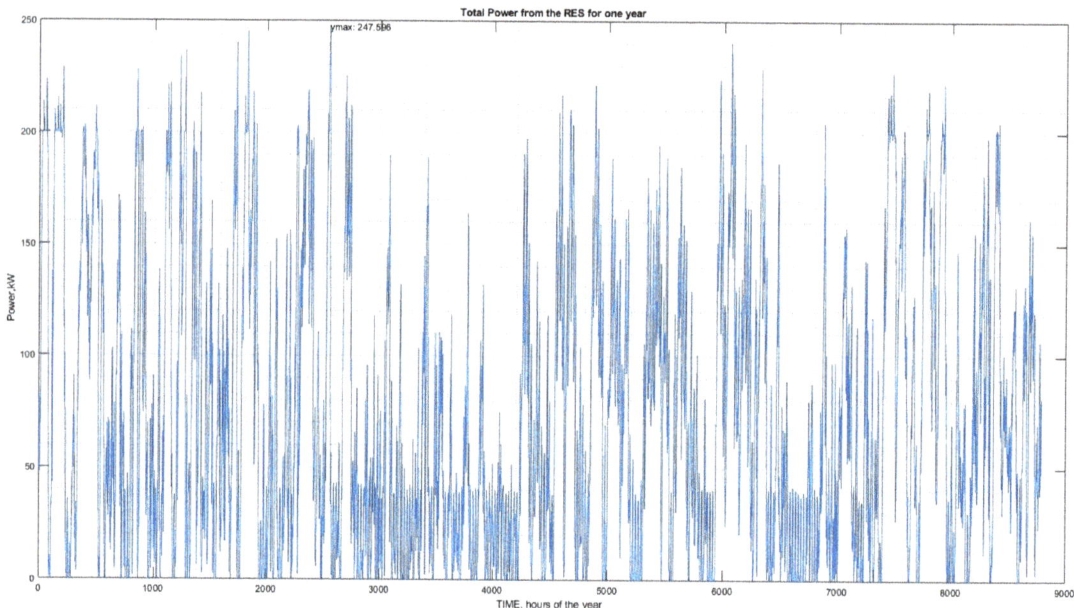

Figure 12. Total power from the RES (WT + PV) for one year (per hour).

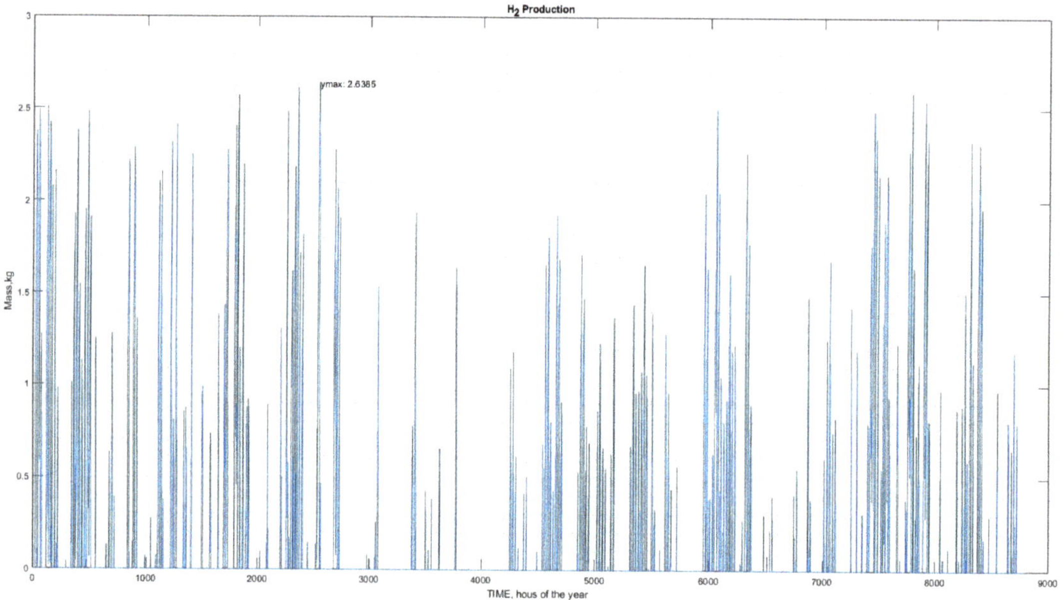

Figure 13. H$_2$ production from the RES (WT + PV) for one year (per hour).

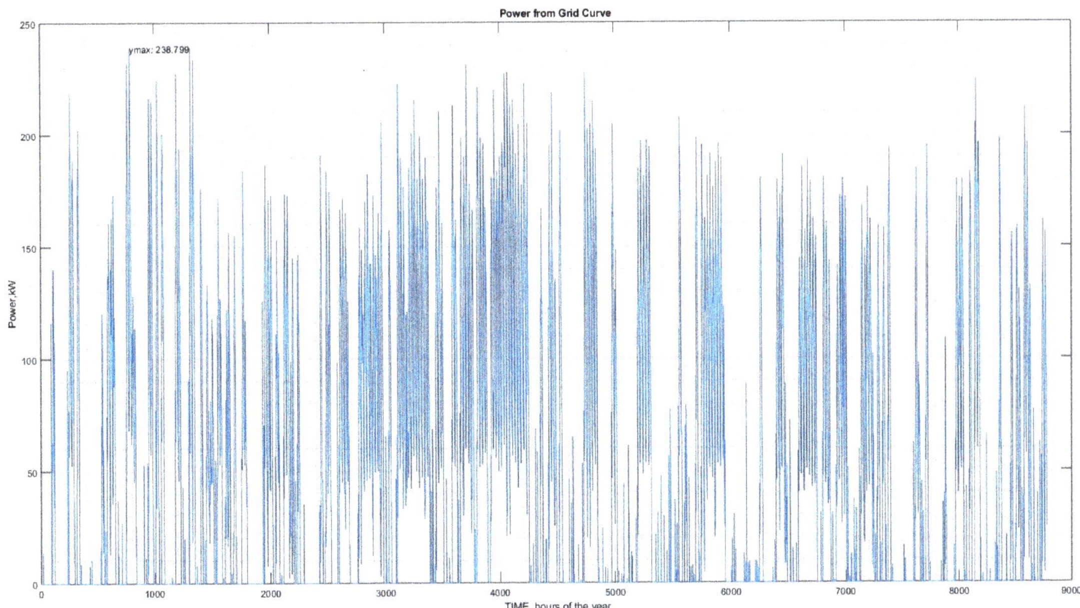

Figure 14. Total power from the grid for one year (per hour).

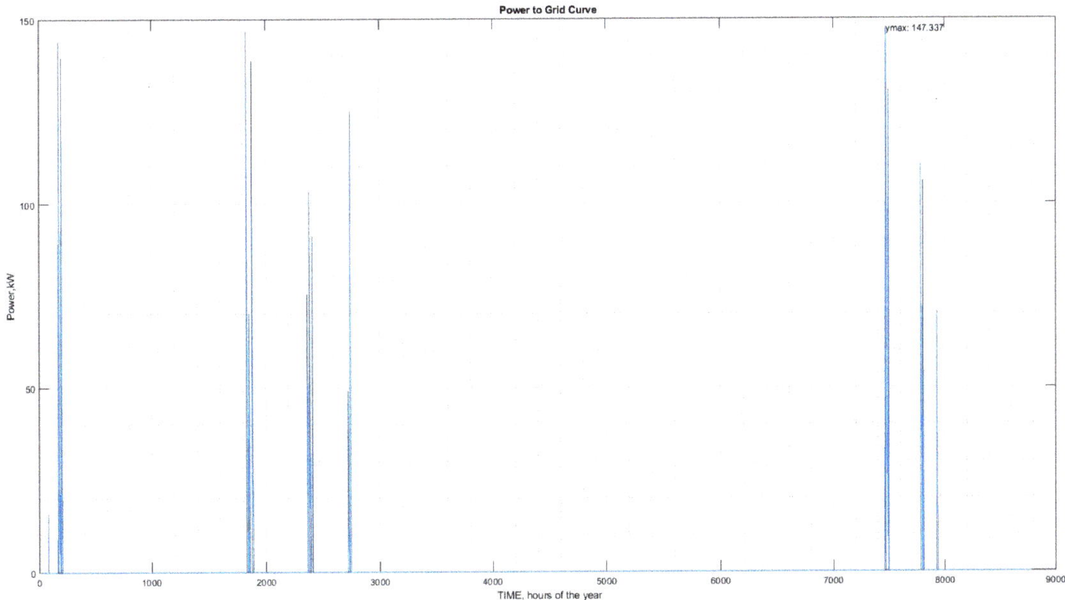

Figure 15. Total power to the grid for one year (per hour).

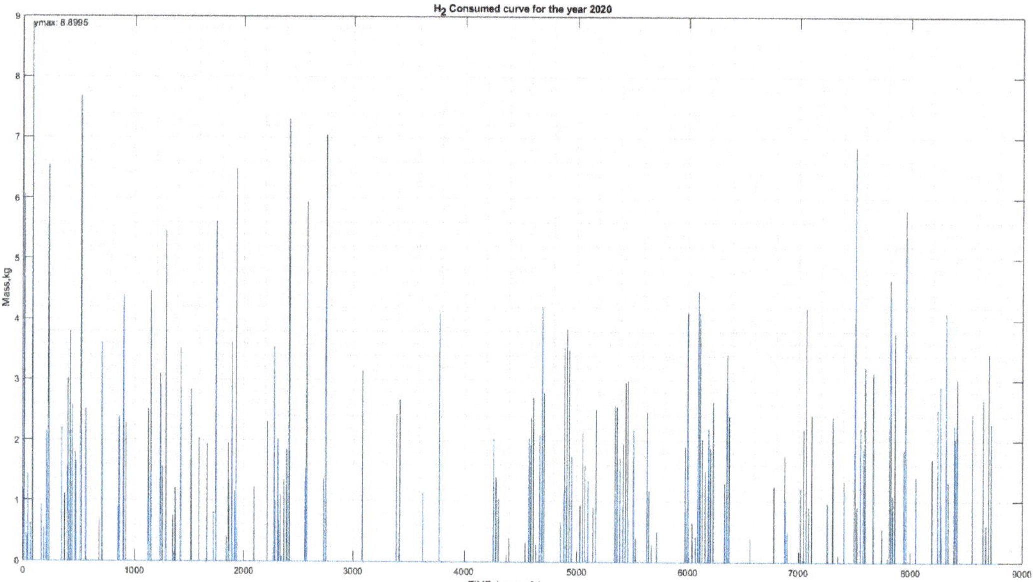

Figure 16. Real hourly H_2 consumption curve for one year.

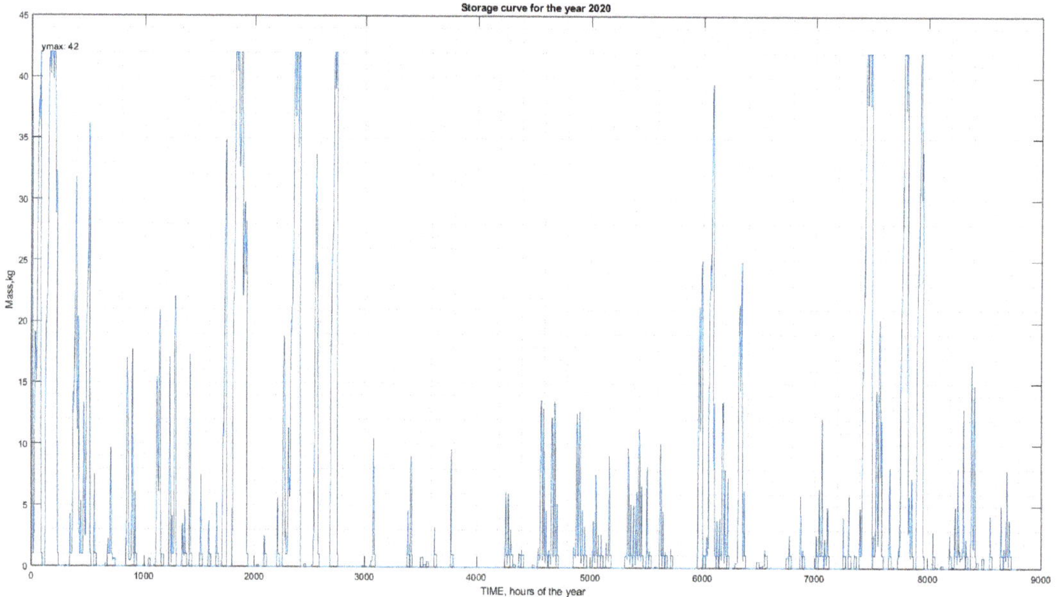

Figure 17. Hourly H_2 storage curve for one year.

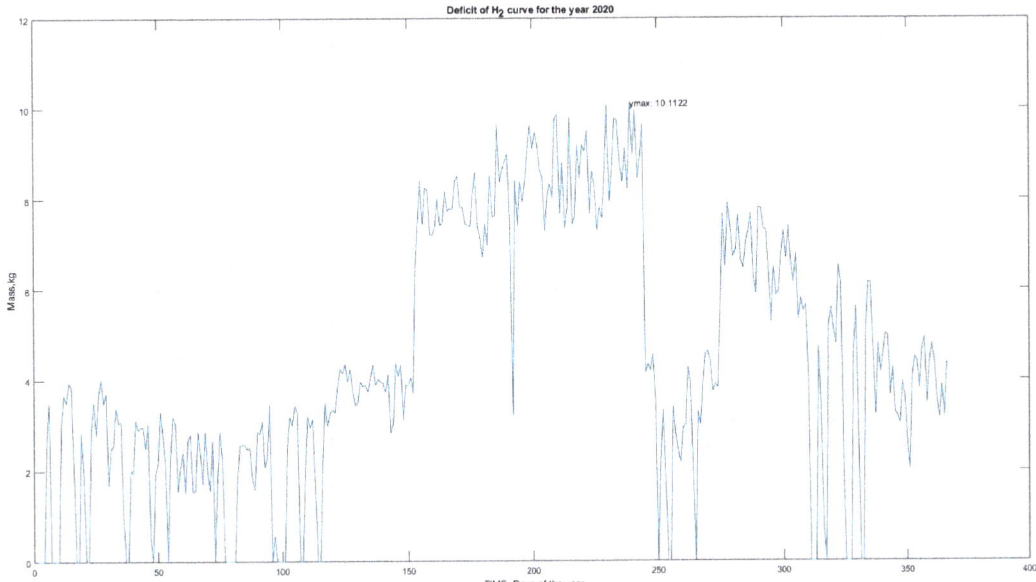

Figure 18. Hourly H$_2$ deficit curve for one year.

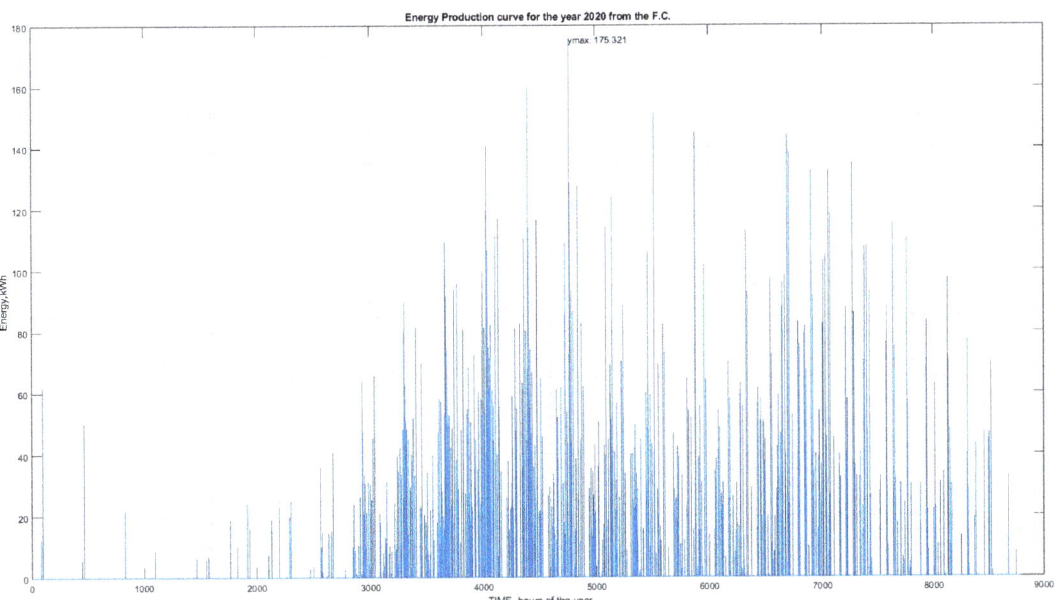

Figure 19. Hourly production curve of fuel cell for one year.

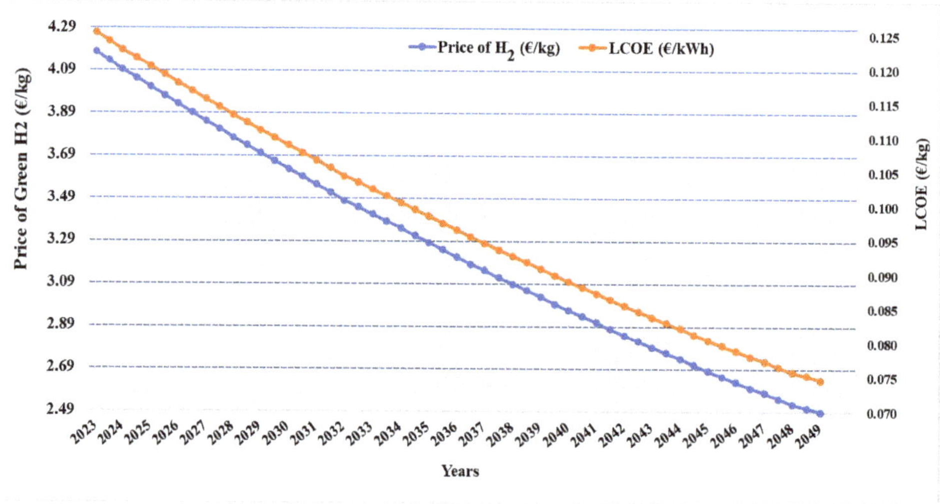

Figure 20. Estimated reduction in the LCOE and price of green hydrogen produced for the proposed hybrid power system with a time horizon of 2050 assuming a uniform reduction in the installation cost (CAPEX) of its components by 2% annually.

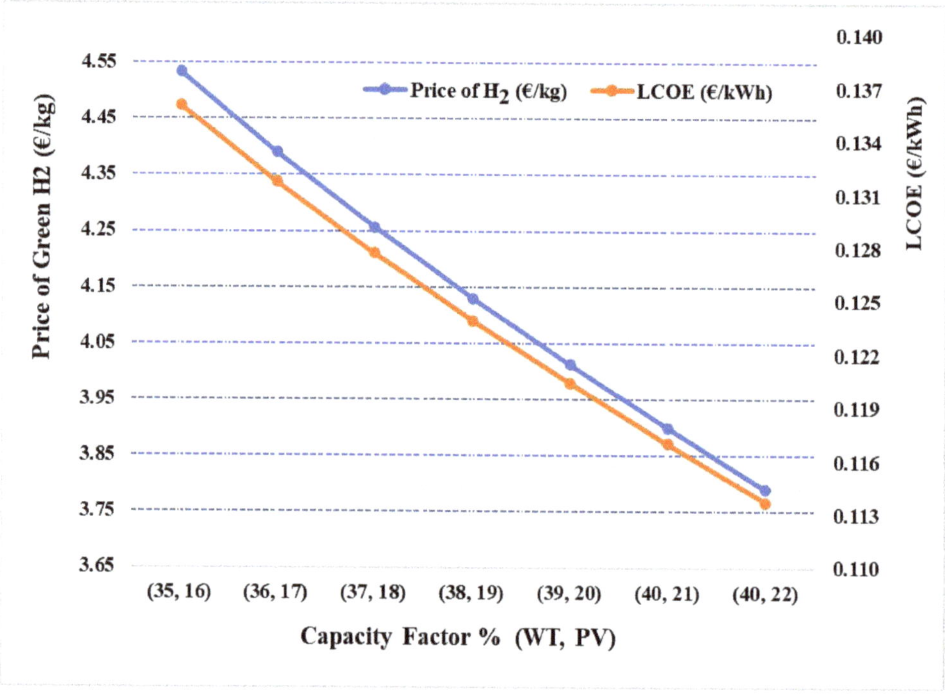

Figure 21. Estimated reduction in the LCOE and price of green hydrogen produced for the proposed hybrid power system through better capacity factor values for both WT (areas with better wind potential) and PV (more southerly areas).

Table 5. Effect of inflation and interest rate on the LCOE price and cost of green produced hydrogen for the proposed hybrid power system.

		Inflation 2%		Inflation 4%		Inflation 6%	
		LCOE (EUR/kWh)	Price of H_2 (EUR/kg)	LCOE (EUR/kWh)	Price of H_2 (EUR/kg)	LCOE (EUR/kWh)	Price of H_2 (EUR/kg)
Initial Interest Rate	7%	0.125	4.177	0.131	4.360	0.137	4.580
Various Interest Rates	4%	0.104	3.467	0.108	3.610	0.113	3.780
	5%	0.111	3.693	0.116	3.850	0.121	4.037
	6%	0.118	3.930	0.123	4.100	0.129	4.303
	7%	0.125	4.177	0.131	4.360	0.137	4.580
	8%	0.133	4.430	0.139	4.630	0.146	4.867
	9%	0.141	4.693	0.147	4.907	0.155	5.160
	10%	0.149	4.963	0.156	5.193	0.164	5.463

An integrated energy system was simulated over a one-year period to assess the feasibility of hydrogen production using renewable energy sources. The analysis revealed fluctuations in hydrogen production ranging from 0 kg to 2.6385 kg per hour due to the intermittent nature of renewable energy sources. However, the excess renewable energy was sufficient to meet the demand of the electrical grid, with the maximum power consumption reaching 238.799 kW and the maximum power supply reaching 147.337 kW. Refueling hydrogen vehicles daily proved challenging due to the vehicles' consumption patterns not aligning with the daily production cycle. Additionally, the fuel cell demonstrated substantial power output variations, reaching a maximum of 175.321 kW. These findings underscore the need for improved energy storage systems to address the intermittency of renewable energy production, as well as the development of more efficient and adaptable hydrogen refueling infrastructure and fuel cells to ensure the consistent supply of hydrogen to vehicles. Overall, the utilization of renewable energy sources for hydrogen production holds promise for a sustainable energy future, but challenges remain that require further technological advancements.

The annual power generated is 714,620.5 kWh and, therefore, the LCOE is 0.1253 EUR/kWh or 4.17 EUR/kgH$_2$ (lower calorific value of H_2, 33.3 kWh/kg). There are similar results from LAZARD and other organizations [37].

What is interesting is the fluctuation in some variables detected using sensitivity analysis. Thus, below, we present charts that have a direct relationship with these fluctuations. The changes are threefold. First of all, the change in the CAPEX reduction of 2% every year as shown in Figure 20; secondly, the change in the capacity factor of the wind turbine and the photovoltaics as shown in Figure 21; and, finally, the change in the price of the electrolyzer as shown in Figure 22.

Finally, a sensitivity analysis was performed, affecting inflation and the initial interest rate, in order to see what would change in the LCOE.

Sensitivity analysis was conducted to evaluate the impact of three key factors on the cost of hydrogen production: the initial capital cost, the efficiency coefficient of the wind turbine and photovoltaic system, and the price of the electrolyser.

The analysis began by examining the influence of the initial capital cost on the hydrogen production cost. A gradual reduction in the initial capital cost by 2% per year led to a substantial decrease in the hydrogen selling price, reaching EUR 2.49/kg. This finding emphasizes the importance of optimizing capital expenditures to enhance hydrogen production cost competitiveness.

Next, the analysis investigated the impact of wind turbine and photovoltaic efficiency coefficients on hydrogen production cost. Incrementally increasing both efficiency coefficients by 1% resulted in a decrease in hydrogen cost from EUR 4.53/kg to EUR 3.79/kg. While the effect was less pronounced compared to the initial capital cost, it still demonstrates the contribution of renewable energy source efficiency in lowering hydrogen production costs.

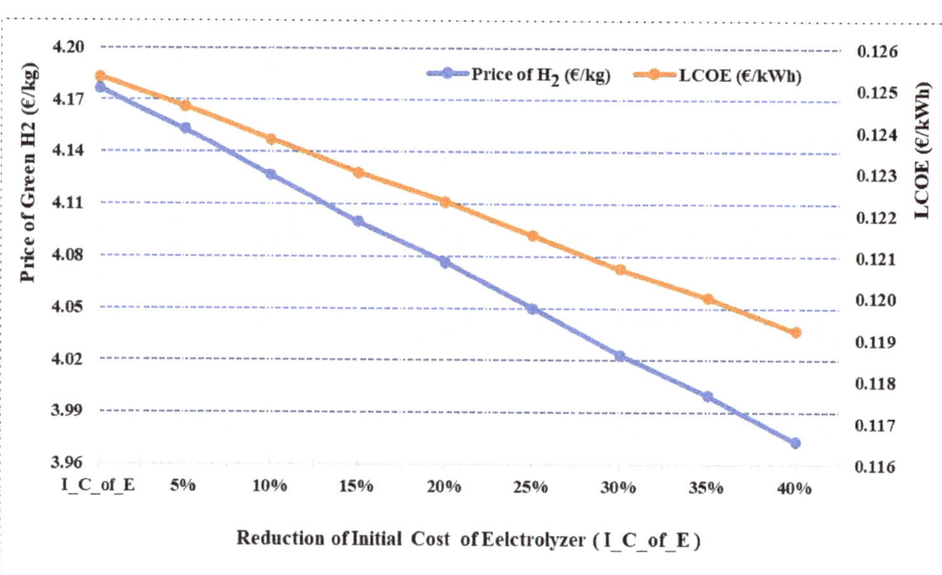

Figure 22. Effect of percentage reduction in electrolyzer cost on the LCOE value for the hybrid case study system.

Finally, the analysis focused on the effect of electrolyser price on the hydrogen production cost. A decrease in electrolyser price resulted in a corresponding decrease in hydrogen production cost, from an initial EUR 4.17/kg to EUR 3.97/kg. This highlights the significance of electrolyser technology developments in driving down hydrogen production costs.

In the latest sensitivity analysis, which was affected by the initial interest rate and inflation, we observed that the lowest LCOE was obtained for the lowest initial interest rate and inflation, which was expected. A good observation is that the LCOE value is more affected by the initial interest rate than by inflation, as shown in Table 5.

In conclusion, the sensitivity analysis revealed that the initial capital cost is the most crucial factor in determining the hydrogen production cost. The efficiency coefficient of renewable energy sources also plays a significant role, while the price of the electrolyser has a moderate impact. The financial aspect of hydrogen production projects is particularly sensitive to changes in interest rates, as shown in the latest sensitivity analysis.

7. Discussion

A hybrid system that combines renewable energy sources (RESs), hydrogen production, storage, and utilization effectively utilizes wind and photovoltaic (PV) generation to produce hydrogen and meet load requirements. Hydrogen production exhibits seasonal patterns, with lower production during low-RES periods and higher production during peak RES periods. The system relies on the grid to supplement energy needs during low-RES and high-hydrogen-consumption periods. The calculated levelized cost of electricity (LCOE) is comparable to the average European electricity price (Eurostat [40]), indicating economic feasibility. The wind turbine exceeds its maximum power due to high wind potential, while the photovoltaic panels contribute significantly to the overall energy production. The hydrogen storage tank occasionally reaches its maximum capacity, suggesting the potential benefit of a larger storage tank. The fuel cell effectively delivers up to 195 kW to the grid. Economic analysis reveals the initial cost and potential cost reductions through component efficiency improvements and capital cost reductions. Sensitivity analysis highlights the impact of various factors on the hydrogen production cost.

8. Conclusions

This study presents a novel hybrid system that integrates renewable energy sources (RESs), hydrogen production, storage, and utilization to address energy demands while minimizing the environmental impact. The system effectively utilizes wind and photovoltaic (PV) generation to produce hydrogen and meet load requirements. The study's key findings include [41–44]:

- Effective utilization of RES: the system successfully exploits RES production to power the hydrogen tank and fulfill load demand. Excess RES energy is used to maximize hydrogen production without environmental pollution.
- Optimized hydrogen production: hydrogen production is maintained within its maximum and minimum limits, ensuring efficient utilization of the storage tank.
- Maximized hydrogen utilization: the produced hydrogen is effectively utilized to charge hydrogen-powered vehicles and supplement fuel cell electrical energy, minimizing reliance on the power grid.
- Adequate system management: the hybrid system demonstrates adequate overall management, effectively balancing energy production, storage, and utilization.

Despite the system's effectiveness, the study acknowledges the limitations of relying solely on the power grid to address energy deficits. Real-world power grids may not always have the capacity to accommodate such deficits, necessitating alternative strategies. Additionally, hydrogen production is observed to be relatively low during the summer months due to high load demand and low wind speeds.

The study concludes with a comprehensive sensitivity analysis, evaluating the impact of cost reductions, capacity factor increases, electrolyzer price reductions, and interest rate and inflation fluctuations on the hydrogen production costs.

Future research directions include incorporating battery storage systems to enhance energy management and expanding the system's scope to larger distribution grids with increased load demands and hydrogen vehicle fleets.

Overall, the study demonstrates the feasibility of a hybrid system that seamlessly integrates RES, hydrogen production, storage, and utilization, providing a promising pathway toward a sustainable and environmentally friendly energy future.

Author Contributions: Conceptualization, A.G.A. and G.A.V.; methodology, P.P.; software, A.G.A. and P.P; validation, A.G.A., P.P., P.V. and G.A.V.; resources, P.V.; writing—original draft preparation, A.G.A. and P.P.; writing—review and editing, A.G.A., P.P. and G.A.V.; supervision, G.A.V.; All authors have read and agreed to the published version of the manuscript.

Funding: This research received no external funding.

Data Availability Statement: Data are contained within the article.

Conflicts of Interest: A.G.A. is employee of Power Public Corporation (PPC S.A.). The other authors declare no conflict of interest in this work.

References

1. Zhou, T.; Francois, B. Modeling and control design of hydrogen production process for an active hydrogen/wind hybrid power system. *Int. J. Hydrogen Energy* **2009**, *34*, 21–30. [CrossRef]
2. Abdallah, A.; Ekekwe, N. Cost and performance of hybrid photovoltaic and wind-powered hydrogen generation systems: A review. *Renew. Sustain. Energy Rev.* **2019**, *101*, 708–723.
3. Hannan, M.A.; Abu, S.M.; Al-Shetwi, A.Q.; Mansor, M.; Ansari, M.N.M.; Muttaqi, K.M.; Dong, Z.Y. Hydrogen energy storage integrated battery and supercapacitor-based hybrid power system: A statistical analysis towards future research directions. *Int. J. Hydrogen Energy* **2022**, *47*, 39523–39548. [CrossRef]

4. Levene, J.; Kroposki, B.; Sverdrup, G. Wind Energy and Production of Hydrogen and Electricity—Opportunities for Renewable Hydrogen. In Proceedings of the 2006 POWER-GEN Renewable Energy and Fuels Technical Conference, Las Vegas, NV, USA, 10–12 April 2006.

5. Becerra, M.; Moran, J.; Jerez, A.; Cepeda, F.; Valenzuela, M. Wind energy potential in Chile: Assessment of a small-scale wind farm for residential clients. *Energy Convers. Manag.* **2017**, *140*, 71–90. [CrossRef]

6. Singh, M. *Dynamic Models for Wind Turbines and Wind Power Plants*; National Renewable Energy Lab. (NREL): Golden, CO, USA, 2011; p. 115.

7. Ulleberg, Ø. The importance of control strategies in PV-hydrogen systems. *Sol. Energy* **2004**, *76*, 323–329. [CrossRef]

8. International Renewable Energy Agency (IRENA). Available online: https://www.irena.org/ (accessed on 15 August 2023).

9. Li, X.; Gao, W.; Feng, Y. A review of hybrid renewable energy systems: A new trend in energy production. *Renew. Sustain. Energy Rev.* **2020**, *132*, 110088.

10. U.S. Department of Energy. 2021. Available online: https://www.energy.gov/eere/ (accessed on 10 July 2023).

11. Bhatt, A.; Chatterjee, P. Hybrid Renewable Energy Systems (HRES) for Sustainable Energy Development. *Energies* **2020**, *13*, 3426.

12. Okunlola, A.; Davis, M.; Kumar, A. The development of an assessment framework to determine the technical hydrogen production potential from wind and solar energy. *Renew. Sustain. Energy Rev.* **2022**, *166*, 112610. [CrossRef]

13. National Renewable Energy Laboratory. Hybrid Wind Photovoltaic Systems. 2020. Available online: https://www.nrel.gov/docs/fy20osti/74251.pdf (accessed on 18 July 2023).

14. Ishaq, H.; Dincer, I. Comparative assessment of renewable energy-based hydrogen production methods. *Renew. Sustain. Energy Rev.* **2021**, *135*, 110192. [CrossRef]

15. Hassan, Q.; Sameen, A.Z.; Salman, H.M.; Jaszczur, M. Large-scale green hydrogen production via alkaline water electrolysis using solar and wind energy. *Int. J. Hydrogen Energy* **2023**, *48*, 34299–34315. [CrossRef]

16. Hassan, N.S.; Jalil, A.A.; Rajendran, S.; Khusnun, N.F.; Bahari, M.B.; Johari, A.; Kamaruddin, M.J.; Ismail, M. Recent review and evaluation of green hydrogen production via water electrolysis for a sustainable and clean energy society. *Int. J. Hydrogen Energy* **2023**. [CrossRef]

17. Nelson, D.B.; Nehrir, M.H.; Wang, C. Unit sizing and cost analysis of stand-alone hybrid wind/PV/fuel cell power generation systems. *Renew. Energy* **2006**, *31*, 1641–1656. [CrossRef]

18. Wijaya, A.; Akbar, H.I.; Winardi, S. Modeling and simulation of a standalone hybrid wind-solar-battery system for hydrogen production. *Int. J. Hydrogen Energy* **2021**, *46*, 3245–3257.

19. Halder, P.; Babaie, M.; Salek, F.; Haque, N.; Savage, R.; Stevanovic, S.; Bodisco, T.A.; Zare, A. Advancements in hydrogen production, storage, distribution and refuelling for a sustainable transport sector: Hydrogen fuel cell vehicles. *Int. J. Hydrogen Energy* **2023**. [CrossRef]

20. MATLAB Program. Available online: https://www.mathworks.com/products/matlab.html (accessed on 20 December 2022).

21. Roche, V.; Laurent, V.; Cardello, G.L.; Jolivet, L.; Scaillet, S. Anatomy of the Cycladic Blueschist Unit on Sifnos Island (Cyclades, Greece). *J. Geodyn.* **2016**, *97*, 62–87. [CrossRef]

22. Zeng, K.; Zhang, D. Recent progress in alkaline water electrolysis for hydrogen production and applications. *Prog. Energy Combust. Sci.* **2010**, *36*, 307–326. [CrossRef]

23. Dekel, D.R. Review of cell performance in anion exchange membrane water electrolysis. *J. Appl. Electrochem.* **2018**, *48*, 427–462.

24. Park, J.; Kim, J.H.; Kim, J.H. Review of high-temperature solid oxide electrolysis for hydrogen production. *Int. J. Hydrogen Energy* **2020**, *45*, 10291–10307. [CrossRef]

25. Narayan, R.; Cheruvally, G. Temperature effects on the efficiency of hydrogen production by water electrolysis—A review. *Int. J. Hydrogen Energy* **2017**, *42*, 19397–19414. [CrossRef]

26. Badwal, S.P.S.; Giddey, S.S.; Munnings, C.; Bhatt, A.I. Review of progress in high-temperature electrolysis and steam electrolysis for hydrogen production. *Int. J. Hydrogen Energy* **2014**, *39*, 10362–10384. [CrossRef]

27. Greiner, C.J.; Korpas, M.; Holen, A.T. A Norwegian case study on the production of hydrogen from wind power. *Int. J. Hydrogen Energy* **2007**, *32*, 1500–1507. [CrossRef]

28. Tar, K. Some statistical characteristics of monthly average wind speed at various heights. *Renew. Sustain. Energy Rev.* **2008**, *12*, 1712–1724. [CrossRef]

29. Vestas Wind Turbine. Available online: https://www.vestas.com/en (accessed on 10 May 2023).

30. Bodie, Z.; Kane, A.; Markus, A. *Investments*; McGraw-Hill Education: New York, NY, USA, 2014.

31. Anastasiadis, A.; Oikonomou, I.; Vokas, G. Optimal Levelised Cost of System Values with increasing Renewable Energy Sources in a Smart Microgrid. *AIP Conf. Proc.* **2019**, *2190*, 020060. [CrossRef]

32. Financial Advisory and Asset Management LAZARD. Available online: https://www.lazard.com/financial-advisory/ (accessed on 20 August 2023).

33. PVGIS Software. Available online: https://joint-research-centre.ec.europa.eu/photovoltaic-geographical-information-system-pvgis_en (accessed on 15 August 2023).

34. Anastasiadis, A.G.; Konstantinopoulos, S.A.; Kondylis, G.P.; Vokas, G.A.; Papageorgas, P. Effect of fuel cell units in the economic and environmental dispatch of a Microgrid with penetration of photovoltaic and microturbine units. *Int. J. Hydrogen Energy* **2017**, *42*, 3479–3486. [CrossRef]

35. Hatziargyriou, N.D.; Anastasiadis, A.G.; Tsikalakis, A.G.; Vasiljevska, J. Quantification of economic, environmental and operational benefits due to significant penetration of Microgrids in a typical LV and MV Greek network. *Eur. Trans. Electr. Power Eur. Trans. Electr. Power* **2011**, *21*, 1217–1237. [CrossRef]
36. Anestis, A.; Georgios, V. Economic Benefits of Smart Microgrids with Penetration of DER and mCHP Units for Non-Interconnected Islands. *Renew. Energy Int. J.* **2019**, *142*, 478–486. [CrossRef]
37. Lazard Site. Available online: https://www.lazard.com/ (accessed on 20 August 2023).
38. International Energy Agency. *World Energy Outlook 2020*; International Energy Agency: Paris, France, 2020.
39. Toyota Mirai Car Site. Available online: https://www.toyota.com/mirai/ (accessed on 12 August 2023).
40. Eurostat. Electricity Price Statistics. Available online: https://ec.europa.eu/eurostat/statisticexplained/index.php?title=Electricity_price_statistics (accessed on 8 September 2023).
41. Hosseini, S.E.; Wahid, M.A. Hydrogen production from renewable and sustainable energy resources: Promising green energy carrier for clean development. *Renew. Sustain. Energy Rev.* **2016**, *57*, 850–866. [CrossRef]
42. Noussan, N.; Raimondi, P.P.; Scita, R.; Hafner, M. The Role of Green and Blue Hydrogen in the Energy Transition—A Technological and Geopolitical Perspective. *Sustainability* **2021**, *13*, 298. [CrossRef]
43. Dincer, I.; Acar, C. Innovation in hydrogen production. *Renew. Sustain. Energy Rev.* **2017**, *42*, 14843–14864. [CrossRef]
44. Nikolaidis, P.; Poullikkas, A. A comparative overview of hydrogen production processes. *Renew. Sustain. Energy Rev.* **2017**, *67*, 597–611. [CrossRef]

Article

A Methodological Approach to the Simulation of a Ship's Electric Power System

Igor P. Boychuk [1], Anna V. Grinek [2], Nikita V. Martyushev [3], Roman V. Klyuev [4,*], Boris V. Malozyomov [5], Vadim S. Tynchenko [6,7,8], Viktor A. Kukartsev [9], Yadviga A. Tynchenko [6,7] and Sergey I. Kondratiev [10]

[1] Department of Higher Mathematics and Physics, Marine Engineering Faculty, Admiral Ushakov Maritime State University, 353918 Novorossiysk, Russia; boychuk@aumsu.ru

[2] Department of Operation of Ship's Electrical Equipment and Automatic Devices, Marine Engineering Faculty, Admiral Ushakov Maritime State University, 353918 Novorossiysk, Russia; grinyokann@gmail.com

[3] Department of Advanced Technologies, Tomsk Polytechnic University, 30, Lenin Ave., 634050 Tomsk, Russia; martjushev@tpu.ru

[4] Department "Technique and Technology of Mining and Oil and Gas Production", Moscow Polytechnic University, 33, B. Semenovskaya Str., 107023 Moscow, Russia

[5] Department of Electrotechnical Complexes, Novosibirsk State Technical University, 20, Karl Marks Ave., 630073 Novosibirsk, Russia; borisnovel@mail.ru

[6] Department of Technological Machines and Equipment of Oil and Gas Complex, School of Petroleum and Natural Gas Engineering, Siberian Federal University, 660041 Krasnoyarsk, Russia; t080801@yandex.ru (Y.A.T.)

[7] Information-Control Systems Department, Institute of Computer Science and Telecommunications, Reshetnev Siberian State University of Science and Technology, 660037 Krasnoyarsk, Russia

[8] Artificial Intelligence Technology Scientific and Education Center, Bauman Moscow State Technical University, 105005 Moscow, Russia

[9] Department of Materials Science and Materials Processing Technology, Polytechnic Institute, Siberian Federal University, 660041 Krasnoyarsk, Russia

[10] Department of Ships Navigation, Faculty of Water Transport Operation and Navigation, Admiral Ushakov Maritime State University, 353918 Novorossiysk, Russia

* Correspondence: kluev-roman@rambler.ru

Citation: Boychuk, I.P.; Grinek, A.V.; Martyushev, N.V.; Klyuev, R.V.; Malozyomov, B.V.; Tynchenko, V.S.; Kukartsev, V.A.; Tynchenko, Y.A.; Kondratiev, S.I. A Methodological Approach to the Simulation of a Ship's Electric Power System. *Energies* **2023**, *16*, 8101. https://doi.org/10.3390/en16248101

Academic Editors: Konstantinos Aravossis and Eleni Strantzali

Received: 30 October 2023
Revised: 4 December 2023
Accepted: 14 December 2023
Published: 16 December 2023

Abstract: Modern ships are complex energy systems containing a large number of different elements. Each of these elements is simulated separately. Since all these models form a single system (ship), they are interdependent. The operating modes of some systems influence others, but at the same time, the work of all the systems should be aimed at fulfilling the basic functions of the ship. The work proposes a methodological approach to combining various systems of ships into a single complex model. This model allows combining models of ship systems of various levels (microlevel, macrolevel, metalevel, megalevel). The work provides examples of models of such multi-level energy systems. These are energy systems composed of an electric generator, a diesel engine, a propeller shaft, and algorithms used for operating the common parts of the ship's electric power system and a piston wear process. Analytical, structural, numerical, and object-oriented models were made for these objects. Each of these particular models describes a limited class of problems, has characteristic properties, and a mathematical structure. The work shows how particular models can be interconnected using a set-theoretic description. Particular models are combined into macrolevel models, whose output parameters are quantities that are by no means related. The macrolevel models are interrelated using control models. Control models belong to the metalevel and allow for assigning settings and response thresholds to algorithms used in automation systems. Such a model (megalevel model) allows, ultimately, investigating the dynamics of the entire system as a whole and managing it.

Keywords: electric power system; ship; mathematical model; SimInTech 2020; simulation; simulator; main distribution board; service life

1. Introduction

To date, modern ships are, in essence, large energy systems having an intricate interaction among their components. Simulating the behavior of complex systems represents a certain art. A mathematical description of a rather complex system, which is a modern ship, does not allow for implementing the ship model in general. The totality of mechanisms, machines, energy sources, heat exchangers, pipelines, and electrical subsystems designed to organize the possibility of ship movement and the transfer of various types of energy to its various consumers is a shipboard power plant. The power plant includes a main power plant, which drives the ship, consisting of the main engine, a ship propulsion, an electric generator, and a shaft line. There are a number of individual energy models, each of which provide a solution to a well-defined range of specific questions concerning the system's behavior. At the same time, each of the models has its own mathematical structure, characteristic mathematical properties, and is suitable for studying a certain class of problems related to the system's functioning. Since all these models form a single system (ship), they are interdependent. The operating modes of some systems influence others. However, at the same time, the work of all the systems should be aimed at performing the main functions of the ship. Therefore, the task of combining all the disparate models of the ship systems within a single model is extremely relevant. The solution to this problem is possible using different types of mathematical description [1].

At present, the need for creating complex mathematical models is steadily growing. Complex simulation represents a complex interdisciplinary analysis and an interaction of design stages that requires new approaches and design tools.

From the viewpoint of a unified position, the full coverage and the simulation of all the ship systems are rather a difficult and, generally speaking, currently unrealizable task. At the same time, considering some systems separately, as well as many separate systems along with the interaction occurring between the parts, implements the so-called integrated approach to simulation. For instance, the mathematical simulation of the processes proceeding in the ship's electrical power and the mechanical system is of major importance in solving the tasks of designing new types of ships, research, management, and reliability assessment.

The multiple connectivity and complexity of such systems complicate the approach to complex simulation. This fact is explained by the difference and the diversity of the physical processes and the mathematical description complexity of these processes.

The modern literature reflects a number of studies conducted in each local area that are of interest when constructing such a large system.

The works [2,3] demonstrate the use of new methods intended for controlling the ship's power consumption, including the hybrid methods based on neuro-fuzzy systems, namely on the ANFIS model. The authors show that the transition to a flexible intelligent system, in comparison with classical control modeling methods, has a greater economic effect, reducing harmful emissions (if a similar approach is applied to fuel systems).

The connection between the input and output parameters of the simulation can be of a reverse nature in the "design-construction-repair" chain. The issues of structural analysis of the shaft line constructions allow for optimizing the construction at the stage of the ship project in terms of vibration problems. The propeller shaft system experiences a complex nature of periodic loadings. For example, the authors in [4,5] use the finite element modal analysis to determine the range of forbidden frequencies of the main engine and to build an optimal scheme for installing bearings. The work [6] presents the results of studying an analytical model of axial vibrations of the shaft line that helps to improve the design of ship vibration isolators.

The creation of a digital twin of mechanical power systems should fully cover all the stages. The results of simulating the behavior scenario of an object of a ship's electric power system (SEPS) are embedded in the process of managing and training the specialists [7–9]. The methods used for accelerating the modeling of the onboard simulation processes of a ship's electric power system using multicore computing are described in [10]. Approaches to improving the performance of software and hardware methods are presented in [11,12].

The geometric simulation followed by the finite element analysis and the introduction of CAD tools significantly optimizes the design activity [13].

The research [14] demonstrates using a finite element model of a synchronous generator to simulate and identify a transient process, spectral analysis of the ship network, defects (identifying the current and voltage spectrum of the faulty generator), and a variant of arranging the diagnostics system of the ship's electric machines.

In the case of complex systems, the application of intelligent systems accompanied by reinforcing learning is effective as shown in [15]. When the model-free control is used (for example, the ship power supply), the constant replenishment of the knowledge about the object allows for predicting and managing under conditions of uncertainty and increasing the safety level [16,17].

The similar diagnostic models of the state of an internal combustion engine that were constructed using neural networks (for example, those mentioned in [18]) can also be used for the ship systems supplemented with approximation dependencies relating to the wear of the piston rings to operating modes.

An analytical assessment of the functioning reliability of the heat engines is hindered due to the difficulty in finding a function that correlates the wear of engine elements with the operating modes [19], unlike, for example, turbo engines.

In addition, the intelligent models that work with a database of the current modes and equipment operation models, so-called digital twins, complex models that display the relationship of the model with real operating conditions, are very relevant [20–22].

In [23–25], the operation reliability of the systems is increased owing to use of intelligent models based on the digital twins. Reliability is enhanced by the introduction of a self-diagnosis system that compares the real and generated data. This approach is being investigated by the example of the engines, bearing systems, batteries, etc. Processing the signals of various individual systems of the ship is quite a difficult task. The combination of a complex structure and a nonstationary operating mode makes the measured signal of the machine multicomponent have a complex, time-varying spectral structure. In this paper, we propose an implementation of the analysis method based on adaptive synchronous demodulation transform (ASDP). The analysis is performed by searching and constructing the demodulation term element based on the analysis of target signals of spectral data [26]. The overall goal is to efficiently analyze the frequency–time characteristics of the target signals. Using adaptive synchronous demodulation and continuous wavelet transform, the method aims to capture the time-varying frequency components of the signal. A three-dimensional matrix facilitates the visualization and interpretation of the time–frequency information [27].

In general, it is worth noting that the tasks of modeling various systems of ships and the movement of the ships themselves [28] are quite numerous, complex, and very heterogeneous.

The general disadvantages of the mentioned literature sources are as follows. The first is that the analytical approaches are used to simplify the analysis, but the applicability of such approaches is questionable. The second is the fact that the presented models are local in nature and they describe particular physical processes. The third is an issue that there is no attempt in combining the models into a multiple-model structure.

Combining the multiple-model complexes to include the above models necessitates understanding the mathematical description of a particular object and can be approximated by a formal model. The model structure of the object or level can be formalized and generated in the desired way in accordance with the modeling problems (research, optimization, or management) [29]. The development of the structures similar to that shown in Figure 1 should be noted to illustrate the complex process of a multi-criterion selection, which requires a constant coordination of the structure levels and a heterogeneous representation of the objects of the ship system.

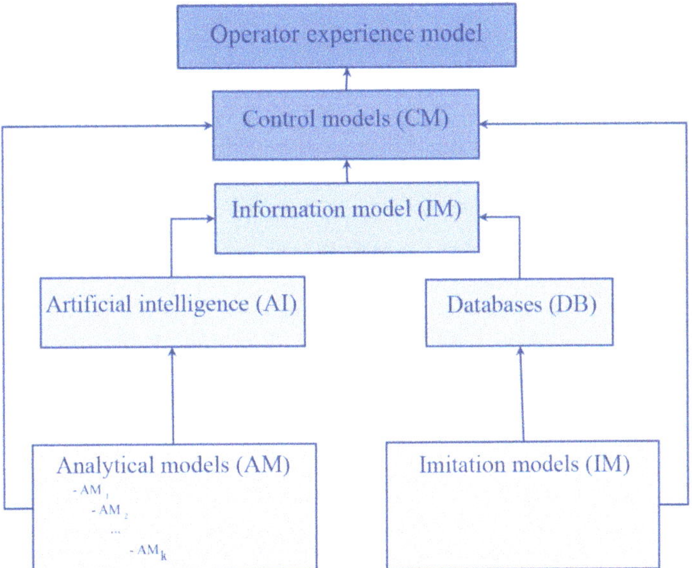

Figure 1. The simulation system structure.

An analysis of the literature shows that to date there is no single model connecting the following ship systems:

- Mechanical system of power plants (propeller shafting);
- Regression experimental models of wear of main engine elements;
- Diagnostic models of the state of the electrical equipment (synchronous generator);
- Models used for controlling the electric power system;
- Algorithmic models used for controlling the main distribution board of a ship.

In this regard, the objective of this study is to develop a methodology for simulating a complex single ship model, which includes a set of subsystems of various levels (microlevel, macrolevel, metalevel, megalevel). The paper shows how it is possible, using the developed approach, to create a digital twin of a ship as a large unified system, rather than a set of separate models. Such a complex model is constructed by the example of the relationship established among mechanical, electric power, algorithmic models and the control model. To achieve the purpose in view, the following tasks must be solved:

1. To develop local mathematical models of various mathematic levels:

   - Model of the stress–strain state of the propeller shafting and natural frequencies;
   - Model of the wear of piston rings in the main engine;
   - Models of the transient process occurring in the electric motor;
   - Model of the electromagnetic processes of a synchronous generator, as well as the identification of defects;
   - Model of controlling the main distribution board.

2. To determine the relationship between modeling levels and service life stages, a unified model of the interaction occurring among the ship systems should be built.

In this study, an approach is proposed that connects the models of the elements of the ship's electric power system with the functional capabilities. Below, the work presents in detail the experience of simulating various ship systems and a mathematical model allowing the combining of the models into one large system.

The main task of the scientific work was to develop a set of mathematical models combined into a system capable of describing in aggregate the electrical and mechanical processes proceeding on a ship of various modes. The paper presents the newly developed, complex, polymodel approach to modeling a ship as a simulation system. The general polymodel approach to the description of a ship as a large system allows for presenting the mathematical models of the objects, the processes, and the dynamics of the system as a whole from a common viewpoint. The partial verification was carried out for the following models: the shafting, generator, the ring wear of the container ship COSCO Long Beach (IMO 9285677) built by HHI in 2004. It should be noted that the ring wear model was originally built using the experimental data obtained from the real ship.

2. Materials and Methods

Technological objects may demonstrate different properties in different situations, and may not have adequate models in terms of classical differential equations. The objects can have a continuously discrete behavior, containing stochastic constituents in the varying parameters. Neglecting these facts does not facilitate the construction of adequate analytical models of the objects.

Simulation modeling provides a mathematical apparatus (a formalized scheme) that generates the desired structure of the object model that meets the purposes of the study, filling this structure with quantitative relationships that describe the relationships among its elements, thus solving various problems of the analysis and selection. The creation of SM, as well as simulation models (simulation-level models), is a complex multi-stage iterative process, whose main peculiarity (as compared to "purely" simulation modeling) consists in the necessity to coordinate different models describing different aspects of the object functioning at each of the research stages (at the conceptual, algorithmic, informational, and software levels).

This work shows a methodological approach to a complex simulation of complex systems and presents the experience of using a wide class of methods and tools intended for simulating elements, the electric power, and the mechanical systems of a ship. Heterogeneous models are related through the apparatus of the set theory. The elements of the corresponding sets (models) and the relations connecting them are determined by the specifics of the system. Such a description of the system allows for an analysis of the structure of the system and its behavior. Various systems containing heterogeneous constituent elements can be investigated by means of this methodology.

The simulation objects in this work are:

- Mechanical systems (a built-up construction of the propeller shaft line consisting of bearings and a system of propeller shafts having a flange connection);
- Process of wear of the piston rings of a diesel engine;
- Dynamic processes occurring in a synchronous generator;
- Current and voltage spectra of a synchronous generator with and without a defect;
- Rules and algorithms of the functioning of the main distribution board of the ship: diesel generators, electrical networks, consumers;
- Short circuit processes proceeding in electrical machines, emergency modes.

These models generally reflect the ship behavior as a large multivariable system.

Different approaches are used to develop local mathematical models depending on the model specifics. The simulation itself is performed using specific software suites and tools of these suites. These models are combined into one system by means of a set-theoretical approach. A mathematical description of each of the models, including a description of the dynamics of each of the systems, is provided in the framework of the approach. In this case, the output parameters of one model are the input parameters of the other.

2.1. The Theoretical-Set Description of the System

A wide range of the objects at various simulation levels was used in these studies:

1. Simulation microlevel:
   - Three-dimensional geometric models of the elements of the propeller shafting and the generator;
   - Forms and frequencies of natural vibrations of mechanical systems;
   - Stress–strain state of the elements of the propeller shafting system;
   - Static and dynamic electromagnetic processes occurring in a synchronous generator.
2. Simulation macrolevel:
   - Generator diagnostic model;
   - Diagnostic model of piston ring wear.
3. Simulation metalevel:
   - Algorithms used for controlling the main distribution board and the ship's electric power;
   - Generator control model;
   - Load models of ship consumers.

The local models presented in the study are combined into one set-theoretical model. The scheme of connecting various systems and their interaction is shown in Figure 2. The purpose of this scheme is to show the mechanisms of the interaction occurring among the models of various levels (microlevel, macrolevel, metalevel, megalevel).

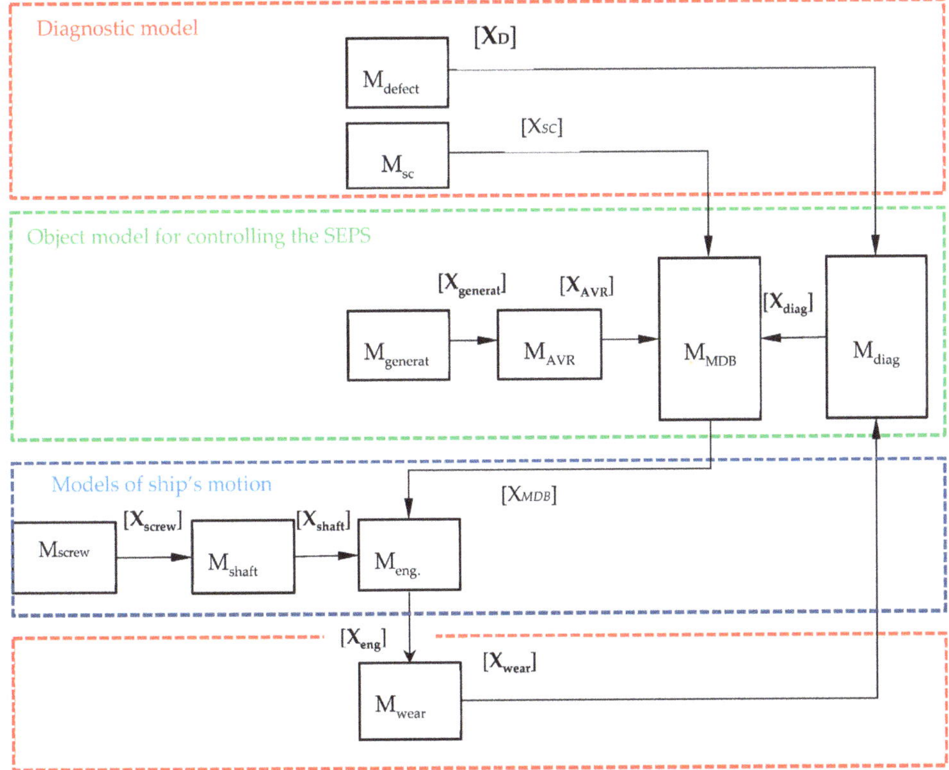

Figure 2. The generalized scheme of the interaction occurring among the models of motion, diagnostics, and system control of the ship.

The following notations are provided in the figure:

$X_D = \begin{bmatrix} I_D, & U_D, & f_D \end{bmatrix}^T$—state vector of the parameters of the synchronous generator model having a mechanical defect: phase currents, phase voltages, and spectral frequency;

$X_{SC} = \begin{bmatrix} I_{SC}, & U_{SC}, & f_{SC} \end{bmatrix}^T$—vector of the parameters of the synchronous generator having a short circuit: phase currents, voltages, and spectral frequency;

$X_{generator} = \begin{bmatrix} I_g, & U_g, & w_g, & \omega_g \end{bmatrix}^T$—vector of synchronous generator parameters: current, voltage, power, and speed;

$X_{AVR} = \begin{bmatrix} I_e, & U_g \end{bmatrix}$—vector of the parameters of the synchronous generator voltage regulator: exciter current and generator voltage;

$X_{MDB} = \begin{bmatrix} P_{inst}, & Q \end{bmatrix}^T$—vector of internal combustion engine parameters: instantaneous power and consumption;

$X_{diag} = \begin{bmatrix} f_{mech.def}, & I_{mech.def}, & U_{mech.def}, & f_{elec.def}, & I_{elec.def}, & U_{elec.def} \end{bmatrix}^T$—vector of the parameters for diagnostic algorithms: frequencies, currents, and voltages associated with a mechanical and electrical defect of the synchronous generator;

$X_{wear} = \begin{bmatrix} I_{rings} \end{bmatrix}$—wear values of the different rings of the engine cylinders;

$X_{shaft} = \begin{bmatrix} \sigma, & \varepsilon, & f_h \end{bmatrix}^T$—vector of the parameters of stresses, displacements, and natural frequencies of the propeller shaft system;

$X_{screw} = \begin{bmatrix} M_{rev}, & M_{stat}, & M_{dyn} \end{bmatrix}^T$—vector of values of the torques of external forces caused by the propeller screw: reversible, dynamic, and static torques.

An example of calculating the parameters of these vectors is shown in Section 3 of this work.

The constructed model represents a deterministic, nonlinear, nonstationary, finite-dimensional, differential, dynamic system. The figure shows the relationship of the models $M_{defect}, M_{sc}, M_{generat}, M_{AVR}, M_{MDB}, M_{diag}, M_{screw}, M_{shaft}, M_{eng}, M_{wear}$, which are part of the generalized model (Figure 1).

The set theory allows for operating with mathematical abstractions and the relationships between their elements. The elements of the corresponding sets and the relations connecting them are determined by the system specifics. There are two sets X, Y, whose elements are associated with the ship. The relationship between the sets is described using binary relations, which are represented by the matrix of incidents (1). From a geometric point of view, the relations define a simplified complex scenario where the elements of the Y set are considered as vertices, and the elements of the X set are simplexes. Therefore, the Y_1 element (generator) is a 0-simplex consisting of the X_4 vertex (wear).

Let the system Σ (ship) contain a set of the investigated objects Y and a set of the states X, where:

Y = {generator, shaft, rings, main distribution board (MDB)},

X = {shaft eccentricity, turn-to-turn fault, modes, harmonics, wear}.

The studied states are the values of the parameters of various ship systems (resonant shaft frequencies, the generator operation frequency, etc.). The incidence matrix (1) shows whether or not there is a relationship between these objects and their states, where 1—there is a connection, 0—there is no connection.

Having an incident matrix Λ such that $(\Lambda)_{ij} = \begin{cases} 1, & (y_i, x_j) \in \lambda \\ 0, & (y_i, x_j) \notin \lambda \end{cases}$, where $x \in X, y \in Y$. In this case:

$$\Lambda = \begin{bmatrix} 1 & 1 & 0 & 1 & 1 \\ 0 & 0 & 1 & 1 & 1 \\ 0 & 0 & 0 & 0 & 1 \\ 1 & 1 & 1 & 1 & 0 \end{bmatrix} \tag{1}$$

The set theory provides an opportunity to operate with mathematical abstractions and relations between their elements. The elements of the corresponding sets and the relations connecting them are determined by the system specifics. There are two X and Y sets, whose elements are related to the ship. The connection between the sets is described by means of binary relations, which are represented by the matrix of incidents (1). Geometrically, the relations define a simplistic complex in which the elements of the Y set are treated as vertices and the elements of the X set are simplexes. In view of this, the element Y_1 (generator) is a 0-simplex consisting of the vertex X_4 (wear and tear). The system dynamics is described by the notion of the image Π. The image Π is a mapping, which attributes a certain number to each simplex, for example, the wear value of the engine rings (Π_0). Since each simplex has a geometric dimension, the image Π is a ranked image (2), i.e., the sum of images Π_i.

A complete graded pattern Π of this system has the form:

$$\Pi = \Pi_0 \oplus \Pi_1 \oplus \Pi_2 \oplus \Pi_3, \tag{2}$$

where:

$$
\begin{aligned}
\Pi_0 &: \{y_3(rings)\} \to wear\ ratio \\
\Pi_1 &: empty \\
\Pi_2 &: \{y_2(shaft)\} \to wear\ ratio \\
\Pi_3 &: \{y_1(generator)\} \to number\ of\ failures \\
&\quad\ \{y_4(MDB)\} \to number\ of\ failures
\end{aligned}
$$

The system dynamics will be described as a change in the Π image at each moment of time [30]. For this system, these changes will be forces acting on the fixed geometry of the complex, specified by certain models.

These models can be classified by the simulating type and by hierarchy. That is, the models can be attributed to certain micro, macro, and metalevels. A complex approach is presented to solve the set tasks. These tasks are solved by means of a multicomponent model complex, which includes blocks containing incoming model classes.

2.2. The Mathematical Description of the Models

The mathematical description of the models $y_1 - y_3$ is internal and can be presented in the general case by a differential equation of the form:

$$M_g = \left\{ \vec{x}(t) \Big| \dot{\vec{x}} = f(\vec{x},t);\ \vec{h}_0(\vec{x}(t_0)) \le O, \vec{h}_1(\vec{x}(t_f)) \le O, \vec{q}(\vec{x},t) \le O \right\}, \tag{3}$$

where $\vec{x}$ is a vector, characterizing the state of motion of the elements; $\vec{q}$, $\vec{h}_0$, $\vec{h}_1$ are known vector functions through which the restrictions to the motion process of the objects and boundary conditions are set. At the same time, the dynamics of the system $f(\cdot,\cdot)$ are set in various forms.

Equation (3) describes the dynamics of the models, which are generally represented by a differential equation. In the equation, the derivatives of the state of the element are on the left side, and the applied load is on the right. For example, the deformation rate of the shaft depends on the applied force load.

The mathematical description of the model y_4 is external and can be presented in the form of the generalized mathematical model of the processes of controlling the MDB functioning:

$$M = \left\{ \vec{u}(t) \Big| \dot{\vec{x}} = f(\vec{x},\vec{u},t);\ \vec{h}_0(\vec{x}(t_0)) \le O, \vec{h}_1(\vec{x}(t_f)) \le O, \vec{q}^{(1)}(\vec{x},\vec{u}) = O, \vec{q}^{(2)}(\vec{x},\vec{u}) < O \right\}, \tag{4}$$

where $\vec{x}, \vec{u}$ are the generalized vectors of the state and control of MDB; $\vec{h}_0$, $\vec{h}_1$ are known vector functions, through which the boundary conditions for the vector $\vec{x}$ are set at time points t_0, t_f; $\vec{q}^{(1)}, \vec{q}^{(2)}$ are vector functions, setting the main space–time, technical, and technological constraints imposed on the functioning process.

Equation (4) represents the control models. In such models, the state of the model elements depends, in addition to the power component, also on the control action. Such models include, for example, the generator controller model.

Therefore, each model has a vector of input and output parameters. The input and output parameters depend on what model type the system is described as a whole. The dynamic models are described by the equations of type (3); the control models are described by the equations of type (4).

As a result, Equation (3) describes the dynamics of the models, which are generally represented by a differential equation. In the equation on its left side, there are derivatives from the state of the element, and the applied load is on the right side. For example, the shaft deformation speed depends on the attached power load.

Equation (4) is a control model. In such models, the state of the model elements depends, except for the power component, also on the controlling exposure. Such models include, for example, the generator regulator model.

In each specified case, for the given objects of the equation, a specific type (for example, Equations (5) and (6)) will describe the dynamics of the generator and the gross. In this work, obtaining optimal management was not a goal. Optimal management is the next stage in the development of this approach.

3. Results

This section presents the results of simulating the various elements of a ship, whose dynamics is described mainly by the equations of the form (2) and (3), considering the specifics of simulating each element. Specific mathematical formulations are provided for each of the considered elements of the system.

The simulation was performed for a ship of the transport fleet such as a tanker. This ship is a system that moves mechanically, whose synchronous generator is the main one. The simulated ship has the main ship low-speed dual-fuel diesel engine manufactured by Hyundai Heavy Industries (Ulsan, Republic of Korea) under a license granted by MAN Energy Solution (COSCO Long Beach (IMO 9285677).

3.1. The Development of a General Scheme of the Interaction among the Models of Various Ship Systems

A complex model of a ship as an energy-mechanical object can be obtained by combining analytical, simulation, logical, and algorithmic models, and it represents a simulation system (SS). Combining the isolated models of individual ship nodes facilitates solving the problem of forming a single model of the functioning of all the ship's systems. This enables performing the basic functions of the ship without failures of individual systems and minimizing emergency situations. In fact, a digital twin of the ship is being built. The block diagram of such a model is shown in Figure 2 [1]. The diagram shows the interaction of different models. At the same time, let us take into account the fact that different models may have different variants of the used solutions (using databases, artificial intelligence). These models transmit data and form a digital twin of the ship (IM). Then, all the necessary parameters of the general condition of the ship are transmitted to the control model (CM). The control model generates the required control actions to achieve the required condition of the ship. It transmits these actions to the ship system controls and to the ship captain. At the same time, the control model (CM) obtains data not only from the ship digital twin (IM), but also directly from low-level models (AM, IM). Examples of the application of this approach to solving particular problems are provided in [13,31,32].

The studies of the local structures and the objects of the ship's electric power system discussed below are still of private nature and describe the physical processes proceeding in the system elements, but can and should be combined into a multiple-model structure as shown in Figure 1.

The simulation system can be generally represented in the form of a specially organized simulation complex consisting of the following elements. The first one refers to the structure of simulation models in its object domain. The second represents the structure of analytical models that provide an accounting of the parameters of modeled objects. The third is a subsystem representing information technology based on a knowledge base, algorithms, and competencies. It is based on expert systems and machine learning. The fourth one is based on the control system and control signal transmission system, which connects all parts of the system and organizes the user interface and dialog with the system operators.

The simulation experience shows that the focused specialization of SS is weakly amenable to the universalization requirements.

At the same time, the hierarchical models are divided into the following levels: a microlevel, a macrolevel, a metalevel, and a megalevel. The connection of the levels and the transition among them are arranged according to certain rules and patterns. Each of the levels has its own solutions concerning the simulation and management of the objects (Figure 3). The microlevel combines the analytical models (AM) shown in Figure 2. An example of such models is the equation of the propeller shaft vibrations, the wear model of the engine piston rings, etc. The macrolevel is simulation models (SM). An example of such models is a generator operation model, a modal analysis of a shaft line accompanied by the determination of natural oscillation frequencies, etc. The metalevel is a control system model (CM); it includes a generator controller model, a voltage regulator model, etc. The megalevel is the termination model that determines the control of the entire system as a whole. This is a simulator model of the main distribution board. More examples of ship systems related to these levels can be found in [13,31,32].

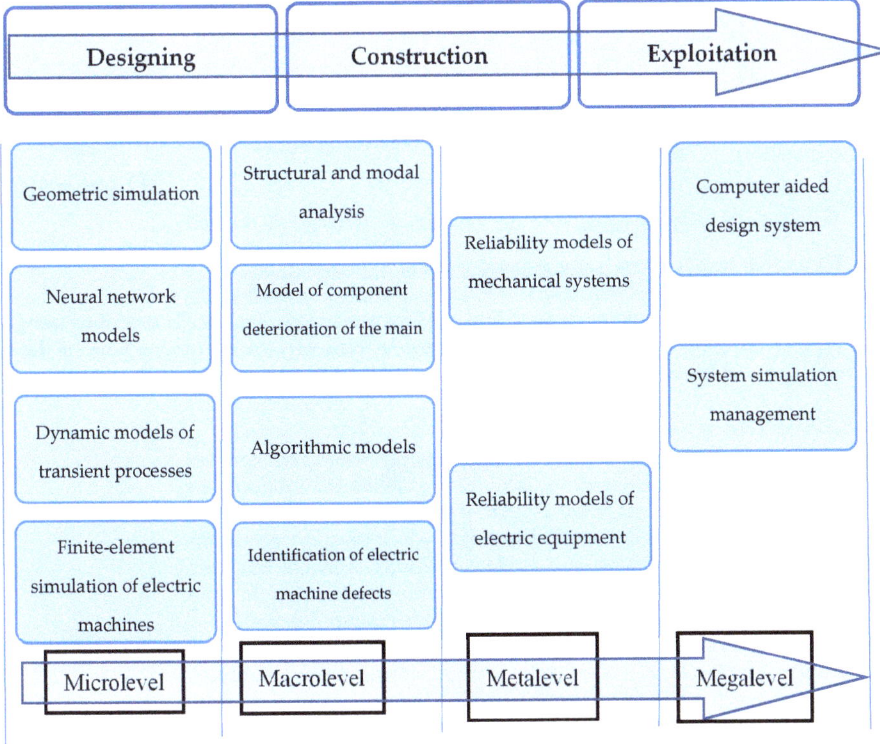

Figure 3. The generalized structure and the composition of the simulation levels of the ship systems.

The design, construction, and operation of the ship are directly related to solving the problem of providing connections and transitions between these levels. Arranging such an interaction represents a difficult task. The labor costs and ship building expenditures will depend on the accuracy of solving this problem. In addition, the interaction between these systems will determine the entire subsequent service life of the ship.

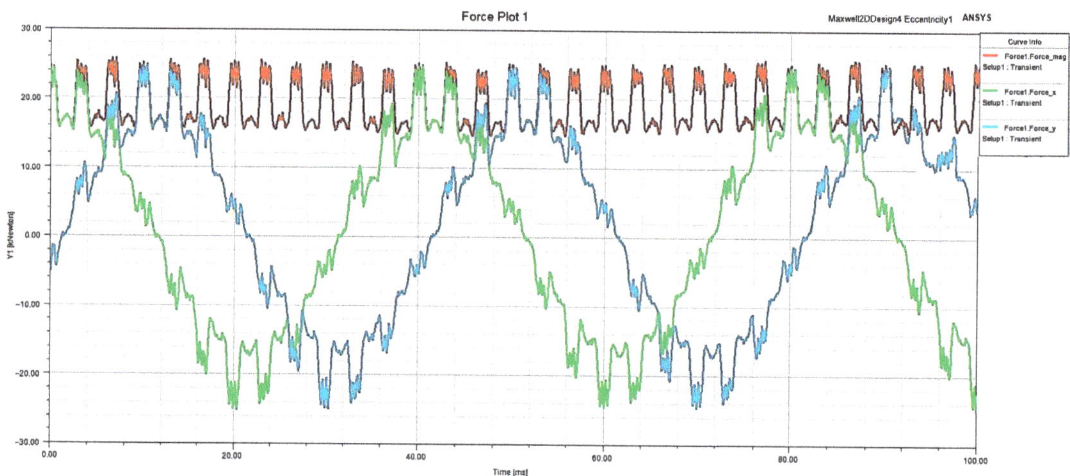

Figure 6. The variation in the force over time of the generator eccentricity.

It is imperative to note that these investigations were conducted through simulation on a synchronous brushless generator. The simulation outputs comprise detailed time oscillograms illustrating the behavior of key parameters such as flux linkage, phase currents, and voltages, as well as forces and torques. This comprehensive analysis provides insights into the dynamic interactions within the generator system, shedding light on the influence of rotor eccentricity on the force vector components and their associated frequencies. The frequency characteristics of the specified generator in idle modes were obtained based on them.

The results of simulating the start-up of the generator and reaching the stationary mode are shown in Figure 7a by the example of changes occurring in phase voltages. The time-and-frequency analysis of the signal was performed using the wavelet transform (Figure 7). The illuminated regions in the wavelet transform coefficients Ca,b, as illustrated in Figure 7b, provide insights into the augmentation of rotor rotation speed. Examination of these coefficients reveals the existence of three distinct frequency domains: the domain associated with the acceleration of speed, the phase leading to the critical speed, and the stable, steady-state mode. A crucial observation is that the intensity of the light areas in the scalogram, corresponding to higher wavelet transform coefficients, directly correlates with the magnitude of energy encapsulated within the frequency component of the series. In essence, the more pronounced the illumination in the scalogram, the greater the energy content of the corresponding frequency component.

This numerical modeling approach provides the ability to identify spectra of voltages, currents, forces, and moments corresponding to both mechanical and electromagnetic defects. Importantly, this model is able to accurately diagnose generator defects caused by rotor eccentricity, even under start-up conditions and in initial operating modes.

With a deeper look at the numerical model, we are able to identify and analyze the spectra of voltages, currents, forces, and moments associated with possible mechanical and electromagnetic faults. This approach allows for predicting and detecting even the smallest deviations associated with generator defects, which makes it unique in the aspect of pre-diagnosis already at the start-up phase and in the initial modes of operation.

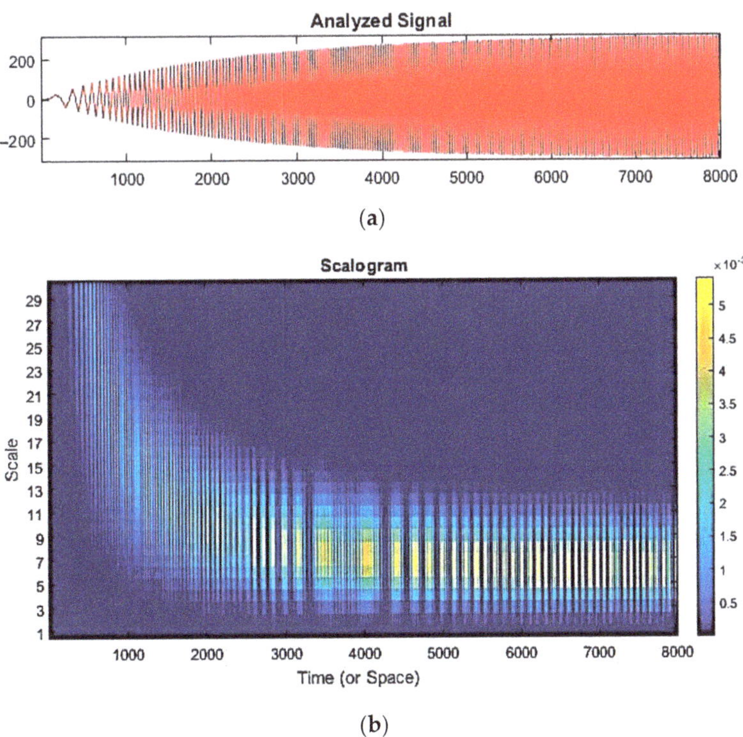

Figure 7. The wavelet transform of the transient process using the Morlet wavelet: (**a**) dynamics of changes in phase voltages, (**b**) scalogram.

Thus, the numerical model provides a high degree of accuracy and sensitivity in detecting rotor-eccentricity-related faults, making it an effective tool for the early diagnosis and prevention of potential generator problems.

3.3. Simulating the Stress–Strain State of the Propeller Shafting and Studying Its Dynamics at the Design Stage

At the design stage, it is important to know the natural resonant frequencies to develop effective design solutions, as well as the purpose of the operating ranges of the main engine [38]. Abnormal vibrations can cause element failures. The task of studying the state and the behavior of composite propulsion plants, where a complex process of interconnected torsional, longitudinal, and transverse vibrations takes place, can be divided into a number of composite problems of statics and dynamics.

In this work, the stress–strain state is calculated and a modal analysis of the constructions is performed using the finite element simulation in the ANSYS environment.

When considering the majority of structural dynamics problems of mechanical systems, the application of spatial discretization using the finite element method allows us to obtain a semi-discrete finite element equation of motion based on the principle of virtual work, and it is as follows:

$$[M]\,\ddot{\vec{u}}\,(t) + [C]\,\dot{\vec{u}}\,(t) + \vec{F}^{\,i}\,(t) = \vec{F}^{\,a}\,(t) \tag{6}$$

where

$[M]$—structural mass matrix;

$\ddot{\vec{u}}\,(t)$—nodal acceleration vector;

$[C]$—structural damping matrix;

$\overrightarrow{\dot{u}}(t)$—nodal velocity vector;

$\overrightarrow{F}^{i}(t)$—internal load vector;

$\overrightarrow{F}^{a}(t)$—applied load vector.

Based on the three-dimensional model of the gross, the finish-element grid was created containing 31,045 nodes. The dynamics of such a system is described by Equation (6). The solution of this equation enables finding the displacement, speed, and acceleration for each node. The solution is revealed using numerical modeling in the ANSYS system. The computing environment implies an ANSYS package.

Equation (6) describes the movement dynamics of the shaft points during the finite element discretization of the shaft line. The system geometry consists of a thrust shaft, intermediate shaft, and propeller shaft (Figure 8). The model parameters are given in Table 1. The shaft line of the ships of the R-1706 series (Samotlor-type tankers) with a diameter of 0.695 m was chosen as a calculation model.

Figure 8. The three-dimensional model of the shafting mechanical system.

Table 1. Model parameters.

Parameter	Thrust Shaft	Intermediate Shaft	Propeller Shaft
Length, mm	1000	1000	1030
Diameter, mm	200	200	200
Volume, mm^3	1.624×10^7	1.624×10^7	1.8308×10^7
Weight, kg	127.49	127.49	143.72

Based on the three-dimensional model, a finite element mesh of the tetrahedral elements was created, having a maximum size of 200 mm, where the number of nodes was 31,045 and the number of elements was 17,839. Bearings set the following boundary conditions: the restriction of movements in a plane perpendicular to the shaft. Bolted connections were "rigid restraints".

The calculation involves solving three problems, which represent statical, modal, and harmonic analyses (Figure 9).

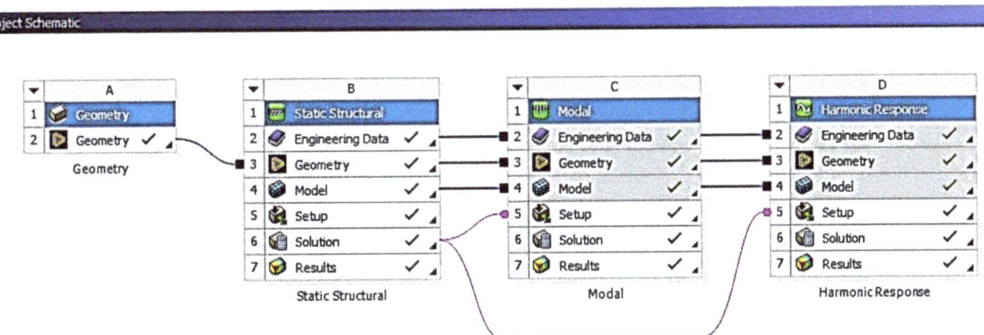

Figure 9. The scheme of the calculation project. (**A**) Design calculation. (**B**) Statistical analysis. (**C**) Preloaded modal calculations. (**D**) Preloaded harmonic calculations.

The geometric and finite element simulation enables the design calculation and a technology design for other ship elements and constructions [39,40].

The statistical analysis of the construction allowed conducting preloaded modal and harmonic calculations.

We used the method of modal analysis to identify frequencies and shapes of natural vibrations of structures. This method can serve as a first step for other forms of dynamic analysis such as transient, harmonic, and spectral analysis. It is important to note that the modal analysis method assumes linearity of the system, and all types of nonlinearities such as nonlinear material behavior, boundary contact conditions, and finite displacements are not considered. Contacts remain either open or closed depending on their initial state, and external forces and damping are assumed to be zero. In the real rotary systems, the bearings are not infinitely rigid. In addition, friction and lubricant add damping to them. The rigidity of the bearings often changes along with the rotation frequency and differs along the coordinate axes. The same applies to damping [41]. In ANSYS, to simulate bearings when calculating the rotary dynamics, there are the COMBI14 or COMBI214 elements, which allows the user to set in each case the desired rigidity and damping coefficients of bearing supports depending on the rotation speed. The description of the complex elastic linkage "shaft-bearing" is based on the elastic interaction model.

The modal analysis at the design stage optimizes the design in terms of resonant frequencies. As a result of the modal analysis, the forms of resonant vibrations of the propeller shaft system and the frequencies that correspond to them were obtained (Figure 10). Based on the simulation results, the amplitude–frequency spectrum of the natural oscillations of the propeller shaft system in various modes was obtained. The forms and the description of the modes are given in [13].

Figure 11 shows the twelfth mode of the natural oscillations of the shaft, as well as the shaft distortion.

After that, a harmonic analysis was conducted to obtain the system frequency response when the efforts were applied. The harmonic analysis showed the presence of high structural stresses near the flanges and platforms on which the bearing was mounted and fixed. The oscillation frequency during operation, according to the calculation, reached 150 Hz, which means that even the first resonant frequency was not reached.

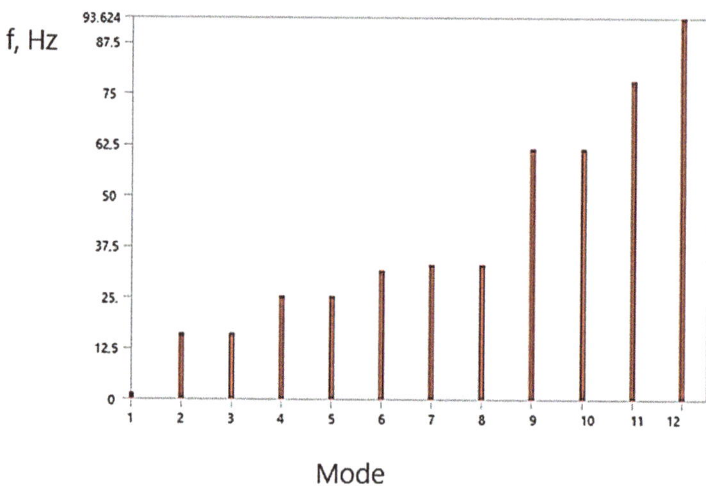

Figure 10. Modal analysis.

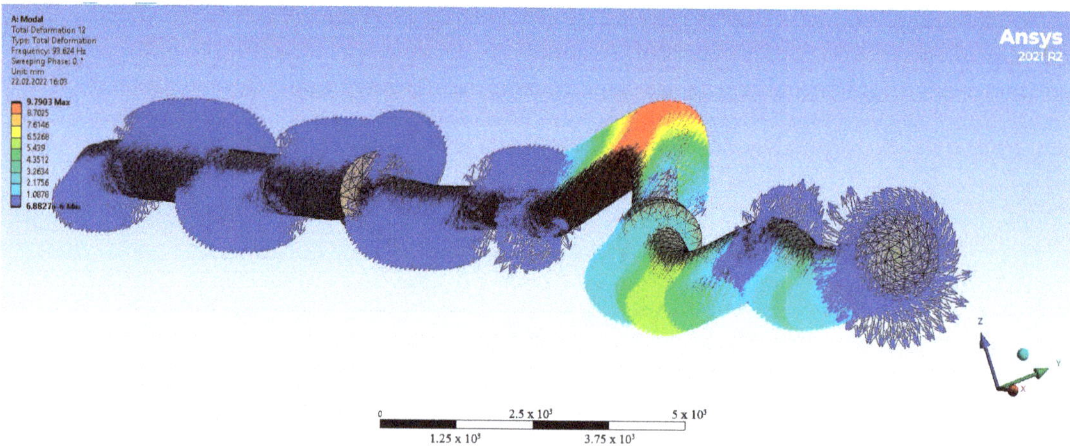

Figure 11. Modal analysis. The twelfth form of natural vibrations of the shaft.

3.4. Developing the Empirical Models of the Wear of Piston Rings of the Main Engine Based on Neural Networks

Describing the complex models that associate the wear of the elements of the ship's electric power system with operating modes is difficult when using analytical methods. In this study, heuristic models based on neural networks, describing the relationship between the wear of diesel engine elements and engine operating modes, are proposed. The MatLab package is used to create a neural network.

The model used for assessing the wear of piston rings has a complex appearance and functions involving many parameters. Ne is the average effective power over the period, kW; ϕrel is the average relative humidity over the period, %; T is the average boost temperature over the period, C; S is the mass content of sulfur in the fuel, %; qm is the specific consumption of the cylinder oil, g/kWh. The surfaces of the rings are worn out as a result of friction against the piston while simultaneously burning the fuel. The inner surface of the rings has several functions: for example, insulation, heat exchange, and reduction in piston vibrations. In this regard, the description of the wear dynamics is a difficult task from the viewpoint of the analytical description. The factor analysis justifies the fact that the most significant factors in such a model are the power and the cylinder oil consumption.

The complexity of this model lies in the fact that the mutual influence of the factors can be ambiguous. Using the neural network for approximating the empirical data is effective in this case. The approximation of a complex relationship existing among the parameters by regression models in the form of a neural network can become a good platform for subsequent "fast" calculations performed for digital ship twins in the real-time mode. Based on the measurement results obtained for the diesel engine, the neural network (Figure 12) was built on the ship (Table 2), associating the wear of the piston rings with the operating modes. The calculations were performed for a tanker shipping crude oil. The main ship low-speed dual-fuel diesel engine, manufactured by Hyundai Heavy Industries (South Korea) under a license granted by MAN Energy Solution COSCO Long Beach (IMO 9285677), was chosen as the object of the calculations. The data on the wear of piston rings were obtained by applying an ultrasonic thickness gauge Elcometer of the 456 B SCALE 1 model used for metallic and non-metallic coatings. The data were obtained from the ship reporting documentation and were associated with the data on engine operating modes.

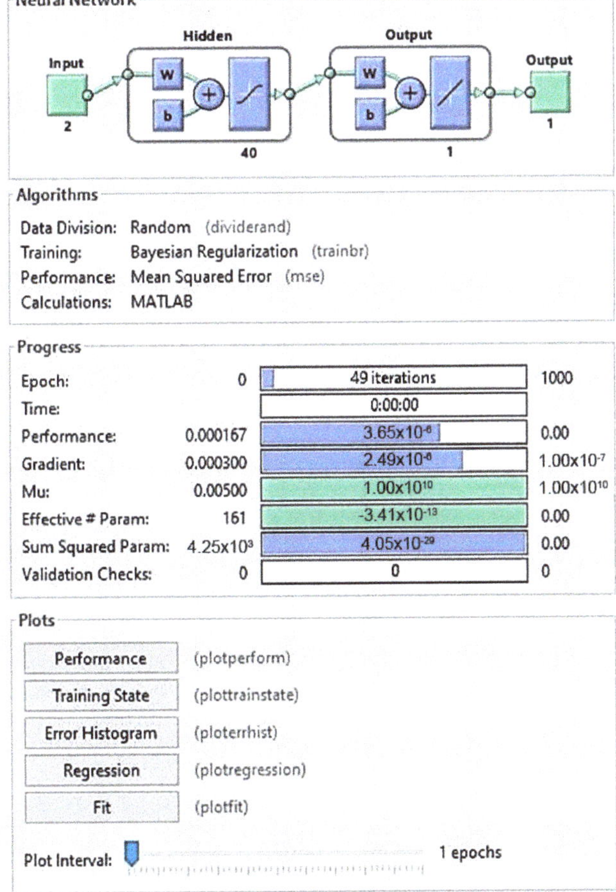

Figure 12. The structure of the approximating neural network.

Table 2. Summary table of the coating wear intensity approximated by the neural network.

Engine Power, %	Specific Consumption of the Cylinder Oil, g/kWh	Wear Intensity, μm/h
54	0.8	0.011023
54	1.2	0.01105
82	0.8	0.01495
82	1.2	0.01015
55	0.85	0.01087
55	0.8	0.01113
56	0.82	0.01105
56	0.85	0.01077
56	0.86	0.01070
57	0.85	0.01083
57	0.87	0.01064
58	0.87	0.01070
58	0.88	0.01059
58	0.91	0.01033
59	0.93	0.01015
61	0.94	0.01007
63	0.96	0.00997
65	0.96	0.01027
66	0.97	0.01026
66	1	0.00978
67	1.01	0.00980
69	1.03	0.00986
69	1.06	0.00978
73	1.08	0.01008
74	1.09	0.01013
75	1.12	0.01032
76	1.15	0.01059
77	1.16	0.01059
77	1.2	0.01123
80	1.2	0.01060

Figure 13 below shows a surface constructed by the neural network, describing the relationship among the wear intensity, power, and the specific consumption of the cylinder oil. The power and the specific oil consumption were measured along the axes of abscissa and ordinate; the wear intensity of the ceramic-metal coating was measured along the applicate axis.

The surface graph shows that there are combinations of extreme values of the power and the cylinder oil consumption, at which the wear is at a maximum.

It is obvious that under the maximum power in the case of the low fuel consumption, there is a slight wear of the rings, whose excess leads to a decrease in the wear. This information should be included in the accompanying documents, the inspection schedule, and the overhaul.

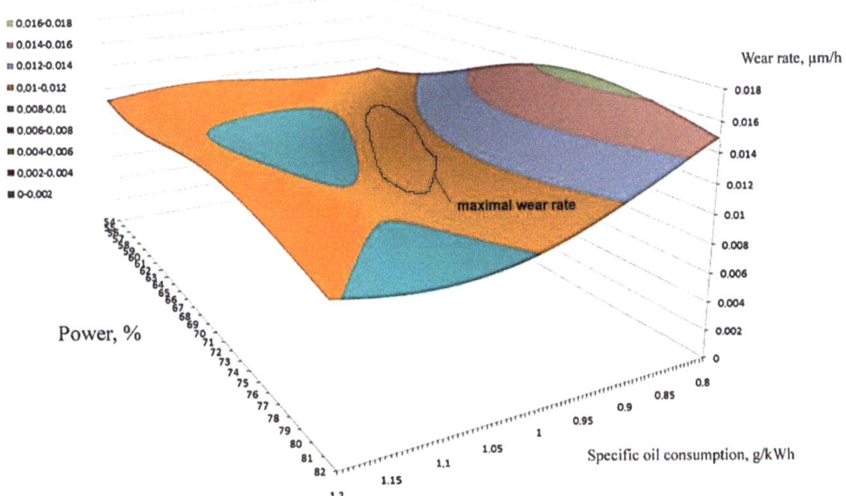

Figure 13. The visualization of the approximating wear surface obtained by the neural network.

The resulting approximating dependence can be expanded by introducing new parameters, such as humidity and temperature, accompanied by an increase in the data set of the measurements and observations.

A similar method can be proposed to replace "heavy" models with "lighter" approximating complex dependencies, for example, the relationship of the models "the stress-strain state of the elements of the propeller shafting design—the probability of failure" in predictive diagnostics systems.

3.5. Simulating a Short Circuit in the Generator

In the context of a synchronous machine with three-phase windings, an excitation circuit, and both longitudinal and transverse damper windings, the set of differential equations governing the equilibrium of electromotive force (EMF) and voltage drops in the synchronous machine's circuits can be expressed as follows:

$$\begin{cases} u_\eta = -\frac{d\psi_\eta}{dt} - R_\eta i_\eta & (\eta = a,b,c) \\ u_f = -\frac{d\psi_f}{dt} - R_f i_f \end{cases} \quad \begin{cases} 0 = -\frac{d\psi_{yd}}{dt} - R_{yd} i_{yd} \\ 0 = -\frac{d\psi_{yq}}{dt} - R_{yq} i_{yq} \end{cases} \tag{7}$$

where

u_η ($\eta = a,\ b,\ c$) and u_f—values of voltage on the phase windings and the excitation windings, respectively;

i_η and i_f—current values of phase winding and field winding, respectively;

ψ_η and ψ_f—flux linkages of phase winding and field winding, respectively;

R_η and R_f—ohmic resistances of phase winding and field winding, respectively;

ψ_{yd} and ψ_{yq}—flux-coupling of the damper windings of the longitudinal and transverse components, respectively;

R_{yd} and R_{yq}—their active resistances; i_{yd} and i_{yq}—current values in damping circuits.

Equation (7) represents the differential equations describing the dynamics of electric machines. The calculation of the transient processes requires a joint solution of these equations supplemented with the expressions of the transformer and load voltage drops.

The dynamic processes proceeding in the electrical machines and networks are simulated to determine the setup variables in the algorithms that connect different levels [22]. An example of reaching the idling mode followed by a three-phase short circuit to one of the systems is presented below. The rotation speed of the generator $\omega t = 1$ corresponds to the nominal one and does not change. At the time of 70 s, a three-phase short circuit occurs to the first three-phase system and the stator current increases taking into account an aperiodic component. Then, the aperiodic and periodic components fade to a steady-state value in accordance with the expressions (since the short circuit occurs in the idle mode, and the transverse component is absent).

Figure 14 demonstrates the dynamic model of a transient process occurring in the electrical circuit at the level of a physical short circuit process. The calculations were carried out for the tanker shipping crude oil.

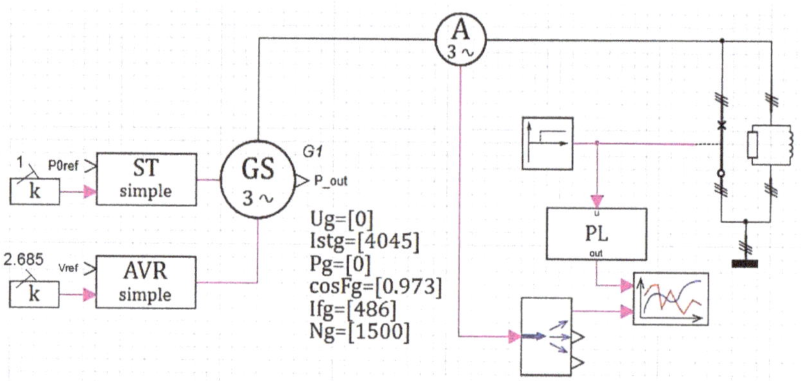

Figure 14. The model of the short circuit dynamics in a synchronous generator system.

To compare the calculation results with the theoretical values, the current of one phase of the generator in the short circuit mode is calculated. The analytical curve of the generator current lasting up to the short circuit moment is represented by a straight line having a constant value of 3600.9 A equal to the reference current (Figures 15 and 16).

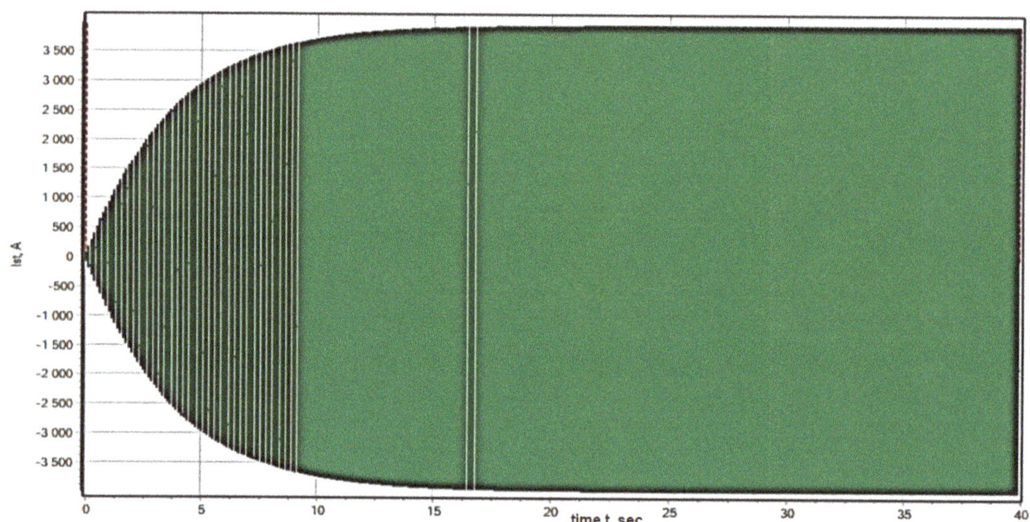

Figure 15. The generator phase current.

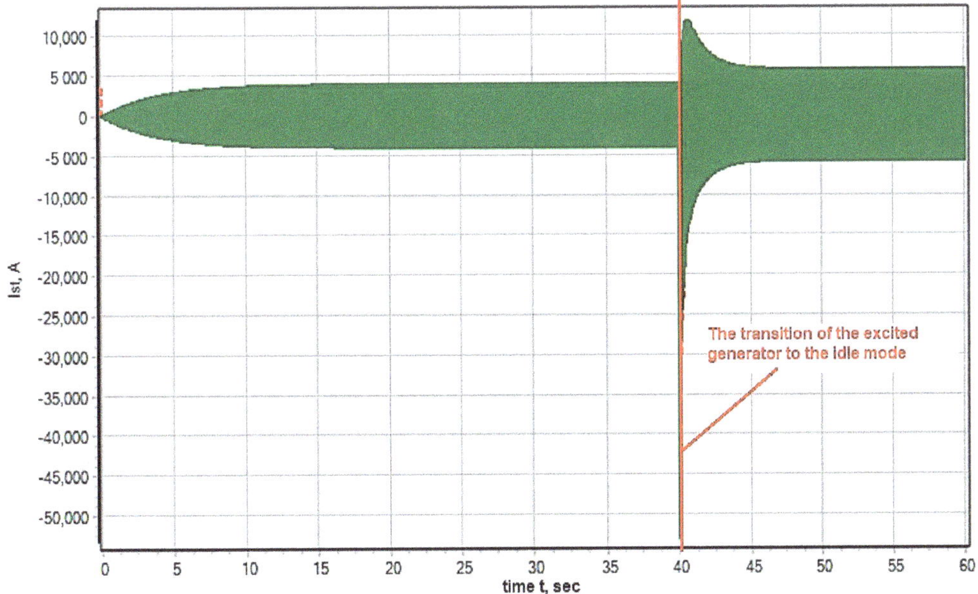

Figure 16. The dynamics of the short circuit current of the generator phase.

That is, at a time interval from 0 to 70 s, the excited generator goes into the idle mode. The generator stator voltage becomes close to the nominal voltage. This model facilitates studying the dynamics of the short circuit current.

3.6. The Object-Oriented Simulation of the Operation Process of the Ship's Electric Power System

Using modern methods to solve the electromagnetic equations, describing the functioning of the electrical machines, provides great opportunities for studying transient processes. The influence of various mechanical factors, electrical factors, and operating conditions: eccentricities, mechanical damages, uneven laying of windings, elevated temperature, can also be considered.

Electric current is transmitted from the main distribution board to consumers via electric networks consisting of wires and distribution devices.

The main elements of the ship's electric power system that make up the object-oriented model are:

(1)   Electric power sources consisting of DC or AC generators and accumulator batteries; additionally, ship's electric power systems have various converters of the current type, its voltage, and frequency;

(2)   Distributing gears including the boards consisting of apparatuses, distributing electric power, and devices used for controlling the operation of electrical installations and monitoring them;

(3)   Electrical networks consisting of cables and wires that transmit electric power from sources to consumers;

(4)   Consumers of electric power, representing various electric motors that convert electric power into mechanical work, as well as devices and apparatuses that convert it into another type of energy: heat, light, electromagnetic, and others.

This work presents a solution to the problem of creating a virtual simulator of a ship's electric power system. The mathematical model of this element belongs to the control models and is described by the equations of type (3).

A similar model facilitates systematically investigating the ship's electric power system, simulating the dynamics, and making a decision on the satisfactory quality of the designing solution.

Figure 17 shows a model of the main distribution board (MDB) of a tanker used to ship oil, implementing the algorithms for controlling diesel generators (DG). The interface of this project consists of "primitives" formed into the constructions that are more complex. SimInTech can restore an animated or static graphic interface of any complexity without special efforts and connects them to the necessary parts of the model [42].

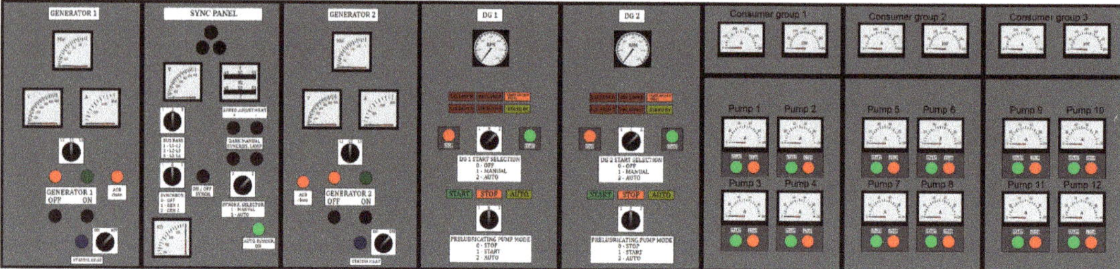

Figure 17. The MDB model visualized in the SimInTech environment, consisting of diesel generators, consumer panels, and loads due to actuators.

The creation of the model required simulating eight sections of MDB, which are typical of a ship having two diesel power stations.

The number of logical and dynamic elements in this project exceeds 400 units, but the total number of elements in the work region amounts to 2400 units (Figures 18 and 19).

The algorithms have been developed for the MDB operation: starting up diesel generators, the introduction into parallel operation. The algorithms provide manual and automatic control by diesel generators, conditions for stoppage, starting up, and switching.

This model works in the real-time mode and performs the tasks that correspond to the real tasks set on the ship MDB. The model performs its functions according to the logic of implementing the required conditions. All the MDB panels correspond to the logic of a real device and polish the solutions of various kinds of tasks:

- Starting up DG in the manual mode, observing all the conditions for starting up;
- Starting up DG in the fully automatic mode, where automation fulfills all the required conditions;
- DG load management, switching on and off the pumps;
- Simulating the short circuit conditions under loading;
- Simulating the conditions for the moisture appearance in the stator windings;
- Simulating the conditions of low pressure of the lubricating oil;
- Simulating the conditions of high temperature of the lubricating oil;
- Simulating the conditions of high temperature of the cooling water;
- Simulating the conditions of low pressure of the cooling water;
- Taking two diesel generators in parallel in the manual and fully automatic modes;
- Demonstrating the work of the protection of active and reactive loads.

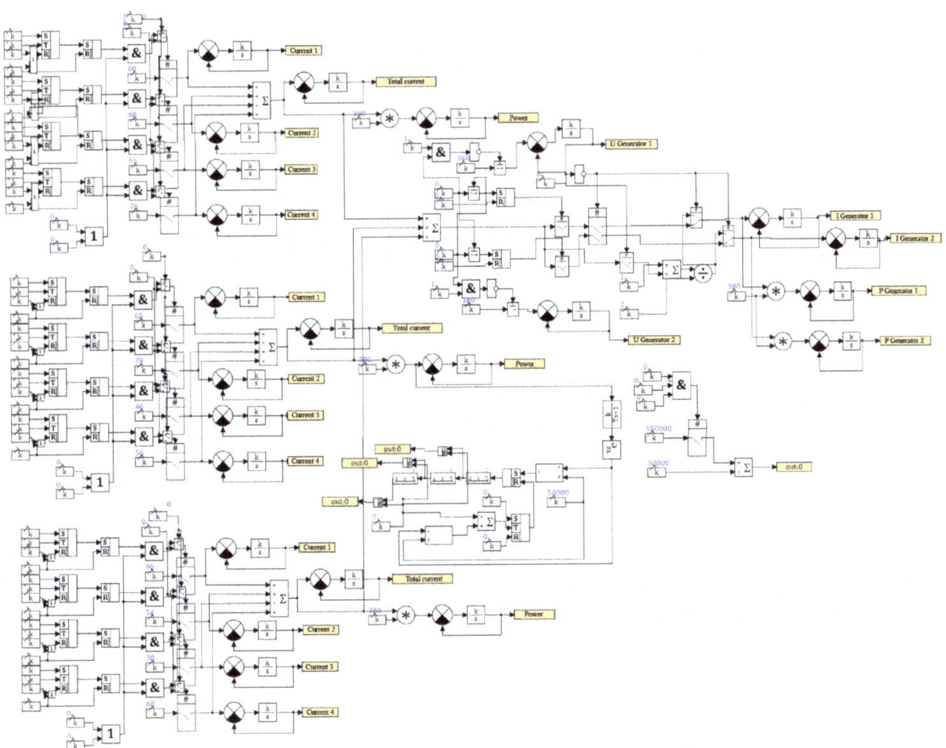

Figure 18. The model of including a group of consumers in SimInTech.

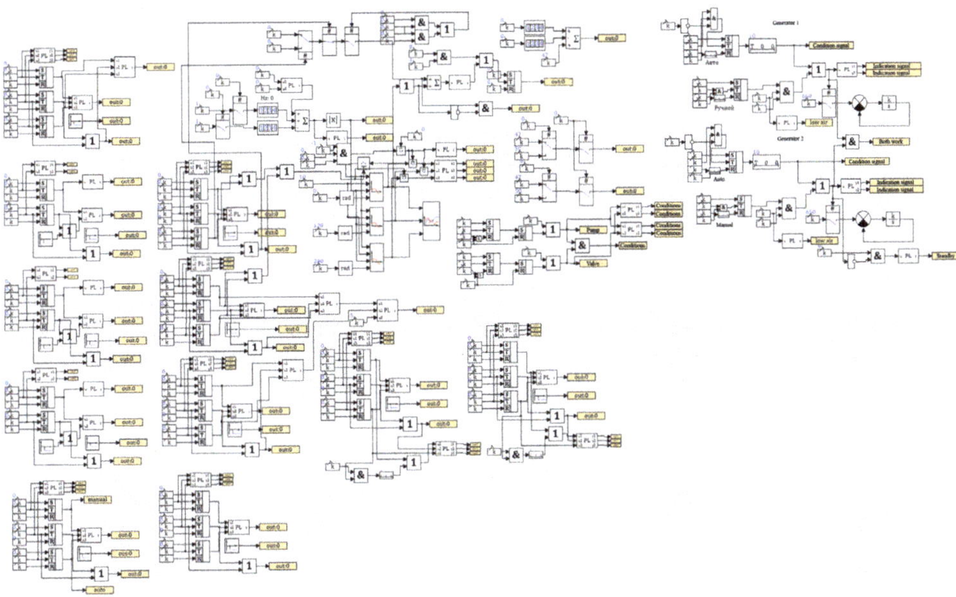

Figure 19. The model of the starting up and synchronization of the diesel generators.

The developed MDB model promotes fully obtaining a stable simulation of the main ship distribution board, performing all possible manipulations with it, and providing a possibility of conducting practical works to familiarize with the object of simulation.

4. Discussion

The model of the engine wear (M_{wear}) associated with its operation modes is the initial one for the MDB operation algorithm (M_{MDB}). The numerical model of the generator ($M_{generat}$) identifies defects and simulates the ship's electric network. The operating frequency ranges of the main engines ($M_{eng.}$), electric propulsion systems ($M_{generat}$), and their design peculiarities (disbalance, geometric inaccuracy, disequilibrium) are determined (M_{defect}) both at the design stage and when eliminating problems arising on the operated ship using the numerical structural models. In addition, if there are measurement data of a real ship, the strength analysis allows for evaluating the fatigue strength and the reliability of a mathematical twin.

A description of the relationship between the operating modes of the equipment and the wear indicators of the engine elements (M_{wear}) using neural models is a promising direction. The modern ship equipment and the development of big data transmission and processing technologies make it possible to develop such an approach in predicting the engine operation processes. In this study, a neural network was used to simulate the wear of piston rings. The volume of the training sample is not as large as the number of the parameters, but this approach seems very promising, since similar models can be continuously supplemented and expanded. The nature of the processes proceeding in the main engines is continuous without a large number of local minima and maxima. That is, the approximating function of the neural networks is effective for simulation.

The information on the resonant frequencies of the propeller shaft system (M_{shaft}) allows for effectively designing the geometry and installing the "screw-propeller shaft–engine" (M_{screw}, M_{shaft}, $M_{eng.}$) system. The Register of Shipping facilitates the calculation of the natural frequencies and the determination of oscillation forms up to the twelfth inclusive.

The value of displacements in the case of different resonant forms of vibrations can be associated with the accuracy and calculation of the wear along the selected directions in the above-mentioned shaft system. The complex load and the evaluation of the shafting fluctuations can be set in the form of several boundary conditions: harmonic disturbances caused by the screw and the main engine, a geometric inaccuracy of structural elements, own weight, and interrelated fluctuations.

Structural, modal, and dynamic analyses of the propeller shafts (M_{shaft}) and other elements promote effective design and technological solutions in the field of reducing loads, vibrations, and resonances, which is the so-called reverse digital design. The digital twin of the shaft elements, simulating harmonic loads, applied according to the results of the measurements taken from a real active ship, is able to calculate and predict the fatigue strength of the structure to a higher accuracy. The variability and cheapness of such calculations by the dimension types and modes form a database for selecting the optimal states of the system as a whole (it should be noted that such calculations must be verified).

MDB (M_{MDB}), in turn, can be represented by algorithmic and dynamic object-oriented models. Hence, on the one hand, based on the micromodels (M_{sc}, M_{defect}, M_{wear}), setpoints and discrimination thresholds in algorithms are assigned, and on the other hand, simulating the transient processes ($M_{generat}$, $M_{eng.}$) makes it possible to build automation systems (M_{AVR}) for MDB (M_{MDB}). The object-oriented description of the dynamic systems simulates and predicts the ship system state or implements optimal control [43–45].

Let us consider the following example for the above-mentioned ships as an illustration of the proposed approach [46–48]. The shaft model M_{shaft} obtained resonant frequencies and an operating frequency of 150 Hz for the shaft oscillations. This engine oscillation frequency corresponded to the engine power of 73% ($M_{eng.}$). This power value was compared with the wear value (Table 2) according to M_{wear}.

The numerical generator model ($M_{generat}$) determined the nature of the transient processes at a rotor rotation speed of 625 rpm at a frequency of 50 Hz. This model $M_{generat}$ provided an opportunity for obtaining the characteristics of transient processes in the presence of a mechanical defect (M_{defect}) in the generator (this was a harmonic component in the spectrum of a high amplitude force having a maximum value at a low frequency). This model $M_{generat}$ also obtained the characteristics of transient processes in the presence of an electrical defect (M_{sc}) or phase short circuit (the current was 15,000 A at 40 s, reaching an idling mode up to 70 s). The characteristics of the operating modes obtained based on the generator model $M_{denerat}$ (current, voltage, power, speed) were input values for the control model M_{AVR}.

The algorithms of installing, starting up, and switching DE combined the X_{sc} vector and X_D, X_{wear} vectors in the MDB model (M_{MDB}) through the diagnostic model M_{diagn}.

The local models and the research results described above, in turn, are already the objects and the states of the already formalized, complex, ranked image of the system. The system image includes the dynamics models, where the boundary and limiting conditions represent a common area for design and functioning tasks.

5. Conclusions

Modern ships are complex systems containing a large number of different elements. In the hierarchy of models, mathematical models of such elements belong to the microlevel and describe mainly the dynamics of various physical processes. At the same time, a mathematical apparatus based on the laws of conservation is used. In particular, this apparatus is represented by, for example, Maxwell's equations, oscillation equations, Navier–Stokes equations, etc. Such models are preset by the equations of type (3). Since such equations describe physical processes of different natures, it is impossible to combine the mathematical models at the microlevel.

The application of microlevel models to specific processes (ring wear, shaft vibrations, generator processes, etc.) enables building macrolevel models, whose output parameters are in no way related quantities. The macrolevel models are connected by means of control models (Equation (4)). The control models belong to the metalevel and allow assigning settings and response thresholds in the algorithms used in automation systems (MDB model). Such a model (megalevel model) promotes, ultimately, the study of the dynamics of the entire system as a whole in terms of a change occurring in the full ranked image of $\prod$ (2) at each moment in time.

The approach (1)–(4) proposed by the authors connects the elements of a complex technical system and investigates its dynamics.

In summary, in this study, an approach to simulating a ship as a relationship of various systems was developed. The mathematical models were created for each of the composite systems and simulation results were presented. Each of the systems described a limited class of problems, had characteristic properties, and a mathematical structure. The models were interconnected using a set-theoretic description. The results of the conducted research can be used in modern intelligent digital platforms for support and decision-making when using maritime transportation, in control systems, and SCADA tracking.

The next step in the development of this work is to set mathematical optimization models and to conduct multi-criteria optimization to determine the optimal parameters of a complex technical system.

Author Contributions: Conceptualization and methodology, R.V.K. and B.V.M.; validation and formal analysis, V.S.T. and V.A.K.; investigation, Y.A.T. and S.I.K.; resources, S.I.K. and N.V.M.; data curation, Y.A.T.; writing—original draft preparation, I.P.B. and A.V.G.; writing—review and editing, I.P.B. and A.V.G.; visualization, V.S.T. and V.A.K.; supervision, R.V.K. and B.V.M.; project administration, N.V.M. All authors have read and agreed to the published version of the manuscript.

Funding: This research received no external funding.

Data Availability Statement: Data are contained within the article.

Conflicts of Interest: The authors declare no conflict of interest.

References

1. Sokolov, B.V.; Yusupov, R.M. Multiple-model description and structure dynamics analysis of space-facilities control systems. *SPIIRAS Proc.* **2010**, *4*, 7–52. [CrossRef]
2. Gaber, M.; Elbanna, S.; El-Dabah, M.; Hamad, M. Intelligent Energy Management System for an all-electric ship based on adaptive neuro-fuzzy inference system. *Energy Rep.* **2021**, *7*, 7989–7998. [CrossRef]
3. Rao, X.; Sheng, C.; Guo, Z.; Yuan, C. A review of online condition monitoring and maintenance strategy for cylinder liner-piston rings of diesel engines. *Mech. Syst. Signal Process.* **2022**, *165*, 108385. [CrossRef]
4. Lai, G.; Liu, J.; Liu, S.; Zeng, F.; Zhou, R.; Lei, J. Comprehensive optimization for the alignment quality and whirling vibration damping of a motor drive shafting. *Ocean. Eng.* **2018**, *157*, 26–34. [CrossRef]
5. Li, C.; Huang, H.; Hua, H. Dynamic modeling and analysis of axial vibration of a coupled propeller and shaft system. *J. Mech. Sci. Technol.* **2016**, *30*, 2953–2960. [CrossRef]
6. Amirudin, A.A.A.; Kamarumtham, K.I.; Haripriyono, A.A.; Yunus, A.A.R.; Suheimy, N.A.S.; Ahmed, Y.A. A review of the dynamic analysis of axial vibrations in marine propulsion shafting system due to propeller excitation. *ASEAN Eng. J.* **2022**, *12*, 19–27. [CrossRef]
7. Xie, X.; He, P.; Wu, D.; Zhang, Z. Vibration attenuation of a propulsion shafting system by electromagnetic forces: Static thrust force balance and harmonic vibration suppression. *Mech. Syst. Signal Process.* **2022**, *179*, 109406. [CrossRef]
8. Dermentzoglou, J.C.; Prousalidis, J.M. Contribution to a detailed modeling and more reliable simulation of a ship's shaft machine. In Proceedings of the 2015 International Conference on Electrical Systems for Aircraft, Railway, Ship Propulsion and Road Vehicles (ESARS), Aachen, Germany, 3–5 March 2015; pp. 1–6.
9. Shi, J.; Amgai, R.; Abdelwahed, S. Modelling of shipboard medium-voltage direct current system for system level dynamic analysis. *IET Electr. Syst. Transp.* **2015**, *5*, 156–165. [CrossRef]
10. Uriarte, F.M.; Hebner, R.E.; Gattozzi, A.L. Accelerating the simulation of shipboard power systems. In Proceedings of the Grand Challenges in Modeling & Simulation (GCMS2011), Part of the 2011 Summer Simulation Multiconference 2011 (Summer Sim2011), The Hague, The Netherlands, 27–30 June 2011.
11. Karavaichenko, M.G.; Gazaleev, L.I. Numerical modeling of a double-walled spherical reservoir. *J. Min. Inst.* **2020**, *245*, 561–568. [CrossRef]
12. Koteleva, N.I.; Korolev, N.A.; Zhukovskiy, Y.L. Identification of the technical condition of induction motor groups by the total energy flow. *Energies* **2021**, *14*, 6677. [CrossRef]
13. Grinek, A.V.; Boychuk, I.P.; Fishenko, A.M.; Savosteenko, N.V.; Gerasimenko, O.N. Investigation of the operation of a ship's synchronous generator based on a numerical model. *J. Phys. Conf. Ser.* **2021**, *2061*, 012004. [CrossRef]
14. Ouroua, A.; Jackson, J.R.; Beno, J.H.; Thompson, R.C.; Schroeder, E. Modeling and simulation of electric ships' power system components and their interaction. In Proceedings of the 2007 Summer Computer Simulation Conference, SCSC 2007, San Diego, CA, USA, 16–19 July 2007; pp. 250–257.
15. Shang, C.; Fu, L.; Bao, X.; Xu, X.; Zhang, Y.; Xiao, H. Energy optimal dispatching of ship's integrated power system based on deep reinforcement learning. *Electr. Power Syst. Res.* **2022**, *208*, 107885. [CrossRef]
16. Gendler, S.G.; Prokhorova, E.A. Assessment of the cumulative impact of occupational injuries and diseases on the state of labor protection in the coal industry. *MIAB. Mining Inf. Anal. Bull.* **2022**, *10–12*, 105–116. [CrossRef]
17. Gendler, S.G.; Gabov, V.V.; Babyr, N.V.; Prokhorova, E.A. Justification of engineering solutions on reduction of occupational traumatism in coal longwalls. *MIAB. Mining Inf. Anal. Bull.* **2022**, *1*, 5–19. [CrossRef]
18. Jones, N.B.; Li, Y.-H. A review of condition monitoring and fault diagnosis for diesel engines. *Tribotest* **2000**, *6*, 267–291. [CrossRef]
19. Kimmich, F.; Schwarte, A.; Isermann, R. Fault detection for modern Diesel engines using signal- and process model-based methods. *Control Eng. Pract.* **2005**, *13*, 189–203. [CrossRef]
20. Djagarov, N.; Grozdev, Z.; Bonev, M. Use of adaptive system stabilizers in ship power systems. In Proceedings of the 2015 IEEE 15th International Conference on Environment and Electrical Engineering (EEEIC), Rome, Italy, 10–13 June 2015; pp. 593–598. [CrossRef]
21. Sivkov, Y.A. Building maritime data hub by using the arduino IoT platform. In Proceedings of the Global Perspectives in MET: Towards Sustainable, Green and Integrated Maritime Transport, Varna, Bulgaria, 11–14 October 2017; pp. 533–540.
22. Shin, Y.J.; Monti, A.; Ponci, F.; Arapostathis, A.; Grady, W.M.; Powers, E.J.; Dougal, R. Virtual power quality analysis for ship power system design. In Proceedings of the 21st IEEE Instrumentation and Measurement Technology Conference (IEEE Cat. No.04CH37510), Como, Italy, 18–20 May 2004; Volume 3, pp. 1758–1763. [CrossRef]
23. Min, H.; Fang, Y.; Wu, X.; Lei, X.; Chen, S.; Teixeira, R.; Zhu, B.; Zhao, X.; Xu, Z. A fault diagnosis framework for autonomous vehicles with sensor self-diagnosis. *Expert Syst. Appl.* **2023**, *224*, 120002. [CrossRef]
24. Huang, N.; Chen, Q.; Cai, G.; Xu, D.; Zhang, L.; Zhao, W. Fault Diagnosis of Bearing in Wind Turbine Gearbox under Actual Operating Conditions Driven by Limited Data With Noise Labels. *IEEE Trans. Instrum. Meas.* **2021**, *70*, 1–10. [CrossRef]
25. Wei, D.; Shengjun, L.; Bo, Y.; Zhengtong, Y.; Mingzhe, L.; Lirong, Y.; Wenfeng, Z. An encoder-decoder fusion battery life prediction method based on Gaussian process regression and improvement. *J. Energy Storage* **2023**, *59*, 106469.

26. Martyushev, N.V.; Malozyomov, B.V.; Sorokova, S.N.; Efremenkov, E.A.; Qi, M. Mathematical Modeling of the State of the Battery of Cargo Electric Vehicles. *Mathematics* **2023**, *11*, 536. [CrossRef]
27. Li, M.; Liu, Y.; Wang, T.; Chu, F.; Peng, Z. Adaptive synchronous demodulation transform with application to analyzing multicomponent signals for machinery fault diagnostics. *Mech. Syst. Signal Process.* **2023**, *191*, 110208. [CrossRef]
28. Chen, H.; Xiong, Y.; Li, S.; Song, Z.; Hu, Z.; Liu, F. Multi-Sensor Data Driven with PARAFAC-IPSO-PNN for Identification of Mechanical Nonstationary Multi-Fault Mode. *Machines* **2022**, *10*, 155. [CrossRef]
29. Yasir, M.; Zhan, L.; Liu, S.; Wan, J.; Hossain, M.S.; Isiacik Colak, A.T.; Liu, M.; Islam, Q.U.; Raza Mehdi, S.; Yang, Q. Instance segmentation ship detection based on improved Yolov7 using complex background SAR images. *Front. Mar. Sci.* **2023**, *10*, 1113669. [CrossRef]
30. Karikov, E.B.; Rubanov, V.G.; Duyun, T.A.; Grinek, A.V. Method for Identification of Complex Control Object of Fractional Order. Patent RF 2,592,464, 2015.
31. Grinek, A.V.; Boichuk, I.P.; Fishchenko, A.M.; Perelygin, D.N.; Alfimova, N.I. Predictive Diagnostics of a Ship's Propeller Shaft Using a Digital Twin. *Russ. Engin. Res.* **2023**, *43*, 99–102. [CrossRef]
32. Martyushev, N.V.; Malozyomov, B.V.; Khalikov, I.H.; Kukartsev, V.A.; Kukartsev, V.V.; Tynchenko, V.S.; Tynchenko, Y.A.; Qi, M. Review of Methods for Improving the Energy Efficiency of Electrified Ground Transport by Optimizing Battery Consumption. *Energies* **2023**, *16*, 729. [CrossRef]
33. Casti, J.L. *Connectivity, Complexity and Catastrophe in Large-Scale Systems*; Wiley-Interscience: New York, NY, USA, 1979.
34. Han, H.; Lee, K.; Park, S. Estimate of the fatigue life of the propulsion shaft from torsional vibration measurement and the linear damage summation law in ships. *Ocean Eng.* **2015**, *107*, 212–221. [CrossRef]
35. Yaghobi, H.; Ansari, K.; Mashhadi, H.R. Analysis of magnetic flux linkage distribution in salient-pole synchronous generator with different kinds of inter-turn winding faults. *Iran. J. Electr. Electron. Eng.* **2011**, *7*, 260–272.
36. Litvinenko, V.S. Correction to: Digital economy as a factor in the technological development of the mineral sector. *Nat. Resour. Res.* **2020**, *29*, 1521–1541. [CrossRef]
37. Seregin, A.S.; Fazylov, I.R.; Prokhorova, E.A. Justification of safe operating conditions for mining transportation machines powered by internal combustion engines using air pollutant emission criterion. *MIAB. Mining Inf. Anal. Bull.* **2022**, *11*, 37–51. [CrossRef]
38. Wang, B.-L.; Gu, W.; Chu, J.-X.; Wu, W.-M.; Guo, Y. Modeling a dual three-phase permanent magnet synchronous motor for electrical propulsion of ships. **2009**, *30*, 347–352. Available online: https://www.researchgate.net/publication/290770963_Modeling_a_dual_three-phase_permanent_magnet_synchronous_motor_for_electrical_propulsion_of_ships (accessed on 1 July 2023).
39. Bashkatov, V.A.; Khudyakov, S.A.; Ignatenko, A.V. Practical confirmation of mechanical balancers effectiveness to reduce vibration of marine main diesel engines. *J. Phys. Conf. Ser.* **2021**, *2061*, 012054. [CrossRef]
40. Guellec, C.; Doudard, C.; Levieil, B.; Jian, L.; Ezanno, A.; Calloch, S. Parametric method for the assessment of fatigue damage for marine shaft lines. *Mar. Struct.* **2023**, *87*, 103325. [CrossRef]
41. Seung, H.; Park, C.; Lee, J.K.; Lim, D.; Kim, Y.Y. Magnetostrictive patch sensor system for battery-less real-time measurement of torsional vibrations of rotating shafts. *J. Sound Vib.* **2018**, *414*, 245–258. [CrossRef]
42. Voronin, V.A.; Nepsha, F.S. Simulation of the electric drive of the shearer to assess the energy efficiency indicators of the power supply system. *J. Min. Inst.* **2020**, *246*, 633–639. [CrossRef]
43. SimInTech. Available online: https://simintech.ru (accessed on 24 January 2023).
44. Boikov, A.; Payor, V. The present issues of control automation for levitation metal melting. *Symmetry* **2022**, *14*, 1968. [CrossRef]
45. Babyr, K.V.; Ustinov, D.A.; Pelenev, D.N. Improving electrical safety of the maintenance personnel in the conditions of incomplete single-phase ground faults. *Bezop. Tr. Promyshlennosti* **2022**, *8*, 55–61. [CrossRef]
46. Vasilyeva, N.V.; Boikov, A.V.; Erokhina, O.O.; Trifonov, A.Y. Automated digitization of radial charts. *J. Min. Inst.* **2021**, *247*, 82–87. [CrossRef]
47. Malozyomov, B.V.; Martyushev, N.V.; Sorokova, S.N.; Efremenkov, E.A.; Qi, M. Mathematical Modeling of Mechanical Forces and Power Balance in Electromechanical Energy Converter. *Mathematics* **2023**, *11*, 2394. [CrossRef]
48. Martyushev, N.V.; Malozyomov, B.V.; Sorokova, S.N.; Efremenkov, E.A.; Valuev, D.V.; Qi, M. Review Models and Methods for Determining and Predicting the Reliability of Technical Systems and Transport. *Mathematics* **2023**, *11*, 3317. [CrossRef]

Article

Techno-Economic Analysis of Combined Gas and Steam Propulsion System of Liquefied Natural Gas Carrier

Muhammad Arif Budiyanto [1,*], Gerry Liston Putra [1], Achmad Riadi [1], Riezqa Andika [2], Sultan Alif Zidane [1], Andi Haris Muhammad [3] and Gerasimos Theotokatos [4]

[1] Department of Mechanical Engineering, Faculty of Engineering, Universitas Indonesia, Kampus Baru UI, Depok 16424, Jawa Barat, Indonesia
[2] Department of Chemical Engineering, Faculty of Engineering, Universitas Indonesia, Kampus Baru UI, Depok 16424, Jawa Barat, Indonesia
[3] Department Marine Systems Engineering, Faculty of Engineering, Hasanuddin University, Makassar 92171, South Sulawesi, Indonesia
[4] Department of Naval Architecture, Ocean and Marine Engineering, University of Strathclyde, Glasgow G4 0LZ, UK; gerasimos.theotokatos@strath.ac.uk
[*] Correspondence: arif@eng.ui.ac.id

Abstract: Various combinations of ship propulsion systems have been developed with low-carbon-emission technologies to meet regulations and policies related to climate change, one of which is the combined gas turbine and steam turbine integrated electric drive system (COGES), which is claimed to be a promising ship propulsion system for the future. The objective of this paper is to perform a techno-economic and environmental assessment of the COGES propulsion system applied to liquefied natural gas (LNG) carriers. A propulsion system design for a 7500 m³ LNG carrier was evaluated through the thermodynamics approach of the energy system. Subsequently, carbon emissions and environmental impact analyses were carried out through a life cycle assessment based on the power and fuel input of the system. Afterwards, a techno-economic analysis was carried out by considering the use of boil-off gas for fuel and additional income from carbon emission incentives. The proposed propulsion system design produces 1832 kilowatts of power for a service speed of 12 knots with the total efficiency of the system in the range of 30.1%. The results of the environmental evaluation resulted an overall environmental impact of 10.01 mPts/s. The results of the economic evaluation resulted in a positive net present value and a logical payback period for investment within 8 years of operation. The impact of this result shows that the COGES has a promising technological commercial application as an environmentally friendly propulsion system. Last, for the economy of the propulsion system, the COGES design has a positive net present value, an internal rate return in the range of 12–18%, and a payback period between 6 and 8 years, depending on the charter rate of the LNG carrier.

Keywords: LNG; COGES; energy system; boil-off gas utilization

Citation: Budiyanto, M.A.; Putra, G.L.; Riadi, A.; Andika, R.; Zidane, S.A.; Muhammad, A.H.; Theotokatos, G. Techno-Economic Analysis of Combined Gas and Steam Propulsion System of Liquefied Natural Gas Carrier. *Energies* **2024**, *17*, 1415. https://doi.org/10.3390/en17061415

Academic Editors: Konstantinos Aravossis and Eleni Strantzali

Received: 30 December 2023
Revised: 5 March 2024
Accepted: 8 March 2024
Published: 15 March 2024

1. Introduction

The International Maritime Organization (IMO) has adopted the IMO Strategy 2023 to reduce ships' greenhouse gas emissions. The strategy is to achieve net-zero greenhouse gas emissions from international shipping in or around 2050 [1,2]. Shipping industries have made various efforts and commitments to meet this emission reduction target. Several technologies continue to be developed to increase decarbonization on ships, including combined marine propulsion [3]. In its current development, there are various combined marine propulsion systems, including a combination of one or more diesel engines, gas engines, gas turbines, steam turbines, and electric propulsion [4].One of the most promising combined marine propulsion systems is a combined gas turbine and steam turbine integrated electric drive system (COGES) [5]. Initially, the COGES concept was introduced as an effort to increase power plant efficiency by integrating the gas cycle and steam cycle in one system [6]. Early in its development, COGES was faced with several technical and economic challenges, but over time, innovation and technological improvements have improved its performance and capabilities [7].

Some advantages of the COGES system are high power density, high thermal efficiency, and low noise and vibration [8,9]. The COGES system is usually applied and is promoted to be used for LNG carriers since it can utilize wasted cargo in the form of boil-off gas (BoG) [9,10]. LNG carriers are ships equipped with dedicated tanks for LNG cargo with various capacities [11,12]. Several LNG ships are classified as very large tankers with a capacity of 125,000–266,000 m^3 [13]; the latest development is small LNG carriers, which have a capacity of 1000–40,000 m^3 [14,15]. Despite the fact that the tanks on LNG carriers are insulated, a small amount of warming occurs, causing the LNG cargo to evaporate as it approaches its boiling point [16,17]. This natural evaporation is unavoidable, and the resultant boil-off gas must be evacuated to keep the tanks' pressure stable [18]. LNG carriers are suitable for LNG distribution between countries operating in international waters [19,20], while for distribution in inter-island areas, it is more appropriate to use small LNG carriers, because regular-sized LNG carriers cannot enter these locations due to low draft [21,22]. To improve the performance of the LNG carrier operating system, a reliable and efficient propulsion system design is required.

Currently, most LNG carriers use steam turbines and diesel dual-fuel engines with a percentage of more than 40% [23]. Looking at overall efficiency, there are several alternative propulsion systems that have been developed and have better efficiency values, such as dual-fuel diesel engines and combined cycles [24,25]. Further studies related to the use of gas turbines include the use of combined cycle steam–gas turbines, with an overall efficiency reaching 50% compared to other systems such as petrol engines or diesel engines [26]. COGES marine power plants are proven to increase cogeneration efficiency on ferries and cruise ships. Compared with reciprocating engines, COGES plants yield cogeneration efficiency gains of 1–5%, with a maximum total efficiency of 51% [27].

The propulsion system of LNG carriers consists of three main parts, namely, the prime mover, the transmission system, and the ship propulsion device [28]. The design of the ship's propulsion system depends on the type of ship, main size, ship speed, stern model, and hull model [29]. The problem that usually arises in the design of the propulsion system is the unmet service speed. The propulsion system of the main propulsion part of the ship is closely related to the thermal power generation cycle [30]. The thermal power generation cycle is a cycle that comes from burning fuel to generate power [31]. From this thermal power generation, the cycle begins with the chemical energy of the fuel being burned so that it becomes thermal energy. Then the thermal energy generated from the combustion is converted using a gas turbine and a steam turbine into mechanical energy. This electrical energy is used to drive the ship's propulsion system and meet the needs of the ship [32].

The COGES system uses a gas turbine to drive a generator and provide electrical power and propulsion according to needs that are regulated by the main switchboard in turn [33]. In this system, the propeller is driven by an electric motor that is controlled by frequency. Then the exhaust gas from the gas turbine is used to raise the steam in the heat recovery steam generator (HRSG) [34,35]. The steam from here drives a steam turbine generator in turn, which also generates electrical energy and feeds into the main switchboard [36]. Many previous studies have discussed the efficiency of HRSG. Effective utilization of waste heat energy can increase power generation efficiency and reduce emissions, either by using dedicated waste heat recovery systems for electricity production or by using it for heating services [37]. Waste heat recovery systems can utilize the remaining heat to generate mechanical/electrical power, which can meet the demand for propulsion and auxiliary services [38].

Ship engine manufacturing companies see the possibility of using waste heat recovery systems to achieve a total efficiency of 60% for the fuel energy used on ships [39]. Other researchers estimate using exergy analysis that fuel savings of 4–16% can be achieved for medium-sized long-haul tankers using waste heat recovery systems. For applications in the maritime industry, previous researchers have compared the organic Rankine cycle, the Kalina cycle, and the steam Rankine cycle for marine waste heat recovery systems, the results being that the organic Rankine cycle has the most significant potential to increase fuel efficiency and the combined cycle offers thermal efficiency [40]. The research examines a combined system encompassing a gas turbine powered by solid oxide, a supercritical carbon dioxide loop, an organic Rankine cycle, and an absorption circulation cycle utilizing ammonia air, indicating that thermal efficiency reaches 67% [41]. The waste heat recovery installation used for the production of saturated steam and electric power for the case of two-stroke and four-stroke engine propulsion plants on merchant ships, as a result of simulations, was carried out by increasing the energy efficiency design index [42]. Application of a waste heat recovery installation system on passenger ships is supplied by a steam power plant, which utilizes waste heat from exhaust gas from the main diesel engine [43].

Waste heat recovery systems can recover up to 10% of the fuel energy from the ship prime mover, resulting in an overall system peak efficiency of 60–65% [44]. Based on the performance data of a two-stroke diesel engine adopted for a crude oil tanker propulsion plant, the performance of the optimized waste heat recovery system was also evaluated by comparing it under off-design engine load conditions in the engine power range between 50% and 100% of the rated maximum continuity [45]. By applying the optimization numerical code to the examined passenger ships, two different sizes of turbogenerators were found, respectively, for retrofit and new design solutions. This more significant amount of steam is essentially due to the full exploitation of the flue gas thermal flow compared to retrofitting solutions, where the dimensions of the existing boiler are already fixed [46]. Another study reviewed four types of waste heat recovery systems, namely, organic Rankine cycles, thermoelectric generators, six-stroke cycles, and development of turbocharger technology [47]. Standard technologies used for waste heat recovery from engines include thermoelectric devices, organic Rankine cycles, and turbocharger systems. By maximizing the potential energy of exhaust gases, engine efficiency and net power can be increased [48]. Many studies have investigated the performance and efficiency of waste recovery from marine combined cycles. On the other hand, only a few studies have identified the potential environmental impacts of life cycle analysis.

This paper aims to investigate the combined gas–electric and steam turbine systems for marine propulsion systems on LNG carriers. The results of this work obtain three things at once: analysis of the COGES performance system on the LNG carrier, environmental impact analysis using life cycle assessment, and techno-economic analysis of HRSG installation. The proposed COGES system using HRSG is to address the limitation of current diesel propulsion systems by increasing energy efficiency and lowering exhaust emissions. The contribution to the results of this research can be alternative propulsion systems for LNG carriers and other commercial vessels, especially in using environmentally friendly fuel by utilizing a waste heat recovery system. In the end, it is hoped that the results of this research can be used as developments to support the GHG reduction target program launched by the IMO for the maritime industry.

2. Research Methodology

In this research, the methodology used is in systematic stages, as shown in Figure 1. The research starts from the design data of the LNG carrier, which consists of the principal dimensions, general arrangement, and ship power predictions. From the ship data, a COGES propulsion system design, including HRSG, was designed considering the ship design. The proposed COGES design was then subjected to thermodynamic analysis and life cycle assessment to determine the performance and environmental impact of the system. Lastly, a techno-economic analysis was carried out to determine the system's feasibility in terms of economic scale. The research methodology relies on secondary data and case studies involving different types of LNG carrier ship propulsion systems as its empirical foundation.

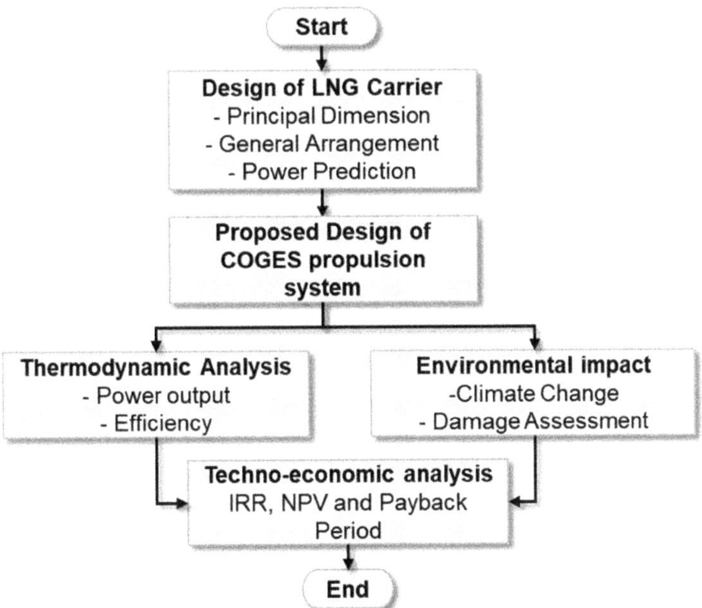

Figure 1. Research stages on the techno-economics of the COGES propulsion system.

2.1. Design Data of the LNG Carrier

The design of the LNG carrier used in this research was based on a design comparison of existing ships with a 7500 m^3 LNG capacity. The design method of the LNG carrier uses a spiral design, which starts with the hull design, the general arrangement, and the calculation of power prediction. The principal dimensions obtained based on the existing basic requirements and standards are shown in Table 1. The general arrangement was designed to ensure that the spaces on the ship were accommodated correctly, including the LNG loading space and engine room layout. The general arrangement of the LNG carrier used is shown in Figure 2. Power prediction calculations use naval architecture software, which provides integrated hull modeling and optimization tools. The results of power prediction on ship speed are shown in Figure 3. Based on the results of hull modeling, the power required for the ship to move at a service speed of 12 knots is around 1832 kW and at a maximum speed of 14 knots is 3377 kW.

Table 1. Principal dimensions of LNG carrier.

Principal Parameters	Dimensions
Length overall	: 117.8 m
Length between perpendicular	: 110.2 m
Beam	: 18.6 m
Depth	: 10.6 m
Draft	: 7.15 m
Service speed	: 12 knots
Cargo tank capacity	: 7500 m^3
Boil-off gas rate	: 0.3%/day
Crew number	: 19

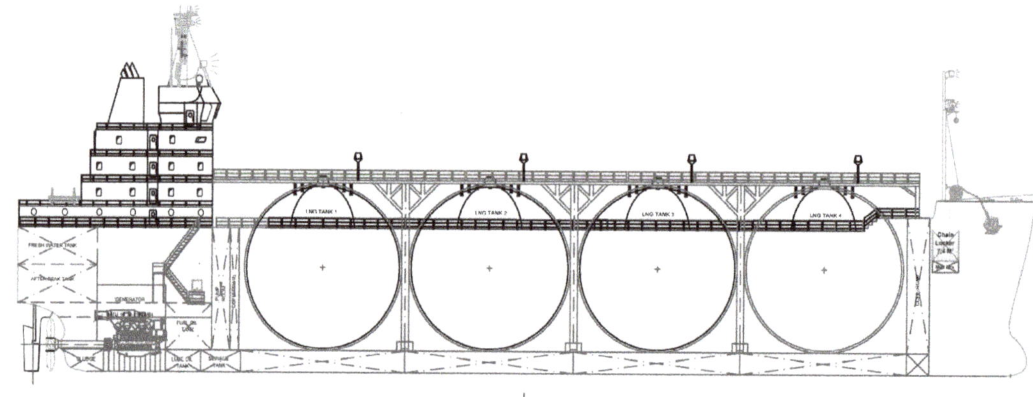

Figure 2. General arrangement of LNG carrier.

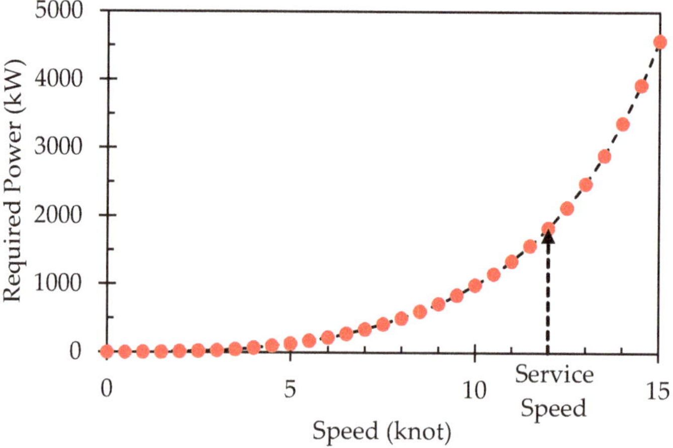

Figure 3. Power requirements of designed LNG carrier.

2.2. Design of COGES Propulsion System

The design of a ship's propulsion system starts from power requirements based on ship speed, determining the parameters and components of the combined gas turbine and steam turbine integrated electric drive system. The COGES system proposed in this research is shown in Figure 4. In this COGES system, the fuel is assumed to come from 100% boil-off gas produced from the LNG cargo tank. The use of boil-off gas as ship fuel makes this system more economical in terms of operations and saves more space on the ship compared to systems that still use diesel engines, both conventional diesel engines and dual-fuel diesel electric (DFDE) propulsion systems. However, in actual conditions, to meet the ship's overall electricity needs, additional auxiliary engines are still needed. This aims to be a safety factor as a source of backup energy for ships.

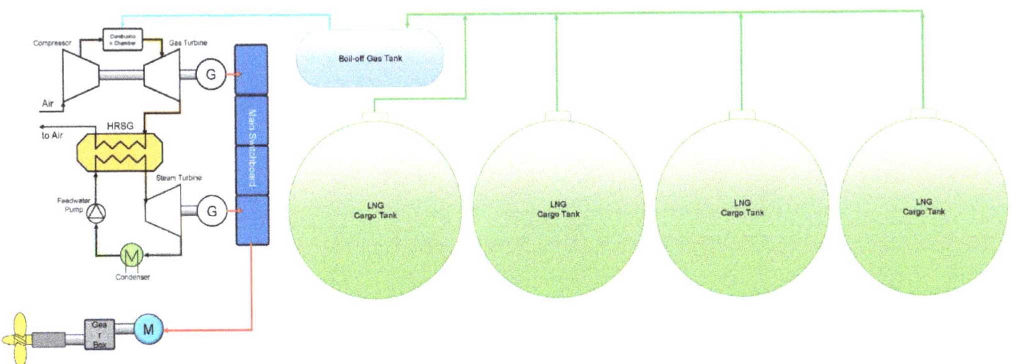

Figure 4. Proposed design of COGES propulsion system.

The fuel used in designing this propulsion system is natural gas that comes from boil-off gas from the LNG cargo being transported. The natural gas load transported has a CH_4 composition of 91.2% and a lower heating value of 49,426.97 kJ/kg. Gas boil-off (*BoG*) is calculated using Equation (1).

$$BoG = \frac{V \times \rho \times BOR \times t}{24} \tag{1}$$

where V is the LNG cargo volume (m^3), ρ is the LNG density (kg/m^3), *BOR* is the boil-off rate (%/day), and t is the shipping time (hours). The gas turbine used in this system has a power specification of 100% load of 5.4 MW, the fuel is natural gas, the exhaust gas temperature (100% load) is 494 °C, the pressure ratio is 13.9, and the exhaust mass flow is 21.3 kg/s. The selection of this gas turbine is based on the total electric load balance requirements of the ship design, as well as considerations of its availability in the market. The steam turbine specifications have an inlet steam pressure of 25 bar, a maximum inlet temperature of 510 °C, and a maximum output of 5000 kW.

In the COGES design used, the heat recovery steam generator (HRSG) used is composed of three components, namely, a superheater, evaporator, and economizer. The working process of these three components becomes a single HRSG operation where the economizer is useful as an initial heater to raise the temperature of the working fluid again from the fluid phase to the saturated liquid phase. Later, the working fluid is processed in the evaporator to become a saturated vapor phase and reheated in the superheater to heat the water in the boiler. The parameter value used in the simulation for the superheater component (ΔT_{hi}) is 30 °C and the evaporator ($\Delta T_{pinchpoint}$) is 25 °C.

The main system-supporting components consist of a boiler feed pump with maximum head specifications of 1200 m, capacity of 70 m^3/h, and isentropic efficiency of 85%. The deaerator component has a capacity specification of 14,000 lbs/h and the inlet pressure (Pin) is 2 bar. The other supporting components are a pump with an isentropic efficiency of 85%, a condenser assumed to have a pressure drop of 0.1 Bar, and a generator with a mechanical efficiency of 90%.

2.3. Thermodynamic Analysis of Designed System

The thermodynamic cycle used in the design of the COGES propulsion system is a combination system between the Brayton cycle and the Rankine cycle. The combined cycle aims to have a higher thermal efficiency value than the cycles used separately. This is possible because gas turbines generally operate at higher temperatures than steam turbine cycles. Thermodynamic analysis of this system uses Cycle-Tempo version 5.1 software and tools program to design, analyze, optimize, and monitor the thermodynamics of the energy system [49].

The thermodynamic analysis scheme of the proposed COGES system is shown in Figure 5. The COGES propulsion system design begins with a gas turbine system where air from the environment enters through the air source at number 3, which is then compressed by compressor number 1 to increase the pressure. Then, the compressed air enters combustion chamber number 5 to be mixed with fuel, namely, natural gas, producing pressurized hot gas to drive gas turbine number 4 to produce energy. Gas turbine number 4 has also been coupled with a generator to produce the required electrical power. The gas produced from the turbine, or what could be called flue gas, exits the gas turbine system through pipe number 5 and is used for the following cycle process.

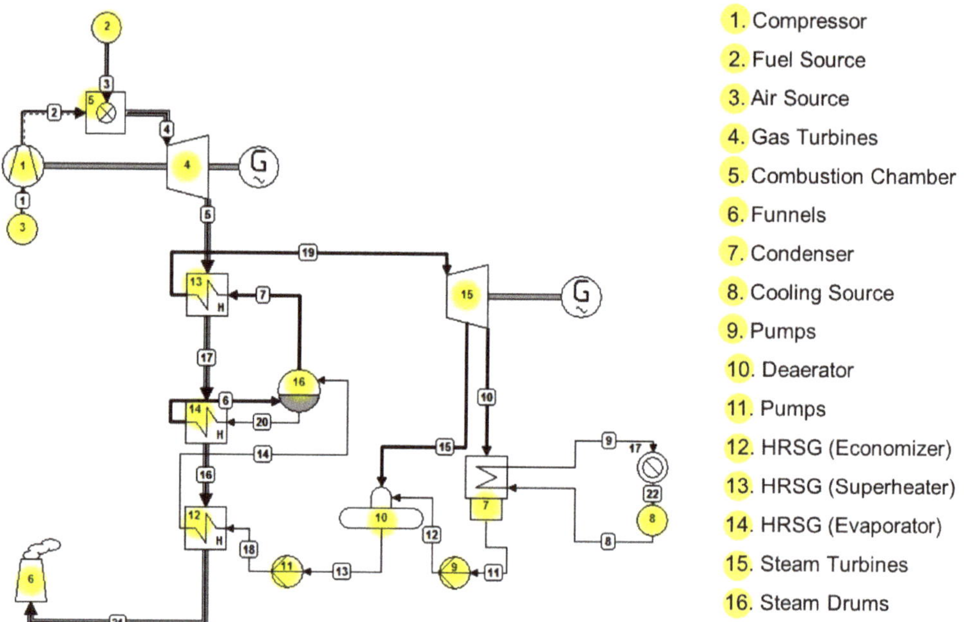

Figure 5. The thermodynamic analysis scheme of the proposed COGES system.

The following essential component in the propulsion system of this design is the heat recovery steam generator, which is an advanced component where the exhaust gas from the turbine is used in the next cycle. This component is the energy source in the steam turbine cycle. This component consists of several heat exchangers that function as superheaters, evaporators, and economizers. Thus, the gas turbine exhaust gas is utilized in stages, starting from the exhaust gas with the highest temperature of around 494 °C, which heats the saturated steam in the superheater. Then the exhaust gas, whose temperature has been reduced to around 444 °C, changes the working fluid phase to steam in the evaporator. The exhaust gas is then used to heat the working fluid that comes from the condenser in the heat exchanger, which acts as an economizer.

Thermodynamic analysis of the gas turbine combination cycle was carried out to determine the power produced by the design system and whether it can meet the ship's propulsion power requirements. The input values and thermodynamic equations for each system are shown in Table 2.

Table 2. Parameter input of thermodynamic analysis.

Cycle	Parameter Input	Thermodynamic Equations
Ambient	T = 298 K P = 1.013 bar	
Gas Turbine Cycle 1g—Compression	T = 298 K P = 1.013 bar m_{air} = 21.3 kg/s m_{fuel} = 0.512 kg/s	$T = T1g \times \left(\frac{P2}{P1}\right)^{\frac{k-1}{k}} = 647$ K $W_{compressor} = m \times Cp \times \Delta T \times \eta s$
Gas Turbine Cycle 2g—Combustion	P = 11.92 bar h = 374 kJ/kg	
Gas Turbine Cycle 3g—Expansion	P = 11.92 bar	$T = T4g \times \left(\frac{P3}{P4}\right)^{\frac{k-1}{k}} = 1373$ K
Gas Turbine Cycle 4g—Heat Rejection	T = 767 K P = 1.013 bar	$W_{gas\,turbine} = m \times Cp \times \Delta T \times \eta s$ $W_{actual} = W_{gas\,turbine} - W_{compressor}$
Steam Turbine Cycle 1s—Water Feeding	T = 297 K h = 100.5 kJ/kg P = 1.0 bar	$W_{pump} = v \times \Delta P / \eta$
Steam Turbine Cycle 2s—Pump Work Input	T = 302 K h = 125.74 kJ/kg $m_{exhaust}$ = 21.812 kg/s	
Steam Turbine Cycle 3s—Heat Addition	T = 558 K h = 1263.1 kJ/kg P = 80 bar	$Q_{exhaust} = Q_{in,\,rankine}$ $m_{exhaust}.\,Cp.\,\Delta T = m_{fluid}\,(h_4 - h_3)$
Steam Turbine Cycle 4s—Heat Addition	T = 568 K h = 2758.7 kJ/kg	
Steam Turbine Cycle 5s—Work Output	T = 834 K h = 3384 kJ/kg	$W_{actual} = W_{gas\,turbine} - W_{compressor}$ $W_{turbine} = m_{fluid}\,(h_5 - h_6) - W_{pump}$
Steam Turbine Cycle 6s—Heat Rejection	T = 302 K h = 2555.6 kJ/kg	

The power produced by the COGES system ($W_{overall}$) is a combination of the total power between the gas turbine cycle and the steam turbine cycle using Equation (2) and the overall system efficiency ($\eta_{overall}$) using Equation (3).

$$W_{overall} = W_{Gas\,Turbine} + W_{Steam\,Turbine} \tag{2}$$

$$\eta_{overall} = \frac{W_{Gas\,Turbine} + W_{Steam\,Turbine}}{Q_{in}} \tag{3}$$

Then, to find out the Q_{in} value, which is the product of the gas fuel mass flow rate and the lower heating value (LHV) of the fuel, the Q_{in} calculation uses Equation (4). Thus, the value of the system's overall efficiency is designed using Equation (5).

$$Q_{in} = m_{fuel} \times LHV \tag{4}$$

$$\eta_{overall} = \frac{W_{Gas\,Turbine} + W_{Steam\,Turbine}}{m_{in} \times LHV} \tag{5}$$

2.4. Environmental Impact Assessment

The method used in environmental impact assessment is eco-indicator 99. By using this method, the environmental impact is expressed in eco-indicator points per unit time (Pts/s or mPts/s). The value of 1 Pt (one point) represents one thousandth of the environmental burden per year of a European resident. Apart from that, the environmental impact can also be in units of $kgCO_2$ per unit of product produced by the system, because CO_2 is a greenhouse gas that has an effect on climate change. Environmental impact assessment is obtained by conducting an emission factor analysis, which uses an approach based on international standards (ISO 14004). With the eco-indicator method, evaluations can be carried out for materials, production processes, transportation processes, energy generation processes, and disposal scenarios. This method is also used to identify each component in the system that has a high environmental impact. Then the effects of environmental impacts are also divided into three categories of damage, including human health, ecosystem quality, and resources. The impact on human health (human health) is in units of DALY (disability-adjusted life years), where 1 DALY is one year of healthy life lost by a person. The impact of environmental ecosystems (ecosystems quality) has $PDFm^2yr$ units (potentially disappeared fraction of species per square meter year); 1 $PDFm^2yr$ means damage to species or ecosystems covering an area of 1 m^2 in one year. The impact of resources has MJ surplus units, where 1 MJ surplus is the basic amount of energy needed to extract a natural resource.

The analysis of greenhouse gas emissions such as CO_2, CH_4, and NO_2 in the designed propulsion system is intended to determine the estimated amount of emissions produced. The analysis is carried out using the tier method, which uses emission factors to fuel consumption data. Analysis using this tier method uses Equation (6).

$$E_{GHG} = (FC \times EC) \tag{6}$$

where FC is the fuel consumed for each fuel type, such as diesel, gasoline, or gas, while EF is the emission factor from the fuel type used for the ship engine. The values used for calculations include CO_2 emission factors for natural gas of 56,100 kg/TJ and fuel consumption for bulk liquid vessels of 14,685 + 0.00079GRT. Then, based on the data that were obtained from the previous discussion, the data are simulated using SimaPro software to obtain an emission analysis using the eco-indicator 99 (H) life-cycle assessment (LCA) method. Then the results of the damage assessment are obtained in several categories such as ecosystems quality, resorts, human events, and climate change. From the results of the damage assessment obtained from the simulation, an analysis is carried out to compare the output emissions produced by the COGES propulsion system with other propulsion systems, such as the diesel propulsion system or the DFDE system.

2.5. Techno-Economic Analysis

The economic theory used to build a propulsion system for ships, especially the COGES combination propulsion system for small LNG carrier 7500 m^3 ships, in this research is to use several aspects of an economic approach, namely, net present value (NPV) in Equation (7), internal rate of return (IRR) in Equation (8), and payback period (PBP) in Equation (9).

$$NPV = \sum_{t=0}^{T} \frac{X_t}{(1+i)^t} \tag{7}$$

$$IRR = \sum_{t=0}^{T} \frac{X_t}{(1+ROR)^t} \tag{8}$$

$$PBP = \sum_{t=0}^{t=POT} X_t = 0 \tag{9}$$

To obtain the above economic values, it is necessary to take into account the capital costs (CAPEX) and operating costs (OPEX). CAPEX in this research focuses on the cost of the design of propulsion system components. The propulsion system proposed in this research is the COGES system, which consists of a gas turbine, steam turbine, HRSG, boiler pump, boiler, condenser, deaerator, and cooling pump. Each component's capital cost uses assumptions based on open study reports and market prices [50]. The cost for each component of the proposed propulsion system is shown in Table 3. OPEX is the operational costs incurred during one year of operation, which consist of fuel costs, fresh water, maintenance costs, lubricating oil, and overhead costs. This study will estimate operating costs according to existing conditions in the field from various sources. Because the proposed propulsion system uses BoG as gas turbine fuel, fuel costs for diesel oil are eliminated.

Table 3. Estimated cost for each component of the proposed propulsion system.

Component	Estimated Cost	Reference
Gas Turbine	USD 55,000,000.00	[51]
Steam Turbine	USD 18,000,000.00	[51]
HRSG	USD 26,000,000.00	[51]
Generator	USD 15,000,000.00	[51]
Hot Water Supply System	USD 7,000,000.00	[51]
Condenser	USD 80,000.00	[51]
Deaerator	USD 800,000.00	[51]
Cooling Pump	USD 8000.00	[51]

To calculate the economic feasibility of the proposed system, it is assumed that income comes from ship charters. The ship is an LNG carrier-type ship, which functions as a charter ship that delivers LNG from resources to places closer to consumers. The type of charter used in this study involves the party carrying out ship operations being the ship owner with a time charter type. According to LNG ship charter rate data from LNG industry sources, the daily charter costs for LNG carrier ships vary greatly, namely, 36,038 USD/day in 2015 and 89,200 USD/day in 2021 [52,53]. This study used several variations of the charter rate to assess whether the proposed propulsion system is feasible. The charter rate variations used are 30,000–70,000 USD/day.

3. Results and Discussion

3.1. Performance of COGES System

Performance analysis of the combined gas turbine cycle propulsion system was conducted to determine the amount of power that can be generated by the design system and whether it can meet the ship's propulsion power needs, as previously calculated. Then, for system performance analysis, Cycle-Tempo applications were used to determine the value of the thermodynamic input in the design system, so that the power output of the combined gas-steam turbine system in this design could be determined. The thermodynamic results obtained at 100% fuel load are shown in Table 4. Based on the simulation results of the COGES system, the total power produced reached 8369 kW under 100% loading conditions.

Table 4. Results of system thermodynamics.

Component	Power (kW)
Gas turbine air compressor	7359.36
Gas turbine	13,085.89
Actual gas turbine	**5726.53**
Pump	2.82
Steam turbine	2645.52
Actual steam turbine	**2642.70**
Total Combined Gas and Steam	**8369.23**

Based on the results shown in Figure 6, it is obtained that the minimum requirement for the ship to be able to move at a service speed of 12 knots is 1832 kW; using COGES, the system needs to work at 24% loading conditions. Meanwhile, if the ship is going to move at its maximum design speed of 14 knots, the required power is 3377 kW using a COGES combination design system at 44% loading conditions. Based on thermodynamic simulation data, the overall efficiency of the COGES system was calculated using

Equation (5), resulting in a maximum efficiency of the system is 30.1%. This aligns with the practical operation of low-power-range power plants, where efficiencies typically range from 25% to 35% [54,55]. From these results, the COGES designed propulsion system can produce greater output power than commonly used factory engines such as diesel engines or dual-fuel engines. The COGES system can produce power output in the range of 15 kWh at a heat input of 50,000 kJ to 42.6 kWh at an input of 150,000 kJ. Meanwhile, the diesel and DFDE engines produce a power output of around 6 kWh at an input of 50,000 kJ to 20 kWh at an input of 150,000 kJ.

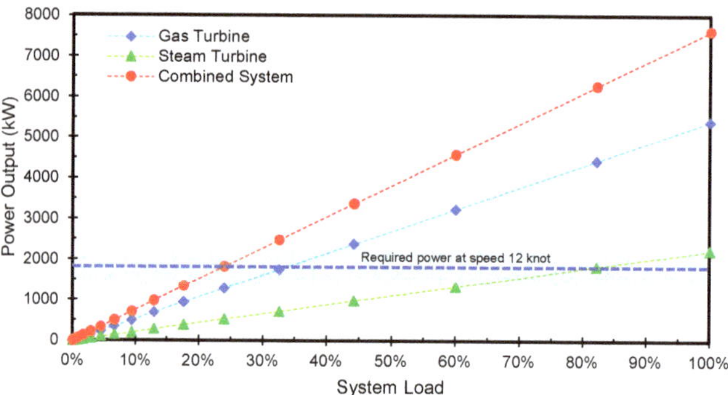

Figure 6. Power output of design propulsion system.

The system running condition depends on fuel availability, namely, boil-off gas produced from the LNG cargo tank. Figure 7 shows the boil-off gas produced with a boil-off rate variation of 0.1–0.3% of the LNG cargo capacity at each ship speed. The boil-off rate of the LNG generally ranges from 0.1% to 0.15%/day for large LNG vessels, while small LNG vessels have a high boil-off rate of between 0.2% and 0.6%/day, depending on the type of tank and the amount of heat introduced [56,57]. The design propulsion system requirements are based on fuel availability; it was found that the design propulsion system using combined gas–steam can meet its needs under conditions of a gas boil-off rate of 0.3%/day at a service speed of 12 knots with a need of 69.5 m^3 and availability of 71.6 m^3 to cover the route cruise according to a plan for 3 days of travel. In conditions of a gas boil-off rate of 0.25%/day, the ship can sail at a constant speed of 11 knots, with the ship's boil-off gas availability still experiencing a positive margin with an availability of 65.1 m^3 with a requirement of 55.9 m^3. In conditions of a gas boil-off rate of 0.2%/day, the ship can sail at a constant speed of 10 knots, with the ship's boil-off gas availability still experiencing a positive margin with an availability of 57.3 m^3 with a requirement of 45.1 m^3. With a gas boil-off rate of 0.15%/day, the ship can sail at a constant speed of 9 knots with the availability of boil-off gas. The ship still experiences a positive margin with an availability of 47.7 m^3 and a need of 36.3 m^3. Then, in the condition of a gas boil-off rate of 0.1%/day, the ship can sail at a constant speed of 8 knots with the availability of boil-off gas; the ship still experiences a positive margin with an availability of 35.8 m^3 with a requirement of 28.7 m^3. The relationship between the availability of boil-off gas fuel is that the faster the ship sails, the shorter the travel time will be, so the availability of boil-off gas will also be less. Still, the system requirements will be more significant by increasing the existing speed.

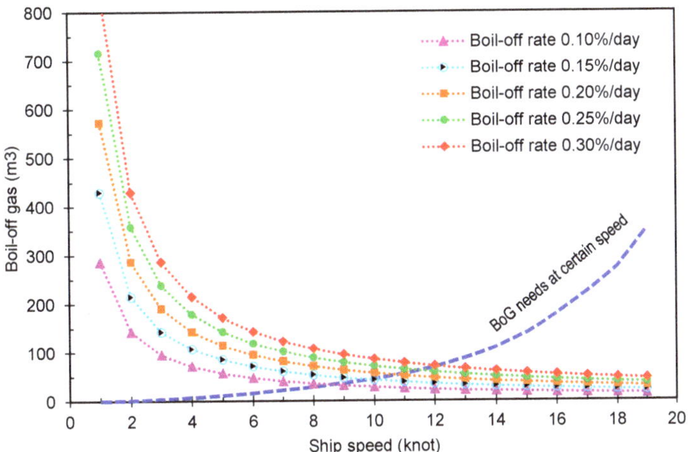

Figure 7. Boil-off gas production at certain ship speeds.

3.2. Analysis of the Environmental Impact of the System

The system's environmental impact was analyzed using the SimaPro software emission simulation application, which aims to obtain the results of emissions released by the designed propulsion system and the comparative propulsion system. Simulations using the SimaPro application were carried out using the IPCC 2013 GWP 100a method, which is a method based on data published by the Intergovernmental Panel on Climate Change that was chosen to provide assessment results for greenhouse gas emissions produced by a cycle in kilograms of CO_2 with global warming potential for the next 100 years front. Referring to the input and output of each component in the COGES system, each component influences the environment, which can be transmitted into an eco-indicator by calculating the total emissions produced by each component. Processing data for each component produces eco-indicators that impact the human health, ecosystem, and resources in the form of PTS, which indicates a representation of the annual environmental load, as seen in Table 5. Then the results of the eco-indicator data are processed to obtain the value of the environmental impact caused by the system in mPts/s units, which is shown in Figure 8.

Table 5. Eco-indicator results from each component of COGES system.

Component	Eco-Indicator (Pts)			
	Ecosystem	Health	Source	Total
Compressor	0.326	10.5	386	396.8
Combustion Chamber	1.2	38.9	1420	1460.1
Turbin Gas	0.535	17.3	634	651.8
HRSG	0.5	16.2	592	608.7
Turbin Steam	0.101	3.26	119	122.4
Boiler	0.243	7.86	288	296.1
Deaerator	0.0307	0.993	36.4	37.4
Feed Pump	0.0316	1.02	37.4	38.4
Condenser	0.204	6.61	242	248.8
Cooling Pump	0.0013	0.043	1.58	1.62

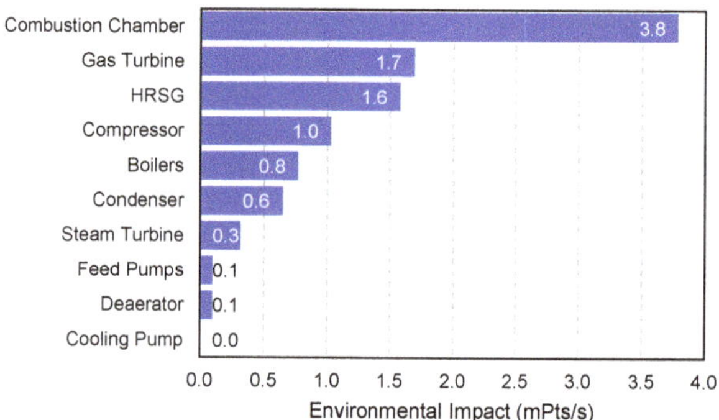

Figure 8. Environmental impact of each COGES component.

In this research, carbon emissions were also investigated for the entire system compared to commonly used engines such as diesel or DFDE. The emission calculation parameters are limited to each system's CO_2 gas emissions and damage assessment. The results of comparing emissions and damage assessments for each system are shown in Table 6. The results of the environmental impact analysis show that the COGES system has the most negligible impact on the environment compared to the diesel propulsion system and the DFDE propulsion system. This can happen because the COGES system only uses natural gas as fuel and does not use any other fossil fuels. This makes the COGES propulsion system more environmentally friendly. From the LCA simulation, it show the climate change contribution of the COGES propulsion system is 0.149 $kgCO_2$ equivalents, compared to the contribution of the diesel propulsion system of 0.314 $kgCO_2$ equivalents and the DFDE system of 0.155 kg CO_2 equivalents. Although not wholly zero-emissions, COGES systems produce relatively lower emissions than conventional diesel engines. Higher efficiency and cleaner fuel can reduce carbon dioxide (CO_2) and nitrogen oxide (NO_x) emissions [58]. The COGES system does not rely on combustion like diesel and DFDE systems, which can lead to lower methane emissions, a potent greenhouse gas [59]. This is in line with other research; from direct measurements on LNG carrier ships, it was found that CO_2 emissions from using gas fuel were lower compared to conventional diesel engines [60].

Table 6. Comparison of emissions and damage assessments for COGES and other systems.

Category	Unit	COGES	Diesel	DFDE
Climate Change	$kgCO_2$ equivalents	0.149	0.314	0.155
Human Health	DALY	0.755×10^{-7}	4.61×10^{-7}	1.64×10^{-7}
Ecosystem Quality	$PDFm^2yr$	0.00102	0.0153	0.00514
Resource	MJ surplus	2.06	4.09	2.744

Apart from the results of CO_2 emissions, a comparison of environmental impacts was obtained using the eco-indicator 99 (H) method. Where environmental impacts affect health, ecosystem quality, and resources for a power output of 1 kWh, the result is that the COGES system has the most negligible influence on damage for the three existing categories. In the health category, only 0.755×10^{-7} DALY or around 2.2 s of healthy life is lost by a person to produce 1 kWh, while the diesel system affects 4.61×10^{-7} DALY or around 14.5 s of healthy life per 1 kWh and the DFDE system affect 1.64×10^{-7} DALYs or around 5.2 s of healthy life per 1 kWh.

3.3. Economic Feasibility of COGES System

The economic feasibility results of the proposed propulsion system are shown in Table 7. Based on the results of the economic study, it can be seen that the ship's charter rate costs greatly influence the NPV value. The COGES system is economically feasible if the minimum charter rate is 50,000 USD/day. With this charter rate, a positive NPV value of around 25 million USD is obtained with an IRR of 12.6% with a payback period of 8 years of operation. With a charter rate of 50,000 USD/day, it is assumed that the profits obtained can exceed the capital costs incurred for the proposed propulsion system during the specified payback period. The estimated capital costs were shown in the previous section in Table 3. With the charter rate variations shown in Table 4, to obtain a positive NPV value, a minimum charter rate of 50,000 USD/day is required. A positive NPV means that the proposed system produces a more significant present value than the initial investment, indicating that the investment can provide good financial returns.

Table 7. Economic feasibility of COGES system.

Charter Rate	Net Present Value (NPV)	Internal Rate Return (IRR)	Payback Period (PBP)
30,000 USD/Day	USD 33,845,188.90	6.18%	13 Years
40,000 USD/Day	USD 4,057,766.87	9.54%	10 Years
50,000 USD/Day	USD 25,585,478.17	12.63%	8 Years
60,000 USD/Day	USD 55,228,723.22	15.55%	7 Years
70,000 USD/Day	USD 84,871,968.26	18.38%	6 Years

The results of the economic feasibility study are analyzed further to obtain recommendations for the maximum value of the discount rate if the investment is made based on the rate of return used to calculate the present value. Figure 9 shows the NPV and discount rate values with variations in the charter rate. To assess whether a proposed system investment is feasible or not, the IRR of a system must be higher than the discount rate; then, the system is considered feasible because the rate of return generated is greater than the discount rate used. From these results, it is found that for the proposed system to be economically feasible, the discount rate value cannot be greater than 13% for a charter rate of 50,000 USD/day. Determining the discount rate for investment, including investment in LNG vessels, includes system risks, LNG market conditions, the life cycle of LNG vessels, and regulations, including government policy. The risk level of the LNG ship system affects the discount rate. If a system has a high level of risk, investors may expect a higher rate of return to compensate for that risk. In general, systems with higher risk require a more significant discount rate to reflect the higher level of risk.

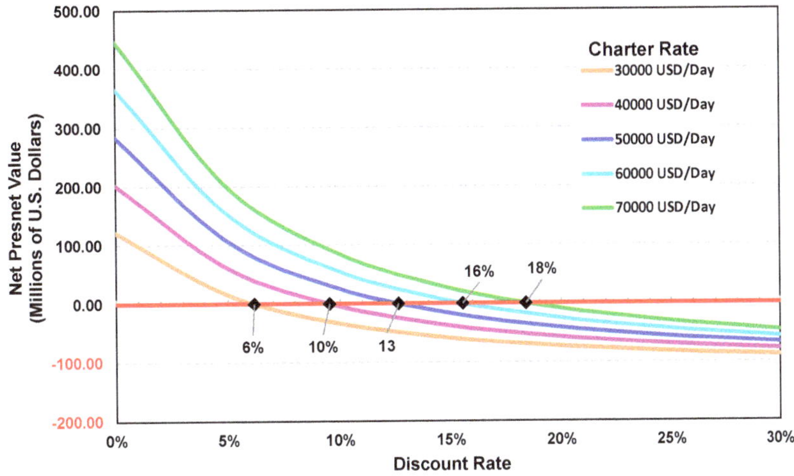

Figure 9. The NPV and discount rate values with variations in the charter rate.

4. Conclusions

In this research, a techno-economic study of the COGES system of LNG carriers was carried out. The study was carried out using a case study of an LNG ship with a capacity of 7500 m^3 with a power requirement at an average speed of 12 knots of 1832 kW. The design of the proposed COGES propulsion system consists of three main parts, namely, the gas turbine, HRSG, and steam turbine, which utilizes boil-off gas as the primary fuel. Based on the results of thermodynamic analysis, the COGES system obtains a total system efficiency of 30.1%, which can achieve the desired power output. An environmental impact assessment compared the life cycle with that of commonly used propulsion systems. The climate change from the COGES propulsion system contribute 0.149 kgCO$_2$ equivalents compared to the contribution of the diesel propulsion system of 0.314 kgCO$_2$ equivalents and the DFDE system of 0.155 kgCO$_2$ equivalents. The latest study carried out an economic study with variations in the charter rate value of LNG ships; it was found that the COGES system is economically feasible if the minimum charter rate is 50,000 USD/day. A positive NPV value of around 25 million USD is obtained with an IRR of 12.6% with a payback period of 8 years of operation. From these results, it can be concluded that the COGES system is feasible from the aspects of performance, environment, and investment, so this system can be used as a convincing alternative for future propulsion systems. The findings support prior studies indicating that the COGES system yields reduced emissions. While not completely emission-free, COGES demonstrates comparatively lower emissions than traditional diesel engines. Enhanced efficiency and the adoption of cleaner fuels contribute to a reduction in carbon emissions, thereby mitigating the environmental footprint of maritime shipping.

Author Contributions: Conceptualization, M.A.B.; methodology, M.A.B. and S.A.Z.; software, S.A.Z.; validation, M.A.B., R.A. and G.T.; formal analysis, M.A.B., G.L.P., A.R., A.H.M. and R.A.; investigation, M.A.B., G.L.P. and A.R.; resources, G.L.P.; data curation, A.R.; writing—original draft preparation, M.A.B. and S.A.Z.; writing—review and editing; M.A.B. and G.T.; funding acquisition, M.A.B. All authors have read and agreed to the published version of the manuscript.

Funding: Q1 International Indexed Publication (PUTI) Grant for Fiscal Year 2023–2024 Number NKB-533/UN2.RST/HKP.05.00/2023 Universitas Indonesia.

Data Availability Statement: Data is unavailable due to privacy.

Acknowledgments: The authors express great appreciation to the Directorate of Research and Development at the Universitas Indonesia.

Conflicts of Interest: The authors declare no conflict of interest.

References

1. International Maritime Organization (IMO). *Initial Imo Strategy on Reduction of Ghg Emissions from Ships*; International Maritime Organization (IMO): London, UK, 2018.
2. Joung, T.-H.; Kang, S.-G.; Lee, J.-K.; Ahn, J. The IMO initial strategy for reducing Greenhouse Gas (GHG) emissions, and its follow-up actions towards 2050. *J. Int. Marit. Safety Environ. Aff. Shipp.* **2020**, *4*, 1–7. [CrossRef]
3. Mallouppas, G.; Yfantis, E.A. Decarbonization in Shipping Industry: A Review of Research, Technology Development, and Innovation Proposals. *J. Mar. Sci. Eng.* **2021**, *9*, 415. [CrossRef]
4. Nuchturee, C.; Li, T.; Xia, H. Energy efficiency of integrated electric propulsion for ships—A review. *Renew. Sustain. Energy Rev.* **2020**, *134*, 110145. [CrossRef]
5. Dotto, A.; Campora, U.; Satta, F. Feasibility study of an integrated COGES-DF engine power plant in LNG propulsion for a cruise-ferry. *Energy Convers. Manag.* **2021**, *245*, 114602. [CrossRef]
6. Merz, C.A.; Pakula, T.J. The Design and Operational Characteristics of a Combined Cycle Marine Powerplant. In Proceedings of the ASME Turbo Expo, Montreal, QC, Canada, 15–19 June 2015.
7. Haglind, F. A review on the use of gas and steam turbine combined cycles as prime movers for large ships. Part II: Previous work and implications. *Energy Convers. Manag.* **2008**, *49*, 3468–3475. [CrossRef]
8. Packalén, S.; Karlsson Nord, N. *Combined Gas- and Steam Turbine as Prime Mover in Marine Applications*; Department of Shipping and Marine Technology, Chalmers University of Technology: Gothenburg, Sweden, 2017.

9. Alzayedi, A.M.T.; Sampath, S.; Pilidis, P. Techno-Environmental Evaluation of a Liquefied Natural Gas-Fuelled Combined Gas Turbine with Steam Cycles for Large Container Ship Propulsion Systems. *Energies* **2022**, *15*, 1764. [CrossRef]

10. Pereira, S.; Almeida, P. Energy Assessment and Comparison of the Use of a Conventional Steam Turbine System in a Lng Carrier Propulsion with the Use of a Coges System. *SSRN Electron. J.* **2022**. [CrossRef]

11. LNGreen: Next-Generation LNG Carrier Concept by DNV GL, HHI, GTT and GasLog n.d. Available online: https://www.dnv.com/news/lngreen-next-generation-lng-carrier-concept-by-dnv-gl-hhi-gtt-and-gaslog-30420 (accessed on 6 December 2023).

12. Pamitran, A.S.; Budiyanto, M.A.; Dandy Yusuf Maynardi, R. Analysis of ISO-tank wall physical exergy characteristic—Case study of LNG boil-off rate from retrofitted dual fuel engine conversion. *Evergreen* **2019**, *6*, 134–142. [CrossRef]

13. Al-Kubaisi, A. An Insight into the World's Largest LNG Ships. In Proceedings of the International Petroleum Technology Conference, Kuala Lumpur, Malaysia, 3–5 December 2008. [CrossRef]

14. Wibisana, L.P.; Budiyanto, M.A. Design and Cost Multi-Objective Optimization of Small-Scale LNG Carriers using the Value Engineering Approach. *Int. J. Technol.* **2021**, *12*, 1288–1301. [CrossRef]

15. Budiyanto, M.A.; Singgih, I.K.; Riadi, A.; Putra, G.L. Study on the LNG distribution to Mobile Power Plants using a Small-Scale LNG Carrier for the case of the Sulawesi region of Indonesia. *Energy Rep.* **2022**, *8*, 374–380. [CrossRef]

16. Yin, L.; Ju, Y. Review on the design and optimization of BOG re-liquefaction process in LNG ship. *Energy* **2022**, *244*, 123065. [CrossRef]

17. Qu, Y.; Noba, I.; Xu, X.; Privat, R.; Jaubert, J.-N. A thermal and thermodynamic code for the computation of Boil-Off Gas—Industrial applications of LNG carrier. *Cryogenics* **2019**, *99*, 105–113. [CrossRef]

18. Ferrín, J.L.; Pérez-Pérez, L.J. Numerical simulation of natural convection and boil-off in a small size pressurized LNG storage tank. *Comput. Chem. Eng.* **2020**, *138*, 106840. [CrossRef]

19. Gerasimov, V.E.; Kuz'Menko, I.F.; Peredel'Skii, V.A.; Darbinyan, R.V. Introduction of technologies and equipment for production, storage, transportation, and use of LNG. *Chem. Pet. Eng.* **2004**, *40*, 31–35. [CrossRef]

20. Budiyanto, M.A.; Riadi, A.; Buana, I.G.N.S.; Kurnia, G. Study on the LNG distribution to mobile power plants utilizing small-scale LNG carriers. *Heliyon* **2020**, *6*, e04538. [CrossRef] [PubMed]

21. APEC. Study on Optimal Use of Small-Scale Shallow-Draft LNG Carriers and FSRUs in the APEC Region APEC Energy Working Group. 2020. Available online: https://www.apec.org/docs/default-source/publications/2020/4/study-on-optimal-use-of-small-scale-shallow-draft-lng-carriers-and-fsrus-in-the-apec-region/220_ewg_study-on-optimal-use-of-small-scale-shallow-draft-lng-carriers-and-fsrus-in-the-apec-region.pdf (accessed on 21 December 2023).

22. Budiyanto, M.A.; Pamitran, A.S.; Yusman, T. Optimization of the Route of Distribution of LNG using Small Scale LNG Carrier: A Case Study of a Gas Power Plant in the Sumatra Region, Indonesia. *Int. J. Energy Econ. Policy* **2019**, *9*, 179–187. [CrossRef]

23. Hongjun, F.; Lei, W.; Bo, Z.; Enshaei, H.; Jiaolong, C. A study on design optimization of LNG power system onboard a dual fueled platform supply vessel. In Proceedings of the 2nd International Conference on Smart & Green Technology for Shipping and Maritime Industries, Glasgow, Scotland, 11–12 July 2019. Available online: https://hdl.handle.net/102.100.100/522683 (accessed on 21 December 2023).

24. Boretti, A. Advantages and Disadvantages of Diesel Single and Dual-Fuel Engines. *Front. Mech. Eng.* **2019**, *5*, 493925. [CrossRef]

25. Budiyanto, M.A.; Nasruddin; Nawara, R. The optimization of exergoenvironmental factors in the combined gas turbine cycle and carbon dioxide cascade to generate power in LNG tanker ship. *Energy Convers. Manag.* **2020**, *205*, 112468. [CrossRef]

26. Cwilewicz, R.; Górski, Z. Prognosis of marine propulsion plants development in view of new requirements concerning marine fuels. *J. Kones* **2014**, *21*, 61–68. [CrossRef]

27. Dotto, A.; Sacchi, R.; Satta, F.; Campora, U. Dynamic performance simulation of combined gas electric and steam power plants for cruise-ferry ships. *Next Energy* **2023**, *1*, 100020. [CrossRef]

28. Fernández, I.A.; Gómez, M.R.; Gómez, J.R.; Insua, B. Review of propulsion systems on LNG carriers. *Renew. Sustain. Energy Rev.* **2017**, *67*, 1395–1411. [CrossRef]

29. Tadros, M.; Ventura, M.; Guedes Soares, C. A nonlinear optimization tool to simulate a marine propulsion system for ship conceptual design. *Ocean Eng.* **2020**, *210*, 107417. [CrossRef]

30. Inal, O.B.; Charpentier, J.F.; Deniz, C. Hybrid power and propulsion systems for ships: Current status and future challenges. *Renew. Sustain. Energy Rev.* **2021**, *156*, 111965. [CrossRef]

31. Kurz, R. Parameter Optimization on Combined Gas Turbine-Fuel Cell Power Plants. *J. Fuel Cell Sci. Technol.* **2005**, *2*, 268–273. [CrossRef]

32. Sulligoi, G.; Vicenzutti, A.; Menis, R. All-electric ship design: From electrical propulsion to integrated electrical and electronic power systems. *IEEE Trans. Transp. Electrif.* **2016**, *2*, 507–521. [CrossRef]

33. Pamik, M.; Nuran, M. The historical process of the diesel electric propulsion system. *Dokuz Eylül Üniversitesi Denizcilik Fakültesi Derg.* **2021**, *13*, 299–316. [CrossRef]

34. Ahmadi, G.; Jahangiri, A.; Toghraie, D. Design of heat recovery steam generator (HRSG) and selection of gas turbine based on energy, exergy, exergoeconomic, and exergo-environmental prospects. *Process. Saf. Environ. Prot.* **2023**, *172*, 353–368. [CrossRef]

35. Nirbito, W.; Budiyanto, M.A.; Muliadi, R. Performance Analysis of Combined Cycle with Air Breathing Derivative Gas Turbine, Heat Recovery Steam Generator, and Steam Turbine as LNG Tanker Main Engine Propulsion System. *J. Mar. Sci. Eng.* **2020**, *8*, 726. [CrossRef]

36. Huan, T.; Hongjun, F.; Wei, L.; Guoqiang, Z. Options and Evaluations on Propulsion Systems of LNG Carriers. In *Propulsion Systems*; IntechOpen: London, UK, 2019. [CrossRef]

37. Lombardi, L.; Carnevale, E.; Corti, A. A review of technologies and performances of thermal treatment systems for energy recovery from waste. *Waste Manag.* **2015**, *37*, 26–44. [CrossRef] [PubMed]

38. Singh, D.V.; Pedersen, E. A review of waste heat recovery technologies for maritime applications. *Energy Convers. Manag.* **2016**, *111*, 315–328. [CrossRef]

39. Díaz-Secades, L.A.; González, R.; Rivera, N.; Montañés, E.; Quevedo, J.R. Waste heat recovery system for marine engines optimized through a preference learning rank function embedded into a Bayesian optimizer. *Ocean Eng.* **2023**, *281*, 114747. [CrossRef]

40. Larsen, U.; Sigthorsson, O.; Haglind, F. A comparison of advanced heat recovery power cycles in a combined cycle for large ships. *Energy* **2014**, *74*, 260–268. [CrossRef]

41. Ouyang, T.; Zhao, Z.; Lu, J.; Su, Z.; Li, J.; Huang, H. Waste heat cascade utilisation of solid oxide fuel cell for marine applications. *J. Clean. Prod.* **2020**, *275*, 124133. [CrossRef]

42. Theotokatos, G.; Livanos, G. Techno-economical analysis of single pressure exhaust gas waste heat recovery systems in marine propulsion plants. *Proc. Inst. Mech. Eng. Part M J. Eng. Marit. Environ.* **2012**, *227*, 83–97. [CrossRef]

43. Buonomano, A.; Del Papa, G.; Francesco Giuzio, G.; Maka, R.; Palombo, A. Advancing sustainability in the maritime sector: Energy design and optimization of large ships through information modelling and dynamic simulation. *Appl. Therm. Eng.* **2023**, *235*, 121359. [CrossRef]

44. Çolak, K.; Ölmez, H.; Saraç, B. Waste heat recovery assessment of triple heat-exchanger usage for ship main engine pre-heating and fresh water generation systems. *Proc. Inst. Mech. Eng. Part M J. Eng. Marit. Environ.* **2023**, *238*, 209–230. [CrossRef]

45. Benvenuto, G.; Trucco, A.; Campora, U. Optimization of waste heat recovery from the exhaust gas of marine diesel engines. *Proc. Inst. Mech. Eng. Part M J. Eng. Marit. Environ.* **2014**, *230*, 83–94. [CrossRef]

46. Altosole, M.; Benvenuto, G.; Campora, U.; Laviola, M.; Trucco, A. Waste Heat Recovery from Marine Gas Turbines and Diesel Engines. *Energies* **2017**, *10*, 718. [CrossRef]

47. Saidur, R.; Rezaei, M.; Muzammil, W.K.; Hassan, M.H.; Paria, S.; Hasanuzzaman, M. Technologies to recover exhaust heat from internal combustion engines. *Renew. Sustain. Energy Rev.* **2012**, *16*, 5649–5659. [CrossRef]

48. Farhat, O.; Faraj, J.; Hachem, F.; Castelain, C.; Khaled, M. A recent review on waste heat recovery methodologies and applications: Comprehensive review, critical analysis and potential recommendations. *Clean. Eng. Technol.* **2022**, *6*, 100387. [CrossRef]

49. Asimptote. Cycle-Tempo. n.d. Available online: https://asimptote.com/cycle-tempo/ (accessed on 31 January 2023).

50. Japan International Cooperation Agency. Financial and Economical Analysis of New Power Plant. 2019. Available online: https://openjicareport.jica.go.jp/pdf/12119806_01.pdf (accessed on 21 December 2023).

51. Japan International Cooperation Agency (JICA). Cost Estimation of the New Power Plant for Construction and Operation. 2004. Available online: https://openjicareport.jica.go.jp/pdf/11750742_19.pdf (accessed on 21 December 2023).

52. Global LNG Shipping—ORders and Deliveries Report | Wood Mackenzie. n.d. Available online: https://www.woodmac.com/reports/lng-global-lng-shipping-orders-and-deliveries-58425985/ (accessed on 21 December 2023).

53. LNG Tanker Average Spot Charter Rate 2021 | Statista n.d. Available online: https://www.statista.com/statistics/1112660/lng-tanker-average-spot-charter-rate/ (accessed on 21 December 2023).

54. Poullikkas, A. An overview of current and future sustainable gas turbine technologies. *Renew. Sustain. Energy Rev.* **2005**, *9*, 409–443. [CrossRef]

55. Banihabib, R.; Assadi, M. The Role of Micro Gas Turbines in Energy Transition. *Energies* **2022**, *15*, 8084. [CrossRef]

56. Harperscheidt, J. LNG as Fuel-Bunkering, Storage and Processing. 2003. Available online: https://www.stg-online.org/onTEAM/shipefficiency/programm/12-Harperscheidt_TGE.pdf (accessed on 21 December 2023).

57. Kim, K.; Park, K.; Roh, G.; Chun, K. Case Study on Boil-Off Gas (BOG) Minimization for LNG Bunkering Vessel Using Energy Storage System (ESS). *J. Mar. Sci. Eng.* **2019**, *7*, 130. [CrossRef]

58. Budiyanto, M.A.; Zidane, S.A.; Putra, G.L.; Riadi, A.; Andika, R.; Theotokatos, G. Performance Analysis of Combined Gas-Electric Steam Turbine System as Main Propulsion for Small-scale LNG Carrier Ships. *Int. J. Technol.* **2023**, *14*, 1093–1102. [CrossRef]

59. Davidson, A.; Wait, N.; Desouza, C.; Marsh, D.; Green, D.; Andrews, P.; Chapman, S. Clean Air Gas Engine (CAGE): Reducing emissions by replacing diesel power in construction. *High Speed Two (HS2) Infrastruct. Des. Constr.* **2023**, *4*, 229–235. [CrossRef]

60. Balcombe, P.; Heggo, D.A.; Harrison, M. Total Methane and CO_2 Emissions from Liquefied Natural Gas Carrier Ships: The First Primary Measurements. *Environ. Sci. Technol.* **2022**, *56*, 9632–9640. [CrossRef]

Article

Active Autonomous Open-Loop Technique for Static and Dynamic Current Balancing of Parallel-Connected Silicon Carbide MOSFETs

Nektarios Giannopoulos, Georgios Ioannidis *, Georgios Vokas and Constantinos Psomopoulos

Department of Electrical and Electronics Engineering, University of West Attica (Ancient Olive Grove Campus), 250 Thivon & P. Ralli Str., 12241 Egaleo, Greece; n.giannopoulos@uniwa.gr (N.G.); gvokas@uniwa.gr (G.V.); cpsomop@uniwa.gr (C.P.)
* Correspondence: gioan@uniwa.gr

Abstract: Silicon carbide (SiC) MOSFETs tend to become one of the main switching elements in power electronics applications of medium- and high-power density. Usually, SiC MOSFETs are connected in parallel to increase power rating. Unfortunately, unequal current sharing between power devices occurs due to mismatches in the technical parameters between devices and the layout of the power circuit. This current imbalance causes different current stress upon power switches, raising concerns about power system reliability. For over a decade, various methods and techniques have been proposed for balancing the currents between parallel-connected SiC MOSFETs. However, most of these methods cannot be implemented unless the deviation between the technical parameters of semiconductor switches is known. This requirement increases the system cost because screening methods are extremely costly and time-consuming. In addition, most techniques aim at suppressing only the transient current imbalance. In this paper, a simple but innovative current balancing technique is proposed, without the need of screening any power device. The proposed technique consists of an open-loop system capable of balancing the currents between two parallel-connected SiC MOSFETs, with the aid of two active gate drivers and an FPGA, actively and independently of the cause. Experimental test results validate that the proposed open-loop method can successfully achieve suppression of current imbalance between parallel-connected SiC MOSFETs, proving its durability and validity level.

Keywords: energy systems; energy system components; parallel-connected silicon carbide (SiC) MOSFETs; active current balancing technique; optimization models

Citation: Giannopoulos, N.; Ioannidis, G.; Vokas, G.; Psomopoulos, C. Active Autonomous Open-Loop Technique for Static and Dynamic Current Balancing of Parallel-Connected Silicon Carbide MOSFETs. *Energies* **2023**, *16*, 7670. https://doi.org/10.3390/en16227670

Academic Editors: Konstantinos Aravossis and Eleni Strantzali

Received: 19 October 2023
Revised: 17 November 2023
Accepted: 18 November 2023
Published: 20 November 2023

1. Introduction

Power electronic systems and applications are considered to be one of the most essential parts of renewable energy sources (RES), electric vehicles (EV), EV chargers, and EV main inverters, reducing the environmental impact. High-capacity, high-temperature, and high-frequency power converters are increasingly demanded to increase power density, reduce costs, and save manpower. For this reason, the most significant features of new technology power converters are minimization of volume, maximization of efficiency, high reliability, and increased durability against short-circuit/overvoltage conditions. To achieve these goals, the transaction from silicon (Si) to wide band gap (WBG)-based semiconductor power switches is of great significance because of their outstanding features compared to Si ones.

Among WBG power switches, SiC MOSFET is considered to be the most promising alternative solution to conventional semiconductor devices in medium- and high-power-density power converter fields. This is attributed to its exceptional characteristics, such as the relatively mature technology, the low cost, and its more stable construction [1,2]. Indeed, the most significant features of SiC MOSFET are the high thermal conductivity and operating temperature capability, the higher breakdown voltage, its superior switching characteristics, the lower on-resistance, the usage of low complexity gate drivers, and the normally-off characteristic. Moreover, SiC MOSFET has no tail current leading to reduced switching losses as well as higher switching frequency [1–3].

Nevertheless, SiC MOSFET offers less current capability than Si ones due to its smaller chip area. This derives from the lower maturity of the manufacturing process of SiC MOSFETs compared to Si ones which includes the lower yield in the wafer as well as the high thermal and mechanical stress in the device. As a result, the current ratings of commercially available discrete SiC MOSFETs with a maximum blocking voltage of 1.2 kV and 1.7 kV are within 120 A and 100 A, respectively. For this reason, the current rating is usually boosted by connecting multiple SiC MOSFET devices in parallel [4–8].

However, the parallelization of SiC MOSFETs introduces the problem of current imbalance which is unpredictable. This results in uneven conduction and switching losses between parallel devices. This, in turn, causes uneven sharing of junction temperature, increasing the risk of SiC MOSFET(s) being led to thermal runaway [9,10]. Also, an over-current and, at the same time, overheating is quite possible. Therefore, it is essential to overcome any possibility of power device failure due to high junction temperature swing by suppressing the current imbalance and retaining distributed heat between power devices as equally as possible [4,11,12].

Current imbalance can be caused by a device package parameter mismatch. For instance, the variation in on-state resistance ($R_{DS\text{-}on}$) causes unequal current sharing during the steady-state, leading to static current imbalance. Moreover, any difference in threshold voltage (V_{th}) and trans-conductance (g_m) leads to uneven current distribution during turn-on and -off intervals, causing dynamic current imbalance [4]. Puschkarsky et al., in [13], experimentally proved the V_{th} instability of high-voltage SiC MOSFETs which may be either short-time or even permanent. Asymmetry of the PCB layout of the power circuit may also affect the current sharing of power devices [14]. Also, both types of imbalances result in unequal temperature rises and electromagnetic interference (EMI), endangering system reliability.

Over the last decade, the current imbalance issue has been addressed by many researchers, proposing techniques and methods capable of minimizing one or both current imbalance types between discrete parallel-connected SiC MOSFETs.

Refs. [4,15,16] proposed several active current balancing methods that make use of current sensors to actively detect current imbalance. Subsequently, an analog controller receives the dynamic imbalance and suppresses it through a gate driver by matching the switching behaviors of the parallel switches.

Most proposed methods suppress static or/and dynamic current imbalance by using passive elements. Refs. [11,17] mitigated dynamic imbalance by adding extra coupled inductance and external same-size gate resistors. As a result, control voltages of parallel devices vary only during transient stages, eliminating the entire transient imbalance. In the same way, ref. [18] eliminated dynamic imbalance with the addition of different-size gate resistors differentiating each gate loop impedance and eliminating turn-on dynamic imbalance. Ref. [19] also deals with turn-on dynamic imbalance by making the gate resistor of the power switch with the smallest V_{th} greater, delaying the charging process of its input capacitance (C_{iss}). In refs. [8,20], static current imbalance was suppressed by adding same-size resistors, serially connected with the drains of the parallel-connected devices. In addition, ref. [8] mitigates both types of imbalances with a differential mode choke. Similarly, refs. [21,22] suppressed the overall current imbalance with the incorporation of a series-connected coupled inductor with the drains of power devices. On the other hand, ref. [23] suggests an alternative way to implement the two aforementioned methods to suppress current imbalance and avoid the disadvantages provoked by the usage of a coupled inductor. Finally, ref. [24] eliminates the entire imbalance by connecting a planar transformer in series with the drains of each power device.

Refs. [1,25–30] proposed novel screening methods for discovering SiC MOSFETs with very close technical parameters, such as $R_{DS\text{-}on}$, V_{th}, and g_m, achieving a balanced current distribution without the requirement of a current imbalance suppression technique. Based on [10,31], the current imbalance, caused by the asymmetrical PCB layout of the power circuit, can be mitigated by lowering the deviations between drain and common source parasitic inductances. Nonetheless, ref. [32] deals with the imbalance, attributed to the asymmetry of the power circuit layout by incorporating a common mode choke to each parallel SiC MOSFET gate loop. This method holds that the current imbalance is limited as the choke mutual inductance becomes larger. Ref. [33] achieved an optimal transient current sharing by reducing the gate resistance, weakening the effect of V_{th} mismatch.

In addition, ref. [34] proposes a gate driver that generates PWMs with different time delays of picoseconds, suppressing dynamic imbalance. Additionally, ref. [35] balances the dynamic imbalance with a multi-stage gate driver with the ability to change the gate resistor during the transient stages. Finally, ref. [36] proposes a gate driver capable of varying the gate voltage to mitigate transient current imbalance.

All the aforementioned methods have the ability to eliminate either static or/and dynamic current imbalance between parallel SiC MOSFETs. On the other hand, most techniques cannot be implemented unless the technical parameters of SiC MOSFETs are known. For this reason, screening processes are needed and conducted with the aid of power device analyzers/curve tracers [20]. Also, ref. [37] proposed a method to monitor the on-resistance of SiC MOSFET. This necessity can be an inhibiting factor for their application in the industry since screening is an extremely costly and time-consuming process [4]. The implementation of the methods, proposed by refs. [4,15,16], can be realized without any screening process since dynamic current imbalance is mitigated with a closed-loop method that actively monitors and suppresses transient imbalance. However, static current imbalance suppression is not addressed. As a result, none of these techniques can minimize the whole current imbalance without knowledge of the device parameter mismatch. Also, the validity of most methods has not been tested under the condition of an asymmetrical PCB layout.

In this paper, an innovative, active, and autonomous open-loop current balancing technique is proposed which addresses the imbalance issue without the necessity of knowing the technical parameters of the power devices. In addition, the proposed technique can eliminate static and dynamic current imbalances actively and irrelevantly of the cause. In Section 2, an analysis to investigate potential strategies for current imbalance suppression is conducted. In Section 3, an analytical description of the structure, functions, and design guidelines of the active current balancing technique is provided. In Section 4, two experimental tests are conducted to verify its effectiveness and efficiency. In Section 5, extension guidelines of the proposed current balancing technique are proposed. Finally, in Section 6, the most important conclusions of this research are presented.

2. Strategies of Static and Dynamic Current Balancing

In this section, an analysis concerning the factors that lead to static and dynamic imbalance is conducted. In addition, an investigation to suppress both current imbalance types between parallel SiC MOSFETs is performed.

2.1. Strategies of Dynamic Current Imbalance Suppresion

As pointed out earlier, device parameter mismatch and parasitic element deviation of the power circuit cause dynamic current imbalance. Based on Equation (1), the technical parameters of the device affect its turn-on delay ($t_{d(on)}$). In this way, during turn-on transience, device parameter mismatch results in different $t_{d(on)}$ between parallel power devices leading to transient current imbalance [16].

$$t_{d(on)} = C_{iss} R_G \, ln\left(\frac{V_{CC}}{V_{CC} - V_{th}}\right) \tag{1}$$

In addition, during turn-on transience and before the drain–source voltage (V_{DS}) begins to fall, the power switch current becomes maximal [16]. At this stage, the power switch is in saturation mode and its current is expressed by Equation (2).

$$\frac{di_D}{dt} = \frac{V_{CC} - V_{th} - i_D/g_m}{L_S + R_G C_{iss}/g_m} \tag{2}$$

i_D is the drain current; V_{CC} is the activating voltage of SiC MOSFET; L_S is the source parasitic inductance of the gate and power loop; di_D/dt is the slew rate of drain current; and R_G is the gate resistance. According to Equation (2), the drain current slew rate is dependent on several factors. Therefore, any deviation of these parameters between parallel-connected SiC MOSFETs can affect transient current sharing causing dynamic current imbalance. Equations (1) and (2) can be similarly written for the turn-off transition.

Based on Equations (1) and (2), the magnitude of R_G affects t_d and di_D/dt [16]. The equivalent circuit of two power devices (dashed line) along with their gate drivers, during transient stages, is depicted in Figure 1. According to Figure 1, each time the SiC MOSFET drivers output V_{CC}, the gate-drain and gate-source capacitances of the two parallel MOSFETs ($C_{GD,i}$ and $C_{GS,i}$, respectively) begin to charge by gate current ($i_{G,i}$) which is expressed in Equation (3). In the preceding symbols and below, wherever an index i appears it refers to a parameter and quantity of MOSFET 1 when $i = 1$ and MOSFET 2 when $i = 2$. Also, $C_{DS,i}$ is the drain–source capacitance of the power device. As shown in Figure 1, $i_{GD,i}$, $i_{GS,i}$, and $i_{DS,i}$ are the currents that conduct $C_{GD,i}$, $C_{GS,i}$, and $C_{DS,i}$, respectively. $L_{G,int,i}$, $L_{D,int,i}$, and $L_{S,int,i}$ represent the parasitic inductances of the pins of power devices mainly caused by the manufacturing technology and production process. $i_{D,i}$ is the drain current conducting the drain and common source parasitic inductances ($L_{D,i}$ and $L_{S,i}$, respectively) which are attributed to the PCB power circuit. $V_{dr,i}$ is the output driving voltage of the gate driver and $u_{GS,i}(t)$ is the gate-source voltage which is applied across $C_{GS,i}$. Finally, $di_{D,i}(t)/dt$ is the slew rate of the drain current.

$$i_{G,i} = i_{GD,i} + i_{GS,i} \tag{3}$$

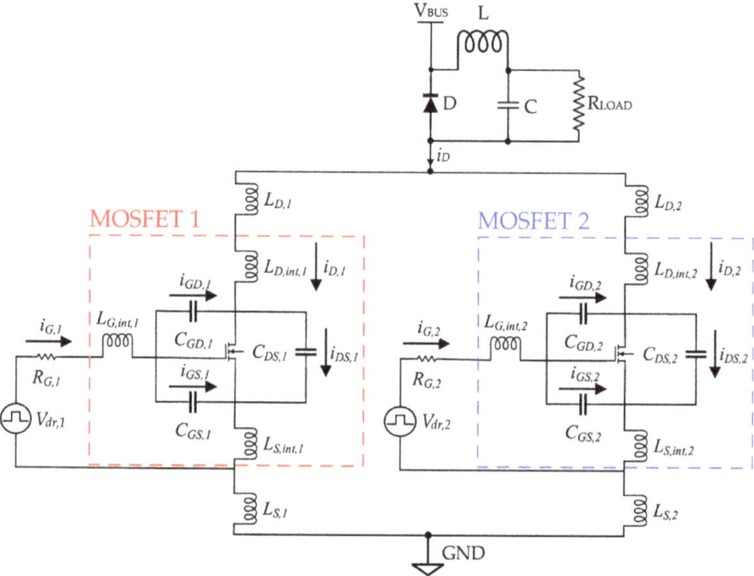

Figure 1. Equivalent circuit of two SiC MOSFETs connected in parallel, during transient stages, incorporated with a DC–DC buck converter.

As for the gate loop parasitic inductance ($L_{G,int,i}$), its effect on $V_{dr,i}$ can be neglected when the gate resistor $R_{G,i}$ is large enough. The Kirchhoff voltage equation for the gate drive current loop $i_{G,i}$ can be written as

$$i_{G,i} = \frac{V_{dr,i} - u_{GS,i}(t) - L_{S,int,i}\frac{di_{D,i}(t)}{dt}}{R_{G,i}} \tag{4}$$

As Equation (4) indicates, $i_{G,i}$ that charges and discharges $C_{GS,i}$, is affected by the $R_{G,i}$ magnitude, influencing the device behavior during transient stages. As a result, modifying the gate current by adjusting $R_{G,i}$ can lead to dynamic current imbalance suppression since di_D/dt and t_d of power devices tend to be synchronized.

According to ref. [34], transient imbalance can be suppressed by adjusting the time delays between power switches. Turn-on and -off delays ($t_{d,on}$ and $t_{d,off}$, respectively) can be directly affected by varying the firing angle (turn-on delay) and duty cycle (turn-off delay) of the PWM signal. During the turn-on interval, the SiC MOSFET which turns on faster is carried by a larger current than the other one. As a result, turn-on imbalance can be minimized by increasing the turn-on delay of the PWM controlling the fastest SiC MOSFET, forcing it to turn on slower. Contrariwise, during the turn-off interval, the power switch that turns off faster is carried by the least current compared to the other switch. The turn-off current imbalance can be reduced by increasing the turn-off delay of the PWM which drives the fastest SiC MOSFET, forcing it to turn off slower. Therefore, transient current imbalance can be suppressed with the proper adjustments of the turn-on and -off delays of the driving signals. Equations (5) and (6) express the total turn-on and -off time delays ($t_{d,on,total}$ and $t_{d,off,total}$, respectively). $t_{d,angle}$ and $t_{d,DC}$ represent the modifications of the firing angle and duty cycle, respectively.

$$t_{d,on,total} = t_{d,on} + t_{d,angle} \tag{5}$$

$$t_{d,off,total} = t_{d,off} + t_{d,DC} \tag{6}$$

To suppress current imbalance during turn-on and -off intervals, each dynamic imbalance requires different modifications and separate control for turn-on and -off intervals. This is attributed to different gate drive strengths and current/voltage waveforms at the drain [16].

2.2. Strategy of Static Current Imbalance Suppression

The equivalent circuit of two power devices during conduction stage, without the gate drivers, is shown in Figure 2. When V_{DS} across SiC MOSFET falls under the difference between the gate-source voltage and threshold voltage ($V_{GS} - V_{th}$), SiC MOSFET is treated as a resistance ($R_{DS-on,i}$, $i = 1,2$), as illustrated in Figure 2. $L_{D,i}$ represents the sum of the parasitic drain inductance and the one that is attributed to the PCB layout or the wiring. In addition, $L_{S,i}$ refers to the source terminal having the same meaning as $L_{D,i}$.

As pointed out earlier, steady-state imbalance is mainly caused by the R_{DS-on} mismatch between power switches. On-resistance of SiC MOSFET shows a positive temperature coefficient (PTC), such as Si MOSFET. In this way, the junction temperature of the power switch carrying the largest current will increase, making its R_{DS-on} greater. Therefore, static current imbalance could automatically be suppressed due to the thermal capability of Si MOSFET R_{DS-on}. However, SiC MOSFET R_{DS-on} shows limited thermal sensitivity compared to Si ones [8]. As a result, the static current imbalance should be addressed in a different manner and without relying on the PTC characteristic of SiC MOSFET R_{DS-on}.

As mentioned above, during the steady-state, SiC MOSFET is equivalent to a resistance. For this reason, its drain current can be calculated by Equation (7) while static current imbalance ($\Delta i_{D,static}$) for two SiC MOSFETs connected in parallel is expressed in Equation (8).

$$i_{DS} = \frac{V_{DS}}{R_{DS\text{-}on}} \tag{7}$$

$$\Delta i_{DS,static} = \frac{V_{DS}\Delta R_{DS\text{-}on}}{R_{DS\text{-}on,1}R_{DS\text{-}on,2}} \tag{8}$$

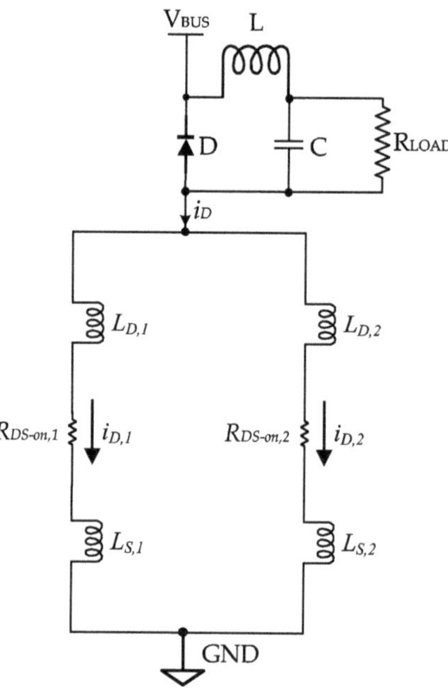

Figure 2. Equivalent circuit of two SiC MOSFETs connected in parallel, during conduction stage, incorporated with a DC–DC buck converter.

According to Equation (8), $\Delta i_{D,static}$ can be mitigated in case the $R_{DS\text{-}on}$ deviation ($\Delta R_{DS\text{-}on} = R_{DS\text{-}on,1} - R_{DS\text{-}on,2}$) between SiC MOSFETs becomes less. Based on [38], when an N-channel FET operates in the linear region, its drain current is expressed by Equation (9) where C_{ox} is the oxide capacitance, μ_n is the electron mobility, and L and W are the length and width of the gate.

$$i_D = \mu_n \, C_{OX} \frac{W}{L} [(V_{GS} - V_{th})V_{DS}] \quad when \quad V_{DS} \ll (V_{GS} - V_{th}) \tag{9}$$

By combining Equations (7) and (9), on-state resistance can be written as shown in Equation (10).

$$R_{DS\text{-}on} = \frac{1}{\mu_n \, C_{OX} \frac{W}{L}(V_{GS} - V_{th})} \tag{10}$$

Based on Equation (10), $R_{DS\text{-}on}$ magnitude can be controlled by V_{GS} while Equation (7) holds that the SiC MOSFET drain current depends on $R_{DS\text{-}on}$ during the steady-state. Therefore, $\Delta i_{D,static}$ can be minimized by properly varying the V_{GS} of the correct power device, leading to static current imbalance elimination.

However, modification of gate-source voltage V_{GS} affects not only static current imbalance but dynamic imbalance as well. According to [20], during the transient stage of the power switch, i_D satisfies the following relationship:

$$i_D = \begin{cases} 0 & V_{GS} < V_{th} \\ g_m(V_{GS} - V_{th}) & V_{th} < V_{GS} < V_{GP} \\ I_L & V_{GS} > V_{GP} \end{cases} \tag{11}$$

where I_L is the load current and V_{GP} is the plateau voltage caused by the Miller effect.

Based on Equation (11), dynamic current sharing is also influenced by V_{GS} difference (ΔV_{GS}) since the drain current is affected by V_{GS} during transient stages as well. However, it is not possible to eliminate the entire imbalance only by modifying V_{GS} because ΔV_{GS} has a different effect on each current imbalance.

For this reason, an efficient strategy is to eliminate static current imbalance by increasing ΔV_{GS} while transient imbalance should be mitigated by combining the two aforementioned transient imbalance suppression strategies.

3. Design of Active Current Balancing Technique

Since the current imbalance can be attributed to various factors, it is impossible to predict its type and magnitude. For this reason, static and dynamic current imbalances should be eliminated independently and regardless of the cause. In addition, the implementation of a current imbalance suppression technique should not require the knowledge of the technical parameters mismatch, current imbalance, and parasitic inductances between parallel SiC MOSFETs or the operating conditions. For this reason, the proposed technique is designed to address the current unbalance issue, fulfilling these requirements.

In this section, the operation principle and structure of the proposed technique are presented in detail. To address the current imbalance issue, the open-loop method is derived from two parts, as illustrated in Figure 3.

1. Gate driver: For every semiconductor device, an active gate driver (AGD) is utilized, capable of controlling power devices, and actively variate V_{GS} and R_G;
2. Digital controller: The PWM signals of the power devices are generated with the use of a digital controller. Additionally, the controller can control the parallel devices and eliminate the static and dynamic current imbalances by imposing the proper variations to several control parameters (V_{GS}, R_G, $t_{d,angle}$, and $t_{d,DC}$). Finally, the modification of the control parameters is realized manually through the digital controller.

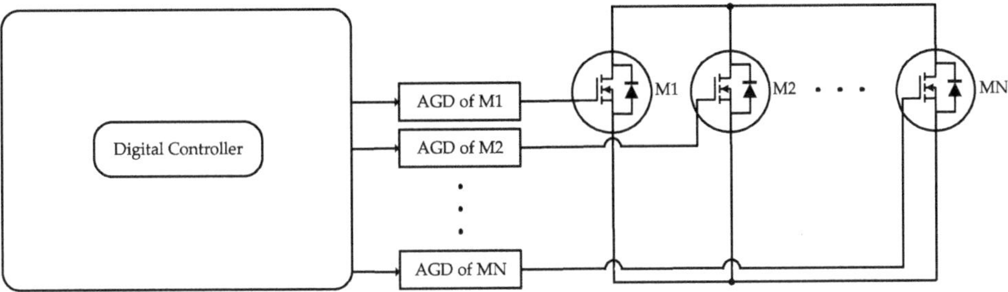

Figure 3. Application structure of the proposed active current balancing technique.

3.1. Capabilities and Structure of the Proposed Active Gate Driver

Based on the analysis of the previous section, all parameters affecting the current imbalance can be controlled by a gate driver and are related to the driving pulse generation source of the SiC MOSFETs. For this reason, an active gate driver is proposed capable of driving power devices and modifying these parameters to eliminate the current imbalance.

However, an active gate driver circuit is mandatory for every parallel-connected semiconductor device to apply different modifications to each SiC MOSFET control parameter.

3.1.1. Operation Principles of the Proposed Active Gate Driver

The proposed active gate driver includes a driving circuit to activate and deactivate the power device with the application of the V_{CC} and V_{EE} control voltages, respectively, as illustrated in Figure 4. Between the V_{CC} and the driving circuit, a forward converter is inserted to provide DC–DC isolation and actively modify the V_{CC} of SiC MOSFET by changing the PWM duty cycle ($\mathrm{PWM_{VCC}}$) and controlling the forward converter switch. Duty cycle control is performed via the digital controller. As a result, the static current imbalance is eliminated by properly adjusting the correct V_{CC}.

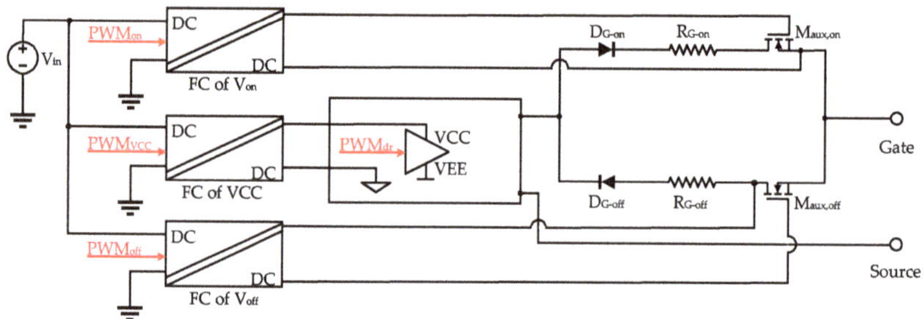

Figure 4. Proposed active gate driver circuit.

As pointed out in Section 2.1, dynamic current imbalance can be suppressed by properly varying the firing angle and duty cycle of the correct $\mathrm{PWM_{dr}}$ signal. However, it is not always possible to eliminate dynamic current imbalance by only changing these two control parameters. Varying the firing angle and duty cycle only affect the turn-on and -off processes of the power switch current, respectively, without influencing their current slopes. To sufficiently suppress dynamic current imbalance, turn-on and -off delays and current slopes should be properly adjusted. In conclusion, a great portion of the dynamic imbalance can be reduced through the variation in turn-on and -off delays. The remaining imbalance can be minimized with the proper adjustment of the gate current of the correct SiC MOSFET. As a result, this current balancing pattern offers the proposed technique the ability to mitigate any current imbalance independently and regardless of the cause.

3.1.2. Design of an Active Gate Driver

Variation in gate current can be achieved by changing the gate resistance size. As illustrated in Figure 4, the branch connected with the gate of SiC MOSFET is derived by two sub-branches which include a resistor ($R_{G\text{-}on}$ and $R_{G\text{-}off}$) connected in series with an auxiliary MOSFET ($M_{aux,on}$ and $M_{aux,off}$). Since the $R_{DS\text{-}on}$ of each MOSFET depends on its gate-source voltage, modifying the V_{GS} of each auxiliary MOSFET changes the entire resistance of each branch. In this way, the turn-on and -off delays and the current slope of the power device are affected. The upper and lower branch control the charging and discharging gate current, respectively. In each branch, a diode ($D_{G\text{-}on}$ and $D_{G\text{-}off}$) is series-connected with their elements to independently control the charging and discharging currents. The control voltages of $M_{aux,on}$ and $M_{aux,off}$ are modified with the aid of two other forward converters with the control of the duty cycles of the $\mathrm{PWM_{on}}$ and $\mathrm{PWM_{off}}$ signals, respectively. In conclusion, each subbranch is treated as a voltage-controlled gate current source.

Therefore, the gate-source voltage of each MOSFET can be modified through the duty cycle, controlling the switch of each forward converter. The duty cycle control of each PWM signal is performed through the digital controller. Finally, the control of the power device along with the forward converter switches requires four different driving pulses for each parallel power device.

3.1.3. Forward Converter Design Guidelines

The elimination of every current imbalance type is influenced by the digital controller's maximum clock frequency. Each PWM signal is generated with the aid of a step-up counter which includes a reset capability. As illustrated in Figure 5, the counter increases by one step for every positive edge of the clock and resets when it reaches a certain value.

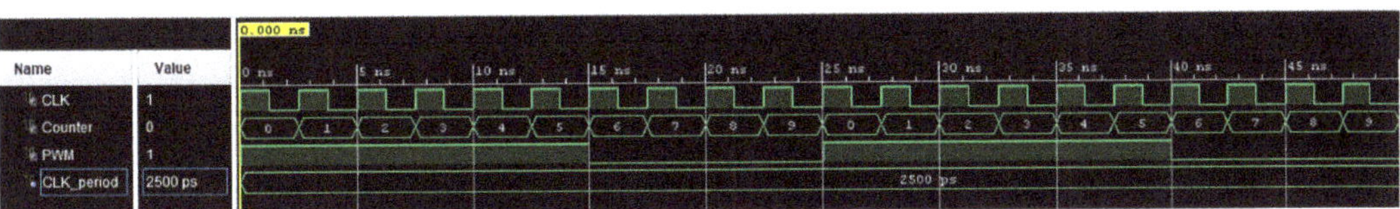

Figure 5. PWM generation strategy.

Therefore, PWM frequency and duty cycle are set based on this function. As depicted in Figure 5, when the counter does not reach a specific value (e.g., lower than 6), PWM turns "high", but when it reaches and exceeds a limit (e.g., 6 or higher), it turns "low". For a specific time period, the count times of the counter increase as the clock frequency becomes higher. In this way, the minimum variations of the turn-on and -off delays are decreased. As a result, the minimization of dynamic current imbalance by adjusting the turn-on and -off delays can become even more efficient. Additionally, the minimum variation step of each gate-source voltage decreases, making the current balancing process even more reliable. The minimum modification step on the control voltage ($V_{var,step}$) can be expressed by Equation (12) where f_{clk}, V_{out}, and f_{sw} are the clock frequency of each counter, the output voltage, and the forward converter switching frequency, respectively. As a result, the minimum variation on the output voltage can be reduced with the usage of a digital controller with a high clock frequency capability while the forward converter frequency (f_{sw}) should be kept low. On the other hand, digital controllers with ultra-high fundamental frequency (f_{clk}) are expensive. In conclusion, the selection of f_{sw} is a trade-off between the current balancing process reliability and the implementation cost of the proposed technique.

$$V_{var,step} = \frac{V_{out} f_{sw}}{f_{clk}} \tag{12}$$

In practice, one crucial matter is the design of the forward converter regarding the reset method of the transformer core. For this reason, a two-switch forward converter topology with two MOSFET switches Q_1 and Q_2 is used, as shown in Figure 6. Both switches are controlled by one gate driver circuit which is derived from the R_{HI}, R_{LI}, D_{BOOT}, C_{BOOT}, and Gate Driver IC elements and simultaneously turns both switches on and off. This method manages to reset the transformer core by using two demagnetization diodes D_1 and D_2. When the switches are turned off, the demagnetization diodes become forward biased and the magnetizing energy in the transformer is returned to the input voltage source (V_{in}) [39].

In addition, R_1 and C_1 and R_2 and C_2 are the snubbing elements connected in parallel with the secondary diodes (D_3 and D_4) to dampen the oscillations that appear across them. These oscillations are attributed to the leakage inductance of the secondary side of the transformer with the capacitor behavior of the diodes when they are blocked. The oscillations take place at the end of the diode conduction.

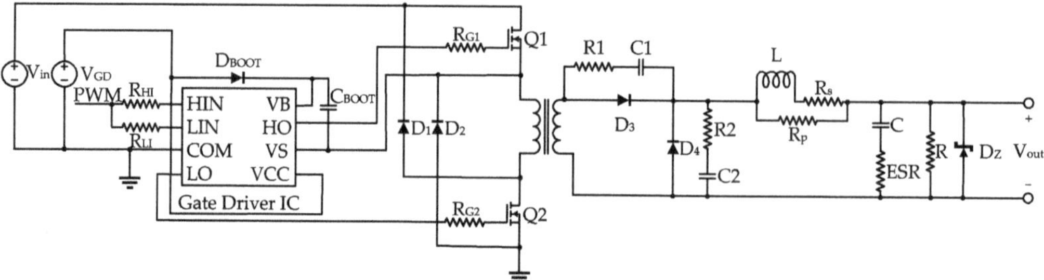

Figure 6. Two-switch forward converter topology.

Another essential matter is the output voltage ripple which is strongly dependent on the inductor (L) and capacitor (C) magnitudes. R_S is the inductor equivalent series resistance and R_p is the parallel resistance correlated with the parallel leakage path across the inductor. Also, ESR is the capacitor series resistance. Each forward converter has a Zener diode (D_Z) connected in parallel with the load resistance (R). According to Equation (13), the reduction in output inductor current ripple (ΔI_{LX}) can be achieved by increasing the switching frequency. However, Equation (12) states that $V_{var,step}$ increases when f_{sw} becomes larger. Additionally, ΔI_{LX} is affected by the inductor size while Equation (14) states that the output voltage ripple (ΔV_{out}) is affected by the capacitor value. Therefore, LC filter values should be properly chosen to reduce ΔV_{out} since ΔI_{LX} and ΔV_{out} can be decreased as the inductor and capacitor increase. However, the inductor affects the converter output voltage because of the voltage drop caused by its parasitic resistance which becomes larger as the inductor size increases. In addition, the output voltage depends on the load current which lowers as the inductor becomes larger. As for the capacitor, its value should be selected concerning the response time of the converter to the duty cycle (D) variations which increase as the capacitor becomes larger. Finally, R should be as large as possible for lowering converter power consumption, taking into account the fact that the R value affects the converter response time.

In conclusion, the forward converter should offer an output voltage with a low ripple. Also, the output voltage should be able to vary in a quite short period of time (mseconds) which is a trade-off between the LC filter and switching frequency.

$$\Delta I_{LX} = \frac{V_{out}(1-D)}{L f_{sw}} \tag{13}$$

$$\Delta V_{out} = \Delta I_{LX}\left(\frac{1}{8\,C\,f_{sw}} + ESR\right) \tag{14}$$

3.2. Functions of the Digital Controller

To address the current imbalance issue, the digital controller includes a number of functions. Initially, the digital controller generates the driving pulses for the control of the SiC MOSFETs with a controllable switching frequency and duty cycle. In addition, the digital controller generates three PWM signals with a fixed frequency and an initial duty cycle which can be modified with the purpose of varying the V_{CC} of power devices and the control voltages of the auxiliary MOSFETs to affect the gate currents. Finally, the duty cycle and turn-on delay of each power switch PWM signal can also be modified.

Current Imbalance Suppression Methodology

To eliminate the entire imbalance, a current balancing methodology should be followed. Before applying any necessary correction to the control parameters of the appropriate power device(s), it is necessary to identify the polarities of the three imbalances.

In the process, all three imbalances should not be suppressed simultaneously but in a specified order, as depicted in Figure 7. If the static current imbalance is larger than a certain value (e.g., 0.1A), the V_{CC} of the SiC MOSFET with the lowest current should start to increase with the purpose of lowering its on-resistance. In case V_{CC} reaches a specific limit and static imbalance remains, the V_{CC} of the SiC MOSFET with the highest current should start to decrease until on-resistances become equal. Otherwise, static current imbalance can also be eliminated by decreasing only the V_{CC} of the SiC MOSFET with the highest current.

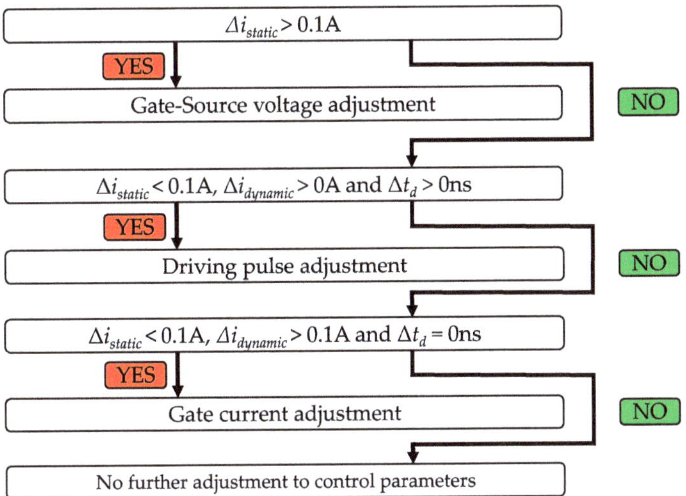

Figure 7. Current balancing strategy flow.

Once static current imbalance becomes less than a certain threshold, if there is dynamic imbalance ($\Delta i_{dynamic}$) a certain balancing order should be executed. Dynamic current balancing can be achieved by forcing the peak currents during turn-on and -off intervals ($\Delta i_{D,on}$ and $\Delta i_{D,off}$) to match. If $\Delta i_{D,on}$ and turn-on delay difference ($\Delta t_{d,on}$) between parallel currents are greater than zero, the turn-on delay ($t_{dl,on}$) of the PWM signal driving the power switch with the largest current during the turn-on transience should begin to increase. Whether $\Delta i_{D,on}$ is eliminated or $\Delta t_{d,on}$ becomes zero, modification of turn-on delay should cease to increase. In case the turn-on imbalance still exists, the charging gate current of the SiC MOSFET, carrying the highest current, should start to become less by decreasing the V_{GS} of the appropriate auxiliary MOSFET until the turn-on peak current difference is minimized.

As for the elimination of the turn-off dynamic imbalance, if $\Delta i_{D,off}$ and turn-off delay difference ($\Delta t_{d,off}$) between parallel currents are greater than zero, the duty cycle of the PWM controlling the power switch with the lowest current during turn-off should start to increase by $t_{dl,off}$. Either $\Delta i_{D,off}$ is minimized or $\Delta t_{d,off}$ becomes zero; the variation in the duty cycle should be ceased. If the turn-off dynamic imbalance remains, the charging gate current of the power device with the least current during the turn-off interval should become less by lowering the V_{GS} of the correct auxiliary MOSFET until the difference between peak currents is minimized.

In any case, the balancing process should always be executed following this pattern. When there is a static current imbalance, the two transient imbalances include both static and dynamic imbalances. Also, the modification of V_{CC} affects not only static imbalance but also dynamic imbalance as well. This may force the dynamic current imbalances to change polarity, especially when the modification of V_{CC} is too large. If this precaution is not taken, it is difficult or even impossible to discover and impose the proper modifications to the control parameters of the correct power device to suppress dynamic imbalances.

4. Test Platform and Experimental Results

4.1. Test Platform

In previous work, the effectiveness and durability of the proposed technique against current imbalance were tested through simulation tests eliminating current imbalance automatically [40,41]. To experimentally verify the effectiveness of the proposed current balancing technique, an experimental test platform is constructed. The structure of the test platform and the proposed current balancing system are depicted in Figure 8. Table 1 lists all the equipment used. The test platform is derived by a DC–DC buck converter with two SiC MOSFETs C2M0080120D (M1 and M2) connected in parallel. As a free-wheeling diode, SiC Schottky E4D20120D is used. The power converter supplies a resistive load while an *LC* filter is used for smoothing the output voltage. The realization of the proposed method includes the digital controller, two current sensors, and two active gate driver circuits which are powered by a separate DC power supply. All the capabilities of the digital controller can be realized with an algorithm and executed with an FPGA (field programmable gate array). For this reason, the Nexys A7-100T FPGA trainer board is used which includes the FPGA Artix-7 offering a clock speed of 500 MHz. Also, an algorithm is written in the VHDL programming language to execute all the digital controller functions. Since two gate drivers are utilized, the FPGA generates eight PWMs for the control of the power devices and the forward converters. The FPGA algorithm utilizes 14 switches (SW), the 7-segment displays, and the pushbuttons of the FPGA board. Table 2 mentions in detail the function of each FPGA switch. The control results of the switching frequency and duty cycle of the SiC MOSFETs as well as the control parameters are displayed in the FPGA 7-segment displays and controlled with the help of the pushbuttons. Finally, the measurement of each drain current is achieved with a surface mount resistor R_{sense} of 100 mΩ and 1 W, connected in series with the source pin of each parallel power device.

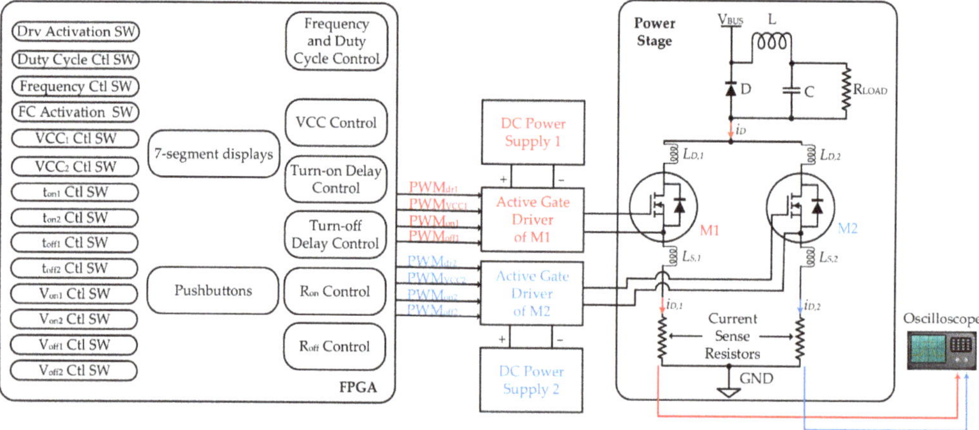

Figure 8. Application of the proposed method for two parallel-connected SiC MOSFET.

Table 1. Equipment used in the experimental tests.

Equipment	Model	Bandwidth	Function
Digital oscilloscope	Keysight MSOX3014A	100 MHz	Capture curves
Current Sense Resistor	LTR10LEZPFLR100	-	Measure I_D
BNC Coaxial cable	141-12BM+	3 GHz	Measure I_D and I_G
BNC Coaxial connector	CONBNC002	1 GHz	Measure I_D and I_G
Voltage probe	Agilent N2862A	150 MHz	Measure V_{GS} and V_{DS}

Table 2. Functions of FPGA Switches.

Switch	Function
Drv Activation SW	Activation driving of SiC MOSFETs
Duty Cycle Ctl SW	Duty cycle control of SiC MOSFETs
Frequency Ctl SW	Switching frequency control of SiC MOSFETs
FC Activation SW	Activation driving of forward converters
VCC_1 Ctl SW	V_{CC} control of M1
VCC_2 Ctl SW	V_{CC} control of M2
t_{on1} Ctl SW	Turn-on delay control of M1
t_{on2} Ctl SW	Turn-on delay control of M2
t_{off1} Ctl SW	Turn-off delay control of M1
t_{off2} Ctl SW	Turn-off delay control of M2
V_{on1} Ctl SW	Control of Gate-Source voltage of $M_{aux,on,1}$
V_{on2} Ctl SW	Control of Gate-Source voltage of $M_{aux,on,2}$
V_{off1} Ctl SW	Control of Gate-Source voltage of $M_{aux,off,1}$
V_{off2} Ctl SW	Control of Gate-Source voltage of $M_{aux,off,2}$

The DC bus power supply (V_{BUS}) of the DC–DC buck converter is implemented using a three-phase full-bridge diode rectifier which is connected in parallel with two capacitors (1500 µF/550 V) to smooth the output voltage of the bridge. The DC bus power supply also consists of a protection system that includes a resistor of 3.3 kΩ/10 W. When the resistor is connected in parallel with the capacitors, it discharges them for safety purposes. As shown in Figure 9, the operation of the entire platform is controlled by a relay (Rel) which is powered by a single-phase AC power supply of 230 V. Once the switch (SW) is closed and the AC supply is ON, the relay connects the three-phase power supply with the rectifier bridge through a normally open (NO) three-phase switch, and at the same time disconnects the discharging resistor from the capacitors through a normally closed (NC) single-phase switch. Otherwise, the relay disconnects the bridge with the three-phase AC supply and connects the discharging resistor with the capacitors. In addition, four electric fuses are used, three in the three-phase AC supply (F1, F2, and F3) and one in the single-phase AC supply (F4), offering overcurrent protection to the experimental platform. The constructed experimental test platform is shown in Figure 10.

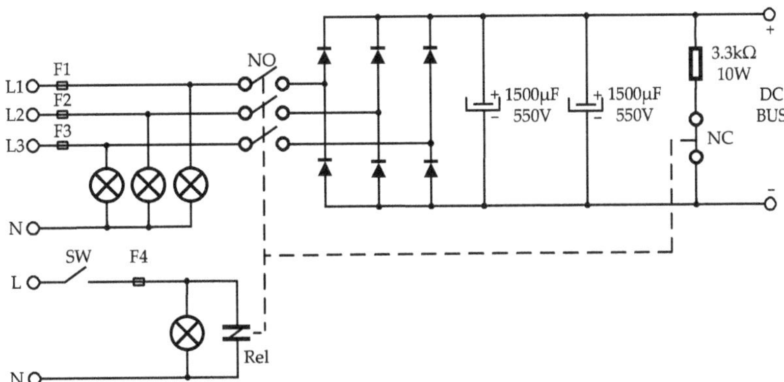

Figure 9. Power supply circuit consists of a three-phase rectifier, a smoothing filter, and a protection circuit.

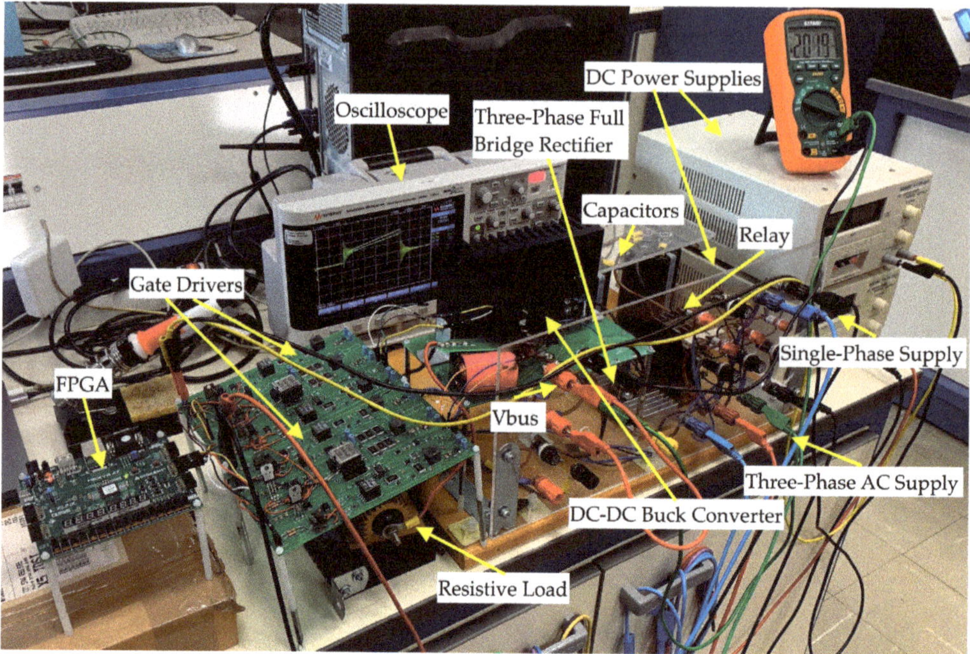

Figure 10. Test platform for the proposed current balancing method.

As for the test conditions, the buck converter operates under a DC bus voltage of 200 V with a frequency and duty cycle of 25 kHz and 25%, respectively, supplying a load of 10 Ω. The initial V_{dr} values for activating and deactivating each parallel SiC MOSFET are 20 V and −5 V, respectively. Finally, $R_{G\text{-}on}$ and $R_{G\text{-}off}$ are equal to 10 Ω while the gate-source voltages of $M_{aux,on}$ and $M_{aux,off}$ are set to 9 V. $V_{ctl\text{-}M,on,i}$ and $V_{ctl\text{-}M,off,i}$ are the output voltages of the forward converter of $M_{aux,on}$ and $M_{aux,off}$, respectively.

The initial values of the activation and deactivation voltages were selected as 20 V and −5 V, respectively, recommended by the datasheet of the utilized SiC MOSFETs C2M0080120D. The manufacturing company (Wolfspeed) has constructed a gate driver (CGD15SG00D2) designated for the driving of that particular SiC MOSFET model. CGD 15SG00D2 uses an isolated DC/DC converter to generate two output voltages of 20 V and −5 V while its output power is 2 W with an efficiency of 86%, meaning that the input power of the driver is 2.3 W. On the other hand, the input voltage (V_{in}) of our proposed gate driver is 17 V to supply all three forward converters and an isolated DC/DC converter that generates the deactivation voltage of 5 V which is always constant. In case the FPGA controls only the forward converter which generates the activation voltage at 20 V, the input current is measured to be 0.15 A. Therefore, the input power of the proposed active gate driver when it supplies constant activation and deactivation voltages is almost 2.5 W, which is very close to the input power of the commercial gate driver (CGD15SG00D2). In case the FPGA controls all forward converters and the output voltages that control auxiliary MOSFETs are 9 V, the overall input current and power of the active gate driver are 0.21 A and 3.6 W, respectively. Therefore, both forward converters and the auxiliary MOSFETs add only 1.1 W of power consumption compared to the previous case in which the proposed active gate driver works in a similar way to the commercial driver. In the worst-case scenario, when the activation voltage reaches 23 V for the suppression of static imbalance, the input current and power are 0.29 A and 4.9 W, respectively.

4.2. Current Sensing System Accuracy

The current measurement of SiC MOSFETs requires sensors of high bandwidth because of their fast-switching speed. Based on [42], surface mount resistors can offer exceptional measurement accuracy since their bandwidth can be on the order of hundreds of megahertz. One of the most important factors affecting their bandwidth is parasitic inductance which is inevitable because of the magnetic field induced by the current conducting the sensor [43]. For this reason, surface mount resistor size should be as low as possible to provide low parasitic inductance in the order of nano or even picohenry. Also, R_{sense} should have low resistance without significantly affecting the current level and offering quite low power losses. On the other hand, one of the pulse current measurement methods that are strongly recommended for the current measurement of WBG devices is the coaxial shunt resistors which can offer MHz or even GHz measurement bandwidth.

To examine the measurement accuracy of the utilized current sensor, two experimental tests are performed, measuring the current of one SiC MOSFET. In the first test, the power switch current is measured with the surface mount resistor of 100 mΩ. In the process, the SiC MOSFET current is measured with the coaxial shunt resistor SDN-414-01 which offers a measurement bandwidth of 400 MHz. The voltage developed across the current sense resistor is illustrated to the oscilloscope with the aid of a BNC coaxial connector and a BNC coaxial cable of quite high bandwidth, as shown in Table 1. Figure 11 depicts and compares the current waveforms of both tests during the conduction stage as well as the turn-on and -off intervals, demonstrating the surface mount resistor measurement accuracy. Compared to the coaxial shunt resistor, the surface mount resistor only shows a measurement delay of around 5 ns during the current ringing stages. However, this time delay difference can be ignored because both waveforms are identical during the rising, conduction, and falling stages of the drain current. In conclusion, for the purposes of this research, the current sense resistor offers high enough measurement accuracy of the drain current of SiC MOSFET.

4.3. Experimental Test Results

In this subsection, the effectiveness of the proposed method is tested by performing two experimental tests. In the first test, a pair of devices is connected in parallel causing current imbalance during steady and dynamic stages which may be attributed to the variation in the technical parameters between SiC MOSFETs. PCB layout of the power circuit is designed to be symmetrical to minimize the length differences between PCB traces to exclude any current imbalance attributed to mismatched parasitic inductances. In the second test, another pair of devices is connected in parallel which originally shows an equal current sharing between power devices. However, the layout was designed to be asymmetrical by connecting the power devices with different gate, drain, and source pin lengths. This leads to static and dynamic current imbalance caused mainly by the mismatch of the drain and common source parasitic inductances. Ref. [31] argues that the mismatch of gate parasitic inductances has an almost negligible effect on dynamic current sharing. The experimental results are further compared under different test conditions (a) without and (b) with the proposed current balancing technique. Figure 12 depicts the drain-source voltage (V_{DS}), developed across power devices, during the conduction stage and the turn-on and -off transitions. The experimental results of the two tests are illustrated in Figures 13–18, depicting the drain currents of the parallel-connected SiC MOSFETs as well as their turn-on and -off gate currents and driving signals. Tables 3 and 4 report the imposed modifications to the control parameters to balance the parallel currents.

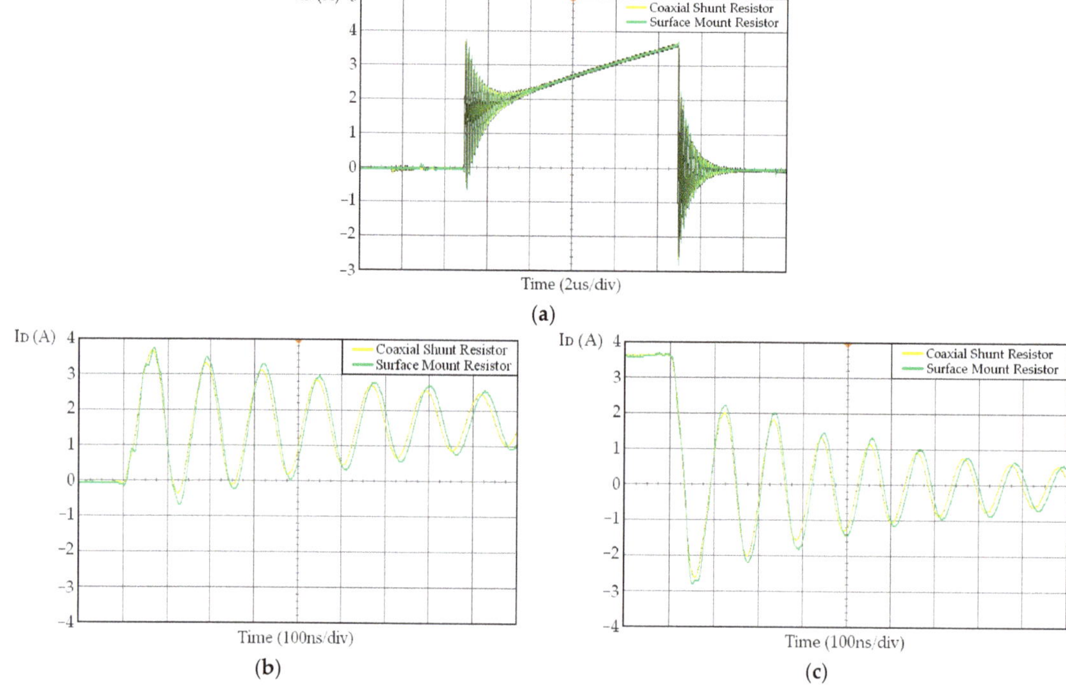

Figure 11. Experimental comparison of current measurement accuracy between the coaxial shunt resistor SDN-414-01 and the surface mount resistor of 100 mΩ during the: (**a**) conduction stage; (**b**) turn-on transition; and (**c**) turn-off transition.

Table 3. Modifications upon control parameters (a) without and (b) with the proposed method of the first experimental test.

a/a	$V_{CC,1}$ (V)	$V_{CC,2}$ (V)	$t_{dl,on,1}$ (ns)	$t_{dl,on,2}$ (ns)	$t_{dl,off,1}$ (ns)	$t_{dl,off,2}$ (ns)	$V_{ctl\text{-}M,on,1}$ (V)	$V_{ctl\text{-}M,on,2}$ (V)	$V_{ctl\text{-}M,off,1}$ (V)	$V_{ctl\text{-}M,off,2}$ (V)
(a)	20	20	0	0	0	0	9	9	9	9
(b)	12.5	23	0	22	0	6	9	9	9	9

Table 4. Modifications upon control parameters (a) without and (b) with the proposed method of the second experimental test.

a/a	$V_{CC,1}$ (V)	$V_{CC,2}$ (V)	$t_{dl,on,1}$ (ns)	$t_{dl,on,2}$ (ns)	$t_{dl,off,1}$ (ns)	$t_{dl,off,2}$ (ns)	$V_{ctl\text{-}M,on,1}$ (V)	$V_{ctl\text{-}M,on,2}$ (V)	$V_{ctl\text{-}M,off,1}$ (V)	$V_{ctl\text{-}M,off,2}$ (V)
(a)	20	20	0	0	0	0	9	9	9	9
(b)	20	17	11	0	4	0	6.5	9	4.8	9

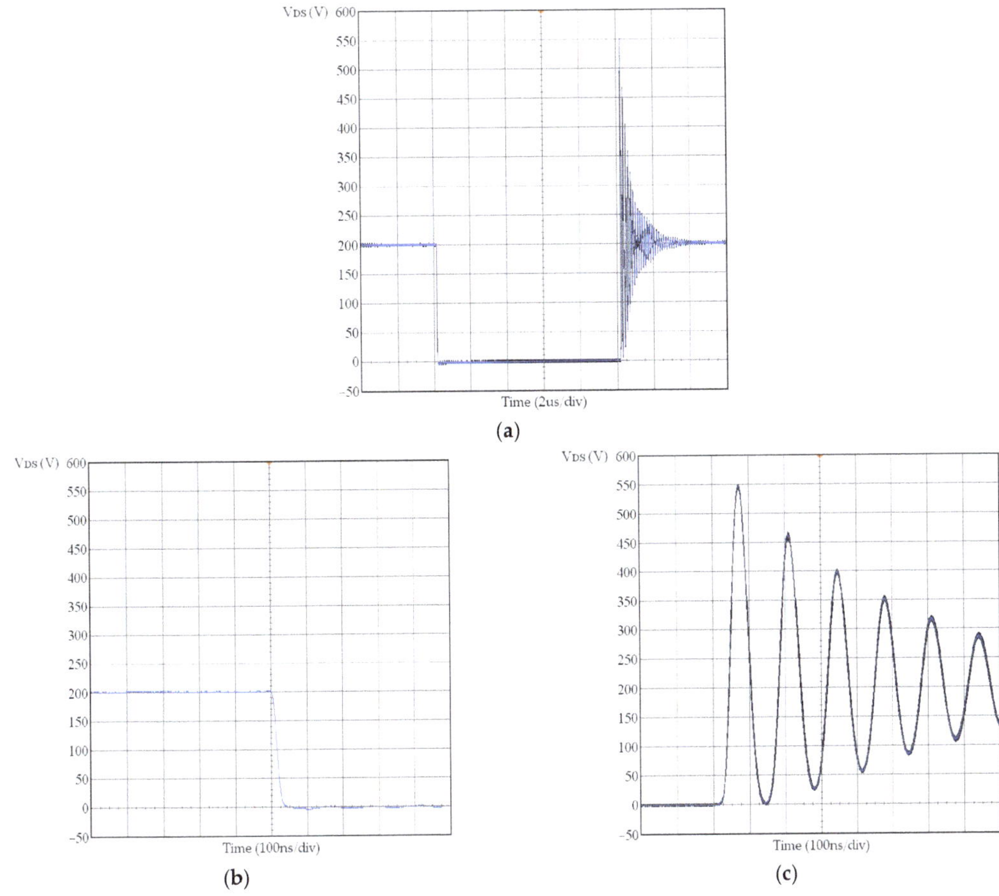

Figure 12. Drain–source voltage during the: (**a**) conduction stage; (**b**) turn-on transition; and (**c**) turn-off transition.

Current imbalance levels (ΔI_D) are mentioned in Tables 5 and 6 for both tests without and with the proposed solution. Based on the experimental results, the current curves and peak currents (I_{Dmax}) between the parallel power devices are almost the same.

Table 5. Comparison of experimental results for the first test.

Condition	Device	Turn-on		Steady-State		Turn-off	
		I_{Dmax} (A)	ΔI_D (A)	I_{Dmax} (A)	ΔI_D (A)	I_{Dmax} (A)	ΔI_D (A)
Without solution	M1	6.2	3.8	4.55	0.9	4.7	1.1
	M2	2.4		3.65		5.8	
With the proposed method	M1	4	0.1	4	0.1	4.2	0.1
	M2	3.9		3.9		4.1	

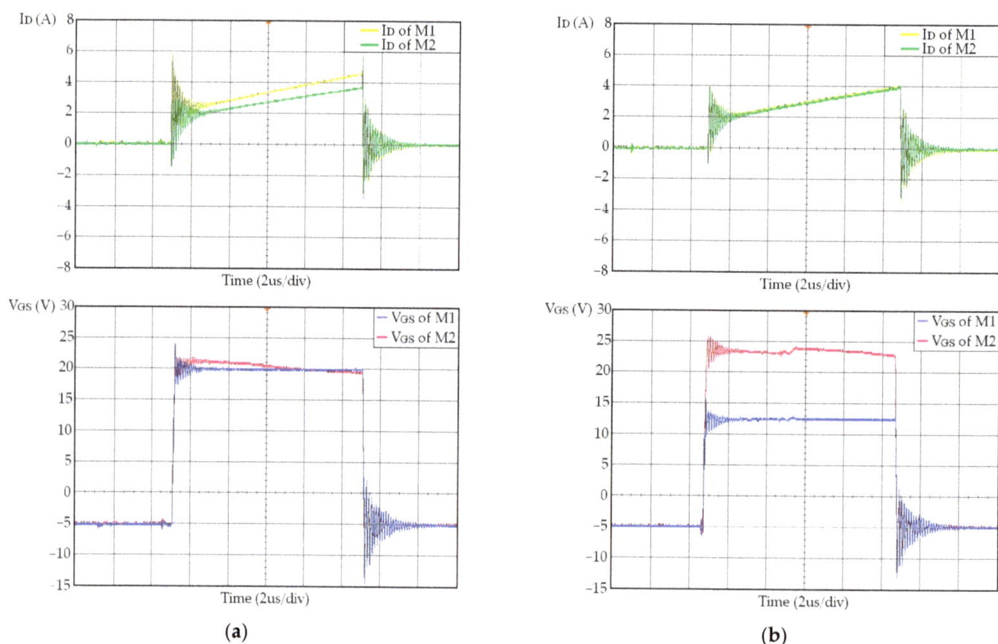

Figure 13. Drain to source currents and driving signals of the first experimental test: (**a**) without; and (**b**) with the proposed current balancing technique.

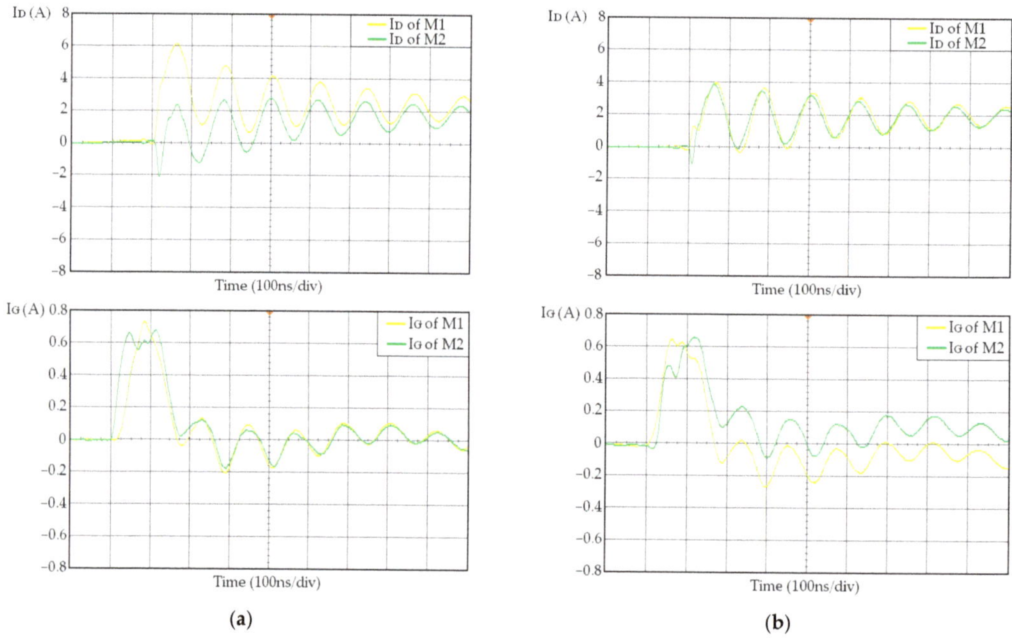

Figure 14. Drain and gate currents of the first experimental test during turn-on transition: (**a**) without; and (**b**) with the proposed current balancing technique.

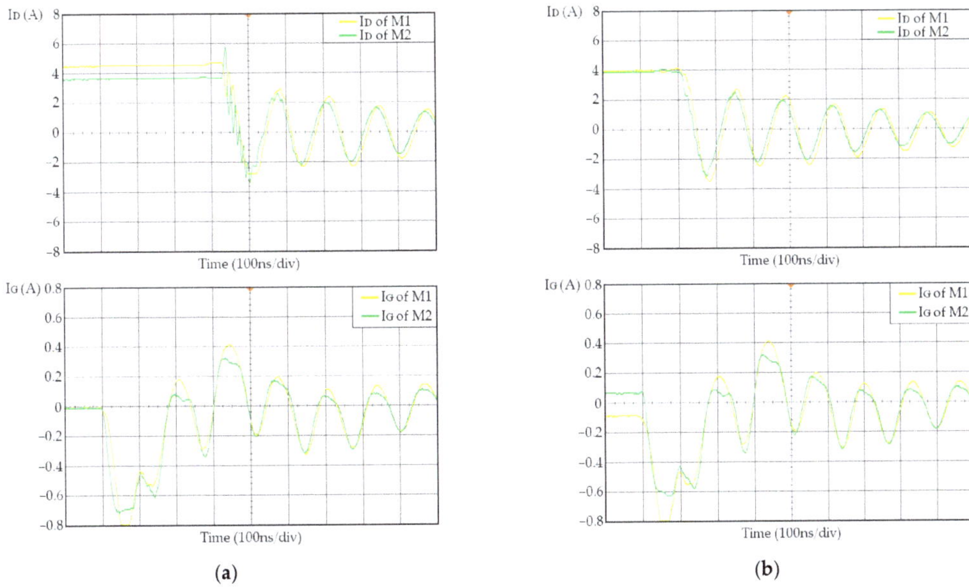

Figure 15. Drain and gate currents of the first experimental test during turn-off transition: (**a**) without; and (**b**) with the proposed current balancing technique.

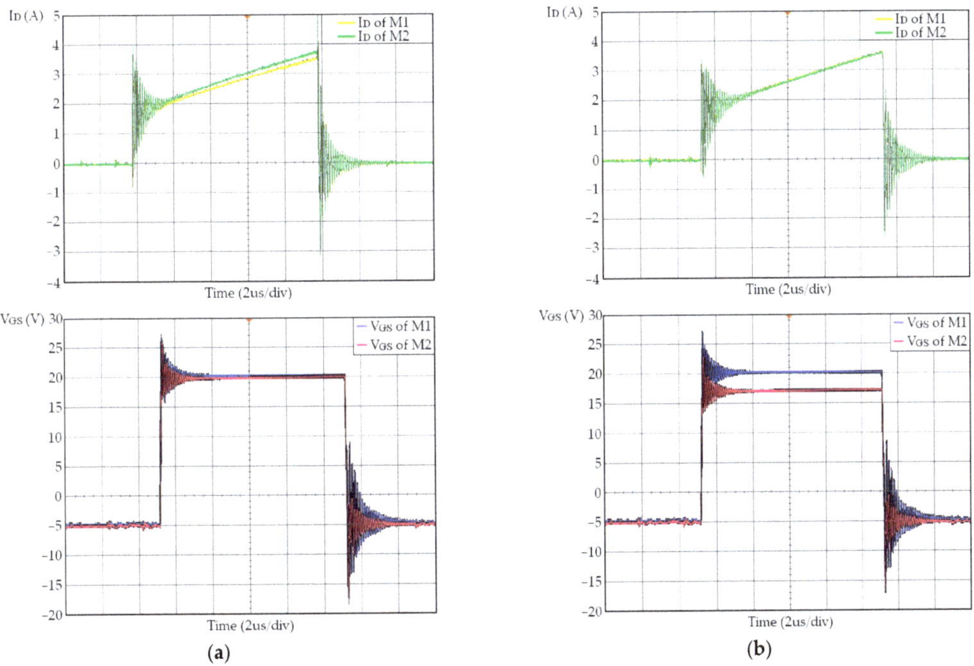

Figure 16. Drain to source currents and driving signals of the second experimental test: (**a**) without; and (**b**) with the proposed current balancing technique.

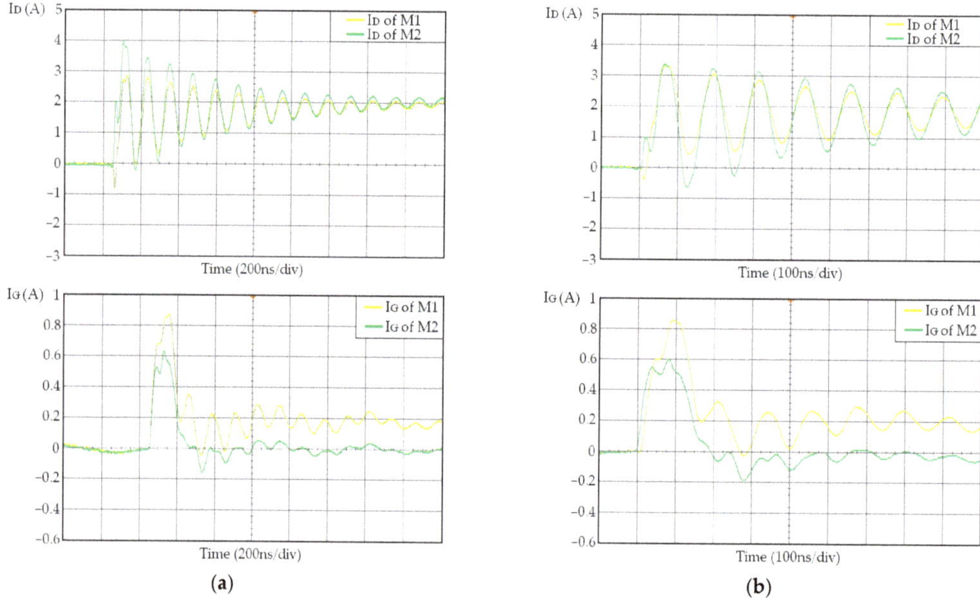

Figure 17. Drain and gate currents of the second experimental test during turn-on transition: (**a**) without; and (**b**) with the proposed current balancing technique.

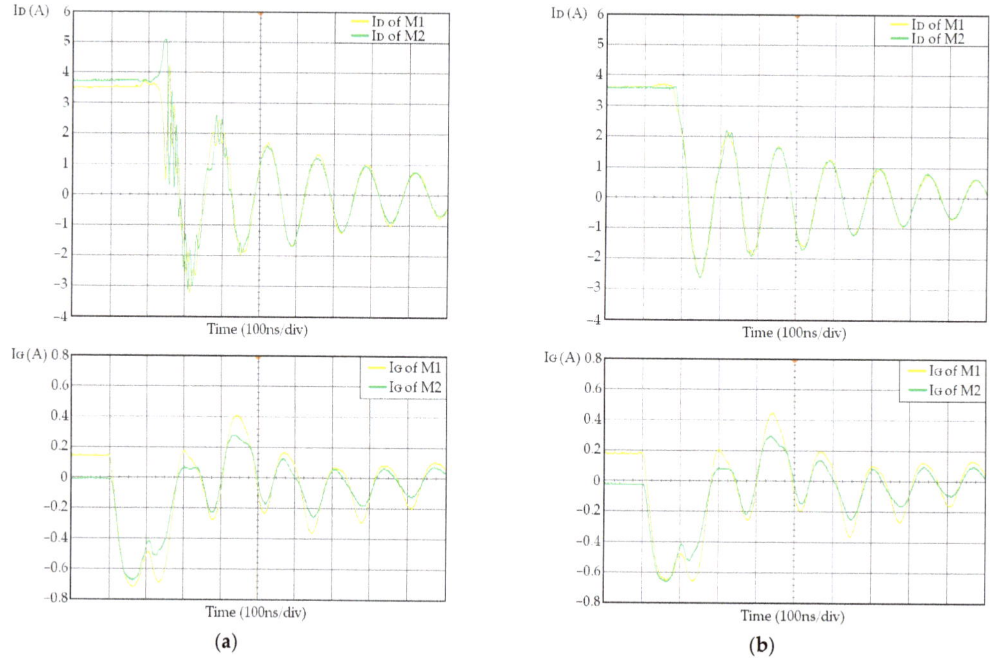

Figure 18. Drain and gate currents of the second experimental test during turn-off transition: (**a**) without; and (**b**) with the proposed current balancing technique.

Table 6. Comparison of experimental results for the second test.

Condition	Device	Turn-on		Steady-State		Turn-off	
		I_{Dmax} (A)	ΔI_D (A)	I_{Dmax} (A)	ΔI_D (A)	I_{Dmax} (A)	ΔI_D (A)
Without solution	M1	2.85	1.15	3.45	0.3	3.7	1.4
	M2	4		3.75		5.1	
With the proposed method	M1	3.3	0.1	3.6	0.05	3.7	0.1
	M2	3.4		3.65		3.6	

In the first test, by implementing the proposed method, turn-on and -off imbalances are reduced from 44% to 1.3% and 10.5% to 1.2%, respectively, while the static imbalance is decreased from 11% to 1.3%. In the second test, turn-on and -off imbalances are reduced from 16.8% to 1.5% and 15.9% to 1.4%, respectively, while the static imbalance is decreased from 4.2% to 0.7% $(((0.05 \times 100)/(3.65 + 3.6)) \times 100\% = 0.7\%)$.

According to the experimental results, before applying the proposed innovative technique, a significant amount of current imbalance is shown between the drain currents during all device stages. However, the proposed method can offer a well-balanced current sharing between SiC MOSFETs by properly adjusting the correct control parameters, proving its current balancing performance against multiple impact factors and promoting the safety of the parallel SiC MOSFETs.

The pair of SiC MOSFETs that were used for the first experimental test was determined through a number of tests, connecting several pairs of SiC MOSFETs in parallel. We have reached the decision to use this particular pair of SiC MOSFETs due to the large static and dynamic current imbalances that occur. Both imbalances may be attributed to the deviation of the technical parameters between parallel devices. Such static current imbalance can only be caused due to the on-resistance difference of the SiC MOSFETs. For this reason, the elimination of the static current imbalance was only possible by driving the parallel devices with such a large V_{CC} difference until both on-resistances of power devices became equal. In the second experimental test, the static current imbalance is much lower compared to the first case and is attributed to the asymmetrical power circuit layout. As a result, the difference between the activation voltages (V_{CC}) of the power devices is much smaller compared to the V_{CC} difference of the first experimental test.

Lowering V_{CC} increases switching and conduction losses because the gate current decreases and on-resistance of SiC MOSFET increases. Therefore, it is important to suppress static current imbalance without increasing conduction losses. Static current imbalance can be minimized by equalizing the on-resistances either by increasing the on-resistance of the SiC MOSFET carrying the highest current or by reducing the on-resistance of the SiC MOSFET carrying the lowest current. However, the first static balancing approach will result in higher conduction losses compared to the second one. For this reason, in the first experimental test, the static current balancing process starts with the increase in V_{CC}, forcing the on-resistance of the SiC MOSFET with the least drain current to become lower. However, the absolute V_{CC} of the SiC MOSFET C2M0080120D is 25 V. As a result, the balancing process should be proceeded by decreasing the V_{CC} of the SiC MOSFET with the highest drain current until the on-resistances of both SiC MOSFETs become equal. The V_{CC} difference may influence dynamic current sharing, but it is compensated with the implementation of the proposed dynamic current balancing methods. As a result, the entire current imbalance is eliminated, retaining balanced switching and conduction power losses. Also, SiC MOSFETs operate under the same temperature stress level because distributed heat between power devices is kept almost equal. Since the drain-source voltage (V_{DS}) of both parallel devices is common, due to their parallel connection, and their drain currents become equal, switching and conduction power losses are balanced.

The turn-on current ringing is attributed to the free-wheeling SiC Schottky diode due to the discharge of its self-capacitance under reverse bias. On the other hand, the turn-off current ringing is caused by the parasitic inductance of the resistive load. The drain–source voltage ringing during turn-off becomes lower with the use of the SiC Schottky diode but it cannot be dramatically reduced. Nevertheless, the switching oscillations of SiC MOSFETs could be eliminated with an RLC snubber [44]. However, the examination of the effectiveness of the proposed method is not affected by the current ringing during the switching intervals.

5. Design the Proposed Current Balancing Technique for Multiple-Device Operation

The scalability of current balancing techniques is very important since increased current capacity is required by medium- and high-power applications. For this reason, more than two SiC MOSFETs should be connected in parallel.

The proposed current balancing system is easy to implement for more than two parallel devices. The implementation design of the proposed method under the multi-device operation scenario is shown in Figure 19, considering N parallel-connected SiC MOSFETs. The extension of the proposed technique and adaption to the scenario with more than two parallel power switches requires one gate driver for every parallel power device. In addition, the algorithm should modify the control parameters for more than two power devices. Since each parallel device requires the generation of four PWM signals, the algorithm of the proposed method should be modified based on the number of parallel devices. Moreover, the number of parallel devices is defined by the number of FPGA switches for the control of the SiC MOSFETs and the adjustment of the balancing parameters of each device independently. Also, the maximum number of parallel SiC devices depends on the number of digital outputs of the constructed FPGA board.

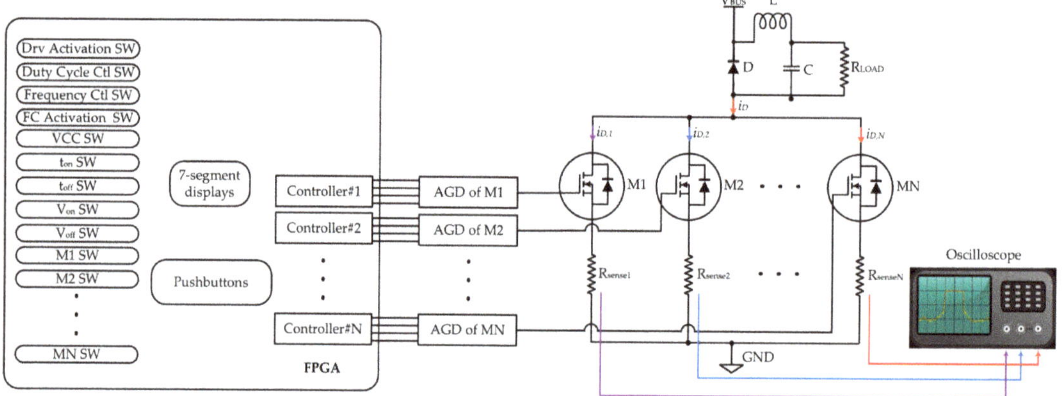

Figure 19. Implementation design of the proposed driving system for multi-device-parallel operation incorporated with a DC–DC buck converter.

6. Conclusions

In this paper, a novel autonomous open-loop current balancing technique for parallel-connected SiC MOSFETs is proposed. Regardless of the operating conditions, the proposed current balancing method can be realized without the need to know the deviation between the technical parameters to minimize current imbalance even if the PCB layout is asymmetrical. An active gate driver capable of manually modifying several control parameters with the help of an FPGA is proposed. As a result, the static current imbalance is eliminated by modifying the gate-source voltage of the proper power switch(es). Moreover, the dynamic current imbalance is suppressed by tuning the gate delays of the power devices through the adjustment of the turn-on delays and duty cycles of the driving pulses. The remaining dynamic imbalance is minimized through the variation in the gate currents. The current imbalance suppression ability of the open-loop technique is validated through experimental tests, demonstrating its performance and effectiveness against the main current imbalance impact factors. The significance of the novel proposed method is huge in terms of efficiency and reliability for renewable energy sources and energy-saving systems.

Author Contributions: Conceptualization, N.G. and G.I.; methodology, N.G.; software, N.G.; validation, G.I., G.V. and C.P.; investigation, N.G.; data curation, N.G.; writing—original draft preparation, N.G.; writing—review and editing, N.G.; supervision, G.I., G.V. and C.P.; project administration, N.G. and G.I. All authors have read and agreed to the published version of the manuscript.

Funding: This research was funded by the Special Account for Research Grants (SARG) of University of West Attica, 6000 Euros per year.

Institutional Review Board Statement: Not applicable.

Informed Consent Statement: Not applicable.

Data Availability Statement: Data is contained within the article.

Conflicts of Interest: The authors declare no conflict of interest.

References

1. Ke, J.; Zhao, Z.; Sun, P.; Huang, H.; Abuogo, J.; Cui, X. New Screening Method for Improving Transient Current sharing of Paralleled SiC MOSFETs. In Proceedings of the 2018 International Power Electronics Conference (IPEC-Niigata 2018-ECCE Asia), Niigata, Japan, 20–24 May 2018; pp. 1125–1130.
2. Xiao, Q.; Yan, Y.; Wu, X.; Ren, N.; Sheng, K. A 10 kV/200A SiC MOSFET module with series-parallel hybrid connection of 1200V/50A dies. In Proceedings of the 2015 IEEE 27th International Symposium on Power Semiconductor Devices & IC's (ISPSD), Hong Kong, China, 10–14 May 2015; pp. 349–352.
3. Helong, L.; Munk-Nielsen, S.; Wang, X.; Maheshwari, R.; Bęczkowski, S.; Uhrenfeldt, C.; Franke, W.T. Influences of Device and Circuit Mismatches on Paralleling Silicon Carbide MOSFETs. *IEEE Trans. Power Electron.* **2016**, *31*, 621–634.
4. Xue, Y.; Lu, J.; Wang, Z.; Tolbert, L.M.; Blalock, B.J.; Wang, F. Active current balancing for parallel-connected silicon carbide MOSFETs. In Proceedings of the 2013 IEEE Energy Conversion Congress and Exposition, Denver, CO, USA, 15–19 September 2013; pp. 1563–1569.
5. Li, H.; Munk-Nielsen, S.; Pham, C.; Bęczkowski, S. Circuit mismatch influence on performance of paralleling silicon carbide MOSFETs. In Proceedings of the 2014 16th European Conference on Power Electronics and Applications, Lappeenranta, Finland, 26–28 August 2014; pp. 1–8.
6. Sadik, D.P.; Colmenares, J.; Peftitsis, D.; Lim, J.K.; Rabkowski, J.; Nee, H.P. Experimental investigations of static and transient current sharing of parallel-connected silicon carbide MOSFETs. In Proceedings of the 2013 15th European Conference on Power Electronics and Applications (EPE), Lille, France, 2–6 September 2013; pp. 1–10.
7. Ishikawa, S.; Isobe, T.; Tadano, H. Current imbalance of parallel connected SiC-MOSFET body diodes. In Proceedings of the 2018 20th European Conference on Power Electronics and Applications (EPE'18 ECCE Europe), Riga, Latvia, 17–21 September 2018; pp. 1–10.
8. Zeng, Z.; Zhang, X.; Zhang, Z. Imbalance Current Analysis and Its Suppression Methodology for Parallel SiC MOSFETs with Aid of a Differential Mode Choke. *IEEE Trans. Ind. Electron.* **2020**, *67*, 1508–1519. [CrossRef]
9. Mukunoki, Y.; Horiguchi, T.; Nishizawa, A.; Konno, K.; Matsuo, T.; Kuzumoto, M.; Hagiwara, M.; Akagi, H. Electro-Thermal Co-Simulation of two Parallel-Connected SiC-MOSFETs Under Thermally-Imbalanced Conditions. In Proceedings of the 2018 IEEE Applied Power Electronics Conference and Exposition (APEC), San Antonio, TX, USA, 4–8 March 2018; pp. 2855–2860.
10. La Mantia, S.; Abbatelli, L.; Brusca, C.; Melito, M.; Nania, M. Design Rules for Paralleling of Silicon Carbide Power MOSFETs. In Proceedings of the PCIM Europe 2017: International Exhibition and Conference for Power Electronics, Intelligent Motion, Renewable Energy and Energy Management, Nuremberg, Germany, 16–18 May 2017; pp. 1–6.
11. Mao, Y.; Miao, Z.; Ngo, K.D.T.; Wang, C.M. Balancing of Peak Currents between Paralleled SiC MOSFETs by Source Impedances. In Proceedings of the 2017 IEEE Applied Power Electronics Conference and Exposition (APEC), Tampa, FL, USA, 26–30 March 2017; pp. 800–803.
12. Yang, X.; Xu, S.; Heng, K.; Wu, X. Distributed Thermal modeling for Power Devices and Modules with Equivalent Heat Flow Path Extraction. *IEEE J. Emerg. Sel. Top. Power Electron.* **2023**. Available online: https://ieeexplore.ieee.org/document/10268446 (accessed on 16 November 2023). [CrossRef]
13. Puschkarsky, K.; Grasser, T.; Aichinger, T.; Gustin, W.; Reisinger, H. Review on SiC MOSFETs High-Voltage Device Reliability Focusing on Threshold Voltage Instability. *IEEE Trans. Electron Devices* **2019**, *66*, 4604–4616. [CrossRef]
14. Haihong, Q.; Ying, Z.; Ziyue, Z.; Dan, W.; Dafeng, F.; Shishan, W.; Chaohui, Z. Influences of circuit mismatch on paralleling silicon carbide MOSFETs. In Proceedings of the 2017 12th IEEE Conference on Industrial Electronics and Applications (ICIEA), Siem Reap, Cambodia, 18–20 June 2017; pp. 556–561.
15. Zhang, Y.; Song, Q.; Tang, X.; Zhang, Y. Gate driver for parallel connection SiC MOSFETs with over-current protection and dynamic current balancing scheme. *J. Power Electron.* **2020**, *20*, 319–328. [CrossRef]

16. Xue, Y.; Lu, J.; Wang, Z.; Tolbert, L.M.; Blalock, B.J.; Wang, F. Active compensation of current unbalance in paralleled silicon carbide MOSFETs. In Proceedings of the 2014 IEEE Applied Power Electronics Conference and Exposition—APEC 2014, Fort Worth, TX, USA, 16–20 March 2014; pp. 1471–1477.

17. Mao, Y.; Miao, Z.; Wang, C.M.; Ngo, K.D.T. Balancing of Peak Currents Between Paralleled SiC MOSFETs by Drive-Source Resistors and Coupled Power-Source Inductors. *IEEE Trans. Ind. Electron.* **2017**, *64*, 8334–8343. [CrossRef]

18. Liu, P.; Yu, R.; Huang, A.Q.; Strydom, J. Turn-on Gate Resistor Optimization for Paralleled SiC MOSFETs. In Proceedings of the 2020 IEEE Applied Power Electronics Conference and Exposition (APEC), New Orleans, LA, USA, 15–19 March 2020; pp. 2810–2814.

19. Du, M.; Ding, X.; Guo, H.; Liang, J. Transient unbalanced current analysis and suppression for parallel-connected silicon carbide MOSFETs. In Proceedings of the 2014 IEEE Conference and Expo Transportation Electrification Asia-Pacific (ITEC Asia-Pacific), Beijing, China, 31 August–3 September 2014; pp. 1–4.

20. Zeng, Z.; Shao, W.; Hu, B.; Kang, S.; Liao, X.; Li, H.; Ran, L. Active Current Sharing of Paralleled SiC MOSFETs by Coupling Inductors. *Proc. Chin. Soc. Electr. Eng.* **2017**, *37*, 2068–2080.

21. Wu, Q.; Wang, M.; Zhou, W.; Wang, X. Current Balancing of Paralleled SiC MOSFETs for a Resonant Pulsed Power Converter. *IEEE Trans. Power Electron.* **2020**, *35*, 5557–5561. [CrossRef]

22. Wu, Q.; Wang, M.; Zhou, W.; Liu, G.; Wang, X. Applying Coupled Inductor to Voltage and Current Balanced Between Paralleled SiC MOSFETs for a Resonant Pulsed Power Converter. In Proceedings of the 2020 IEEE Applied Power Electronics Conference and Exposition (APEC), New Orleans, LA, USA, 15–19 March 2020; pp. 2192–2198.

23. Ding, S.; Wang, P.; Wang, W.; Xu, D.; Blaabjerg, F. Current Sharing Behavior of Parallel Connected Silicon Carbide MOSFETs Influenced by Parasitic Inductance. In Proceedings of the 2019 10th International Conference on Power Electronics and ECCE Asia (ICPE 2019—ECCE Asia), Busan, Republic of Korea, 27–30 May 2019.

24. Sarfejo, M.D.; Allahyari, H.; Bahrami, H.; Afifi, A.; Zadeh, M.A.L.; Yavari, E.; Jalise, M.G. A Passive Compensator for Imbalances in Current Sharing of Parallel-SiC MOSFETs Based on Planar Transformer. *IET Power Electron.* **2021**, *14*, 2400–2412. [CrossRef]

25. Ke, J.; Zhao, Z.; Sun, P.; Huang, H.; Abuogo, J.; Cui, X. Chips Classification for Suppressing Transient Current Imbalance of Parallel-Connected Silicon Carbide MOSFETs. *IEEE Trans. Power Electron.* **2020**, *35*, 3963–3972. [CrossRef]

26. Ke, J.; Zhao, Z.; Zou, Q.; Peng, J.; Chen, Z.; Cui, X. Device Screening Strategy for Balancing Short-Circuit Behavior of Paralleling Silicon Carbide MOSFETs. *IEEE Trans. Device Mater. Reliab.* **2019**, *19*, 757–765. [CrossRef]

27. Liu, Y.; Dai, X.; Jiang, X.; Zeng, Z.; Qi, F.; Liu, Y.; Ke, P.; Wang, Y.; Wang, J. A New Screening Method for Alleviating Transient Current Imbalance of Paralleled SiC MOSFETs. In Proceedings of the 2020 IEEE 1st China International Youth Conference on Electrical Engineering (CIYCEE), Wuhan, China, 1–4 November 2020.

28. Abuogo, J.O.; Zhao, Z. Machine learning approach for sorting SiC MOSFET devices for paralleling. *J. Power Electron.* **2020**, *20*, 329–340. [CrossRef]

29. Abuogo, J.O.; Zao, Z.; Ke, J. Linear regression model for screening SiC MOSFETs for paralleling to minimize transient current imbalance. In Proceedings of the 5th International Conference on Electrical Engineering, Control and Robotics (EECR 2019), Guangzhou, China, 12–14 January 2019.

30. Zhao, B.; Sun, P.; Yu, Q.; Cai, Y.; Cai, J.; Zhao, Z. Chip Screening Method for Suppressing Current Imbalance of Parallel-Connected SiC MOSFETs. In Proceedings of the 2021 4th International Conference on Energy, Electrical and Power Engineering (CEEPE), Chongqing, China, 23–25 April 2021; pp. 19–25.

31. Zhao, C.; Wang, L.; Zhang, F. Effect of Asymmetric Layout and Unequal Junction Temperature on Current Sharing of Paralleled SiC MOSFETs with Kevin-Source Connection. *IEEE Trans. Power Electron.* **2020**, *35*, 7392–7404. [CrossRef]

32. Liu, J.; Zheng, Z. Switching Current Imbalance Mitigation for Paralleled SiC MOSFETs Using Common-mode Choke in Gate Loop. In Proceedings of the 2020 IEEE Energy Conversion Congress and Exposition (ECCE), Detroit, MI, USA, 11–15 October 2020; pp. 705–710.

33. Wang, G.; Mookken, J.; Rice, J.; Schupbach, M. Dynamic and static behavior of packaged silicon carbide MOSFETs in paralleled applications. In Proceedings of the 2014 IEEE Applied Power Electronics Conference and Exposition—APEC 2014, Fort Worth, TX, USA, 16–20 March 2014; pp. 1478–1483.

34. Luedecke, C.; Krichel, F.; Laumen, M.; De Doncker, R.W. Balancing the Switching Losses of Paralleled SiC MOSFETs Using an Intelligent Gate Driver. In Proceedings of the PCIM Asia 2020: International Exhibition and Conference for Power Electronics, Intelligent Motion, Renewable Energy and Energy Management, Shanghai, China, 16–18 November 2020.

35. Lüdecke, C.; Aghdaei, A.; Laumen, M.; Doncker, R.W.D. Balancing the Switching Losses of Paralleled SiC MOSFETs Using a Stepwise Gate Driver. In Proceedings of the 2021 IEEE Energy Conversion Congress and Exposition (ECCE), Vancouver, BC, Canada, 10–14 October 2021; pp. 5400–5406.

36. Wei, Y.; Sweeting, R.; Hossain, M.M.; Mhiesan, H.; Mantooth, A. Variable Gate Voltage Control for Paralleled SiC MOSFETs. In Proceedings of the 2020 IEEE Workshop on Wide Bandgap Power Devices and Applications in Asia (WiPDA Asia), Suita, Japan, 23–25 September 2020.

37. Chen, Z.; Chen, C.; Huang, A.Q. Driver Integrated Online Rds-on Monitoring Method for SiC Power Converters. In Proceedings of the 2022 IEEE Energy Conversion Congress and Exposition (ECCE), Detroit, MI, USA, 9–13 October 2022.

38. Fu, J. Fundamentals of On-Resistance in Load Switches. TI., Application Report SLVA771. Available online: https://www.ti. com/lit/an/slva771/slva771.pdf?ts=1624372175829&ref_url=https%253A%252F%252Fwww.google.com%252F (accessed on 22 June 2021).
39. Choragudi, V.S.A.K. Analysis and Design of Pulse-Width Modulated Two-Switch Forward DC-DC Converter for Universal Laptop Adapter. Master's Thesis, Wright State University, India, Ohio, 2011.
40. Giannopoulos, N.; Ioannidis, G.C.; Psomopoulos, C.S. Active Auto-Suppression Current Unbalance Technique for Parallel-Connected Silicon Carbide MOSFETs. *Electronics* **2022**, *11*, 445. [CrossRef]
41. Giannopoulos, N.; Ioannidis, G.; Vokas, G.; Psomopoulos, C. Autonomous Active Current Balancing Method for Parallel-Connected Silicon Carbide MOSFETs. In Proceedings of the AIP Conference, Tmrees20, Athens, Greece, 25–27 June 2020.
42. Zhang, W. Current Sensor for Wide Bandgap Devices Dynamic Characterization. Master's Thesis, University of Tennessee, Knoxville, TN, USA, 2019.
43. Zhang, W.; Zhang, Z.; Wang, F. Review and Bandwidth Measurement of Coaxial Shunt Resistors for Wide-Bandgap Devices Dynamic Characterization. In Proceedings of the 2019 IEEE Energy Conversion and Exposition (ECCE), Baltimore, MD, USA, 29 September–3 October 2019; pp. 3259–3264.
44. Li, J.; Yang, X.; Xu, M.; Wang, X.; Heng, K. A novel inductively coupled RLC damping scheme for eliminating switching oscillations of SiC MOSFET. *IET Power Electron.* **2023**, *16*, 1486–1498. [CrossRef]

MDPI
St. Alban-Anlage 66
4052 Basel
Switzerland
www.mdpi.com

Energies Editorial Office
E-mail: energies@mdpi.com
www.mdpi.com/journal/energies